Lecture Notes in Computer Science 8476

Commenced Publication in 1973
Founding and Former Series Editors:
Gerhard Goos, Juris Hartmanis, and Jan van Leeuwen

T0236355

Edward A. Hirsch   Sergei O. Kuznetsov
Jean-Éric Pin   Nikolay K. Vereshchagin (Eds.)

Computer Science - Theory and Applications

9th International Computer Science Symposium
in Russia, CSR 2014
Moscow, Russia, June 7-11, 2014
Proceedings

 Springer

Volume Editors

Edward A. Hirsch
Steklov Institute of Mathematics at St. Petersburg
Russian Academy of Sciences, St. Petersburg, Russia
E-mail: hirsch@pdmi.ras.ru

Sergei O. Kuznetsov
National Research University - Higher School of Economics, Moscow, Russia
E-mail: skuznetsov@hse.ru

Jean-Éric Pin
CNRS and University Paris Diderot, Paris, France
E-mail: jean-eric.pin@liafa.univ-paris-diderot.fr

Nikolay K. Vereshchagin
Moscow State University, Russia
E-mail: ver@mech.math.msu.su

ISSN 0302-9743   e-ISSN 1611-3349
ISBN 978-3-319-06685-1   e-ISBN 978-3-319-06686-8
DOI 10.1007/978-3-319-06686-8
Springer Cham Heidelberg New York Dordrecht London

Library of Congress Control Number: 2014936745

Typesetting: Camera-ready by author, data conversion by Scientific Publishing Services, Chennai, India

Printed on acid-free paper

Springer is part of Springer Science+Business Media (www.springer.com)

Preface

The 9th International Computer Science Symposium in Russia (CSR 2014) was held during June 7–11, 2014, in Moscow, hosted by the Moscow Center for Continuous Mathematical Education. It was the ninth event in the series of regular international meetings following CSR 2006 in St. Petersburg, CSR 2007 in Ekaterinburg, CSR 2008 in Moscow, CSR 2009 in Novosibirsk, CSR 2010 in Kazan, CSR 2011 in St. Petersburg, CSR 2012 in Nizhny Novgorod, and CSR 2013 in Ekaterinburg.

The opening lecture was given by Shafi Goldwasser and six other invited plenary lectures were given by Mark Braverman, Volker Diekert, Martin Grohe, Benjamin Rossman, Alexei Semenov, and Igor Walukiewicz.

This volume contains the accepted papers and abstracts of the invited talks. The scope of the proposed topics for the symposium is quite broad and covers a wide range of areas in theoretical computer science and its applications. We received 76 papers in total, and out of these the Program Committee selected 27 papers for presentation at the symposium and for publication in the proceedings.

As usual, Yandex provided the Best Paper Awards. The recipients of these awards were selected by the Program Committee, but due to potential conflicts of interest, the procedure of selecting award winners did not involve the PC chair and was chaired by Edward A. Hirsch. The winners are:

- Best Paper Award: Akinori Kawachi, Benjamin Rossman and Osamu Watanabe, "The Query Complexity of Witness Finding"
- Konrad Schwerdtfeger, The Connectivity of Boolean Satisfiability: Dichotomies for Formulas and Circuits.

The reviewing process was organized using the EasyChair conference system created by Andrei Voronkov. We would like to acknowledge that this system helped greatly to improve the efficiency of the committee work.

The following satellite events were co-located with CSR 2014:

- Workshop on Current Trends in Cryptology (CTCrypt)
- Extremal Graph Theory
- New Directions in Cryptography
- Program Semantics, Specification and Verification (PSSV 2014)

We are grateful to our sponsors:

- Russian Foundation for Basic Research
- Higher School of Economics (HSE)
- Yandex
- Talksum
- IPONWEB
- CMA Small Systems AB
- Dynasty Foundation

We also thank the local organizers: Alexander Kulikov, Daniil Musatov, Vladimir Podolskii, Alexander Smal, Tatiana Starikovskaya, Mikhail Raskin, Mikhail Andreev, Anton Makhlin.

March 2014

Edward A. Hirsch
Sergei O. Kuznetsov
Jean-Éric Pin
Nikolay K. Vereshchagin

Organization

CSR 2014 was organized by:

- Moscow Center for Continuous Mathematical Education
- Steklov Institute of Mathematics at St. Petersburg, Russian Academy of Sciences
- National Research University Higher School of Economics

Program Committee Chair

Jean-Éric Pin   LIAFA, CNRS and University of Paris-Diderot,
                             France

Program Committee

Eric Allender   Rutgers, USA
Andris Ambainis   University of Latvia, Latvia
Christel Baier   TU Dresden, Germany
Petra Berenbrink   Simon Fraser University, Canada
Mikolaj Bojanczyk   University of Warsaw, Poland
Andrei A. Bulatov   Simon Fraser University, Canada
Victor Dalmau   Pompeu Fabra University, Spain
Manfred Droste   University of Leipzig, Germany
Zoltan Esik   University of Szeged, Hungary
Fedor Fomin   University of Bergen, Norway
Edward A. Hirsch   Steklov Institute of Mathematics at
                               St. Petersburg, Russia
Gregory Kucherov   CNRS and University of Marne-la-Vallée,
                               France
Michal Kunc   Masaryk University, Czech Republic
Leonid Libkin   University of Edinburgh, UK
Konstantin Makarychev   Microsoft Research, Redmond, USA
Kurt Mehlhorn   Max Planck Institute, Germany
Georg Moser   University of Innsbruck, Austria
Alexander Okhotin   University of Turku, Finland
Giovanni Pighizzini   University of Milan, Italy
Jean-Éric Pin   LIAFA, CNRS and University of Paris-Diderot,
                               France

Alexander Razborov	University of Chicago, USA, and Steklov Mathematical Institute, Russia
Michel Rigo	University of Liège, Belgium
Nicole Schweikardt	University of Frankfurt, Germany
Jacobo Toran	University of Ulm, Germany
Mikhail Volkov	Ural Federal University, Russia
Carsten Witt	TU Denmark, Denmark

Symposium Co-chairs

Nikolai K. Vereshchagin	MCCME and Moscow State University, Russia
Edward A. Hirsch	Steklov Institute of Mathematics at St. Petersburg, Russia
Sergei O. Kuznetsov	School of Applied Mathematics and Information Science, Higher School of Economics, Russia

Organizing Committee

Alexander Kulikov	Steklov Institute of Mathematics at St. Petersburg, Russia
Daniil Musatov	Moscow Institute of Physics and Technology, Russia
Vladimir Podolskii	Steklov Institute of Mathematics, Moscow, Russia
Alexander Smal	Steklov Institute of Mathematics at St. Petersburg, Russia
Tatiana Starikovskaya	School of Applied Mathematics and Information Science, Higher School of Economics, Russia

Steering Committee

Anna Frid	Sobolev Institute of Mathematics, Russia
Edward A. Hirsch	Steklov Institute of Mathematics at St. Petersburg, Russian Academy of Sciences, Russia
Juhani Karhumäki	University of Turku, Finland
Ernst W. Mayr	Technical University of Munich, Germany
Alexander Razborov	University of Chicago, USA, and Steklov Mathematical Institute, Russia
Mikhail Volkov	Ural Federal University, Russia

External Reviewers

Amano, Kazuyuki
Ananichev, Dmitry
Autebert, Jean-Michel
Avanzini, Martin
Balkova, Lubimora
Barash, Mikhail
Barto, Libor
Batu, Tugkan
Ben-Sasson, Eli
Berlinkov, Mikhail
Björklund, Henrik
Blais, Eric
Bliznets, Ivan
Boasson, Luc
Boigelot, Bernard
Boyar, Joan
Braverman, Mark
Brihaye, Thomas
Broadbent, Christopher
Buchin, Kevin
Carayol, Arnaud
Carton, Olivier
Castro, Jorge
Chailloux, Andre
Charlier, Emilie
Choffrut, Christian
Clemente, Lorenzo
Cording, Patrick Hagge
Crama, Yves
Crochemore, Maxime
de Wolf, Ronald
Di Summa, Marco
Doerr, Benjamin
Doerr, Carola
Drange, Paal Groenaas
Dubslaff, Clemens
Duchêne, Eric
Dvorak, Zdenek
Dziembowski, Stefan
Eidelman, Yuli
Elbassioni, Khaled
Esik, Zoltan
Felgenhauer, Bertram

Friedetzky, Tom
Friedman, Luke
Galántai, Aurél
Gambette, Philippe
Gasarch, William
Gastin, Paul
Grenet, Bruno
Groote, Jan Friso
Guerraoui, Rachid
Gurevich, Yuri
Hansen, Kristoffer Arnsfelt
Hatami, Pooya
Hirvensalo, Mika
Hlineny, Petr
Holzer, Markus
Huang, Shenwei
Huschenbett, Martin
Hüffner, Falk
Iemhoff, Rosalie
Immerman, Neil
Itsykson, Dmitry
Ivan, Szabolcs
Jansen, Bart M.P.
Jeż, Artur
Jitsukawa, Toshiaki
Jones, Mark
Jukna, Stasys
Kabanets, Valentine
Kaiser, Tomas
Karhumäki, Juhani
Kartzow, Alexander
Kerenidis, Iordanis
Khachay, Mikhail
Koebler, Johannes
Komusiewicz, Christian
Kozik, Marcin
Kral, Daniel
Kratsch, Dieter
Krebs, Andreas
Krohn, Erik
Kulikov, Alexander
Kullmann, Oliver
Kuske, Dietrich

Kuznets, Roman
Kötzing, Timo
Laud, Peeter
Lauria, Massimo
Leupold, Peter
Loff, Bruno
Lohrey, Markus
Lokshtanov, Daniel
Löding, Christof
Magdon-Ismail, Malik
Malcher, Andreas
Malod, Guillaume
Matulef, Kevin
Mayr, Richard
Meckes, Mark
van Melkebeek, Dieter
Mercas, Robert
Mereghetti, Carlo
Milnikel, Robert
Misra, Neeldhara
Monmege, Benjamin
Müller, David
Nagy, Benedek
Navarro Perez, Juan Antonio
Nebel, Markus
Nederlof, Jesper
Niedermeier, Rolf
Nordstrom, Jakob
Ouaknine, Joel
Palano, Beatrice
Pilipczuk, Marcin
Pilipczuk, Michał
Plaskota, Leszek
Quaas, Karin
Rahonis, George
Rampersad, Narad

Rasin, Oleg
Razgon, Igor
Rytter, Wojciech
Saarela, Aleksi
Sau, Ignasi
Schaper, Michael
Seki, Shinnosuke
Serre, Olivier
Shallit, Jeffrey
Shpilrain, Vladimir
Sokolov, Dmitry
Srba, Jiri
Stacho, Ladislav
Steinberg, Benjamin
Thapper, Johan
Thierauf, Thomas
Tirthapura, Srikanta
Tsur, Dekel
Tygert, Mark
van Leeuwen, Erik Jan
Velner, Yaron
Vialette, Stéphane
Villard, Jules
Walen, Tomasz
Wang, Kaishun
Weil, Pascal
Wilke, Thomas
Xia, Jianlin
Xie, Ning
Yakaryilmaz, Abuzer
Yamamoto, Masaki
Zakov, Shay
Zdanowski, Konrad
Zhang, Shengyu
Zimand, Marius

Invited Talks

Error-Correction for Interactive Computation

Mark Braverman

Department of Computer Science, Princeton University

Abstract. Classical error-correcting codes deal with the problem of data transmission over a noisy channel. There are efficient error-correcting codes that work even when the noise is adversarial. In the interactive setting, the goal is to protect an entire conversation between two (or more) parties from adversarial errors. The area of interactive error correcting codes has experienced a substantial amount of activity in the last few years. In this talk we will introduce the problem of interactive error-correction and discuss some of the recent results.

Finding All Solutions of Equations
in Free Groups and Monoids with Involution

Volker Diekert[1], Artur Jeż[2,4,*], and Wojciech Plandowski[3]

[1] Institut für Formale Methoden der Informatik, University of Stuttgart, Germany
[2] Institute of Computer Science, University of Wroclaw, Poland
[3] Max Planck Institute für Informatik, Saarbrcken, Germany
[4] Institute of Informatics, University of Warsaw, Poland

Abstract. The aim of this paper is to present a PSPACE algorithm which yields a finite graph of exponential size and which describes the set of all solutions of equations in free groups and monoids with involution in the presence of rational constraints. This became possible due to the recently invented recompression technique of the second author.
He successfully applied the recompression technique for pure word equations without involution or rational constraints. In particular, his method could not be used as a black box for free groups (even without rational constraints). Actually, the presence of an involution (inverse elements) and rational constraints complicates the situation and some additional analysis is necessary. Still, the recompression technique is powerful enough to simplify proofs for many existing results in the literature. In particular, it simplifies proofs that solving word equations is in PSPACE (Plandowski 1999) and the corresponding result for equations in free groups with rational constraints (Diekert, Hagenah and Gutiérrez 2001). As a byproduct we obtain a direct proof that it is decidable in PSPACE whether or not the solution set is finite.[1]

---

[*] Supported by Humboldt Research Fellowship for Postdoctoral Researchers.
[1] A full version of the present paper with detailed proofs can be found on arXiv.

Algorithmic Meta Theorems
for Sparse Graph Classes

Martin Grohe

RWTH Aachen University, Germany
grohe@informatik.rwth-aachen.de

Abstract. Algorithmic meta theorems give efficient algorithms for classes of algorithmic problems, instead of just individual problems. They unify families of algorithmic results obtained by similar techniques and thus exhibit the core of these techniques. The classes of problems are typically defined in terms of logic and structural graph theory. A well-known example of an algorithmic meta theorem is Courcelle's Theorem, stating that all properties of graphs of bounded tree width that are definable in monadic second-order logic are decidable in linear time.

This paper is a brief and nontechnical survey of the most important algorithmic meta theorems.

The Lattice of Definability. Origins, Recent Developments, and Further Directions

Alexei Semenov[1,2,3], Sergey Soprunov[2], and Vladimir Uspensky[1]

[1] Moscow State University
[2] Dorodnicyn Computing Center of the Russian Academy of Sciences
[3] Moscow Pedagogical State University
alsemenov@umail.ru, soprunov@mail.ru, vau30@list.ru

Abstract. The paper presents recent results and open problems on classes of definable relations (definability spaces, reducts, relational algebras) as well as sources for the research starting from the XIX century. Finiteness conditions are investigated, including quantifier alternation depth and number of arguments width. The infinite lattice of definability for integers with a successor function (a non ω-categorical structure) is described. Methods of investigation include study of automorphism groups of elementary extensions of structures under consideration, using Svenonius theorem and a generalization of it.

Table of Contents

Finding All Solutions of Equations
in Free Groups and Monoids with Involution

Volker Diekert[1], Artur Jeż[2,4,*], and Wojciech Plandowski[3]

[1] Institut für Formale Methoden der Informatik, University of Stuttgart, Germany
[2] Institute of Computer Science, University of Wroclaw, Poland
[3] Max Planck Institute für Informatik, Saarbrücken, Germany
[4] Institute of Informatics, University of Warsaw, Poland

Abstract. The aim of this paper is to present a PSPACE algorithm which yields a finite graph of exponential size and which describes the set of all solutions of equations in free groups and monoids with involution in the presence of rational constraints. This became possible due to the recently invented recompression technique of the second author.

He successfully applied the recompression technique for pure word equations without involution or rational constraints. In particular, his method could not be used as a black box for free groups (even without rational constraints). Actually, the presence of an involution (inverse elements) and rational constraints complicates the situation and some additional analysis is necessary. Still, the recompression technique is powerful enough to simplify proofs for many existing results in the literature. In particular, it simplifies proofs that solving word equations is in PSPACE (Plandowski 1999) and the corresponding result for equations in free groups with rational constraints (Diekert, Hagenah and Gutiérrez 2001). As a byproduct we obtain a direct proof that it is decidable in PSPACE whether or not the solution set is finite.[1]

Introduction

A word equation is a simple object. It consists of a pair (U, V) of words over constants and variables and a solution is a substitution of the variables by words in constants such that U and V are identical words. The study of word equations has a long tradition. Let *WordEquation* be the problem to decide whether a given word equation has a solution. It is fairly easy to see that WordEquation reduces to Hilbert's 10th Problem (in Hilbert's famous list presented in 1900 for his address at the International Congress of Mathematicians). Hence in the mid 1960s the Russian school of mathematics outlined the roadmap to prove undecidability of Hilbert 10 via undecidability of WordEquation. The program failed in the sense that Matiyasevich proved Hilbert's 10th Problem to be undecidable in 1970, but by a completely different method, which employed number theory. The missing piece in the proof of the undecidability of Hilbert's 10th Problem

---

* Supported by Humboldt Research Fellowship for Postdoctoral Researchers.
[1] A full version of the present paper with detailed proofs can be found on arXiv.

E.A. Hirsch et al. (Eds.): CSR 2014, LNCS 8476, pp. 1–15, 2014.
© Springer International Publishing Switzerland 2014

was based on methods due to Robinson, Davis, and Putnam [20]. On the other hand, in 1977 Makanin showed in a seminal paper [17] that WordEquation is decidable! The program went a different way, but its outcome were two major achievements in mathematics. Makanin's algorithm became famous since it settled a long standing problem and also because his algorithm had an extremely complex termination proof. In fact, his paper showed that the existential theory of equations in free monoids is decidable. This is close to the borderline of decidability as already the $\forall\exists^3$ positive theory of free monoids is undecidable [7]. Furthermore Makanin extended his results to free groups and showed that the existential and positive theories in free groups are decidable [18, 19]. Later Razborov was able in [26] (partly shown also in [27]) to describe the set of all solutions for systems of equations in free groups (see also [14] for a description of Razborov's work). This line of decidability results culminated in the proof of Tarski's conjectures by Kharlampovich and Myasnikov in a series of papers ending in [15]. In particular, they showed that the theory of free groups is decidable. In order to prove this fundamental result the description of all solutions of an equation in a free group is crucial. Another branch of research was to extend Makanin's result to more general algebraic structures including free partially commutative monoids [21, 5], free partially commutative monoids with involution, graph groups (also known as right-angled Artin groups) [6], graph products [4], and hyperbolic groups [28, 2]. In all these cases the existential theory of equations is decidable. Proofs used the notion of *equation with rational constraints*, which was first developed in the habilitation of Schulz, see [29]. A concept which is also used throughout in the present paper.

In parallel to these developments there were drastic improvements in the complexity of deciding word equations. It is fairly easy to see that the problem is NP-hard. Thus, NP is a lower bound. First estimations for the time complexity on Makanin's algorithm for free monoids led to a tower of several exponentials, but it was lowered over time to EXPSPACE in [9]. On the the other hand it was shown in [16] that Makanin's scheme for solving equations in free groups is not primitive recursive. (Already in the mid 1990 this statement was somehow puzzling and counter-intuitive, as it suggested a strange crossing of complexities: The existential theory in free monoids seemed to be easier than the one in free groups, whereas it was already known at that time that the positive theory in free monoids is undecidable, but decidable in free groups.) The next important step was done by Plandowski and Rytter, whose approach [25] was the first essentially different than Makanin's original solution. They showed that Lempel-Ziv encodings of minimal solution of word equations leads to an exponential compression (if the solution itself is at least exponential in the length of the equation). Moreover the compression turned out be extremely simple. As a consequence they formulated the still valid conjecture that WordEquation is NP-complete. Following the idea to keep the equation and possible solution in compressed form and employing a novel type of factorisations Plandowski showed that WordEquation is in PSPACE, i.e., it can be solved in polynomial space and exponential time [22]. His method was quite different from Makanin's approach and more symmetric. In particular, it could be also used to generate all solutions of a given word equation [23], however, this required non-trivial extensions of the original method.

Using Plandowski's method Gutiérrez showed that satisfiability of equations in free groups is in PSPACE [10], which led Diekert, Hagenah and Gutiérrez to the result that the existential theory of equations with rational constraints in free groups is PSPACE-complete [3]. To date these are still the best complexity results. Since this proof generalised Plandowski's satisfiability result [22], it is tempting to also extend the generator of all solutions [23]. Indeed, Plandowski claimed that his method applies also to free groups with rational constraints, but he found a gap in his generalization [24].

However in 2013 another substantial progress in solving word equations was done due to a powerful recompression technique due to Jeż [13]. His new proof that WordEquation is in PSPACE simplified the existing proofs drastically. In particular, this approach could be used to describe the set of all solutions rather easily, so the previous construction of Plandowski [23] was simplified as well.

What was missing however was the extension to include free monoids with involution and therefore free groups and another missing block was the the presence of rational constraints. Both extensions are the subject of the present paper.

We first follow the approach of [3] how to transform the set of all solutions of an equation with rational constraints in polynomial time into a set of all solutions of an equation with regular constraints over a free monoid with involution. Starting at that point we show the existence of a PSPACE-transducer which produces a finite graph (of exponential size) which describes all solutions and which is nonempty if and only if the equation has at least one solution. Moreover, the graph also encodes whether or not there are finitely many solutions, only. The technique of recompression simplifies thereby [3] and it yields the important new feature that we can describe all solutions.

1   Preliminaries

Word Equations. Let A and Ω be two finite disjoint sets, called *the alphabet of constants* and *the alphabet of variables* (or *unknowns*), respectively. For the purpose of this paper A and Ω are endowed with an *involution*, which is is a mapping $^-$ such that $\overline{\overline{x}} = x$ for all elements. In particular, an involution is a bijection. Since, the identity mapping is an involution, there is no harm in thinking that all sets come with an involution. If M is a monoid, then we additionally require $\overline{xy} = \overline{y}\,\overline{x}$ for all $x, y \in M$. This applies in particular to a free monoid Σ^* over a set with involution: For a word $w = a_1 \cdots a_m$ we thus have $\overline{w} = \overline{a_m} \cdots \overline{a_1}$. If $\overline{a} = a$ for all $a \in \Sigma$ then $\overline{w}$ simply means to read the word from right-to-left.

A *word equation* is a pair (U, V) of words over $A \cup \Omega$, usually denoted by $U = V$. A *solution* σ of a word equation $U = V$ is a substitution σ of unknowns in Ω by words over constants, such that the replacement of unknowns by the substituted words in U and in V give the same word. Moreover, as we work with involutions we additionally demand that the solution satisfies $\sigma(\overline{X}) = \overline{\sigma(X)}$ for all $X \in \Omega$.

Example 1. Let $\Omega = \{X, Y, \overline{X}, \overline{Y}\}$ and $A = \{a, b\}$ with $b = \overline{a}$. Then $XabY = YbaX$ behaves as a word equation without involution One of its solutions is the substitution $\sigma(X) = bab$, $\sigma(Y) = babab$. Under this substitution we have $\sigma(X)ab\sigma(Y) = bababbabab = \sigma(Y)ba\sigma(X)$. It can be proved that the solution set of the equation $XabY = YbaX$ is closely related to Sturmian words [12].

The notion of word equation immediately generalizes to a system of word equations $U_i = V_i$ for some index set I. In this case a solution σ must satisfy all $U_i = V_i$ simultaneously. Next, we consider constraints. Let $\mathcal{C}$ be a class of formal languages, then a system of *word equations with constraints in $\mathcal{C}$* is given by a finite list $(U_i, V_i)_i$ of word equations and a finite list of constraints of type $X \in L$ (resp. $X \notin L$) where $X \in \Omega$ and $L \subseteq A^*$ with $L \in \mathcal{C}$. For a solution we now additionally demand that $\sigma(X) \in L$ (resp. $\sigma(X) \notin L$) for all constraints.

Here, we focus on rational and recognizable (or regular) constraints and we assume that the reader is familiar with basic facts in formal language theory. The classes of rational and recognizable subsets are defined for every monoid M [8], and they are incomparable, in general. *Rational* sets (or languages) are defined inductively as follows. All finite subsets of M are rational. If $L_1, L_2 \subseteq M$ are rational, then the union $L_1 \cup L_2$, the concatenation $L_1 \cdot L_2$, and the generated submonoid L_1^* are rational. A subset $L \subseteq M$ is called *recognizable*, if there is a homomorphism ρ to some finite monoid E such that $L = \rho^{-1}\rho(L)$. We also say that ρ (or E) *recognizes* L in this case. Kleene's Theorem states that in finitely generated free monoids both classes coincide, and we follow the usual convention to call a rational subset of a free monoid *regular*. If M is generated by some finite set $\Gamma \subseteq M$ (as it always the case in this paper) then every rational set is the image of a regular set L under the canonical homomorphism from Γ^* onto M; and every recognizable set of M is rational. (These statements are trivial consequences of Kleene's Theorem.) Therefore, throughout we assume that a rational (or regular) language is specified by a nondeterministic finite automaton, *NFA* for short.

Equations with Rational Constraints over Free Groups.

By $F(\Gamma)$ we denote the free group over a finite set Γ. We let $A = \Gamma \cup \Gamma^{-1}$. Set also $\overline{x} = x^{-1}$ for all $x \in F(\Gamma)$. Thus, in (free) groups we identify x^{-1} and $\overline{x}$. By a classical result of Benois [1] rational subsets of $F(\Gamma)$ form an effective Boolean algebra. That is: if L is rational and specified by some NFA then $F(\Gamma) \setminus L$ is rational; and we can effectively find the corresponding NFA. There might be an exponential blow-up in the NFA size, though. This is the main reason to allow negative constraints $X \notin L$, so we can avoid explicit complementation. Let us highlight another point: The singleton $\{1\} \subseteq F(\Gamma)$ is, by definition, rational. Hence the set $F(\Gamma) \setminus \{1\}$ is rational, too. Therefore an inequality $U \neq V$ can be handled by a new fresh variable X and writing $U = XV$ & $X \in F(\Gamma) \setminus \{1\}$ instead of $U \neq V$.

Proposition 1 ([3]). *Let $F(\Gamma)$ be a free group and $A = \Gamma \cup \Gamma^{-1}$ be the corresponding set with involution as above. There is polynomial time transformation which takes as input a system S of equations (and inequalities) with rational constraints over $F(\Gamma)$ and outputs a system of word equations with regular constraints S' over A which is solvable if and only if S is solvable in $F(\Gamma)$.*

More precisely, let $\varphi : A^* \to F(\Gamma)$ be the canonical morphism of the free monoid with involution A^* onto the free $F(\Gamma)$. Then the set of all solutions for S' is mapped via $\sigma' \mapsto \varphi \circ \sigma'$ onto the set of all solutions of S.

A system of word equations $(U_1, V_1), \ldots, (U_s, V_s)$ is equivalent to a single equation $(U_1 a \cdots U_s a U_1 b \cdots U_s b, V_1 a \cdots V_s a V_1 b \cdots V_s b)$ where a, b are a fresh constants with $a \neq b$. Hence, Proposition 1 shows that the problem of deciding the satisfiability of a system of equations and inequalities (with rational constraints) in a free group or to find all solutions can be efficiently reduced to solving the corresponding task for word equations with regular constraints in a free monoids with involution.

Input Size. The input size for the reduction is given by the sum over the lengths of the equations and inequalities plus the size of Γ plus the sum of the number of states of the NFAs in the lists for the constraints. The measure is accurate enough with respect to polynomial time and or space. For example note that if an NFA has n states then the number of transitions is bounded by $2n|\Gamma|$. Note also that $|\Gamma|$ can be much larger than the sum over the lengths of the equations and inequalities plus the sum of the number of states of the NFAs in the lists for the constraints. Recall that we encode $X \neq 1$ by a rational constraint, which introduces an NFA with $2|\Gamma| + 1$ states. Since $|\Gamma|$ is part of the input, this does not cause any problem.

Word Equations with Regular Constraints and with Involution. Consider a list of k regular languages $L_i \subseteq \Sigma^*$ each of them being specified by some NFA with n_i states. The disjoint union of these automata yields a single NFA with $n_1 + \cdots + n_k$ states which accepts all L_i by choosing appropriate initial and final sets for each L_i. Let $n = n_1 + \cdots + n_k$. We may assume that the NFA has state set $\{1, \ldots n\}$. Then each letter $a \in A$ defines a Boolean $n \times n$ matrix $\tau(a)$ where the entry (p, q) is 1 if (p, a, q) is a transition and 0 otherwise. This yields a homomorphism $\tau : A^* \to \mathbb{B}^{n \times n}$ such that τ recognizes L_i for all $1 \leq i \leq k$. Moreover, for each i there is a row vector $I_i \in \mathbb{B}^{1 \times n}$ and a column vector $F_i \in \mathbb{B}^{n \times 1}$ such that we have $w \in L_i$ if and only if $I_i \cdot \tau(w) \cdot F_i = 1$.

For a matrix P we let P^T be its transposition. There is no reason that $\tau(\overline{a}) = \tau(a)^T$, hence τ is not necessarily a homomorphism which respects the involution. So, as done in [3], we let $\mathbb{M}_{2n} \subseteq \mathbb{B}^{2n \times 2n}$ denote the following monoid with involution:

$$\mathbb{M}_{2n} = \left\{ \left(\begin{smallmatrix} P & 0 \\ 0 & Q \end{smallmatrix} \right) \mid P, Q \in \mathbb{B}^{n \times n} \right\} \text{ with } \overline{\left(\begin{smallmatrix} P & 0 \\ 0 & Q \end{smallmatrix} \right)} = \left(\begin{smallmatrix} Q^T & 0 \\ 0 & P^T \end{smallmatrix} \right).$$

Define $\rho(a) = \left(\begin{smallmatrix} \tau(a) & 0 \\ 0 & \tau(\overline{a})^T \end{smallmatrix} \right)$. Then the homomorphism $\rho : A^* \to \mathbb{M}_{2n}$ respects the involution. Moreover ρ recognizes all L_i and $\overline{L_i} = \{\overline{w} \mid w \in L_i\}$.

In terms of matrices, each constraint $X \in L$ or $X \notin L$ translates to restriction of possible values of $\rho(\sigma(X))$. On the other hand, if $\rho(\sigma(X))$ is one of the values allowed by all constraints, $\sigma(X)$ satisfies all constraints. Thus, as a preprocessing step our algorithm guesses the $\rho(\sigma(X))$, which we shall shortly denote as $\rho(X)$ and in the following we are interested only in solution for which $\rho(\sigma(X)) = \rho(X)$ and $\rho(\overline{X}) = \overline{\rho(X)}$. Note that, as ρ is a function, each solution of the original system corresponds to a solution for an exactly one such guess, thus we can focus on generating the solutions for this restricted problem.

Moreover, we always assume that the involution on Ω is without fixed points. This is no restriction and avoids some case distinctions. We now give a precise definition of the problem we are considering now:

Definition 1. *An equation E with constraints is a tuple $E = (A, \Omega, \rho; U = V)$ containing the following items:*

- *An alphabet of constants with involution A.*
- *An alphabet of variables with involution without fixed points Ω.*
- *A mapping $\rho : A \cup \Omega \to \mathbb{M}_{2n}$ such that $\overline{\sigma(X)} = \sigma(\overline{X})$ for all $x \in A \cup \Omega$.*
- *The word equation $U = V$ where $U, V \in (A \cup \Omega)^*$.*

A solution of E is a homomorphism $\sigma : (A \cup \Omega)^* \to A^*$ by leaving the letters from A invariant such that the following conditions are satisfied:

$$\sigma(U) = \sigma(V)\,,$$
$$\overline{\sigma(X)} = \sigma(\overline{X}) \ \text{ for all } X \in \Omega,$$
$$\rho\sigma(X) = \rho(X) \ \text{ for all } X \in \Omega.$$

The input size of E is given by $\|E\| = |A| + |\Omega| + |UV| + n$.

In the following, we denote the size of the instance by n, i.e., by the same letter as the size of the matrix. This is not a problem, as we can always increase the size of the matrix.

Equations during the Algorithm. During the procedure we will create various other equations and introduce new constants. Still, the original alphabet A never changes and new constants shall represent words in A^*. As a consequence, we will work with equations over $B \cup \Omega$, where B is the smallest alphabet containing A and all constants in $UV\overline{UV}$. In this setting a solution σ assigns words from B to variables. To track the meaning of constants from $B \setminus A$ we additionally require that a solution has a morphism $h : B \mapsto A^*$, which is constant on A. Then given an equation $U = V$ the $h(\sigma(U))$ corresponds to a solution of the original equation. Note that $|B| \le |A| + 2|UV|$ and we therefore we can ignore $|B|$ for the complexity.

A *weight* of a solution (σ, h) of an an equation (U, V) is

$$\mathrm{w}(\sigma, h) = |U| + |V| + 2n \sum_{X \in \Omega} |h(\sigma(X))| \ .$$

Note that we implicitly assume here that if X does not occur in the equation then $\sigma(X) = \epsilon$. Each next equation in the sequence will have a smaller weight, which ensures that we do not cycle.

Two solutions (σ_1, h) and (σ_2, h) of (U, V) that satisfy $h(\sigma_1(X)) = h(\sigma_2(X))$ for each variable X represent the same solution of the original equation and so in some sense are equivalent. We formalise this notion in the following way: for an equation $U = V$ the solution (σ_1, h) is a *simpler equivalent* of the solution (σ_2, h), written as $(\sigma_1, h) \preceq (\sigma_2, h)$ if for each variable X the $\sigma_1(X)$ is obtained from $\sigma_2(X)$ by replacing some letters $b \in B$ by $h(b)$. It follows straight from the definition that if $(\sigma_1, h) \preceq (\sigma_2, h)$ then σ_1 and σ_2 have the same weight.

Note that h is a technical tool used in the analysis, it is not stored, nor transformed by the algorithm, nor it is used in the graph representation of all solutions.

2 Graph Representation of All Solutions

In this section we give an overview of the graph representation of all solutions and the way such a representation is generated. By an *operator* we denote a function that transforms substitutions (for variables). All our operators are rather simple: $\sigma'(X)$ is usually obtained from $\sigma(X)$ by morphisms, appending/prepending letters, etc. In particular, they have a polynomial description. We usually denote them by φ and their applications as $\varphi[\sigma]$.

Recall that the input equation has size n, so in particular it has at most n variables. We say that a word equation (U, V) (with constraints) is *strictly proper* if $U, V \in (B \cup \Omega)^*$, in total U and V have at most n occurrences of variables and $|U| + |V| \le cn^2$ (a possible constant is $c = 28$ as we will see later); an equation is *proper* if $|U| + |V| \le 2cn^2$. The idea is that strictly proper equations satisfy the desired upper-bound and proper equations are some intermediate equations needed during the computation, so they can be a bit larger. Note that the input equation is strictly proper.

The main technical result of the paper states that:

Lemma 1. *Suppose that (U_0, V_0) is a strictly proper equation with $|U_0|, |V_0| > 0$ and let it have a solution (σ_0, h_0). Then there exists a sequence of proper equations (U_0, V_0), (U_1, V_1), ..., (U_m, V_m), where $m > 0$, and families of operators $\Phi_1, \Phi_2, \ldots, \Phi_m$ such that*

- *(U_m, V_m) is strictly proper;*
- *if (σ_i, h_i) is a solution of (U_i, V_i) then there is $\varphi_{i+1} \in \Phi_{i+1}$ and a solution (σ_{i+1}, h_{i+1}) of (U_{i+1}, V_{i+1}) such that $\sigma_i = \varphi_{i+1}[\sigma_{i+1}]$ and $h_i(\sigma_i(U_i)) = h_{i+1}(\sigma_{i+1}(U_{i+1}))$ furthermore, $\mathrm{w}(\sigma_i, h_i) > \mathrm{w}(\sigma_{i+1}, h_{i+1})$;*
- *if (σ_{i+1}, h_{i+1}) is a solution of (U_{i+1}, V_{i+1}) and $\varphi_{i+1} \in \Phi_{i+1}$ then there is h_i such that (σ_i, h_i) is a solution of (U_i, V_i), where $\sigma_i = \varphi_{i+1}[\sigma_{i+1}]$;*
- *each family Φ_i as well as operator $\varphi_i \in \Phi_i$ have polynomial-size description.*

Given (U_0, V_0), all such sequences (for all possible solutions) can be produced in PSPACE.

The exact definition of allowed families of operators Φ is deferred to Section 3, for the time being let us only note that Φ has polynomial description (which can be read from (U_i, V_i) and (U_{i+1}, V_{i+1})), may be infinite and its elements can be efficiently listed, (in particular, it can be tested, whether Φ is empty or not).

Clearly, for an equation in which both U_i and V_i are the same constant, the solution is entirely described by regular languages in order to satisfy the regular constraints. By a simple preprocessing we can actually ensure that all $X \in \Omega$ are present in the equation. In this case the transformation implies that if $\sigma(U_i) = \sigma(V_i) \in B$ then there is exactly one solution that assigns ϵ to each variable. In this way all solutions of the input equation (U, V) are obtained by a path from (U_0, V_0) to some (U_i, V_i) satisfying with $\sigma(U_i) = \sigma(V_i) \in B$: the solution of (U, V) is a composition of operators from the families of operators on the path applied to the solution of (U_i, V_i).

Using Lemma 1 one can construct in PSPACE a graph like representation of all solutions of a given word equation: for the input equation $U = V$ we construct a directed graph $\mathcal{G}$ which has nodes labelled with a proper equations. Then for each strictly proper equation (U_0, V_0) such that $|U_0|, |V_0| > 1$ we use Lemma 1 to list all possible sequences for (U_0, V_0). For each such sequence $(U_0, V_0), (U_1, V_1), \ldots, (U_m, V_m)$ we put the edges $(U_0, V_0) \rightarrow (U_1, V_1)$, $(U_1, V_1) \rightarrow (U_2, V_2) \ldots, (U_{m-1}, V_{m-1}) \rightarrow (U_m, V_m)$ and annotate the edges with the appropriate family of operators. We lastly remove the nodes that are not reachable from the starting node and those that do not have a path to an ending node.

Note that there may be several ways to obtain the same solution, using different paths in the constructed graph.

3   Compression Step

In this section we describe procedures that show the claim of Lemma 1. In essence, given a word equation (U, V) with a solution σ we want to *compress* the word $\sigma(U)$ directly on the equation, i.e., without the knowledge of the actual solution. These compression steps replace the ab-blocks, as defined later in this section. To do this, we sometimes need to modify the equation (U, V).

The crucial observation is that a properly chosen sequence of such compression guarantees that the obtained equation is strictly proper (assuming (U, V) was), see Lemma 7.

Inverse Operators. Given a (nondeterministic) procedure transforming the equation $U = V$ we say that this procedure *transforms the solutions*, if based on the nondeterministic choices and the input equation we can define a family of operators Φ such that:

– For any solution (σ, h) of $U = V$ there are some nondeterministic choices that lead to an equation $U' = V'$ such that $(\varphi[\sigma'], h) \preceq (\sigma, h)$ for some solution (σ', h') of the equation $U' = V'$ and some operator $\varphi \in \Phi$. Furthermore, $h(\sigma(U)) = h'(\sigma'(U))$.

– for every equation $U' = V'$ that can be obtained from $U = V$ and any its solution (σ', h') and for every operator $\varphi \in \Phi$ there is h such that $(\varphi[\sigma'], h)$ is a solution of $U = V$ and $h(\sigma(U)) = h'(\sigma'(U))$.

Note that both $U' = V'$ and Φ depend on the nondeterministic choices, so it might be that for different choices we can transform $U = V$ to $U' = V'$ (with Φ') and to $U'' = V''$ (with a family Φ'').

We also say that the equation $U = V$ with its solution (σ, h) are *transformed into* $U' = V'$ with (σ', h') and that Φ is the *corresponding family of inverse operators*. In many cases, Φ consists of a single operator φ, in such case we call it the *corresponding inverse operator*, furthermore, in some cases φ does not depend on $U = V$, nor on the nondeterministic choices.

ab-blocks. In an earlier paper using the recompression technique [13] there were two types of compression steps: compression of pairs ab, where $a \neq b$ were two different constants, and compression of maximal factor a^ℓ (i.e., ones that cannot be extended to the right, nor left). In both cases, such factors were replaced with a single fresh constant, say c. The advantage of such compression steps was that the replaced factors were non-overlapping, in the sense that when we fixed a pair or block to be compressed, each constant in a word w belongs to at most one replaced factor.

We would like to use similar compression rules also for the case of monoids with involution, however, one needs to take into the account that when some string w is replaced with a a constant c, then also $\overline{w}$ should be replaced with $\overline{c}$. The situation gets complicated, when some of letters in w are fixed point for the involution, i.e., $\overline{a} = a$. In the worst case, when $\overline{a} = a$ and $\overline{b} = b$ the occurrences of ab and $\overline{ab} = ba$ are overlapping, so the previous approach no longer directly applies. (If we start with an equation over a free group then, in the involution has no fixed points in A, but fixed points are produced during the algorithm. They cannot be avoided in our approach, see below.)

Still, the problem can be resolved by replacing factors from a more general class (for a fixed pair of constants ab).

Definition 2. *Depending on a and b, ab-blocks are*

1. *If $a = b$ then there are two types of ab-blocks: a^i for $i \geq 2$ and $\overline{a}^i$ for $i \geq 2$.*
2. *If $a \neq b$, $\overline{a} \neq a$ and $\overline{b} \neq b$ then ab and $\overline{ba}$ are the two types of ab-blocks.*
3. *If $a \neq b$, $\overline{a} = a$ and $\overline{b} \neq b$ then ab, $\overline{ab} = \overline{b}a$ and $\overline{b}ab$ are the three types of ab-blocks.*
4. *If $a \neq b$, $\overline{a} \neq a$ and $\overline{b} = b$ then ab, $\overline{ab} = b\overline{a}$ and $ab\overline{a}$ are the three types of ab-blocks.*
5. *If $a \neq b$, $\overline{a} = a$ and $\overline{b} = b$ then $(ba)^i$ for $i \geq 1$, $a(ba)^i$ for $i \geq 1$, $(ba)^i b$ for $i \geq 1$ and $(ab)^i$ for $i \geq 1$ are the four types of ab-blocks.*

An occurrence of ab-block in a word is an ab-factor, it is maximal, if it is not contained in any other ab-factor.

For a fixed ab block s the s-reduction of the word w is the word w' in which all maximal factors s ($\overline{s}$) are replaced by a new constant c_s ($\overline{c}_s$, respectively). The inverse function is s-expansion.

Observe that s-reduction introduces new constants to B, we extend ρ to it in a natural way, keeping in mind that if s is replaced with c then $\bar{s}$ is replaced with $\bar{c}$. We let $c = \bar{c}$ if and only if $s = \bar{s}$ Note that in this way letters may become fixed point for the involution. For example, $a\bar{a}$ is an $a\bar{a}$-block for $a \neq \bar{a}$. If $a\bar{a}$ is compressed into c then $c = \bar{c}$. It might be that after s-reduction some letter c in the solution is no longer in B (as it was removed from the equation). In such a case we replace c in the solution by $h(c)$, this is described in larger detail later on.

The following fact is a consequence of the definitions of maximal ab-blocks.

Lemma 2. *For any word $w \in B^*$ and two constants $a, b \in B$, maximal ab-blocks in w do not overlap.*

As a consequence of Lemma 2 we can extend the s-reduction to sets S of ab-blocks in a natural way. Such S-reduction words is well-defined and defines a function on B^*. Clearly, S-expansion is a function.

The s-reduction is easy, if all s factors are wholly contained within the equation or within substitution for a variable. It looks non-obvious, when part of some factor s is within the substitution for the variable and part in the equation. Let us formalise those notions: for a word equation (U, V) we say that an ab-factor is *crossing* in a solution σ if it does not come from U (V, respectively), nor from any $\sigma(X)$ nor $\sigma(\overline{X})$ for an occurrence of a variable X; ab is *crossing* in a solution σ, if some ab-factor is crossing. Otherwise ab is *non-crossing* in σ.

By guessing all $X \in \Omega$ with $\sigma(X) = \epsilon$ (and removing them) we can always assume that $\sigma(X) \neq \epsilon$ for all X. In this case crossing ab's can be alternatively characterized.

Lemma 3. *If we have $\sigma(X) \neq \epsilon$ for all X then ab is crossing in σ if and only if one of the following holds:*

- aX, *for an unknown X, occurs in U or V and $\sigma(X)$ begins with b or*
- $\bar{b}X$, *for an unknown X, occurs in U or V and $\sigma(X)$ begins with $\bar{a}$ or*
- Xb, *for an unknown X, occurs in U or V and $\sigma(X)$ ends with a or*
- $X\bar{a}$, *for an unknown X, occurs in U or V and $\sigma(X)$ ends with $\bar{b}$ or*
- XY *or $\overline{YX}$, for unknowns X, Y, occurs in U or V and $\sigma(X)$ ends with b while $\sigma(Y)$ begins with a.*

Since a crossing word ab can be associated with an occurrence of a variable X, it follows that the number of crossing words is linear in the number of occurrences of variables.

Lemma 4. *Let (U, V) be a proper equation such that $\sigma(X) \neq \epsilon$ for all X. Then there are at most $4n$ different crossing words in σ.*

Reduction for Non-crossing ab. When ab is non-crossing in the solution σ we can make the reduction for all ab-blocks that occur in the equation (U, V) on $\sigma(U)$ simply by replacing each ab-factor in U and V. The correctness follows from the fact that each maximal occurrence of ab-block in $\sigma(U)$ and $\sigma(V)$ comes either wholly from U (V, respectively) or from $\sigma(X)$ or $\sigma(\overline{X})$. The former are replaced by our procedure and the latter are replaced implicitly, by changing the solution. Thus it can be shown that the solutions of the new and old equation are in one-to-one correspondence.

Algorithm 1. CompNCr(U, V, ab) Reduction for a non-crossing ab

---

1: $S \leftarrow$ all maximal ab-blocks in U and V
2: **for** $s \in S$ **do**
3:     let c_s be a fresh constant
4:     **if** $s = \overline{s}$ **then**
5:         let $\overline{c_s}$ denote c_s
6:     **else**
7:         let $\overline{c_s}$ a fresh constant
8:     replace each maximal factor s ($\overline{s}$) in U and V by c_s ($\overline{c_s}$, respectively)
9:     set $\rho(c_s) \leftarrow \rho(s)$ and $\rho(\overline{c_s}) \leftarrow \rho(\overline{s})$
10: **return** (U, V)

We should define an inverse operator for CompNCr(U, V, ab) as well as h'. For the latter, we extend h to new letters c_s simply by $h'(c_s) = h(s)$ and keeping all other values as they were. For the former, let S be the set of all maximal ab-blocks in (U, V) and let CompNCr(U, V, ab) replace $s \in S$ by c_s then $\varphi_{\{c_s \to s\}_{s \in S}}$ is defined as follows: in each $\sigma(X)$ it replaces each c_s by s, for all $s \in S$. Note that $\varphi_{\{c_s \to s\}_{s \in S}}$ is the S-expansion.

Lemma 5. *Let ab be non-crossing in a solution σ of an equation (U, V). Let compute a set of ab-blocks S and replace $s \in S$ by c_s. Then CompNCr(U, V, ab) transforms (U, V) with (σ, h) to (U', V') with (σ', h'), where h' is defined as above. Furthermore, define $\varphi_{\{c_s \to s\}_{s \in S}}$ as above; it is the inverse operator.*

If at least one factor was replaced in the equation then $\mathrm{w}(\sigma'(U')) < \mathrm{w}(\sigma(U))$.

Proof. We define a new solution σ' by simply replacing each maximal factor $s \in S$ by c_s. This is not all, as it might be that this solution uses letters that are no longer in (U', V'). In such a case we replace all such letters c with $h(c)$. Concerning the weight, $h'(\sigma'(X)) = h(\sigma(X))$ but if at least one factor was replaced, $|U'| + |V'| < |U| + |V|$.

Reduction for Crossing ab. Since we already know how to compress a non-crossing ab, a natural way to deal with a crossing ab is to "uncross" it and then compress using CompNCr. To this end we pop from the variables the whole parts of maximal ab-blocks which cause this block to be crossing. Afterwards all maximal ab-blocks are noncrossing and so they can be compressed using CompNCr(U, V, ab)

As an example consider an equation $aaXaXaX = aXaYaYaY$. (For simplicity without constraints.) It is easy to see that all solutions are of the form

$X \in a^{\ell_X}$ and $Y = a^{\ell_Y}$, for an arbitrary k. After the popping this equation is turned into $a^{3\ell_X+4} = a^{\ell_X+3\ell_Y}$, for which aa is noncrossing. Thus solution of the original equation corresponds to the solution of the Diophantine equation: $3\ell_X + 4 = \ell_X + 3\ell_Y + 4$. This points out another idea of the popping: when we pop the whole part of block that is crossing, we do not immediately guess its length, instead we treat the length as a parameter, identify ab-blocks of the same length and only afterwards verify, whether our guesses were correct. The verification is formalised as a linear system of Diophantine equations. Each of its solutions correspond to one "real" lengths of ab-blocks popped from variables.

Note that we still need to calculate the transition of the popped ab-block, which depends on the actual length (i.e., on particular solution ℓ_X or r_X). However, this ab block is long because of repeated ab (or $\overline{ba}$). Now, when we look at $\rho(ab), \rho(ab)^2, \ldots$ then starting from some (at most exponential) value it becomes periodic, the period is also at most exponential. More precisely, if $P \in \mathbb{M}_{2n}$ is a matrix, then we can compute in PSPACE an idempotent power $P^p = P^{2p}$ with $p \leq 2^{n^2}$. Thus, if ℓ is a parameter we can guess (and fix) the remainder $r \equiv \ell \bmod p$ with $0 \leq r < p$. We can guess if $r < \ell$ and in this case we substitute the parameter ℓ by $c \cdot p + r$ and we view c as the new integer parameter with the constraint $c \geq 1$. This can be written as an Diophantine equation and added to the constructed linear Diophantine system which has polynomial size if coefficients are written in binary. We can check solvability (and compute a minimal solution) in NP, see e.g., [11]. (For or a more accurate estimation of constants see [3]).

Let D be the system created by CompCr. In this case there is no single operator, rather a family Φ_D, and its elements $\varphi_{D,\{\ell_X,r_X\}_{X \in \Omega}}$ are defined using a solution $\{\ell_X, r_X\}_{X \in \Omega}$ of D. Given an arithmetic expression with integer parameters $\{x_X, y_X\}_{X \in \Omega}$, by $e_i[\{\ell_X, r_X\}_{X \in \Omega}]$ we denote its evaluation on values $x_X = \ell_X$ and $y_X = r_X$. In order to obtain $\varphi_{D,\{\ell_X,r_X\}_{X \in \Omega}}[\sigma](X)$ we first replace each letter c_{e_i} with appropriate type of ab-block of length $e_i[\{\ell_X, r_X\}_{X \in \Omega}]$. Afterwards, we prepend ab block of length ℓ_X and append ab-block of length r_X (in both cases we need to take into the account the types). Concerning h', we extend h to new letters by setting $h'(c_{e_i}) = h(e_i[\{\ell_X, r_X\}_{X \in \Omega}])$.

Lemma 6. *Let (U, V) have a solution σ. Let S be the set of all ab-blocks in U, V or crossing in σ. CompCr(U, V, ab) transforms (U, V) with (σ, h) into an equation (U', V') with (σ', h'). If at least one ab-factor was replaced then $\mathrm{w}(\sigma') < \mathrm{w}(\sigma')$. Furthermore, the family Φ_D is the corresponding family of operators.*

Main Transformation. The crucial property of TransformEq is that it uses equation of bounded size, as stated in the following lemma. Note that this *does not* depend on the non-deterministic choices of TransformEq.

Lemma 7. *Suppose that (U, V) is a strictly proper equation. Then during TransformEq the (U, V) is a proper equation and after it it is strictly proper.*

Algorithm 2. CompCr(U, V, ab) Compression of ab-blocks for a crossing ab

---
1: **for** $X \in \Omega$ **do**
2:     **if** $\sigma(X)$ begins with ab-block s **then**   ▷ Guess
3:         $p \leftarrow$ period of $\rho(ab)$ in $\mathbb{M}_{2n}$ ▷ The same as period of $\overline{ab}$
                                                       ▷ Guess and verify
4:         guess $\rho_s = \rho(s)$ and ρ_X such that $\rho(X) = \rho_s \rho_X$
5:         let x_X denote the length of s ▷ An integer parameter, not number
6:         add to D constraint on length   ▷ Ensure that $\rho_s = \rho(s)$
7:         replace each X with sX, set $\rho(X) \leftarrow \rho_X$
8:         **if** $\sigma(X) = \epsilon$ and $\rho(X) = \rho(\epsilon)$ **then**   ▷ Guess
9:             remove X from the equation
10:    Perform symmetric actions on the end of X
11: let $\{\mathcal{E}_1, \ldots, \mathcal{E}_k\}$ be maximal ab-blocks in (U, V) (read from left to right)
12: **for** each $\mathcal{E}_i$ **do**
13:    let $e_i \leftarrow |\mathcal{E}_i|$ ▷ Arithmetic expression in $\{x_X, y_X\}_{X \in \Omega}$
14: partition $\{\mathcal{E}_1, \ldots, \mathcal{E}_k\}$, each part has ab-blocks of the same type   ▷ Guess
15: **for** each part $\{\mathcal{E}_{i_1}, \ldots, \mathcal{E}_{i_{k_p}}\}$ **do**
16:    **for** each $\mathcal{E}_{i_j} \in \{\mathcal{E}_{i_1}, \ldots, \mathcal{E}_{i_{k_p}}\}$ **do**
17:       add an equation $e_{i_j} = e_{i_{j+1}}$ to D ▷ Ignore the meaningless last equation
18: verify created system D ▷ In NP
19: **for** each $\mathcal{E}_i = \{E_{i_1}, \ldots, E_{i_{k_p}}\}$ **do**
20:    let $c_{e_{i_1}} \in B$ be an unused letter
21:    **for** each $E_{i_j} \in \mathcal{E}_i$ **do**
22:       replace every E_{i_j} by $c_{e_{i_1}}$

Proof. Consider, how many letters are popped into the equation during Trans-formEq. For a fixed ab, CompCr may introduce long ab-blocks at sides of each variable, but then they are immediately replaced with one letter, so we can count them as one letter (and in the meantime each such popped prefix and suffix is represented by at most four constants). Thus, $2n$ letters are popped in this way. There are at most $4n$ crossing pairs, see Lemma 4, so in total $8n^2$ letters are introduced to the equation.

Consider a constant initially present in the equation, say a, which is not followed by a variable and is not the last letter in U or V. When equation has size m, there are at least $m - 2n - 2$ such letters. Thus this a is followed by a letter, say b, and so ab is in P and we tried to compress the maximal ab-factor containing this ab. The only reason, why we failed is that one of a, b was already compressed, as part of a different factor. Thus, if a (not initially followed by a variable, nor the last letter in U and V) was not compressed during TransformEq, then the two following constants were. The left of those constants was present initially at the equation, so at least $(m-2n-2)/3$ initial constants were removed from the equation. Hence the result.

Algorithm 3. TransformEq(U, V)

1: $P \leftarrow$ list of explicit ab's in U, V.
2: $P' \leftarrow$ crossing ab's   ▷ Done by guessing first and last letters of each $\sigma(X)$,
   $|P'| \leq 4n$
3: $P \leftarrow P \setminus P'$
4: **for** $ab \in P$ **do**
5:     CompNCr(U, V, ab)
6: **for** $ab \in P$ **do**
7:     CompCr(U, V, ab)
8: **return** (U, V)

Using the results above we obtain the following theorem:

Theorem 1. *There is PSPACE-transducer for following problem.*

Input: *A system of word equations with rational constraints over a finitely generated free group (resp. over a finitely generated free monoid with involution). Output: A finite labeled directed graph such that the labelled paths describe all solutions. The graph is is empty if and only if the system has no solution.*

Moreover, it can be decided in PSPACE whether the input system with rational constraints has a finite number of solutions.

References

[1] Benois, M.: Parties rationelles du groupe libre. C. R. Acad. Sci. Paris, Sér. A 269, 1188–1190 (1969)
[2] Dahmani, F., Guirardel, V.: Foliations for solving equations in groups: free, virtually free and hyperbolic groups. J. of Topology 3, 343–404 (2010)
[3] Diekert, V., Gutiérrez, C., Hagenah, C.: The existential theory of equations with rational constraints in free groups is PSPACE-complete. Inf. and Comput. 202, 105–140 (2005); Conference version in Ferreira, A., Reichel, H. (eds.) STACS 2001. LNCS, vol. 2010, pp. 170–182. Springer, Heidelberg (2001)
[4] Diekert, V., Lohrey, M.: Word equations over graph products. IJAC 18(3), 493–533 (2008)
[5] Diekert, V., Matiyasevich, Y., Muscholl, A.: Solving word equations modulo partial commutations. TCS 224, 215–235 (1999); Special issue of LFCS 1997
[6] Diekert, V., Muscholl, A.: Solvability of equations in free partially commutative groups is decidable. International Journal of Algebra and Computation 16, 1047–1070 (2006); Journal version of Orejas, F., Spirakis, P.G., van Leeuwen, J. (eds.) ICALP 2001. LNCS, vol. 2076, pp. 543–554. Springer, Heidelberg (2001)
[7] Durnev, V.G.: Undecidability of the positive $\forall\exists^3$-theory of a free semi-group. Sibirsky Matematicheskie Jurnal 36(5), 1067–1080 (1995) (in Russian); English translation: Sib. Math. J. 36(5), 917–929 (1995)
[8] Eilenberg, S.: Automata, Languages, and Machines, vol. A. Academic Press, New York (1974)
[9] Gutiérrez, C.: Satisfiability of word equations with constants is in exponential space. In: Proc. 39th FOCS 1998, pp. 112–119. IEEE Computer Society Press, Los Alamitos (1998)

[10] Gutiérrez, C.: Satisfiability of equations in free groups is in PSPACE. In: Proc. 32nd STOC 2000, pp. 21–27. ACM Press (2000)

[11] Hopcroft, J.E., Ulman, J.D.: Introduction to Automata Theory, Languages and Computation. Addison-Wesley (1979)

[12] Ilie, L., Plandowski, W.: Two-variable word equations. Theoretical Informatics and Applications 34, 467–501 (2000)

[13] Jeż, A.: Recompression: a simple and powerful technique for word equations. In: Portier, N., Wilke, T. (eds.) STACS. LIPIcs, vol. 20, pp. 233–244. Schloss Dagstuhl-Leibniz-Zentrum fuer Informatik, Dagstuhl (2013)

[14] Kharlampovich, O., Myasnikov, A.: Irreducible affine varieties over a free group. II: Systems in triangular quasi-quadratic form and description of residually free groups. J. of Algebra 200(2), 517–570 (1998)

[15] Kharlampovich, O., Myasnikov, A.: Elementary theory of free non-abelian groups. J. of Algebra 302, 451–552 (2006)

[16] Kościelski, A., Pacholski, L.: Complexity of Makanin's algorithm. J. Association for Computing Machinery 43(4), 670–684 (1996)

[17] Makanin, G.S.: The problem of solvability of equations in a free semigroup. Math. Sbornik 103, 147–236 (1977) English transl. in Math. USSR Sbornik 32 (1977)

[18] Makanin, G.S.: Equations in a free group. Izv. Akad. Nauk SSR, Ser. Math. 46, 1199–1273 (1983) English transl. in Math. USSR Izv. 21 (1983)

[19] Makanin, G.S.: Decidability of the universal and positive theories of a free group. Izv. Akad. Nauk SSSR, Ser. Mat. 48, 735–749 (1984) (in Russian); English translation in: Math. USSR Izvestija 25, 75–88 (1985)

[20] Matiyasevich, Y.: Hilbert's Tenth Problem. MIT Press, Cambridge (1993)

[21] Matiyasevich, Y.: Some decision problems for traces. In: Adian, S., Nerode, A. (eds.) LFCS 1997. LNCS, vol. 1234, pp. 248–257. Springer, Heidelberg (1997)

[22] Plandowski, W.: Satisfiability of word equations with constants is in PSPACE. J. Association for Computing Machinery 51, 483–496 (2004)

[23] Plandowski, W.: An efficient algorithm for solving word equations. In: Proc. 38th STOC 2006, pp. 467–476. ACM Press (2006)

[24] Plandowski, W.: (unpublished, 2014)

[25] Plandowski, W., Rytter, W.: Application of Lempel-Ziv encodings to the solution of word equations. In: Larsen, K.G., Skyum, S., Winskel, G. (eds.) ICALP 1998. LNCS, vol. 1443, pp. 731–742. Springer, Heidelberg (1998)

[26] Razborov, A.A.: On Systems of Equations in Free Groups. PhD thesis, Steklov Institute of Mathematics (1987) (in Russian)

[27] Razborov, A.A.: On systems of equations in free groups. In: Combinatorial and Geometric Group Theory, pp. 269–283. Cambridge University Press (1994)

[28] Rips, E., Sela, Z.: Canonical representatives and equations in hyperbolic groups. Inventiones Mathematicae 120, 489–512 (1995)

[29] Schulz, K.U.: Makanin's algorithm for word equations — Two improvements and a generalization. In: Schulz, K.U. (ed.) IWWERT 1990. LNCS, vol. 572, pp. 85–150. Springer, Heidelberg (1992)

Algorithmic Meta Theorems
for Sparse Graph Classes

Martin Grohe

RWTH Aachen University, Germany
grohe@informatik.rwth-aachen.de

Abstract. Algorithmic meta theorems give efficient algorithms for classes of algorithmic problems, instead of just individual problems. They unify families of algorithmic results obtained by similar techniques and thus exhibit the core of these techniques. The classes of problems are typically defined in terms of logic and structural graph theory. A well-known example of an algorithmic meta theorem is Courcelle's Theorem, stating that all properties of graphs of bounded tree width that are definable in monadic second-order logic are decidable in linear time.

This paper is a brief and nontechnical survey of the most important algorithmic meta theorems.

Introduction

It is often the case that a wide range of algorithmic problems can be solved by essentially the same technique. Think of dynamic programming algorithms on graphs of bounded tree width or planar graph algorithms based on layerwise (or outerplanar) decompositions. In such situations, it is natural to try to find general conditions under which an algorithmic problem can be solved by these techniques—this leads to *algorithmic meta theorems*. However, it is not always easy to describe such conditions in a way that is both mathematically precise and sufficiently general to be widely applicable. Logic gives us convenient ways of doing this. An early example of an algorithmic meta theorem based on logic is Papadimitriou and Yannakakis's [40] result that all optimisation problems in the class MAXSNP, which is defined in terms of a fragment of existential second-order logic, admit constant-ratio polynomial time approximation algorithms.

Besides logic, most algorithmic meta theorems have structural graph theory as a second important ingredient in that they refer to algorithmic problems restricted to specific graph classes. The archetypal example of such a meta theorem is Courcelle's Theorem [3], stating that all properties of graphs of bounded tree width that are definable in monadic second-order logic are decidable in linear time.

The main motivation for algorithmic meta theorems is to understand the core and the scope of certain algorithmic techniques by abstracting from problem-specific details. Sometimes meta theorems are also crucial for obtaining new algorithmic results. Two recent examples are

E.A. Hirsch et al. (Eds.): CSR 2014, LNCS 8476, pp. 16–22, 2014.

- a quadratic-time algorithm for a structural decomposition of graphs with excluded minors [27], which uses Courcelle's Theorem;
- a logspace algorithm for deciding wether a graph is embeddable in a fixed surface [16], which uses a logspace version of Courcelle's Theorem [14].

Furthermore, meta theorems often give a quick and easy way to see that certain problems can be solved efficiently (in principle), for example in linear time on graphs of bounded tree width. Once this has been established, a problem-specific analysis may yield better algorithms. However, an implementation of Courcelle's Theorem has shown that the direct application of meta theorems can yield competitive algorithms for common problems such as the dominating set problem [37].

In the following, I will give an overview of the most important algorithmic meta theorems, mainly for decision problems. For a more thorough introduction, I refer the reader to the surveys [26,28,33].

Meta Theorems for Monadic Second-Order Logic

Recall Courcelle's Theorem [3], stating that all properties of graphs of bounded tree width that are definable in monadic second-order logic (MSO) are decidable in linear time. Courcelle, Makowsky, and Rotics [5] generalised this result to graph classes of bounded clique width, whereas Kreutzer and Tazari showed in a series of papers [34,35,36] that polylogarithmic tree width is a necessary condition for meta theorems for MSO on graph classes satisfying certain closure conditions like being closed under taking subgraphs (also see [23]).

In a different direction, Elberfeld, Jakoby, and Tantau [14,15] proved that all MSO-definable properties of graphs of bounded tree width can be decided in logarithmic space and all MSO-definable properties of graphs of bounded tree depth can be decided in AC^0.

Meta Theorems for First-Order Logic

For properties definable in first-order logic (FO), we know meta theorems on a much larger variety of graph classes. Before I describe them, let me point out an important difference between FO-definable properties and MSO-definable properties. In MSO, we can define NP-complete problems like 3-COLOURABILITY. Thus if we can decide MSO-definable properties of a certain class of graphs in polynomial time, this is worth noting. But the range of graph classes where we can hope to achieve this is limited. For example, the MSO-definable property 3-COLOURABILITY is already NP-complete on the class of planar graphs. On the other hand, all FO-definable properties of (arbitrary) graphs are decidable in polynomial time (even in uniform AC^0). When proving meta theorems for FO, we are usually interested in linear-time algorithms or polynomial-time algorithms with a small fixed exponent in the running time, that is, in fixed-parameter tractability (see, for example, [10,19]). When we say that NP-complete problems like MINIMUM DOMINATING

SET are definable in FO, we mean that for each k there is an FO-sentence ϕ_k stating that a graph has a dominating set of size at most k. Thus we define (the decision version of) the DOMINATING SET problem by a family of FO-sentences, and if we prove that FO-definable properties can be decided in linear time on a certain class of graphs, this implies that DOMINATING SET parameterized by the size of the solution is fixed-parameter tractable on this class of graphs. By comparison, we can define 3-COLOURABILITY by a single MSO-sentence, and if we prove that MSO-definable properties can be decided in linear time on a certain class of graphs, this implies that 3-COLOURABILITY can be decided in linear time on this class of graphs.

After this digression, let us turn to the results. The first notable meta theorem for deciding FO-definable properties, due to Seese [41], says that FO-definable properties of bounded-degree graphs can be decided in linear time. Frick and Grohe [21] gave linear-time algorithms for deciding FO-definable properties of planar graphs and all apex-minor-free graph classes and $O(n^{1+\epsilon})$ algorithms for graph classes of bounded local tree width. Flum and Grohe [18] proved that deciding FO-definable properties is fixed-parameter tractable on graph classes with excluded minors, and Dawar, Grohe, and Kreutzer [8] extended this to classes of graphs locally excluding a minor. Dvořák, Král, and Thomas [13] proved that FO-definable properties can be decided in linear time on graph classes of bounded expansion and in time $O(n^{1+\epsilon})$ on classes of locally bounded expansion. Finally, Grohe, Kreutzer, and Siebertz [30] proved that FO-definable properties can be decided in linear time on nowhere dense graph classes. Figure 1 shows the containment relation between all these and other sparse graph classes. Nowhere dense classes were introduced by Nešetřil and Ossona de Mendez [38,39] (also see [29]) as a formalisation of classes of "sparse" graphs. They include most familiar examples of sparse graph classes like graphs of bounded degree and planar graphs. Notably, classes of bounded average degree or bounded degeneracy are not necessarily nowhere dense.

The meta theorem for FO-definable properties of nowhere dense classes is optimal if we restrict our attention to graph classes closed under taking subgraphs: if $\mathcal{C}$ is a class of graphs closed under taking subgraphs that is somewhere dense (that is, not nowhere dense), then deciding FO-properties of graphs in $\mathcal{C}$ is as hard as deciding FO-properties of arbitrary graphs, with respect to a suitable form of reduction [13,33]. Thus under the widely believed complexity-theoretic assumption FPT $\neq$ AW[$*$], which is implied by more familiar assumptions like the exponential time hypothesis or FPT $\neq$ W[1], deciding FO-definable properties of graphs from $\mathcal{C}$ is not fixed-parameter tractable.

There are a few meta theorems for FO-definable properties of graph classes that are somewhere dense (and hence not closed under taking subgraphs). Ganian et al. [24] give quasilinear-time algorithms for certain classes of interval graphs. By combining the techniques of [5] and [21], it can easily be shown that deciding FO-definable properties is fixed-parameter tractable on graphs of bounded local rank width (see [26]). It is also easy to prove fixed-parameter tractability for classes of unbounded, but slowly growing degree [11,25].

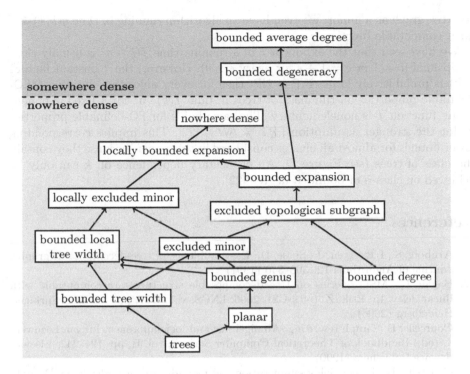

Fig. 1. Sparse graph classes

Counting, Enumeration, and Optimisation Problems

Many of the meta theorems above have variants for counting and enumeration problems, where we are given a formula $\phi(x_1,\ldots,x_k)$ with free variables and want to compute the number of tuples satisfying the formula in a given graph or compute a list of all such tuples, and also for optimisation problems. (See, for example, [1,2,4,6,7,9,12,11,17,20,31,32].)

Uniformity

In this paper, we stated all meta theorems in the form: for every property definable in a logic L on a class $\mathcal{C}$ of graphs there is an $O(n^c)$ algorithm. Here n is the number of vertices of the input graph, and the exponent c is a small constant, most often 1 or $(1+\epsilon)$. However, all these theorems hold in a uniform version of the form: there is an algorithm that, given an L-sentence ϕ and a graph $G \in \mathcal{C}$, decides whether ϕ holds in G in time $f(k) \cdot n^c$, where k is the length of the sentence ϕ and f is some computable function (the exponent c remains the same). For families of classes constrained by an integer parameter ℓ, such as the classes of all graphs of tree width at most ℓ or the classes of all graphs that exclude an

ℓ-vertex graph as a minor, we even have an algorithm running in time $g(k, \ell) \cdot n^c$, for a computable function g.

We have seen that the exponent c in a running time $f(k) \cdot n^c$ is usually close to optimal ($c = 1$ or $c = 1 + \epsilon$ for every $\epsilon > 0$). However, the "constant factor" $f(k)$ is prohibitively large: if $P \neq NP$, then for every algorithm deciding MSO-definable properties on the class of trees in time $f(k) \cdot n^c$ for a fixed exponent c, the function f is nonelementary. The same holds for FO-definable properties under the stronger assumption $FPT \neq AW[*]$ [22]. This implies corresponding lower bounds for almost all classes considered in this paper, because they contain the class of trees (see Figure 1). An elementary dependence on k can only be achieved on classes of bounded degree [22].

References

1. Arnborg, S., Lagergren, J., Seese, D.: Easy problems for tree-decomposable graphs. Journal of Algorithms 12, 308–340 (1991)
2. Bagan, G.: MSO queries on tree decomposable structures are computable with linear delay. In: Ésik, Z. (ed.) CSL 2006. LNCS, vol. 4207, pp. 167–181. Springer, Heidelberg (2006)
3. Courcelle, B.: Graph rewriting: An algebraic and logic approach. In: van Leeuwen, J. (ed.) Handbook of Theoretical Computer Science, vol. B, pp. 194–242. Elsevier Science Publishers (1990)
4. Courcelle, B.: Linear delay enumeration and monadic second-order logic. Discrete Applied Mathematics 157(12), 2675–2700 (2009)
5. Courcelle, B., Makowsky, J., Rotics, U.: Linear time solvable optimization problems on graphs of bounded clique width. Theory of Computing Systems 33(2), 125–150 (2000)
6. Courcelle, B., Makowsky, J., Rotics, U.: On the fixed-parameter complexity of graph enumeration problems definable in monadic second-order logic. Discrete Applied Mathematics 108(1-2), 23–52 (2001)
7. Courcelle, B., Mosbah, M.: Monadic second-order evaluations on tree-decomposable graphs. Theoretical Computer Science 109, 49–82 (1993)
8. Dawar, A., Grohe, M., Kreutzer, S.: Locally excluding a minor. In: Proceedings of the 22nd IEEE Symposium on Logic in Computer Science, pp. 270–279 (2007)
9. Dawar, A., Grohe, M., Kreutzer, S., Schweikardt, N.: Approximation schemes for first-order definable optimisation problems. In: Proceedings of the 21st IEEE Symposium on Logic in Computer Science, pp. 411–420 (2006)
10. Downey, R., Fellows, M.: Fundamentals of Parameterized Complexity. Springer (2013)
11. Durand, A., Schweikardt, N., Segoufin, L.: Enumerating first-order queries over databases of low degree. In: Proceedings of the 33rd ACM Symposium on Principles of Database Systems (2014)
12. Durand, A., Grandjean, E.: First-order queries on structures of bounded degree are computable with constant delay. ACM Transactions on Computational Logic 8(4) (2007)
13. Dvořák, Z., Král, D., Thomas, R.: Deciding first-order properties for sparse graphs. In: Proceedings of the 51st Annual IEEE Symposium on Foundations of Computer Science, pp. 133–142 (2010)

14. Elberfeld, M., Jakoby, A., Tantau, T.: Logspace versions of the theorems of bod-laender and courcelle. In: Proceedings of the 51st Annual IEEE Symposium on Foundations of Computer Science, pp. 143–152 (2010)
15. Elberfeld, M., Jakoby, A., Tantau, T.: Algorithmic meta theorems for circuit classes of constant and logarithmic depth. In: Dürr, C., Wilke, T. (eds.) Proceedings of the 29th International Symposium on Theoretical Aspects of Computer Science. LIPIcs, vol. 14, pp. 66–77. Schloss Dagstuhl, Leibniz-Zentrum fuer Informatik (2012)
16. Elberfeld, M., Kawarabayashi, K.I.: Embedding and canonizing graphs of bounded genus in logspace. In: Proceedings of the 46th ACM Symposium on Theory of Computing (2014)
17. Flum, J., Frick, M., Grohe, M.: Query evaluation via tree-decompositions. Journal of the ACM 49(6), 716–752 (2002)
18. Flum, J., Grohe, M.: Fixed-parameter tractability, definability, and model checking. SIAM Journal on Computing 31(1), 113–145 (2001)
19. Flum, J., Grohe, M.: Parameterized Complexity Theory. Springer (2006)
20. Frick, M.: Generalized model-checking over locally tree-decomposable classes. Theory of Computing Systems 37(1), 157–191 (2004)
21. Frick, M., Grohe, M.: Deciding first-order properties of locally tree-decomposable structures. Journal of the ACM 48, 1184–1206 (2001)
22. Frick, M., Grohe, M.: The complexity of first-order and monadic second-order logic revisited. Annals of Pure and Applied Logic 130, 3–31 (2004)
23. Ganian, R., Hliněný, P., Langer, A., Obdrlek, J., Rossmanith, P., Sikdar, S.: Lower bounds on the complexity of MSO1 model-checking. In: Dürr, C., Wilke, T. (eds.) Proceedings of the 29th International Symposium on Theoretical Aspects of Computer Science. LIPIcs, vol. 14, pp. 326–337. Schloss Dagstuhl, Leibniz-Zentrum fuer Informatik (2012)
24. Ganian, R., Hliněný, P., Král', D., Obdržálek, J., Schwartz, J., Teska, J.: FO model checking of interval graphs. In: Fomin, F.V., Freivalds, R., Kwiatkowska, M., Peleg, D. (eds.) ICALP 2013, Part II. LNCS, vol. 7966, pp. 250–262. Springer, Heidelberg (2013)
25. Grohe, M.: Generalized model-checking problems for first-order logic. In: Ferreira, A., Reichel, H. (eds.) STACS 2001. LNCS, vol. 2010, pp. 12–26. Springer, Heidelberg (2001)
26. Grohe, M.: Logic, graphs, and algorithms. In: Flum, J., Grädel, E., Wilke, T. (eds.) Logic and Automata – History and Perspectives. Texts in Logic and Games, vol. 2, pp. 357–422. Amsterdam University Press (2007)
27. Grohe, M., Kawarabayashi, K., Reed, B.: A simple algorithm for the graph minor decomposition – logic meets structural graph theory. In: Proceedings of the 24th Annual ACM-SIAM Symposium on Discrete Algorithms, pp. 414–431 (2013)
28. Grohe, M., Kreutzer, S.: Methods for algorithmic meta theorems. In: Grohe, M., Makowsky, J. (eds.) Model Theoretic Methods in Finite Combinatorics. Contemporary Mathematics, vol. 558, pp. 181–206. American Mathematical Society (2011)
29. Grohe, M., Kreutzer, S., Siebertz, S.: Characterisations of nowhere denese graphs. In: Seth, A., Vishnoi, N. (eds.) Proceedings of the 32nd IARCS Annual Conference on Foundations of Software Technology and Theoretical Computer Science. LIPIcs, vol. 24, pp. 21–40. Schloss Dagstuhl, Leibniz-Zentrum fuer Informatik (2013)
30. Grohe, M., Kreutzer, S., Siebertz, S.: Deciding first-order properties of nowhere dense graphs. In: Proceedings of the 46th ACM Symposium on Theory of Computing (2014)

31. Kazana, W., Segoufin, L.: First-order query evaluation on structures of bounded degree. Logical Methods in Computer Science 7(2) (2011)
32. Kazana, W., Segoufin, L.: Enumeration of first-order queries on classes of structures with bounded expansion. In: Proceedings of the 32nd ACM Symposium on Principles of Database Systems, pp. 297–308 (2013)
33. Kreutzer, S.: Algorithmic meta-theorems. In: Esparza, J., Michaux, C., Steinhorn, C. (eds.) Finite and Algorithmic Model Theory. London Mathematical Society Lecture Note Series, ch. 5, pp. 177–270. Cambridge University Press (2011)
34. Kreutzer, S., Tazari, S.: Lower bounds for the complexity of monadic second-order logic. In: Proceedings of the 25th IEEE Symposium on Logic in Computer Science, pp. 189–198 (2010)
35. Kreutzer, S., Tazari, S.: On brambles, grid-like minors, and parameterized intractability of monadic second-order logic. In: Proceedings of the 21st Annual ACM-SIAM Symposium on Discrete Algorithms, pp. 354–364 (2010)
36. Kreutzer, S.: On the parameterised intractability of monadic second-order logic. In: Grädel, E., Kahle, R. (eds.) CSL 2009. LNCS, vol. 5771, pp. 348–363. Springer, Heidelberg (2009)
37. Langer, A., Reidl, F., Rossmanith, P., Sikdar, S.: Evaluation of an mso-solver. In: Proceedings of the 14th Meeting on Algorithm Engineering & Experiments, pp. 55–63 (2012)
38. Nešetřil, J., Ossona de Mendez, P.: On nowhere dense graphs. European Journal of Combinatorics 32(4), 600–617 (2011)
39. Nešetřil, J., Ossona de Mendez, P.: Sparsity. Springer (2012)
40. Papadimitriou, C., Yannakakis, M.: Optimization, approximation, and complexity classes. Journal of Computer and System Sciences 43, 425–440 (1991)
41. Seese, D.: Linear time computable problems and first-order descriptions. Mathematical Structures in Computer Science 6, 505–526 (1996)

The Lattice of Definability.
Origins, Recent Developments,
and Further Directions

Alexei Semenov[1,2,3], Sergey Soprunov[2], and Vladimir Uspensky[1]

[1] Moscow State University
[2] Dorodnicyn Computing Center of the Russian Academy of Sciences
[3] Moscow Pedagogical State University
alsemenov@umail.ru, soprunov@mail.ru, vau30@list.ru

Abstract. The paper presents recent results and open problems on classes of definable relations (definability spaces, reducts, relational algebras) as well as sources for the research starting from the XIX century. Finiteness conditions are investigated, including quantifier alternation depth and number of arguments width. The infinite lattice of definability for integers with a successor function (a non ω-categorical structure) is described. Methods of investigation include study of automorphism groups of elementary extensions of structures under consideration, using Svenonius theorem and a generalization of it.

Keywords: Definability, definability space, reducts, Svenonius theorem, quantifier elimination, decidability, automorphisms.

> *"Mathematicians, in general, do not like*
> *to operate with the notion of definability;*
> *their attitude towards this notion*
> *is one of distrust and reserve."*
> Alfred Tarski [Tar4]

1 Introduction. The Basic Definition

One of the "most existential" problems of humanity is "How to define something through something being understood yet". It sounds even more important than "What is Truth?" [John 18:38].

Let us recollect our understanding of the problem in the context of modern mathematics.

Language (of definition):

- Logical symbols including connectives, (free) variables: x_0, x_1, ..., quantifiers and equality $=$.
- Names of relations (sometimes, names of objects and operations as well). The set of these names is called signature. Each name has its number of arguments (its arity).
- Formulas.

E.A. Hirsch et al. (Eds.): CSR 2014, LNCS 8476, pp. 23–38, 2014.

Structure of signature Σ is a triple $\langle D, \Sigma, \mathsf{Int} \rangle$. Here D is a set (mostly countable infinite in this paper) called the *domain* of the structure, Int is an *interpretation* that maps every n-ary name of relation to n-ary relation on D, in other words a subset of D^n.

If the structure of signature Σ is given then every formula in the language with this signature defines a relation on D.

The most common language uses quantifiers over D. But sometimes we consider other options like quantifiers over subsets of D (monadic language), over finite subsets of D (weak monadic language), or no quantifiers at all (quantifier-free language).

Let us fix a domain D and any set of relations S on D. Then we can take any finite subset of S and give names to its elements. We have now a structure and interpretation for elements of its signature as was given beforehand. Any formula in the constructed language defines a relation on D. We call any relation on D obtained in this way *definable* in S (over D).

All relations definable in S constitute *closure* of S. Any closed set of relations is called *definability space* (over D).

Any definability space S has its *group of automorphisms* $\mathsf{Aut}(S)$ i. e. permutations on D that preserve all relations from the space S.

All definability spaces over D constitute the *lattice of definability* with natural lattice operations. Evidently lattices for different (countable infinite) D are isomorphic. In other words we have one lattice only.

Investigation of this lattice is the major topic of our paper. In particular we consider lattice of subspaces of a given definability space. The subspaces are called also *reducts* of the space.

We shall consider finitely generated spaces mostly.

If a set of generators for a space is given we can define the theory of the space. Choosing a system of generators is like choosing a coordinate system for a linear space. For finitely generated spaces such properties as decidability of their theory is invariant for the choosing different sets of generators.

Today we feel that the concept of definability space (independently of formal definition and terminology) is very basic and central for mathematical logic and even for mathematics in general. As we will see from the next chapter it was in use as long as the development of the very discipline of mathematical logic was happening. Nevertheless the major results concerning it, including precise definitions and fundamental theorems were obtained quite late and paid much less respect than those concerning notions of "Truth" and "Provability".

We try to keep our text self-contained and introduce needed definitions.

2   The History and Modern Reflections

In our short historical survey we use materials from [BuDa, Smi, Hod].

We try to trace back original sources and motivations. In some important cases understanding of problems and meaning of notions were changed considerably over time. It is important to consider original ideas along with their maturity 30 years and even much later. As we will see the scene of events was pretty much international.

2.1 Relations, Logic, Languages. Early Approaches of XIX Century

Our central notion of an assignment satisfying a formula is implicit in George Peacock [Pea] and explicit in George Boole [Boo], though without a precise notion of "formula" in either case.

In 1860 – 1890-s Frege developed understanding of relations and quantifiers [Fre, Fre1].

Peirce established the fundamental laws of the calculus of classes and created the theory of relatives. Essentially it was the definability issue. Starting with his 1870 paper [Pei], Peirce presented the final form of his first-order logic in his 1885 paper [Pei1]. Pierce's theory was the frame that made possible the proof by Leopold Löwenheim of the first metamathematical theorem in his [Löw]. Löwenheim proved that every first-order expression [Zählausdrücke] is either contradictory or already satisfiable in a countable infinite domain (see [Sko]). So, the major concepts of semantics were used by Löwenheim as well as Thoralf Skolem, but were not explicitly presented in their papers.

Schröder proposed the first complete axiomatization of the calculus of classes and expanded considerably the calculus and the theory of relatives [Sch, Sch1].

2.2 Automorphisms. Isomorphisms

With the appearance of Klein's Erlangenprogramm in 1872 [Kle] it became clear that automorphism groups are useful means of studying mathematical theories.

The word "isomorphism" appeared in the definition of categoricity bt Huntington [Hun]. There he says that "special attention may be called to the discussion of the notion of isomorphism between two systems, and the notion of a sufficient, or categorical, set of postulates".

Alfred Tarski in his paper "What are Logical Notions?", presented first in 1963 [Tar4] explains the subject of logic as study of "everything" up to permutations: "I shell try to extend his [Klein's] method beyond geometry and apply it also to logic ... I use the term "notion" in a rather loose and general sense ... Thus notions include individuals, classes of individuals, relations on individuals".

2.3 How to Define Major Mathematical Structures? Geometry and Numbers. 1900-s. The Width

At the end of XIX century Italian (Giuseppe Peano, Alessandro Padoa, Mario Pieri, ...) and German (Gotlob Frege, Moritz Pasch, David Hilbert, ...) mathematicians tried to find "the best" set of primitive notions for Geometry and Arithmetic considered as deductive systems. This was about "how to define something through something".

In August 1900 The First International Congress of Philosophy [ICP] followed by the Second International Congress of Mathematicians [ICM] met in Paris.

At the mathematical congress Hilbert presented his list of problems [Hil], some of which became central to mathematical logic, Padoa gave two talks on the axiomatizations of the integers and of geometry.

At the philosophical congress Russell read a paper on the application of the theory of relations to the problem of order and absolute position in space and time. The Italian school of Peano and his disciples contributed papers on the logical analysis of mathematics. Peano and Burali-Forti spoke on definitions, Pieri spoke on geometry considered as a purely logical system. Padoa read his famous essay containing the "logical introduction to any theory whatever", where he states:

> "To prove that the system of undefined symbols is irreducible with respect to the system of unproved propositions [axioms] it is necessary and sufficient to find, for any undefined symbol, an interpretation of the system of undefined symbols that verifies the system of unproved propositions and that continues to do so if we suitably change the meaning of only the symbol considered."

Pieri formulated about 1900 and completed in his 1908 "Point and Sphere" memoir, a full axiomatization of Euclidean geometry based solely on the undefined notions point and equidistance of two points from a third point [Pie].

Tarski's undefined notions were point and two relations: congruence of two point pairs and betweenness of a triple. Tarski and Adolf Lindenbaum [LiTa] showed that in the first-order context, Pieri's selection of equidistance as the sole undefined relation for Euclidean geometry was optimal. No family of binary relations, however large, can serve as the sole undefined relations.

We considered the problem of minimization of maximal number of arguments in generators of a given definability space.

Definition 1. *Let a definability space S is given. Its* width *is the minimal n such as S can be generated by relations with n or less arguments.*

Theorem 1. [Sem] *There are definability spaces of any finite or countable width.*

Huntington and Oswald Veblen were part of a group of mathematicians known as the American Postulate Theorists. Huntington was concerned with providing "complete" axiomatizations of various mathematical systems, such as the theory of the algebra of logic and the theory of real numbers. In 1935 Hungtington published [Hun1] "Inter-relations among the four principal types of order", where he says:

> "The four types of order whose inter-relations are considered in this paper may be called, for brevity, (1) serial order; (2) betweenness; (3) cyclic order; and (4) separation."

These "four types of order" will play special role in the further developments discussed in the present paper.

2.4   The Exact Formulation of Definability

Indirectly the notion of truth and even more indirectly definability were present from beginning of 1900-s and even earlier. For example he word "satisfy" in this context may be due to Huntington (for example in [Hun2]). We mentioned works of Löwenheim and Skolem.

But only the formal ("usual" inductive) definition of truth by Tarski gives the modern (model-theoretic) understanding of semantics of a formula as a relation over a domain [Tar].

Complications in understanding today of Tarski and Lindenbaum meaning of Padoa's method (relevant for our considerations) are discussed in [Hod1].

2.5   Elimination of Quantifiers

In the attempts to describe meaning of logical formulas and to obtain "decidability" (in the sense "to prove or disprove") versions of quantifier elimination were developed in 1910 − 1930-s. Remarkable results were published in the end of 1920-s.

C. H. Langford used this method in 1927 [Lan, Lan1] to prove decidability of the theory of dense order without a first or last element.

Mojżesz Presburger [Pre] proved elimination of quantifiers for the additive theory of the integers.

Not using the formal (Tarski-style) definition Skolem illustrated in 1929 an elimination [Sko1] for the theory of order and multiplication on the natural numbers. The complete proof was obtained by of Mostowski in 1952 [Mos].

Tarski himself announced in 1931 a decision procedure for elementary algebra and geometry (published however only in 1948, see [Tar1]).

Elimination of quantifiers was considered as a part of introducing semantics. A natural modern definition appealing to finite signature was not used. In fact, both Presburger and Tarski structures DO NOT permit elimination of quantifiers in this sense. But in these cases you can either choose using operations and terms in atomic formulas, or take a finite set of generators, then every formula can be effectively transform to an equivalent of a limited quantifier depth (the number of quantifier changes).

Let S be a definability space generated by a finite set of relations F. Consider a quantifier hierarchy of subsets of S: $F_0, F_1, \ldots$. Here F_0 is a quantifier-free (Boolean) closure of F, for every $i = 0, 1, \ldots F_{i+1}$ is obtained from F_i by taking all projections of its relations (adding of existential quantifiers) and then getting Boolean closure. (An alternative definition can be given by counting quantifier alternations.) The hierarchy can be of infinite length if F_{i+1} differs of F_i for all i, or finite length n − minimal for which $F_{n+1} = F_n$.

Here are several well-known examples. We indicate informally the structure and the − length of the hierarchy for it:

− $\langle \mathbb{Q}; < \rangle$ − 0.
− Dense order $[0, 1]$ − 0 (if we include these elements into signature).

- $\langle \mathbb{Z}; +1 \rangle$ – 1.
- Presburger arithmetic – 1. Linear forms, congruences module m can be introduced via existential quantifiers. Extensions of + with rapidly growing functions [Sem1].
- Tarski algebra – 1. Again, polynomials can be explained with existential quantifiers only.
- Skolem arithmetic – 1.
- Multiple successor arithmetic (automata proofs) – 1.
- Arithmetic of + and × – infinity.

A priory the length of the hierarchy for the space S can depend of the choice of (finite) F.

Problem 1. Can the hierarchy length be really different for different choices of F?

Definition 2. *The* depth *of a definability space is the minimal (over all finite sets of generators) length of the quantifier hierarchy for it.*

In [Sem] a problem on existing of other options was formulated. The answer was obtained in 2010:

Theorem 2. [SeSo] *There are spaces of arbitrary finite or infinite depth.*

Problem 2. Are there "natural" examples of "big" (2, 3, 4, ...) finite depth?

Problem 3. What is the depth of Rabin Space [Rab], [Muc]?

2.6   Decidability

Decidability in the sense of existing an algorithm to decide is a statement (closed formula) true or false was a key question of study. For example, Tarski result on the field of reals implies the decidability of Geometry. The decidability results for multiple successor arithmetic led Elgot and Rabin to the following problem

Problem 4. [ElRa] Does there exist a structure with maximally decidable theory?

We say that a finitely generated definability space has a *maximally decidable theory* iff its theory is decidable and any greater finitely generated definability space does not have a decidable theory.

Soprunov proved in [Sop] (using forcing arguments) that every space in which a regular ordering is definable is not maximal. A partial ordering $\langle B; < \rangle$ is said to be *regular* if for every $a \in B$ there exist distinct elements $b_1, b_2 \in B$ such that $b_1 < a$, $b_2 < a$, and no element $c \in B$ satisfies both $c < b_1$ and $c < b_2$. As a corollary he also proved that there is no maximal decidable space if we use weak monadic language for definability instead of our standard language.

In [BeCe], Bès and Cégielski consider a weakening of the Elgot – Rabin question, namely the question of whether all structures M whose theory is decidable can be expanded by some constant in such a way that the resulting structure still has a decidable theory. They answer this question negatively by proving that there exists a structure M with a decidable theory (even monadic theory) and such that any expansion of M by a constant has an undecidable theory.

In [BeCe1] they indicate a sufficient condition for a space with decidable theory no to be maximal.

In our context it is natural to consider also decidability of elements of a definability space. Of course we need a "constructivisation" of the domain D. For example, we can take natural numbers as it.

Definition 3. *We call a space* decidable *if all its elements are decidable. We call a finitely generated space* uniformly decidable *if there is an algorithm providing a decision procedure for any formula (using the generators) and any vector of its arguments.*

Problem 5. [Sem], 2003. Are there spaces of arbitrary finite or infinite depth with decidable theory?

Problem 6. Are there decidable and uniformly decidable spaces of arbitrary finite or infinite depth?

Problem 7. Does there exist a maximal decidable structure?

We say that a finitely generated definability space is *maximal decidable* iff it's decidable but any greater finitely generated definability space is not decidable.

As it was shown in [Sem1] there is an unary predicate R for which the space generated by $+$, R on the domain of natural numbers is decidable, but not uniformly, and has an undecidable theory.

3   General Fundamental Theorems on Definability vs. Provability and Automorphisms. 1950-s

Buchi and Danhof [BuDa] outlined the transition between end of 1930-s and end of 1950-s:

> "At this time it might have seemed that most of the basic problems of elementary axiom systems were solved. A more careful observer however, upon reading the papers of Tarski [Tar2, Tar3], might have wondered about the existence of general theorems which would explain elementary definability as the above theorems explain the basic properties of elementary logical consequence.
>
> One such theorem, the completeness, in the sense of definability, of elementary logic was proved by Beth in 1953 [Bet]. In 1959 Svenonius [Sve] published a further result on elementary definability. Just as with the earlier results of Beth and Craig, logicians seem slow in recognizing Svenonius' theorem as a basic tool in the theory of definability, perhaps because it is not generally known to be available."

These results are generally considered as realization of Padoa's idea (or "method").

Let Σ is a signature, we say that $M' = \langle D', \Sigma, \mathsf{Int}' \rangle$ is an *extension* of $M = \langle D, \Sigma, \mathsf{Int} \rangle$ if D is a subset of D' and $\mathsf{Int}(R)$ is the restriction of $\mathsf{Int}'(R)$ on D for any $R \in \Sigma$.

We say that M' is an *elementary extension* of M if the previous condition holds for any definable relation, i.e. if R is definable in M relation, then R is the restriction on D of the relation, definable in M' by the same formula.

In our context Svenonius' theorem is the most useful tool. Here is its suitable formulation.

Theorem 3. (Svenonius Theorem) *Let M — countable structure with signature Σ^+ and let $\Sigma \subset \Sigma^+, R \in \Sigma^+$. The following statements are equivalent:*

(i) Relation R belongs to closure of Σ in M,

(ii) For any M' countable elementary extension of M and any permutation of the domain of M' which preserves Σ, preserves R.

The idea here is to use an additional structure to the original one and consider its elementary extensions. The additional structure narrows the class of extensions and makes the extensions more comprehensible, so we can find the needed automorphism.

In fact, we can use one universe only in a modification of the theorem as was shown in [SeSo1].

By $\mathcal{F}$ we denote the set of everywhere defined functions $f \colon \mathbb{N} \to D$. If R is n-ary relation on D and φ is a mapping $\mathcal{F} \to \mathcal{F}$ then we say that φ *almost preserves* R if $\{i \mid R(f_1(i), \ldots, f_n(i)) \not\equiv R(\varphi(f_1)(i), \ldots, \varphi(f_n)(i))\}$ is finite for any $f_1, \ldots, f_n$ in $\mathsf{Dom}(\varphi)$.

Theorem 4. (CH) *Let S be a definability space. The following conditions are equivalent:*

(1) Relation $R \in S$,

(2) any permutation φ on $\mathcal{F}$ which almost preserves all relations from S almost preserves R.

The remarkable feature of this form of Svenonius Theorem is that the condition (2) is purely combinatorial, not appealing to any logical language.

4   The Definability Lattice

Numerous results were devoted to the study of specific definability spaces. For example, Inan Korec in [Kor] surveyed different natural generation sets for the definability space generated by addition and multiplication of integers.

Cobham — Semenov's theorem [Sem2] states that nontrivial intersection of spaces generated by automata working in different bases should be exactly the space generated by $+$. (This will be considered later in the context of self-definability of Muchnik.)

4.1 Authomorphisms and Galois Correspondence. ω-categoricity

As we see in Svenonius theorem the authomorphism group is an important object in the study of definability spaces.

The *symmetric group* $\mathsf{Sym}(D)$ on a set D is the group consisting of all permutations of D.

There is a natural topology on the symmetric group, we mean the topology of pointwise convergence: a basis of neighborhoods of an element consists of all permutations that coincide with the element on a finite set.

It's easy to see that for spaces S and T we have $S \subseteq T \Rightarrow \mathsf{Aut}(S) \supseteq \mathsf{Aut}(T)$ and that automorphism groups for spaces are closed. So, we can call groups corresponding to reducts of a space S supergroups of $\mathsf{Aut}(S)$.

Groups for different spaces can coincide.

An ω-*categorical* structure is one for which all countable structures that are elementary equivalent to it are isomorphic to it.

For ω-categorical structures definability subspaces are in one-to-one correspondence with closed automorphism groups, so $S \subseteq T$ iff $\mathsf{Aut}(S) \supseteq \mathsf{Aut}(T)$, i.e the correspondence between definability spaces and their automorphism groups is an antitone Galois connection.

It immediately follows from Svenonius theorem, but in the special case of ω-categoricity it may be concluded from so called Engeler – Ryll-Nardzewski – Svenonius Theorem (see e. g. [Hod2]).

4.2 The Rational Order. Homogeneous Structures

We start with a case of the most famous definability space where all subspaces were discovered first. This result describing the lattice of subspaces of $\langle \mathbb{Q}; < \rangle$ was obtained by Claude Frasnay in 1965 [Fra]. All subspaces of rational order are given by the following descriptions:

- One may view the ordering up to reversal, and so obtain a (ternary) linear Betweenness relation B on $\mathbb{Q}$, where $B(x; y, z)$ holds if and only if $y < x < z$ or $z < x < y$.
- Alternatively, by bending the rational line into a Circle one obtains a natural (ternary) circular ordering K on $\mathbb{Q}$; here, $K(x, y, z) \iff (x < y < z) \vee (y < z < x) \vee (z < x < y)$.
- The latter too may be viewed up to reversal, to obtain the (quaternary) Separation relation S: $S(x, y; z, w)$ if the points x, y in the circular ordering separate the points z, w.

The remarkable fact is that these are exactly the structures that in axiomatic form were described by Huntington in 1935 [Hun1] (as was mentioned above).

The structure $\langle \mathbb{Q}; < \rangle$ is ω-categorical. The method of proof for this is "back-and-forth" argument discovered by Huntington (not Cantor) [Hun3]. In fact the proof shows that $\langle \mathbb{Q}; < \rangle$ is homogeneous in the following sense.

Definition 4. *A structure M is* homogeneous *if every isomorphism between its finite substructures extends to an automorphism of M.*

This definition is a generalization of its "group counterpart".

Definition 5. *A permutation group is* homogeneous *iff any finite subset of its domain can be translated to any other subset of the same cardinality with an element of the group.*

It's obvious, that if $\mathsf{Aut}(S)$ is homogeneous, then the structure is homogeneous as well. Actually not only the structure $\langle \mathbb{Q}; < \rangle$ is homogeneous, but also the group $\mathsf{Aut}(\langle \mathbb{Q}; < \rangle)$ is homogeneous.

Peter Cameron [Cam1] showed that there are just four homogeneous nontrivial groups of permutations on a countable set. As the corollary we get, that in the case of $\langle \mathbb{Q}; < \rangle$ apart from $\mathsf{Aut}(\langle \mathbb{Q}; < \rangle)$ and $\mathsf{Sym}(\mathbb{Q})$, there are just three homogeneous groups. The first is the group of all permutations of $\mathbb{Q}$ which either preserve the order or reverse it. The second is the group of all permutations which preserve the cyclic relation "$x < y < z$ or $y < z < x$ or $z < x < y$"; this corresponds to taking an initial segment of $\mathbb{Q}$ and moving it to the end. The third is the group generated by these other two: it consists of those permutations which preserve the relation "exactly one of x, y lies between z and w".

All countable homogeneous structures are ω-categorical, if they have a finite signature or signature finite for any fixed number of variables. For ω-categorical structures homogeneity is equivalent to quantifier elimination. All reducts of $\langle \mathbb{Q}; < \rangle$ are homogeneous and have quantifier elimination.

A good source for information related to homogeneous structures is [Mac].

4.3   The Random Graph. Thomas Conjecture

Our next example is one more remarkable homogeneous structure.

Definition 6. *We call a countable graph* random *iff given two finite disjoint sets U, V of vertices, there exists a vertex z joined to every vertex in U and to no vertex in V.*

This Is called "Alice's Restaurant Property". The term was coined by Peter Winkler [Win], in reference to a popular song by Arlo Guthrie. The refrain of the song "You can get anything you want at Alice's restaurant" catches the spirit of this property.

Any two random graphs are isomorphic. The proof is similar to the isomorphism proof for every two countable dense unlimited orders (the $\mathbb{Q}$ case). The term "random" can be explained by the following property:

> If a graph X on a fixed countable vertex set is chosen by selecting edges independently at random with probability $1/2$ from the unordered pairs of vertices, then $\mathsf{Prob}(X=R) = 1$.

An explicit construction of R in [Rad]:

> The set of vertices is $\mathbb{N}$, and x is connected to y if and only if the x-th digit in the base 2 expansion of y is equal to 1 or vice versa.

Here are the subspaces of the random graph R

Let $R^{(k)}$ be the k-ary relation that contains all k-tuples of pairwise distinct elements $x_1, \ldots, x_k$ in V such that the number of (undirected) edges between those elements is odd.

$R(a, b)$ – "(ab) is an edge in R"; $R^{(3)}$;$R^{(4)}$;$R^{(5)}$;Sym — equality
This description is given in [Tho].

It easy to see that structure of $R^{(3)}$ is not homogeneous and does not have quantifier elimination.

Simon Thomas proved obtained this description in [Tho1]. and suggested the following conjecture:

If M is a finitely generated homogeneous structure then M has finitely many reducts.

Problem 8. Verify Thomas conjecture.

4.4   Further Examples

In order to verify Thomas conjecture the superposition of two homogeneous structures: $\langle \mathbb{Q}; < \rangle$ and random graph $\langle G; E \rangle$ was considered in [BoPiPo]. They presented a complete classification of the reducts of this random ordered graph up to equivalence. It was shown that without counting obvious reducts $\langle D; <, E \rangle$ and $\langle D; = \rangle$ there are precisely 42 such reducts.

In [JuZi] was described a complete lattice of the reducts of expansion of the structure $\langle \mathbb{Q}; < \rangle$ by a constant. This expansion can be considered as expansion by three unary predicates: "$x < a$"; "$x = a$"; and "$x > a$". Actually in this paper different expansions of $\langle \mathbb{Q}; < \rangle$ by unary predicates that have quantifier elimination were studied. They classified the reducts of such expansions and showed that there are only finitely many such. In particular it shows that in the simplest case: expansion of rational numbers by two convex subsets (a cut of the rational numbers) there are exactly 53 reducts, generated by the 5 standard reducts on the elements of the cut as well as permutations preserving, swapping and mixing elements of the cut.

Let us mention the example of an ω-categorical structure, which shows that the condition of quantifier elimination in the Thomas' Conjecture is necessary: [AhZi] describes infinitely many reducts of a doubled infinite-dimensional projective space over binary field (F_2).

4.5   Not ω-categorical Spaces. Integers with Successor – Depth 1

We don't know too much about the reducts of not ω-categorical structure. Answering the dual question to Thomas' one [BoMa] constructs an example of not ω-categorical structure with the finite reducts lattice — actually the lattice contains only two items. This example is based on tree of valency three structure.

Another (more simple) example was demonstrated in the [KaSi]. Answering a question from [BoMa] they show that the structure $\langle \mathbb{Q}; S(x, y, z) \rangle$, where

$S(x, y, z) \equiv (z{=}(x{+}y)/2)$ (or, the same, the structure $\langle \mathbb{Q}; f(x, y, z)\rangle$ where $f(x, y, z) = x{-}y{+}z$) admits no definable reduct. Though Svenonious theorem is not used explicitly in the proof, the approach is rather similar. They note that the structure $\langle \mathbb{Q}^{<\omega}; +\rangle$ is the saturated elementary extension of the $\langle \mathbb{Q}; +\rangle$, so it's enough to consider permutations of the structure $\langle \mathbb{Q}^{<\omega}; +\rangle$ only. Now the fact that $\mathsf{Aut}(\langle \mathbb{Q}^{<\omega}; f\rangle)$ is maximal closed nontrivial subgroup (proved in the same paper) is used.

The structure $\langle \mathbb{Z}; +1\rangle$ — integer numbers with the successor relation is not ω-categorical, and has depth 1. For any natural number n we define spaces by their generators

"$x_1{-}x_2 = n$" — A_n,
"$x_1{-}x_2 = x_3{-}x_4 = n \vee x_1{-}x_2 = x_3{-}x_4 = -n$" — B_n, and
"$|x_1{-}x_2| = n$" — C_n.

Theorem 5. [SeSo2] *Any subspaces of $\langle \mathbb{Z}; +1\rangle$ is A_n or B_n or C_n for a natural n.*
$A_n \succ B_n \succ C_n$ for any n and if $n \neq m$ then $A_n \succ A_m$, $B_n \succ B_m$, $C_n \succ C_m$ iff n is a divisor of m.

Problem 9. Describe the lattice of subspaces for $\langle \mathbb{N}; +1\rangle$.

Problem 10. Describe the lattice of subspaces for natural numbers with multiple successors.

We leave out the researches on the reducts of the field of real [MaPe, Pet] and complex [MaPi] numbers.

4.6   Decidability of the Lattice Problems. Muchnik's Self-definability

A natural algorithmic problem for an algebraic structure of definability lattice is does an element of a space (given by a formula in or case) belong to a subspace generated by a given set of elements? Positive and negative results on this for homogeneous structures were obtained in [BoPiTs].

Andrei Muchnik in his work [Muc1] introduced the following

Definition 7. *A definability space S is called* self-definable *iff there is a finite signature (set of generators) Σ for S and sequence of formulas $F_1, \ldots, F_n, \ldots$ such that for any $n = 1, 2, \ldots$*
    1. F_n is a closed formula in signature $\Sigma \cup \{P\}$, where P is an n-ary symbol
    2. F_n is true iff we take as interpretation of P an element from S.

He proved

Theorem 6. *The space $\langle \mathbb{N}; +\rangle$ is self-definable.*

He writes:

"Unfortunately, we do not know any other examples of nice self-definable structures.

Structures with unsolvable elementary theory are usually mutually inter-
pretable with the arithmetic of addition and multiplication of integers,
the non-self-definability of which is proved in [Add] (using category arguments and [Tan] using measure arguments).

We believe that the structure formed by algebraic real numbers (with
addition and multiplication) is not self-definable; however, a formal proof
is missing (and seems to be rather complicated).

(Note that it is easy to prove that the structure formed by all real
numbers with addition and multiplication is not self-definable. Indeed,
let us assume that $\Phi(A)$ is true if and only if A is definable. Now we
replace $A(x)$ by $x = y$. The new formula $\Phi'(y)$ is true if and only if y
is algebraic. But we can eliminate quantifiers in $\Phi'(y)$ and get a finite
union of segments. So we come to a contradiction.)"

Problem 11. Give more examples of structures with self-definability property.

References

[Add]   Addison Jr., J.W.: The undefinability of the definable. Notices Amer. Math.
         Soc. 12, 347 (1965)
[AhZi]   Ahlbrandt, G., Ziegler, M.: Invariant subgroups of $^V V$. J. Algebra 151(1),
         26–38 (1992)
[BeCe]   Bès, A., Cégielski, P.: Weakly maximal decidable structures. RAIRO - The-
         oretical Informatics and Applications 42(1), 137–145 (2008)
[BeCe1] Bès, A., Cégielski, P.: Nonmaximal decidable structures. Journal of Mathe-
         matical Sciences 158(5), 615–622 (2009)
[Bet]   Beth, E.W.: On Padoa's method in the theory of definition. Indag. Math. 15,
         330–339 (1953)
[BoMa]   Bodirsky, M., Macpherson, D.: Reducts of structures and maximal-closed
         permutation groups. arXiv:1310.6393. (2013)
[Boo]   Boole, G.: The mathematical analysis of logic. Philosophical Library (1847)
[BoPiPo] Bodirsky, M., Pinsker, M., Pongrácz, A.: The 42 reducts of the random or-
         dered graph. arXiv:1309.2165 (2013)
[BoPiTs] Bodirsky, M., Pinsker, M., Tsankov, T.: Decidability of definability. In:
         26th Annual IEEE Symposium on Logic in Computer Science (LICS). IEEE
         (2011)
[BuDa]   Buchi, J.R., Danhof, K.J.: Definability in normal theories. Israel Journal of
         Mathematics 14(3), 248–256 (1973)
[Cam]   Cameron, P.J.: Aspects of infinite permutation groups. London Mathematical
         Society Lecture Note Series 339, 1 (2007)
[Cam1]   Cameron, P.J.: Transitivity of permutation groups on unordered sets. Math-
         ematische Zeitschrift 148(2), 127–139 (1976)
[ElRa]   Elgot, C.C., Rabin, M.O.: Decidability and Undecidability of Extensions of
         Second (First) Order Theory of (Generalized) Successor. J. Symb. Log. 31(2),
         169–181 (1966)
[Fra]   Frasnay, C.: Quelques problèmes combinatoires concernant les ordres totaux
         et les relations monomorphes. Annales de l' institut Fourier 15(2). Institut
         Fourier (1965)

[Fre]   Frege, G.: Begriffsschrift, eine der arithmetischen nachgebildete Formel-
         sprache des reinen Denkens. Halle. (1879); van Heijenoort J. (trans.) Be-
         griffsschrift, a formula language, modeled upon that of arithmetic, for pure
         thought. From Frege to Gödel: A Source Book in Mathematical Logic, 3–82
         (1879-1931)
[Fre1]   Frege, G.: Grundgesetze der Arithmetik, Jena: Verlag Hermann Pohle, Band
         I/II. The Basic Laws of Arithmetic, by M. Furth. U. of California Press,
         Berkeley (1964)
[Hil]   Hilbert, D.: Mathematische probleme. Nachrichten von der Gesellschaft
         der Wissenschaften zu Göttingen, Mathematisch-Physikalische Klasse 1900,
         253–297 (1900)
[Hod]   Hodges, W.: Model Theory (Draft July 20, 2000),
         http://wilfridhodges.co.uk/history07.pdf
[Hod1]   Hodges, W.: Tarski on Padoa's method (2007),
         http://wilfridhodges.co.uk/history06.pdf
[Hod2]   Hodges, W.: Model theory, Encyclopedia of Mathematics and its Applica-
         tions, vol. 42. Cambridge University Press, Cambridge (1993)
[Hun]   Huntington, E.V.: The Fundamental Laws of Addition and Multiplication in
         Elementary Algebra. The Annals of Mathematics 8(1), 1–44 (1906)
[Hun1]   Huntington, E.V.: Inter-Relations Among the Four Principal Types of Order.
         Transactions of the American Mathematical Society 38(1), 1–9 (1935)
[Hun2]   Huntington, E.V.: A complete set of postulates for the theory of absolute con-
         tinuous magnitude. Transactions of the American Mathematical Society 3(2),
         264–279 (1902)
[Hun3]   Huntington, E.V.: The continuum as a type of order: an exposition of the
         model theory. Ann. Math. 6, 178–179 (1904)
[ICM]   Duporcq, E. (ed.): Compte rendu du deuxième Congrès international des
         mathématiciens: tenu à Paris du 6 au 12 août 1900: procès-verbaux et com-
         munications. Gauthier-Villars (1902)
[ICP]   International Congress of Philosophy. 1900–1903. Bibliothèque du Congrès
         International de Philosophie. Four volumes. Paris: Librairie Armand Colin
[JuZi]   Junker, M., Ziegler, M.: The 116 reducts of $(\mathbb{Q}, <, a)$. Journal of Symbolic
         Logic, 861–884 (2008)
[KaSi]   Kaplan, I., Simon, P.: The affine and projective groups are maximal.
         arXiv:1310.8157 (2013)
[Kle]   Klein, F.: Vergleichende betrachtungen über neuere geometrische forsuchun-
         gen. A. Deichert (1872)
[Kor]   Korec, I.: A list of arithmetical structures complete with respect to the first-
         order definability. Theoretical Computer Science 257(1), 115–151 (2001)
[Lan]   Langford, C.H.: Some theorems on deducibility. Annals of Mathematics Sec-
         ond Series 28(1/4), 16–40 (1926–1927)
[Lan1]   Langford, C.H.: Theorems on Deducibility (Second paper). Annals of Math-
         ematics, Second Series 28(1/4), 459–471 (1926–1927)
[LiTa]   Lindenbaum, A., Tarski, A.: Über die Beschränktheit der Ausdrucksmittel
         deduktiver Theorien. Ergebnisse eines Mathematischen Kolloquiums, fasci-
         cule 7 (1934–1935) (Engl. trans.: On the Limitations of the Means of Ex-
         pression of Deductive Theories. In: Corcoran, J. (ed.) Alfred Tarski: Logic,
         Semantics, Metamathematics, Hackett, Indianapolis, 384–392 (1935))
[Löw]   Löwenheim, L.: Über möglichkeiten im relativkalkül. Mathematische An-
         nalen 76(4), 447–470 (1915)

[Mac]   Macpherson, D.: A survey of homogeneous structures. Discrete Mathematics 311(15), 1599–1634 (2011)

[MaPe]   Marker, D., Peterzil, Y.A., Pillay, A.: Additive reducts of real closed fields. The Journal of Symbolic Logic, 109–117 (1992)

[MaPi]   Marker, D., Pillay, A.: Reducts of $(C, +, \cdot)$ which contain $+$. Journal of Symbolic Logic, 1243–1251 (1990)

[Mos]   Mostowski, A.: On direct products of theories. Journal of Symbolic Logic, 1–31 (1952)

[Muc]   Muchnik, A.A.: Games on infinite trees and automata with dead-ends. A new proof for the decidability of the monadic second order theory of two successors. Bull. EATCS 48, 220–267 (1992)

[Muc1]   Muchnik, A.A.: The definable criterion for definability in Presburger arithmetic and its applications. Theoretical Computer Science 290(3), 1433–1444 (2003)

[Pea]   Peacock, G.: Report on the recent progress and present state of certain branches of analysis. British Association for the Advancement of Science (1833)

[Pei]   Peirce, C.S.: Description of a notation for the logic of relatives, resulting from an amplification of the conceptions of Boole's calculus of logic. Memoirs of the American Academy of Arts and Sciences, 317–378 (1873)

[Pei1]   Peirce, C.S.: On the algebra of logic: A contribution to the philosophy of notation. American Journal of Mathematics 7(2), 180–196 (1885)

[Pet]   Peterzil, Y.: Reducts of some structures over the reals. Journal of Symbolic Logic 58(3), 955–966 (1993)

[Pie]   Pieri, M.: La geometria elementare istituita sulle nozioni 'punto' é 'sfera'. Memorie di Matematica e di Fisica della Società Italiana delle Scienze 15, 345–450 (1908)

[Pre]   Presburger, M.: Über die Vollständigkeit eines gewissen Systems der Arithmetik ganzer Zahlen, in welchem die Addition als einzige Operation hervortritt. Sprawozdanie z 1 Kongresu Matematyków Krajow Slowianskich, Ksiaznica Atlas. pp. 92-10 (Translated: On the completeness of a certain system of arithmetic of whole numbers in which addition occurs as the only operation. History and Philosophy of Logic 12, 225–233 (1930))

[Rab]   Rabin, M.O.: Decidability of second-order theories and automata on infinite trees. Transactions of the American Mathematical Society 141, 1–35 (1969)

[Rad]   Rado, R.: Universal graphs and universal functions. Acta Arithmetica 9(4), 331–340 (1964)

[Sch]   Schröder, E.: On Pasigraphy. Its Present State and the Pasigraphic Movement in Italy. The Monist 9(1), 44–62 (1898)

[Sch1]   Schröder, E.: Vorlesungen über die Algebra der Logik, Volumes 1 to 3. Teubner, Leipzig. Reprinted by Chelsea, New York (1966)

[Sem]   Semenov, A.L.: Finiteness Conditions for Algebras of Relations. Trudy Matematicheskogo Instituta im. V.A. Steklova 242, 103–107 (2003); English version: Proceedings of the Steklov Institute of Mathematics 242, 92–96 (2003)

[Sem1]   Semenov, A.L.: On certain extensions of the arithmetic of addition of natural numbers. Izvestiya: Mathematics 15(2), 401–418 (1980)

[Sem2]   Semenov, A.L.: Predicates that are regular in two positional systems are definable in Presburger arithmetic. Siberian Math. J. 18(2), 403–418 (1977)

[SeSo]   Semenov, A., Soprunov, S.: Finite quantifier hierarchies in relational alge-
          bras. Proceedings of the Steklov Institute of Mathematics 274(1), 267–272
          (2011)
[SeSo1]   Semenov, A.L., Soprunov, S.F.: Remark on Svenonius theorem.
          arXiv:1301.2412 (2013)
[SeSo2]   Semenov, A.L., Soprunov, S.F.: Lattice of relational algebras definable in
          integers with successor. arXiv:1201.4439 (2012)
[Sko]   Skolem, T.: Logisch-kombinatorische Untersuchungen über die Erfullbarkeit
          oder Beweisbarkeit mathematischer Sdtze nebst einem Theorem über dichte
          Mengen. Videnskapsselskapets skrifter. I. Matematisk-naturvidenskabelig
          klasse 4 (1920)
[Sko1]   Skolem, T.: Über gewisse Satzfunktionen in der Arithmetik. Skrifter utgit av
          Videnskapsselskapet i Kristiania, I. klasse 7 (1930)
[Smi]   Smith, J.T.: Definitions and Nondefinability in Geometry. The American
          Mathematical Monthly 117(6), 475–489 (2010)
[Sop]   Soprunov, S.: Decidable expansions of structures. Vopr. Kibern. 134, 175–179
          (1988) (in Russian)
[Sve]   Svenonius, L.: A theorem on permutations in models. Theoria 25(3), 173–178
          (1959)
[Tan]   Tanaka, H.: Some results in the effective descriptive set theory. Publications
          of the Research Institute for Mathematical Sciences 3(1), 11–52 (1967)
[Tar]   Tarski, A.: The Concept of Truth in Formalized Languages. In: Alfred Tarski:
          Logic, Semantics, Metamathematics. Trans. J. H. Woodger, second edition
          ed. and introduced by John Corcoran, Hackett, Indianapolis, 152–278 (1983)
[Tar1]   Tarski, A.: A Decision Method for Elementary Algebra and Geometry Re-
          port R-109 (second revised edn.). The Rand Corporation, Santa Monica, CA
          (1951)
[Tar2]   Tarski, A.: Der Wahrheitsbegriff in den formalisierten Sprachen. Studia
          Philosophica 1 (1935); reprinted in Tarski 2, 51–198 (1986)
[Tar3]   Tarski, A.: Einige methodologifche Unterfuchungen über die Definierbarkeit
          der Begriffe. Erkenntnis 5(1), 80–100 (1935)
[Tar4]   Tarski, A.: What are logical notions? History and Philosophy of Logic 7(2),
          143–154 (1986)
[Tho]   Thomas, S.: Reducts of random hypergraphs. Annals of Pure and Applied
          Logic 80(2), 165–193 (1996)
[Tho1]   Thomas, S.: Reducts of the random graph. Journal of Symbolic Logic 56(1),
          176–181 (1991)
[Win]   Winkler, P.: Random structures and zero-one laws. Finite and infinite com-
          binatorics in sets and logic, pp. 399–420. Springer, Netherlands (1993)

Counting Popular Matchings
in House Allocation Problems

Rupam Acharyya, Sourav Chakraborty, and Nitesh Jha

Chennai Mathematical Institute
Chennai, India
{rupam,sourav,nj}@cmi.ac.in

Abstract. We study the problem of counting the number of *popular matchings* in a given instance. McDermid and Irving gave a poly time algorithm for counting the number of popular matchings when the preference lists are strictly ordered. We first consider the case of ties in preference lists. Nasre proved that the problem of counting the number of popular matching is #P-hard when there are ties. We give an FPRAS for this problem.

We then consider the popular matching problem where preference lists are strictly ordered but each house has a capacity associated with it. We give a *switching graph characterization* of popular matchings in this case. Such characterizations were studied earlier for the case of strictly ordered preference lists (McDermid and Irving) and for preference lists with ties (Nasre). We use our characterization to prove that counting popular matchings in capacitated case is #P-hard.

1 Introduction

A *popular matching problem* instance I comprises a set $\mathcal{A}$ of *agents* and a set $\mathcal{H}$ of *houses*. Each agent a in $\mathcal{A}$ ranks (numbers) a subset of houses in $\mathcal{H}$ (lower rank specify higher preference). The ordered list of houses ranked by $a \in \mathcal{A}$ is called a's *preference list*. For an agent a, let E_a be the set of pairs (a, h) such that the house h appears on a's preference list. Define $E = \cup_{a \in \mathcal{A}} E_a$. The problem instance I is then represented by a bipartite graph $G = (\mathcal{A} \cup \mathcal{H}, E)$. A *matching* M of I is a matching of the bipartite graph G. We use $M(a)$ to denote the house assigned to agent a in M and $M(h)$ to denote the agent that is assigned house h in M. An agent *prefers* a matching M to a matching M' if (i) a is matched in M and unmatched in M', or (ii) a is matched in both M and M' but a prefers the house $M(a)$ to $M'(a)$. Let $\phi(M, M')$ denote the number of agents that prefer M to M'. We say M is *more popular than* M' if $\phi(M, M') > \phi(M', M)$, and denote it by $M \succ M'$. A matching M is called *popular* if there exists no matching M' such that $M' \succ M$.

The popular matching problem was introduced in [5] as a variation of the stable marriage problem [4]. The idea of popular matching has been studied extensively in various settings in recent times [1,14,12,10,8,11,13], mostly in the context where only one side has preference of the other side but the other side

E.A. Hirsch et al. (Eds.): CSR 2014, LNCS 8476, pp. 39–51, 2014.
© Springer International Publishing Switzerland 2014

has no preference at all. We will also focus on this setting. Much of the earlier work focuses on finding efficient algorithms to output a popular matching, if one exists.

The problem of counting the number of "solutions" to a combinatorial question falls into the complexity class #P. An area of interest that has recently gathered a certain amount of attention is the problem of counting stable matchings in graphs. The Gale-Shapely algorithm [4] gives a simple and efficient algorithm to output a stable matching, but counting them was proved to be #P-hard in [6]. Bhatnagar, Greenberg and Randall [2] showed that the random walks on the *stable marriage lattice* are slowly mixing, even in very restricted versions of the problem. [3] gives further evidence towards the conjecture that there may not exist an FPRAS at all for this problem.

Our motivation for this study is largely due to the similarity of structures between stable matchings and popular matchings (although no direct relationship is known). The interest is further fueled by the existence of a linear time algorithm to exactly count the number of popular matchings in the standard setting [12]. We look at generalizations of the standard version - preferences with ties and houses with capacities. In the case where preferences could have ties, it is already known that the counting version is #P-hard [13]. We give an FPRAS for this problem. In the case where houses have capacities, we prove that the counting version is #P-hard. While the FPRAS for the case of ties is achieved via a reduction to a well known algorithm, the #P-hardness for the capacitated case is involving, making it the more interesting setting of the problem.

We now formally describe the different variants of the popular matching problem (borrowing the notation from [14]) and also describe our results alongside.

House Allocation Problem (HA). These are the instances $G = (\mathcal{A} \cup \mathcal{H}, E)$ where the preference list of each agent $a \in \mathcal{A}$ is a linear order. Let $n = |\mathcal{A}| + |\mathcal{H}|$ and $m = |E|$. In [1], Abraham et al. give a complete characterization of popular matchings in an HA instance, using which they give an $O(m + n)$ time algorithm to check if the instance admits a popular matching and to obtain the largest such matching, if one exists. The question of counting popular matchings was first addressed in [12], where McDermid et al. give a new characterization by introducing a powerful structure called the *switching graph* of an instance. The switching graph encodes all the popular matchings via *switching paths* and *switching cycles*. Using this structure, they give a linear time algorithm to count the number of popular matchings.

House Allocation Problem with Ties (HAT). An instance $G = (\mathcal{A} \cup \mathcal{H}, E)$ of HAT can have applicants whose preference list contains ties. For example, the preference list of an agent could be $[h_3, (h_1, h_4), h_2]$, meaning, house h_3 gets rank 1, houses h_1 and h_4 get a tied rank 2 and house h_2 gets the rank 3. A characterization for popular matchings in HAT was given in [1]. The characterization is used to give an $O(\sqrt{n}m)$ time algorithm to solve the maximum cardinality popular matching problem. We outline their characterization briefly in Section 2 where we consider the problem of counting popular matchings in HAT. In [13], Nasre gives a proof of #P-hardness of this problem. We give an FPRAS for this problem by reducing it to the problem of counting perfect matchings in a bipartite graph.

Capacitated House Allocation Problem (CHA). A popular matching instance in CHA has a *capacity* c_i associated with each house $h_i \in \mathcal{H}$, allowing at most c_i agents to be matched to house h_i. The preference list of each agent is strictly ordered. A characterization for popular matchings in CHA was given in [14], along with an algorithm to find the largest popular matching (if one exists) in time $O(\sqrt{C}n_1 + m)$, where $n_1 = |\mathcal{A}|$, $m = |E|$ and C is the total capacity of the houses. In Section 3, we consider the problem counting popular matchings in CHA. We give a switching graph characterization of popular matchings in CHA. This is similar to the switching graph characterization for HA in [12]. Our construction is also motivated from [13], which gives a switching graph characterization of HAT. We use our characterization to prove that it is #P-Complete to compute the number of popular matchings in CHA.

Remark. A natural reduction exists from a CHA instance $G = (\mathcal{A} \cup \mathcal{H}, E)$ to an HAT instance. The reduction is as follows. Treat each house $h_i \subset \mathcal{H}$ with capacity c as c different houses $h_i^1, \ldots, h_i^c$ of unit capacity, which are always tied together and appear together wherever h_i appears in any agent's preference list. Let the HAT instance thus obtained be G'. It is clear that every popular matching of G is a popular matching of G'. Hence, for example, an algorithm which finds a maximum cardinality popular matching for HAT can be used to find a maximum cardinality popular matching for the CHA instance G. In the context of counting, it is important to note that one popular matching of G may translate to many popular matchings in G'. It is not clear if there is a useful map between these two sets that may help in obtaining either hardness or algorithmic results for counting problems.

2   Counting in House Allocation Problem with Ties

In this section we consider the problem of counting the number of popular matchings in House Allocation problem with Ties (HAT). We first describe the characterization given in [1] here using similar notations. Let $G = (\mathcal{A} \cup \mathcal{H}, E)$ be an HAT instance. For any agent $a \in \mathcal{A}$, let $f(a)$ denote the set of first choices of a. For any house $h \in \mathcal{H}$, define $f(h) := \{a \in \mathcal{A}, f(a) = h\}$. A house h for which $f(h) \neq \phi$ is called an f-house. To simplify the definitions, we add a unique last-resort house $l(a)$ with lowest priority for each agent $a \in A$. This forces every popular matching to be an applicant complete matching.

Definition 1. (Section 3.1 in [1]). *The **first choice graph** of G is defined to be $G_1 = (\mathcal{A} \cup \mathcal{H}, E_1)$, where E_1 is the set of all rank one edges.*

Lemma 1. (Lemma 3.1 in [1]). *If M is a popular matching of G, then $M \cap E_1$ is a maximum matching of G_1.*

Let M_1 be any maximum matching of G_1. The matching M_1 can be used to identify the houses h that are always matched to an agent in the set $f(h)$. In this direction, we observe that M_1 defines a partition of the vertices $\mathcal{A} \cup \mathcal{H}$ into three disjoint sets - *even*, *odd* and *unreachable*: a vertex is *even* (resp. *odd*) if there is an even (resp. odd) length alternating path from an unmatched vertex (with respect to M_1) to v; a vertex v is *unreachable* if there is no alternating path from an unmatched vertex to v. Denote the sets *even*, *odd* and *unreachable* by $\mathcal{E}$, $\mathcal{O}$ and $\mathcal{U}$ respectively. The following is a well-known theorem in matching theory [9].

Lemma 2 (Gallai-Edmonds Decomposition). *Let G_1 and M_1 define the partition $\mathcal{E}$, $\mathcal{O}$ and $\mathcal{U}$ as above. Then,*

(a) The sets $\mathcal{E}$, $\mathcal{O}$ and $\mathcal{U}$ are pairwise disjoint, and every maximum matching in G_1 partitions the vertices of G_1 into the same partition of even, odd and unreachable vertices.

(b) In any maximum matching of G_1, every vertex in $\mathcal{U}$ is matched with another vertex in $\mathcal{U}$, and every vertex in $\mathcal{O}$ is matched with some vertex in $\mathcal{E}$. No maximum matching contains an edge between a vertex in $\mathcal{O}$ and a vertex in $\mathcal{O} \cup \mathcal{U}$. The size of a maximum matching is $|\mathcal{O}| + |\mathcal{U}|/2$.

(c) G_1 contains no edge connecting a vertex in $\mathcal{E}$ with a vertex in $\mathcal{U}$.

We show the decomposition of G_1 in Figure 1, where we look at the bipartitions of $\mathcal{U}$, $\mathcal{O}$, and $\mathcal{E}$ into their left and right parts, denoted by subscripts l and r respectively. Since G_1 only contained edges resulting from first-choices, every house in $\mathcal{U}_r$ and $\mathcal{O}_r$ is an f-house. From Lemma 2, each such house $h \in \mathcal{U}_r \cup \mathcal{O}_r$ is matched with an agent in $f(h)$ in every maximum matching of G_1, and correspondingly in every popular matching of G (Lemma 1).

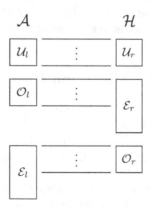

Fig. 1. Gallai-Edmonds decomposition of the first-choice graph of G

For each agent a, define $s(a)$ to be a's most preferred house(s) in $\mathcal{E}_r$. Note that $s(a)$ always exists after the inclusion of last-resort houses $l(a)$. The following is proved in [1].

Lemma 3. (Lemma 3.5 in [1]). *A matching M is popular in G if and only if*

1. *$M \cap E_1$ is a maximum matching of G_1, and*
2. *for each applicant a, $M(a) \in f(a) \cup s(a)$.*

The following hardness result is from [13].

Lemma 4. (Theorem 3 in [13]). *Counting the number of popular matchings in HAT is #P-hard.*

We now give an FPRAS for counting the number of popular matchings in the case of ties. As before, let $G = (\mathcal{A} \cup \mathcal{H}, E)$ be our HAT instance. We assume that that G admits at least one popular matching (this can be tested using the characterization). We reduce our problem to the problem of counting perfect matchings in a bipartite graph. We start with the first-choice graph G_1 of G, and perform a Gallai-Edmonds decomposition of G_1 using any maximum matching of G_1. In order to get a perfect matching instance, we extend the structure obtained from Gallai-Edmonds decomposition described in Figure 1. Let $\mathcal{F}$ be the set of f-houses and $\mathcal{S}$ be the set of s-houses. We make use of the following observations in the decomposition.

— Every agent in $\mathcal{U}_l$ and $\mathcal{O}_l$ gets one of their first-choice houses in every popular matching.
— $\mathcal{E}_r$ can be further partitioned into the following sets:
 - $\mathcal{E}_r^f := \{h \in \mathcal{F} \cap \overline{\mathcal{S}}, h \in \mathcal{E}_r\}$,
 - $\mathcal{E}_r^s := \{h \in \overline{\mathcal{F}} \cap \mathcal{S}, h \in \mathcal{E}_r\}$,
 - $\mathcal{E}_r^{f/s} := \{h \in \mathcal{F} \cap \mathcal{S}, h \in \mathcal{E}_r\}$, and
 - $\mathcal{E}_r^{\star} := \{h \notin \mathcal{F} \cup \mathcal{S}, h \in \mathcal{E}_r\}$.

$\mathcal{O}_l$ can only match with houses in $\mathcal{E}_r^f \cup \mathcal{E}_r^{f/s}$ in every popular matching.

These observations are described in Figure 2(a).

Next, we observe that every agent in $\mathcal{E}_l$ that is already not matched to a house in $\mathcal{O}_r$, must match to a house in $\mathcal{E}_r^s \cup \mathcal{E}_r^{f/s}$. We facilitate this by adding all edges $(a, s(a))$ for each agent in $\mathcal{E}_l$. Finally, we add a set of dummy agent vertices $\mathcal{D}$ on the left side to balance the bipartition. The size of $\mathcal{D}$ is $|\mathcal{A}| - (|\mathcal{H}| - |\mathcal{E}_r^{\star}|)$. This difference is non-negative as long as the preference-lists of agents are complete. We make the bipartition $(\mathcal{D}, \mathcal{E}_r^f \cup \mathcal{E}_r^{f/s} \cup \mathcal{E}_r^s)$ a complete bipartite graph by adding the appropriate edges. This allows us to move from one popular matching to another by switching between first and second-choices and, among second choices of agents. Finally, we remove set $\mathcal{E}_r^{\star}$ from the right side. The new structure is described in Figure 2(b). Denote the new graph by G'.

Lemma 5. *The number of popular matchings in G is $|D|!$ times the number of perfect matchings in G'.*

Proof. Consider a perfect matching M of G'. Let the matching M' be obtained by removing from M all the edges coming out of the set $\mathcal{D}$. Observe that $M' \cap E_1$

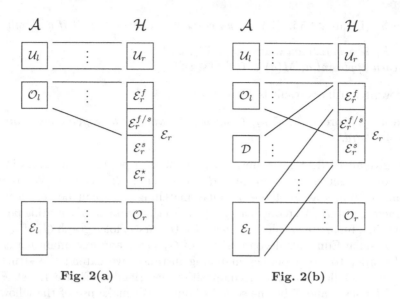

Fig. 2. Reduction to a perfect-matching instance by extending the Gallai-Edmonds decomposition of G_1

is a maximum matching of G_1. This is because the sets $\mathcal{U}_l$, $\mathcal{O}_l$ and $\mathcal{O}_r$ are always matched in M' (or else M would not be a perfect matching of G') and that the size of a maximum matching in G_1 is $(|\mathcal{U}_l| + |\mathcal{O}_l| + |\mathcal{O}_r|)$ by Lemma 2. Also, each agent in $\mathcal{A}$ is matched to either a house in $\mathcal{F}$ or in $\mathcal{S}$ by the construction of graph G'. Using Lemma 3, we conclude that M is a popular matching of G. Finally, observe that every popular matching in M in G can be augmented to a perfect matching of G' by adding exactly $|\mathcal{D}|$ edges. This follows again from Lemma 2 and Lemma 3.   □

We now make use of the following result of Jerrum et al. from [7].

Lemma 6. (Theorem 1.1 in [7]). *There exists an FPRAS for the problem of counting number of perfect matchings a bipartite graph.*

From Lemma 5 and Lemma 6, we have the following.

Theorem 1. *There exists an FPRAS for counting the number of popular matchings in the House Allocation problem with Ties.*

3   Counting in Capacitated House Allocation Problem

In this section, we consider the structure of popular matchings in Capacitated House Allocation problem (CHA). A CHA instance I consists of agents $\mathcal{A}$ and houses $\mathcal{H}$. Let $|\mathcal{A}| = n$ and $|\mathcal{H}| = m$. Let $c : \mathcal{H} \to \mathbb{Z}_{>0}$ be the capacity function for houses. Each agent orders a subset of the houses in a strict order creating its *preference list*. The preference list of $a_i \in \mathcal{A}$ defines a set of edges E_i from a_i to houses in $\mathcal{H}$. Define $E = \cup_{i \in [n]} E_i$. The problem instance I can then be represented by a bipartite graph $G = (\mathcal{A} \cup \mathcal{H}, E)$.

For the instance I, a *matching* M is a subset of E such that each agent appears in at most one edge in M and each house h appears in at most $c(h)$ edges in M. The definitions of *more popular than* relationship between two matchings and *popular matching* is same as described earlier in Section 1.

We now outline a characterization of popular matchings in CHA from [14]. As before, denote by $f(a)$ the first choice of an agent $a \in \mathcal{A}$. A house which is the first choice of at least one agent is called an f-house. For each house $h \in \mathcal{H}$, define $f(h) = \{a \in \mathcal{A}, f(a) = h\}$. For each agent $a \in \mathcal{A}$, we add a unique last-resort house $l(a)$ with least priority and capacity 1.

Lemma 7. (Lemma 1 in [14]) *If M is a popular matching then for each f-house h, $|M(h) \cap f(h)| = \min\{c(h), |f(h)|\}$.*

For each agent $a \in \mathcal{A}$, define $s(a)$ to be the highest ranked house h on a's preference list such that one of the following is true:

- h is not an f-house, or,
- h is an f-house but $h \neq f(a)$ and $|f(h)| < c(h)$.

Notice that $s(a)$ always exists after the inclusion of last-resort houses $l(a)$. The following lemma gives the characterization of popular matchings in G.

Lemma 8. (Theorem 1 in [14]) *A matching M is popular if and only if*

1. *for every f-house $h \in \mathcal{H}$,*
   - *if $|f(h)| \leq c(h)$, then every agent in $f(h)$ is matched to the house h,*
   - *else, house h is matched to exactly $c(h)$ agents, all belonging to $f(h)$,*
2. *M is an agent complete matching such that for each agent $a \in \mathcal{A}$, $M(a) \in \{f(a), s(a)\}$.*

3.1 Switching Graph Characterization of CHA

We now give a *switching graph* characterization of popular matchings for instances from this class. Our results are motivated from similar characterizations for HA in [12] and for HAT in [13]. A switching graph for an instance allows us to move from one popular matching to another by making well defined walks on the switching graph.

Consider a popular matching M of an instance G of CHA. The switching graph of G with respect to M is a directed weighted graph $G_M = (\mathcal{H}, E_M)$, with the edge set E_M defined as follows. For every agent $a \in \mathcal{A}$,

- add a directed edge from M(a) to $\{f(a), s(a)\} \setminus M(a)$,
- if $M(a) = f(a)$, assign a weight of -1 on this edge, otherwise assign a weight of $+1$.

Associated with the switching graph G_M, we have an *unsaturation degree* function $u_M : \mathcal{H} \to \mathbb{Z}_{\geq 0}$, defined $u_M(h) = c(h) - |M(h)|$. A vertex h is called *saturated* if its unsaturation degree is 0, i.e. $u_M(h) = 0$. If $u_M(h) > 0$, h is called *unsaturated*. We make use of the following terminology in the foregoing discussion. We now describe some useful properties of the switching graph G_M.

▷ *Property 1:* Each vertex h can have out-degree at most $c(h)$.
  Proof. Each edge is from a matched house to an unmatched house and since the house h has a maximum capacity $c(h)$, it can only get matched to at most $c(h)$ agents.   □

▷ *Property 2:* Let M and M' be two different popular matchings in G and let G_M and $G_{M'}$ denote the switching graphs respectively. For any vertex house h, the number of -1 outgoing edges from h is invariant across G_M and $G_{M'}$. The number of $+1$ incoming edges on h is also invariant across G_M and $G_{M'}$.
  Proof. From Lemma 7, in any popular matching, each f-house h is matched to exactly $min\{|f(h)|, c(h)\}$ agents and this is also the number of outgoing edges with weight -1. A similar argument can be made for $+1$ weighted incoming edges.   □

▷ *Property 3:* No $+1$ weighted edge can end at an unsaturated vertex.
  Proof. If a $+1$ weighted edge is incident on a vertex h, this means that the house h is an f-house for some agent a that is still not matched to it in M. But if h is unsaturated then it still has some unused capacity. The matching M' obtained by just promoting a to h is popular than M, which is a contradiction.   □

▷ *Property 4:* There can be no incoming -1 weighted edge on a saturated vertex if all its outgoing edges have weight -1.
  Proof. A -1 weighted edge on a vertex h implies that the house h is an s-house for some agent a. But if h is saturated with all outgoing edges having a weight of -1, then all the capacity of h has been used up by agents who had h as their first choice. But by definition, h can not be an s-house for any other agent.   □

▷ *Property 5:* For a given vertex h, if there exists at least one $+1$ weighted incoming edge, then all outgoing edges are of weight -1 and there can be no -1 weighted incoming edge on h.
  Proof. Let agent a correspond to any $+1$ weighted incoming edge. Suppose h has an outgoing $+1$ edge ending at a vertex h' and agent a' corresponds to this edge. We can promote agents a and a' to their first choices and demote any agent which is assigned house h'. This leads to a matching more popular than M. Hence all outgoing edges from h must be of weight -1. Further, Property 3 and Property 4 together imply that there can be no incoming edge on h of weight -1.   □

Switching Moves. We now describe the operation on the switching graph which takes us from one popular matching to another. We make use of the following terminology with reference to the switching graph G_M. Note that the term "path" ("cycle") implies a "directed path" ("directed cycle"). A "+1 edge"("−1 edge") means an "edge with weight +1" ("edge with weight −1").

- A path is called an *alternating path* if it starts with a +1 edge, ends at a −1 edge and alternates between +1 and −1 edges.
- A *switching path* is an alternating path that ends at an unsaturated vertex.
- A *switching cycle* is an even length cycle of alternating −1 and +1 weighted edges.
- A *switching set* is a union of edge-disjoint switching cycles and switching paths, such that at most k switching paths end a vertex of unsaturation degree k.
- A *switching move* is an operation on G_M by a switching set S in which, for every edge e in S, we reversed the direction of e and flip the weight of e ($+1 \leftrightarrow -1$).

Observe that every *valid* switching graph inherently implies a matching (in the context of CHA) of G.

Let $G_M = (\mathcal{H}, E_M)$ and $G_{M'} = (\mathcal{H}, E_{M'})$ be the switching graphs associated with popular matchings M and M' of the CHA instance $G = (\mathcal{A} \cup \mathcal{H}, E)$. Observe that the underlying undirected graph of G_M and $G_{M'}$ are same. We have the following.

Theorem 2. *Let S be the set of edges in G_M that get reversed in $G_{M'}$. Then, S is a switching set for G_M.*

We prove this algorithmically in stages.

Lemma 9. *Every directed cycle in S is a switching cycle of G_M.*

Proof. Let C be any cycle in S. From Property 5 of switching graphs, we know that no vertex in C can have an incoming edge and an outgoing edge of same weight +1. Similarly, since S is the set of edges in G_M which have opposite directions and opposite weights in $G_{M'}$, we observe that S can not contain any vertex with incoming and outgoing edges both having weight −1 (again from Property 5). This forces the weights of cycle C to alternate between +1 and −1. Moreover, this alternation forces the cycle to be of even length.

At this stage we apply the following algorithm to the set S.

Reduction(S):
1. while (*there exists a switching cycle C in S*):
   let $S := S \setminus C$
2. while (S *is non-empty*):
   (a) *find a longest path P in S which alternates between weights +1 and −1*
   (b) *let $S := S \setminus P$*

At the end of every iteration of the while loop in Step 1, Lemma 9 still holds true. We now prove a very crucial invariant of the while loop in Step 2.

Lemma 10. *In every iteration of the* while *loop in Step 2 of the algorithm* Reduction, *the longest path in step 2(a) is a switching path for* G_M.

Proof. Let us denote the stages of the run of algorithm Reduction by t. Initially, at $t = 0$, before any of the while loops run, S is exactly the difference of edges in E_M and E'_M. Let the while loop in Step 1 runs t_1 times and the while loop in Step 2 runs t_2 times.

Let the current stage be $t = t_1 + i$. Let P be the maximal path in step 2(a) at this stage. We show that P starts with an edge of weight $+1$. For contradiction, let (h_i, h_j) be an edge of weight -1 and that this is the first edge of path P. Let a_{ij} be the agent associated with the edge (h_i, h_j).

The Property 5 of switching sets precludes any incoming edge of weight -1 on the vertex h_i. Hence, no switching path could have ended at h_i at any stage $t < t_1 + i$. Similarly, no switching cycle with an incoming edge -1 was incident on h_i at an earlier stage.

Let us assume that there were r cycles that were incident at h_i at $t = 0$. At stage $t = t_1 + i$, let the number of outgoing -1 edges be m. Hence at $t = 0$, h_i had r incoming $+1$ edges and $r + m$ outgoing -1 edges. But this would also imply that at $t = 0$, h_i had $r + m$ incoming $+1$ edges in $G_{M'}$. This contradicts Property 2, requiring the number of incoming $+1$ edges to be constant in the switching graphs corresponding to different popular matchings.

A similar argument can be made for the fact that the path P can only end at an edge with weight -1 and that P ends at an unsaturated vertex.

The following theorem establishes the characterization for popular matchings in CHA.

Theorem 3. *If* G_M *is the switching graph of the CHA instance* G *with respect to a popular matching* M, *then*
(i) every switching move on G_M *generates another popular matching, and*
(ii) every popular matching of G *can be generated by a switching move on* M.

Proof.

(i) We verify that the new matching generated by applying a switching move on G_M satisfies the characterization in Lemma 8. Call the new switching graph $G_{M'}$ and the associated matching M'. First, observe that M' is indeed an agent complete matching since $G_{M'}$ still has a directed edge for each agent in $\mathcal{A}$. Next, each agent a is still matched to $f(a)$ or $s(a)$ as the switching move either reverses an edge of G_M or leaves it as it is. Finally, for each house h, $f(h) \subseteq M'(h)$ if $|f(h)| < c(h)$ and $|M'(h)| = c(h)$ with $M'(h) \subseteq f(h)$ otherwise. This is true because $|M'(h)| = |M(h)|$, by the definition of switching moves.
(ii) This is implied by Theorem 2.

3.2   Hardness of Counting

In this section we prove the #P-hardness of counting popular matchings in CHA. We reduce the problem of counting the number of matchings in a bipartite graph to our problem.

Let $G = (A \cup B, E)$ be a bipartite matching instance in which we want to count the number of matchings. From G we create a CHA instance I such that the number of popular matchings of I is same as the number of matchings of G.

Observe that a description of a switching graph gives the following information about its instance:

- the set of agents A,
- for each agent $a \in \mathcal{A}$, it gives $f(a)$ and $s(a)$, and
- for each s-house or f-house h, the unsaturation degree gives the capacity $c(h)$.

Using this information, we can create the description of the instance I so that it meets our requirement. For simplicity, we assume G to be connected (as isolated vertices do not affect the count). We orient all the edges of G from A to B and call the directed graph $G' = (A \cup B, E')$. Using G', we construct a graph S, which will be the switching graph.

Let $|A| = n_1$, $|B| = n_2$ and $|E| = m$. S is constructed by augmenting G'. We keep all the vertices and edges of G' in S and assign each edge a weight of -1. Further, for each vertex $u \in A$, add a copy u' and add a directed edge from u' to u, and assign a weight of $+1$ to the edge. Call the new set of vertices A'. The sets A' and B contain s-houses and the set A contains f-houses. We label every vertex in A' and A as *saturated* and for each vertex v in B, we label v as *unsaturated* with *unsaturation degree* 1. Hence, the switching graph S has $2n_1 + n_2$ vertices and $n_1 + m$ edges.

The CHA instance I corresponding to the switching graph S has $2n_1 + n_2$ houses and $n_1 + m$ agents. Each agent has a preference list of length 2 that is naturally defined by the weight of edges in S.

Let the popular matching represented by S be M_ϕ. This corresponds to the empty matching of G. Every non-empty matching of G can be obtained by a switching move on S. We make this more explicit in the following theorem.

Theorem 4. *The number of matchings in G is same as the number of popular matchings in I.*

Proof. We prove this by showing that each matching in G corresponds to a unique set of edge disjoint switching paths in the switching graph S of I.

Consider a matching M of G and let $(u, v) \in M$. We look at the length 2 directed path in S that is obtained by extending (u, v) in the reverse direction: $u' \to u \to v$ with $u' \in A'$. It's easy to see that this is a switching path for I. Moreover, the set of switching paths obtained from any matching of G forms a valid switching set (as every pair of such paths arising from a matching are always edge disjoint).

For the converse, observe that S can only have switching paths of length 2 and it has no switching cycles. An edge disjoint set of such paths corresponds to a matching of G. By the definition of S, it's easy to see every matching in M can be obtained by a switching set of S.

Conclusions and Acknowledgement: We obtained an FPRAS for the #P-hard problem of counting popular matchings where instances could have ties. We presented a switching graph characterization for Capacitated House Allocation problem. Though our motivation for studying this structure was to prove a hardness result for counting popular matchings in CHA, the characterization may itself be of importance to many other problems of interest. This also completes the picture of House Allocation problems to a wider extent as such characterizations were only known for HA and HAT instances. We believe that this structure could be used to give an FPRAS for the case of CHA. This remains an open question.

We thank Meghana Nasre for fruitful discussions. We also thank anonymous reviewers for their input.

References

1. Abraham, D.J., Irving, R.W., Kavitha, T., Mehlhorn, K.: Popular matchings. SIAM J. Comput. 37(4), 1030–1045 (2007)
2. Bhatnagar, N., Greenberg, S., Randall, D.: Sampling stable marriages: why spouse-swapping won't work. In: SODA, pp. 1223–1232 (2008)
3. Chebolu, P., Goldberg, L.A., Martin, R.A.: The complexity of approximately counting stable matchings. In: Serna, M., Shaltiel, R., Jansen, K., Rolim, J. (eds.) APPROX 2010. LNCS, vol. 6302, pp. 81–94. Springer, Heidelberg (2010)
4. Gale, D., Shapley, L.S.: College admissions and the stability of marriage. The American Mathematical Monthly 69(1), 9–15 (1962)
5. Gärdenfors, P.: Match making: assignments based on bilateral preferences. Behavioral Science 20(3), 166–173 (1975)
6. Irving, R.W., Leather, P.: The complexity of counting stable marriages. SIAM J. Comput. 15(3), 655–667 (1986)
7. Jerrum, M., Sinclair, A., Vigoda, E.: A polynomial-time approximation algorithm for the permanent of a matrix with non-negative entries. In: STOC, pp. 712–721 (2001)
8. Kavitha, T., Mestre, J., Nasre, M.: Popular mixed matchings. Theor. Comput. Sci. 412(24), 2679–2690 (2011)
9. Lovász, L., Plummer, M.D.: Matching theory. North-Holland Mathematics Studies, vol. 121. North-Holland Publishing Co., Amsterdam (1986), Annals of Discrete Mathematics, 29
10. Mahdian, M.: Random popular matchings. In: ACM Conference on Electronic Commerce, pp. 238–242 (2006)
11. McCutchen, R.M.: The least-unpopularity-factor and least-unpopularity-margin criteria for matching problems with one-sided preferences. In: Laber, E.S., Born-stein, C., Nogueira, L.T., Faria, L. (eds.) LATIN 2008. LNCS, vol. 4957, pp. 593–604. Springer, Heidelberg (2008)

12. McDermid, E., Irving, R.W.: Popular matchings: structure and algorithms. J. Comb. Optim. 22(3), 339–358 (2011)
13. Nasre, M.: Popular matchings: Structure and cheating strategies. In: STACS, pp. 412–423 (2013)
14. Sng, C.T.S., Manlove, D.: Popular matchings in the weighted capacitated house allocation problem. J. Discrete Algorithms 8(2), 102–116 (2010)

Vertex Disjoint Paths in Upward Planar Graphs

Saeed Akhoondian Amiri[1], Ali Golshani[2],
Stephan Kreutzer[1], and Sebastian Siebertz[1]

[1] Technische Universität Berlin, Germany
{saeed.akhoondianamiri,stephan.kreutzer,sebastian.siebertz}@tu-berlin.de
[2] University of Tehran, Iran
ali.golshani@ut.ac.ir

Abstract. The k-vertex disjoint paths problem is one of the most studied problems in algorithmic graph theory. In 1994, Schrijver proved that the problem can be solved in polynomial time for every fixed k when restricted to the class of planar digraphs and it was a long standing open question whether it is fixed-parameter tractable (with respect to parameter k) on this restricted class. Only recently, Cygan et al. [5] achieved a major breakthrough and answered the question positively. Despite the importance of this result, it is of rather theoretical importance. Their proof technique is both technically extremely involved and also has a doubly exponential parameter dependence. Thus, it seems unrealistic that the algorithm could actually be implemented. In this paper, therefore, we study a smaller but well studied class of planar digraphs, the class of upward planar digraphs which can be drawn in a plane such that all edges are drawn upwards. We show that on this class the problem (i) remains NP-complete and (ii) problem is fixed-parameter tractable. While membership in FPT follows immediately from [5]'s general result, our algorithm is very natural and has only singly exponential parameter dependence and linear dependence on the graph size, compared to the doubly exponential parameter dependence and much higher polynomial dependence on the graph size for general planar digraphs. Furthermore, our algorithm can easily be implemented, in contrast to the algorithm in [5].

1 Introduction

Computing vertex or edge disjoint paths in a graph connecting given sources to sinks is one of the fundamental problems in algorithmic graph theory with applications in VLSI-design, network reliability, routing and many other areas. There are many variations of this problem which differ significantly in their computational complexity. If we are simply given a graph (directed or undirected) and two sets of vertices S, T of equal cardinality, and the problem is to compute $|S|$ pairwise vertex or edge disjoint paths connecting sources in S to targets in T, then this problem can be solved efficiently by standard network flow techniques.

A variation of this is the well-known *k-vertex disjoint paths problem*, where the sources and targets are given as lists $(s_1, \ldots, s_k)$ and $(t_1, \ldots, t_k)$ and the problem

E.A. Hirsch et al. (Eds.): CSR 2014, LNCS 8476, pp. 52–64, 2014.

is to find k vertex disjoint paths connecting each source s_i to its corresponding target t_i. The k-disjoint paths problem is NP-complete in general and remains NP-complete even on planar undirected graphs (see [10]).

On undirected graphs, it can be solved in polynomial time for any fixed number k of source/target pairs. This was first proved for the 2-disjoint paths problems, for instance in [18,19,21,14], before Robertson and Seymour proved in [16] that the problem can be solved in polynomial-time for every fixed k. In fact, they proved more, namely that the problem is *fixed-parameter tractable* with parameter k, that is, solvable in time $f(k) \cdot |G|^c$, where f is a computable function, G is the input graph, k the number of source/target pairs and c a fixed constant (not depending on k). See e.g. [7] for an introduction to fixed-parameter tractability.

For directed graphs the situation is quite different (see [1] for a survey). Fortune et al. [8] proved that the problem is already NP-complete for $k = 2$ and hence the problem cannot be expected to be fixed-parameter tractable on directed graphs. They cannot even be expected to be fixed-parameter tractable on acyclic digraphs, as shown by Slivkins [20]. However, on acyclic digraphs the problem can be solved in polynomial time for any fixed k [8].

In [12], Johnson et al. introduced the concept of *directed tree-width* as a directed analogue of undirected tree-width for directed graphs. They showed that on classes of digraphs of bounded directed tree-width the k-disjoint paths problem can be solved in polynomial time for any fixed k. As the class of acyclic digraphs has directed tree-width 1, Slivkins' result [20] implies that the problem cannot be expected to be fixed-parameter tractable on such classes.

Given the computational intractability of the directed disjoint paths problem on many classes of digraphs, determining classes of digraphs on which the problem does become at least fixed-parameter tractable is an interesting and important problem. Using colour coding techniques, the problem can be shown to become fixed-parameter tractable if the length of the disjoint paths is bounded. This has, for instance, been used to show fixed-parameter tractability of the problem on classes of bounded *DAG-depth* [9]. In 1994, Schrijver [17] proved that the directed k-disjoint paths problem can be solved in polynomial time for any fixed k on planar digraphs, using a group theoretical approach and it was a long standing open question whether it is fixed-parameter tractable on this restricted class. Only recently, Cygan et al. [5] achieved a major breakthrough and answered the question positively. Despite the importance of this result (and the brilliance of their proof), it is of rather theoretical importance. Their proof technique is based on irrelevant vertices in sequences of concentric cycles of alternating orientation and is both technically extremely involved and also has a doubly exponential parameter dependence. It also uses the algebraic tools used by Schrijver, which involves in particular a polynomial time algorithm for checking the existence of a solution in a fixed cohomology class. The currently best known algorithms for this have an impractical polynomial time running time. Thus, it seems unrealistic that the algorithm could actually be implemented.

In this paper, therefore, we study a smaller class of planar digraphs, the class of *upward planar digraphs*. These are graphs that have a plane embedding such that every directed edge points "upward", i.e. each directed edge is represented by a curve that is monotone increasing in the y direction. Upward planar digraphs are very well studied in a variety of settings, in particular in graph drawing applications (see e.g. [2]). In contrast to the problem of finding a planar embedding for a planar graph, which is solvable in linear time, the problem of finding an upward planar embedding is NP-complete in general [11]. Much work has gone into finding even more restricted classes inside the upward planar class that allow to find such embeddings in polynomial time [4,3,15].

By definition, upward planar graphs are acyclic graphs. Hence, by the above results, the k-vertex disjoint paths problem can be solved in polynomial time on upward planar graphs for any fixed k. As a first result in this paper we show that the problem remains NP-complete on upward planar graphs, i.e. that we cannot hope to find a general polynomial-time algorithm. Our construction even shows that the problem is NP-complete on directed grid graphs.

Our second result is that the problem is fixed-parameter tractable with respect to parameter k on the class of upward planar digraphs if we are given an upward planar graph together with an upward planar embedding. We present a linear time algorithm that has single exponential parameter dependency. The idea of our algorithm is straight forward but the proof of its correctness requires some work.

2   Preliminaries

By $\mathbb{N}$ we denote the set of non-negative integers and for $n \in \mathbb{N}$, we write $[n]$ for the set $\{1, \ldots, n\}$. We assume familiarity with the basic concepts from (directed) graph theory, planar graphs and graph drawings and refer the reader to [1,2,6] for more details. For background on parameterized complexity theory we refer the reader to [7].

An embedding of a graph $G = (V, E)$ in the real plane is a mapping φ that maps vertices $v \in V$ to points $\varphi_v \in \mathbb{R}^2$ and edges $e = (u, v) \in A$ to continuous functions $\varphi_e : [0, 1] \to \mathbb{R}^2$ such that $\varphi_e(0) = \phi_u$ and $\varphi_e(1) = \varphi_v$. A plane embedding is an embedding such that $\varphi_e(z) = \varphi_{e'}(z')$ if $z, z' \in \{0, 1\}$ for all $e \neq e' \in E$. An upward plane embedding is a plane embedding such that every edge is drawn "upward", i.e. for all edges $e \in A$, if $\varphi_e(z) = (x, y)$, $\varphi_e(z') = (x', y')$ and $z' > z$, then $y' \geq y$. An *upward planar graph* is a graph that has an upward plane embedding. To improve readability, we will draw all graphs in this paper from left to right, instead of upwards.

The k-vertex disjoint paths problem on upward planar graphs is the following problem.

VERTEX DISJOINT PATHS ON UPWARD PLANAR GRAPHS (UPPLAN-VDPP)

  Input: An upward planar graph G together with an upward plane embedding, $(s_1, t_1), \ldots, (s_k, t_k)$.
  Problem: Decide whether there are k pairwise internally vertex disjoint paths $P_1, \ldots, P_k$ linking s_i to t_i, for all i.

---

3   NP-Completeness of UpPlan-VDPP

This section is dedicated to the proof of one of our main theorems:

Theorem 3.1. UPPLAN-VDPP *is NP-complete.*

Before we formally prove the theorem, we give a brief and informal overview of the proof structure. The proof of NP-completeness is by a reduction from SAT, the satisfiability problem for propositional logic, which is well-known to be NP-complete [10]. On a high level, our proof method is inspired by the NP-completeness proof in [13] but the fact that we are working in a restricted class of planar graphs requires a number of changes and additional gadgets.

Let $\mathcal{V} = \{V_1, \ldots, V_n\}$ be a set of variables and $\mathcal{C} = \{C_1, \ldots, C_m\}$ be a set of clauses over the variables from $\mathcal{V}$. For $1 \leq i \leq m$ let $C_i = \{L_{i,1}, L_{i,2}, \ldots, L_{i,n_i}\}$ where each $L_{i,t}$ is a literal, i.e. a variable or the negation thereof. We will construct an upward planar graph $G_{\mathcal{C}} = (V, E)$ together with a set of pairs of vertices in $G_{\mathcal{C}}$ such that $G_{\mathcal{C}}$ contains a set of pairwise vertex disjoint directed paths connecting each source to its corresponding target if, and only if, $\mathcal{C}$ is satisfiable. The graph $G_{\mathcal{C}}$ is roughly sketched in Fig. 1.

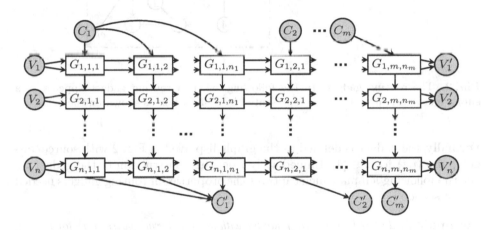

Fig. 1. Structure of the graph $G_{\mathcal{C}}$

We will have the source/target pairs $(V_i, V_i') \in V^2$ for $i \in [n]$ and $(C_j, C_j') \in V^2$ for $j \in [m]$, as well as some other source/target pairs inside the gadgets $G_{i,j,t}$ that guarantee further properties. As the picture suggests, there will be two possible paths from V_i to V_i', an upper path and a lower path and our construction will ensure that these paths cannot interleave. Any interpretation of the variable V_i will thus correspond to the choice of a unique path from V_i to V_i'. Furthermore, we will ensure that there is a path from C_j to C_j' if and only if some literal is interpreted such that C_j is satisfied under this interpretation.

We need some additional gadgets which we describe first to simplify the presentation of the main proof.

Routing Gadget: The rôle of a routing gadget is to act as a planar routing device. It has two incoming connections, the edges e_t from the top and e_l from the left, and two outgoing connections, the edges e_b to the bottom and e_r to the right. The gadget is constructed in a way that in any solution to the disjoint paths problem it allows for only two ways of routing a path through the gadget, either using e_t and e_b or e_l and e_r.

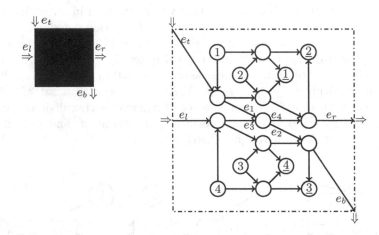

Fig. 2. The routing gadget. In the following, when a routing gadget appears as a subgadget in a figure, it will be represented by a black box as shown on the left.

Formally, the gadget is defined as the graph displayed in Fig. 2 with source/target pairs $(i, \underline{i})$ for $i \in [4]$. Immediately from the construction of the gadget we get the following lemma which captures the properties of routing gadgets needed in the sequel.

Lemma 3.2. *Let R be a routing gadget with source/target pairs $(i, \underline{i})$ for $i \in [4]$.*

1. *There is a solution of the disjoint paths problem in R.*
2. *Let $P_1, \ldots, P_4$ be any solution to the disjoint paths problem in R, where P_i links vertex i to $\underline{i}$. Let $H := R \setminus \bigcup_{i=1}^{4} P_i$.*
   (a) *H neither contains a path which goes through e_t to e_r nor a path which goes through e_l to e_b.*
   (b) *H either contains a unique path P which goes through e_t to e_b or a unique path P' in H which goes through e_l to e_r, and not both.*

Crossing Gadget: A crossing gadget has two incoming connections to its left via the vertices H^{in} and L^{in} and two outgoing connections to its right via the vertices H^{out} and L^{out}. Furthermore, it has one incoming connection at the top via the vertex T and outgoing connection at the bottom via the vertex B. Intuitively, we want that in any solution to the disjoint paths problem, there is exactly one path P going from left to right and exactly one path P' going from top to bottom. Furthermore, if P enters the gadget via H^{in} then it should leave it via H^{out} and if it enters the gadget via L^{in} then it should leave it via L^{out}. Of course, in a planar graph there cannot be such disjoint paths P, P' as they must cross at some point. We will have to split one of the paths, say P, by removing the outward source/sink pair and introducing two new source/sink pairs, one to the left of P' and one to its right.

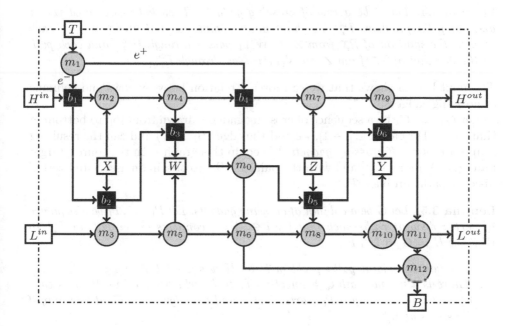

Fig. 3. The crossing gadget

Formally, the gadget is defined as the graph displayed in Fig. 3. The following lemma follows easily from Lemma 3.2.

Lemma 3.3. *Let G be a crossing gadget.*

1. *There are uniquely determined vertex disjoint paths P_1 from H^{in} to W, P_2 from T to B and P_3 from X to Y. Let $H := G \setminus \bigcup_{i=1}^{3} P_i$. Then H contains a path from Z to H^{out} but it does not contain a path from Z to L^{out}.*
2. *There are uniquely determined vertex disjoint paths Q_1 from L^{in} to W, Q_2 from T to B and Q_3 from X to Y. Let $H := G \setminus \bigcup_{i=1}^{3} Q_i$. Then H contains a path from Z to L^{out} but it does not contain a path from Z to H^{out}.*

The next lemma shows that we can connect crossing gadgets in rows in a useful way. It follows easily by induction from Lemma 3.3.

Let $G_1, \ldots, G_s$ be a sequence of crossing gadgets drawn from left to right in in that order. We address the inner vertices of the gadgets by their names in the gadget equipped with corresponding subscripts, e.g., we write H_1^{in} for the vertex H^{in} of gadget G_1. For each $j \in [s-1]$, we add the edges $(H_j^{out}, H_{j+1}^{in})$ and $(L_j^{out}, L_{j+1}^{in})$ and call the resulting graph a *row of crossing gadgets*. We equip this graph with the source/target pairs $(X_j, Y_j), (Z_j, W_{j+1})$ for $j \in [s-1]$ to obtain an associated vertex disjoint paths problem $\mathcal{P}_r$ (the subscript r stands for row). Denote by $\mathcal{P}_r^+$ the problem $\mathcal{P}_r$ with additional source/target pair (H_1^{in}, W_1) and by $\mathcal{P}_r^-$ the problem $\mathcal{P}_r$ with additional source/target pair (L_1^{in}, W_1).

Lemma 3.4. *Let G be a row of crossing gadgets. Then both associated vertex disjoint paths problems $\mathcal{P}_r^+, \mathcal{P}_r^-$ have unique solutions. For all $i \in [t-1]$, each path in the solution of $\mathcal{P}_r^+$ from Z_i to W_{i+1} passes through H_{i+1}^{in} and each path in the solution of $\mathcal{P}_r^-$ from Z_i to W_{i+1} passes through L_{i+1}^{in}.*

The next lemma shows that we can force a relation between rows and columns of crossing gadgets.

Let $G_1, \ldots, G_t$ be a sequence of crossing gadgets drawn from top to bottom in that order. For each $i \in [t-1]$, we add the edge (B_i, T_{i+1}) and call the resulting graph a *column of crossing gadgets*. We equip this graph with the source/target pairs (X_i, Y_i) for $i \in [t]$ and with the pair (T_1, B_t) to obtain an associated vertex disjoint paths problem $\mathcal{P}$.

Lemma 3.5. *Let G be a column of crossing gadgets. Let $P_1, \ldots, P_t$ be a sequence of vertex disjoint paths such that for $i \in [t]$, P_i connects either H_i^{in} or L_i^{in} to W_i. Let $H := G \setminus \bigcup_{i=1}^t P_i$.*

1. *The vertex disjoint paths problem $\mathcal{P}$ on H has a solution.*
2. *There is a unique path Q connecting T_1 to B_t which for all $i \in [t]$ uses edge e^+ in G_i if and only if P_i starts at H_i^{in} and the edge e^- in G_i if and only if P_i starts at L_i^{in}.*

Note that the paths P_i as stated in the lemma exist and they are uniquely determined by Lemma 3.3.

We are now ready to construct a vertex disjoint paths instance for any SAT instance $\mathcal{C}$.

Definition 3.6. *Let $\mathcal{C}$ be a SAT instance over the variables $\mathcal{V} = \{V_1, \ldots, V_n\}$ and let $\{C_1, \ldots, C_m\}$ be its set of clauses. For $j \in [m]$ let $C_j = \{L_{j,1}, L_{j,2}, \ldots, L_{j,n_j}\}$, where each $L_{j,s}$ is a literal, i.e. a variable or the negation thereof.*

1. *The graph $G_\mathcal{C}$ is defined as follows.*
   - *For each variable $V \in \mathcal{V}$ we introduce two vertices V and V'.*
   - *For each clause $C \in \mathcal{C}$ we introduce two vertices C and C'.*
   - *For each variable V_i and each literal $L_{j,t}$ in clause j we introduce a crossing gadget $G_{i,j,t}$.*
   - *For $i \in [n]$ we add the edges $(V_i, H_{i,1,1}^{in})$, $(V_i, L_{i,1,1}^{in})$, $(H_{i,m,n_m}^{out}, V_i')$ and $(L_{i,m,n_m}^{out}, V_i')$.*

  – For $j \in [m], t \in [n_j]$ we add the edges $(C_j, T_{1,j,t})$ and $(B_{n,j,t}, C'_j)$.
  – Finally, we delete the edge e^+ for all $i \in [n], j \in [m], t \in [n_j]$ in $G_{i,j,t}$ if $L_{j,t}$ is a variable and the edge e^- if it is a negated variable.

  We draw the graph G_C as shown in Fig. 1.
2. We define the following vertex disjoint paths problem $\mathcal{P}_C$ on G_C. We add all source/target pairs that are defined inside the routing gadgets. Furthermore:
  – For $i \in [n], j \in [m], t \in [n_j - 1]$, we add the pairs
    • $(V_i, W_{i,1,1})$,
    • (Z_{i,m,n_m}, V'_i),
    • $(X_{i,j,t}, Y_{i,j,t})$ and
    • $(Z_{i,j,t}, W_{i,j,t+1})$.
  – For $i \in [n], j \in [m-1]$, we add the pairs $(Z_{i,j,n_j}, W_{i,j+1,1})$.
  – For $j \in [m]$, we add the pairs (C_j, C'_j).

The proof of the following theorem is based on the fact that in our construction, edge e^+ is present in gadget $G_{i,j,t}$, if and only if C_j does not contain variable V_i negatively and e^- is present in gadget $G_{i,j,t}$, if and only if C_j does not contain variable V_i positively (especially, both edges are present if the clause does not contain the variable at all). In particular, every column contains exactly one gadget where one edge is missing. Now it is easy to conclude with Lemma 3.4 and Lemma 3.5.

Theorem 3.7. *Let C be a SAT-instance and let $\mathcal{P}_C$ be the corresponding vertex disjoint paths instance on G_C as defined in Definition 3.6. Then C is satisfiable if and only if $\mathcal{P}_C$ has a solution.*

It is easily seen that the presented reduction can be computed in polynomial time and this finishes the proof of Theorem 3.1.

If we replace the vertices C_i and C'_i with directed paths, then it is easy to convert the graph G_C to a directed grid graph, i.e. a subgraph of the infinite grid. This implies that the problem is NP-complete even on upward planar graphs of maximum degree 4.

4   A Linear Time Algorithm for Fixed k

In this section we prove that the k-disjoint paths problem for upward planar digraphs can be solved in linear time for any fixed value of k. In other words, the problem is fixed-parameter tractable by a linear time parameterized algorithm.

Theorem 4.1. *The problem UPPLAN-VDPP can be solved in time $\mathcal{O}(k! \cdot k \cdot n)$, where $n := |V(G)|$.*

For the rest of the section we fix a planar upward graph G together with an upward planar embedding and k pairs $(s_1, t_1), \ldots, (s_k, t_k)$ of vertices. We will not distinguish notationally between G and its upward planar embedding. Whenever we speak about a vertex v on a path P we mean a vertex $v \in V(G)$ which is contained in P. If we speak about a *point on the path* we mean a point $(x, y) \in \mathbb{R}^2$

which is contained in the drawing of P with respect to the upward planar drawing of G. The algorithm is based on the concept of a path in G being *to the right* of another path which we define next.

Definition 4.2. *Let P be a path in an upward planar drawing of G. Let (x, y) and (x', y') be the two endpoints of P such that $y \leq y'$, i.e. P starts at (x, y) and ends at (x', y'). We define*

$$right(P) := \{(u, v) \in \mathbb{R}^2 : y \leq v \leq y' \text{ and } u' < u \text{ for all } u' \text{ such that } (u', v) \in P\}$$
$$left(P) := \{(u, v) \in \mathbb{R}^2 : y \leq v \leq y' \text{ and } u' > u \text{ for all } u' \text{ such that } (u', v) \in P\}.$$

The next two lemmas follow immediately from the definition of upward planar drawings.

Lemma 4.3. *Let P and Q be vertex disjoint paths in an upward planar drawing of G. Then either $right(P) \cap Q = \emptyset$ or $left(P) \cap Q = \emptyset$.*

Lemma 4.4. *Let P be a directed path in an upward planar drawing of a digraph G. For $i = 1, 2, 3$ let $p_i := (x_i, y_i)$ be distinct points in P such that $y_1 < y_2 < y_3$. Then p_1, p_2, p_3 occur in this order on P.*

Definition 4.5. *Let P and Q be two vertex disjoint paths in G.*

1. *A point $p = (x, y) \in \mathbb{R}^2 \setminus P$ is to the right of P if $p \in right(P)$. Analogously, we say that $(x, y) \in \mathbb{R}^2 \setminus P$ is to the left of P if $p \in left(P)$.*
2. *The path P is to the right of Q, denoted by $Q \prec P$ if there exists a point $p \in P$ which to the right of some point $q \in Q$. We write $\prec^*$ for the transitive closure of $\prec$.*
3. *If $\mathcal{P}$ is a set of pairwise disjoint paths in G, we write $\prec_{\mathcal{P}}$ and $\prec^*_{\mathcal{P}}$ for the restriction of $\prec$ and $\prec^*$, resp., to the paths in $\mathcal{P}$.*

We show next that for every set $\mathcal{P}$ of pairwise vertex disjoint paths in G the relation $\prec^*$ is a partial order on $\mathcal{P}$. Towards this aim, we first show that $\prec$ is irreflexive and anti-symmetric on $\mathcal{P}$.

Lemma 4.6. *Let $\mathcal{P}$ be a set of pairwise disjoint paths in G.*

1. *The relation $\prec_{\mathcal{P}}$ is irreflexive.*
2. *The relation $\prec_{\mathcal{P}}$ is anti-symmetric, i.e. if $P_1 \prec_{\mathcal{P}} P_2$ then $P_2 \not\prec_{\mathcal{P}} P_1$ for any $P_1, P_2 \in \mathcal{P}$.*

Proof. The first claim immediately follows from the definition of $\prec$. Towards the second statement, suppose there are $P_1, P_2 \in \mathcal{P}$ such that $P_1 \prec_{\mathcal{P}} P_2$ and $P_2 \prec_{\mathcal{P}} P_1$.

Hence, for $j = 1, 2$ and $i = 1, 2$ there are points $p^i_j = (x^i_j, y^i_j)$ such that $p^i_j \in P_i$ and $x^1_1 < x^2_1$, $y^1_1 = y^2_1$ and $x^1_2 > x^2_2$, $y^1_2 = y^2_2$. W.l.o.g. we assume that $y^1_1 < y^1_2$. Let $Q \subseteq P$ be the subpath of P from p^1_1 to p^1_2, including the endpoints. Let $Q_1 := \{(x^1_1, z) : z < y^1_1\}$ and $Q_2 := \{(\vec{x}^1_2, z) : z > y^1_2\}$ be the two lines parallel to the y-axis going from p^1_1 towards negative infinity and from p^1_2 towards infinity.

Then $Q_1 \cup Q \cup Q_2$ separates the plane into two disjoint regions R_1 and R_2 each containing a point of P_2. As P_1 and P_2 are vertex disjoint but p_1^2 and p_2^2 are connected by P_2, P_2 must contain a point in Q_1 or Q_2 which, on P_2 lies between p_1^2 and p_2^2. But the y-coordinate of any point in Q_1 is strictly smaller than y_1^2 and y_2^2 whereas the y-coordinate of any point in Q_2 is strictly bigger than y_1^2 and y_2^2. This contradicts Lemma 4.4.   □

We use the previous lemma to show that $\prec_{\mathcal{P}}^*$ is a partial order for all sets $\mathcal{P}$ of pairwise vertex disjoint paths.

Lemma 4.7. *Let $\mathcal{P}$ be a set of pairwise vertex disjoint directed paths. Then $\prec_{\mathcal{P}}^*$ is a partial order.*

Proof. By definition, $\prec_{\mathcal{P}}^*$ is transitive. Hence we only need to show that it is anti-symmetric for which, by transitivity, it suffices to show that $\prec_{\mathcal{P}}^*$ is irreflexive.

To show that $\prec_{\mathcal{P}}^*$ is irreflexive, we prove by induction on k that if $P_0, \ldots, P_k \in \mathcal{P}$ are paths such that $P_0 \prec_{\mathcal{P}} \cdots \prec_{\mathcal{P}} P_k$ then $P_k \not\prec_{\mathcal{P}} P_0$. As for all $P \in \mathcal{P}$, $P \not\prec_{\mathcal{P}} P$, this proves the lemma.

Towards a contradiction, suppose the claim was false and let k be minimum such that there are paths $P_0, \ldots, P_k \in \mathcal{P}$ with $P_0 \prec_{\mathcal{P}} \cdots \prec_{\mathcal{P}} P_k$ and $P_k \prec_{\mathcal{P}} P_0$. By Lemma 4.6, $k > 1$.

Let $R := \bigcup_{i=0}^{k-2} right(P_i)$. Note that $k - 2 \geq 0$, so R is not empty. Furthermore, as for all P, Q with $P \prec Q$, $right(P) \cap right(Q) \neq \emptyset$, R is a connected region in $\mathbb{R}^2$ without holes. Let $L := \bigcup_{i=1}^{k-1} left(P_i)$. Again, as $k > 1$, $L \neq \emptyset$ and L is a connected region without holes.

As $P_{k-2} \prec_{\mathcal{P}} P_{k-1}$, we have $L \cap R \neq \emptyset$ and therefore $L \cup R$ separates the plane into two unbounded regions, the upper region T and the lower region B.

The minimality of k implies that $P_i \not\prec_{\mathcal{P}} P_k$ for all $i < k - 1$ and therefore $R \cap P_k = \emptyset$. Analogously, as $P_k \not\prec_{\mathcal{P}} P_i$ for any $i > 0$, we have $L \cap P_k = \emptyset$. Hence, either $P_k \subseteq B$ or $P_k \subseteq T$. W.l.o.g. we assume $P_k \subseteq B$. We will show that $left(P_0) \cap B = \emptyset$.

Suppose there was a point $(x, y) \in P$ and some $x' < x$ such that $(x', y) \in B$. This implies that $y < v$ for all $(u, v) \in L$. But this implies that B is bounded by $right(P_0)$ and L contradicting the fact that $right(P_{k-1}) \cap B \neq \emptyset$.   □

We have shown so far that $\prec^*$ is a partial order on every set of pairwise vertex disjoint paths.

Remark 4.8. Note that if two paths $P, Q \in \mathcal{P}$ are incomparable with respect to $\prec_{\mathcal{P}}^*$ then one path is strictly above the other, i.e. $(right(P) \cup left(P)) \cap (right(Q) \cup left(Q)) = \emptyset$. This is used in the next lemma.

Definition 4.9. *Let $s, t \in V(G)$ be vertices in G such that there is a directed path from s to t. The right-most s-t-path in G is an s-t-path P such that for all s-t-paths P', $P \subseteq P' \cup right(P')$.*

Lemma 4.10. *Let $s, t \in V(G)$ be two vertices and let P be a path from s to t in an upward planar drawing of G. If P' is an s-t path such that $P' \cap right(P) \neq \emptyset$ then there is an s-t path Q such that $Q \subseteq P \cup right(P)$ and $Q \cap right(P) \neq \emptyset$.*

Proof. If $P' \subseteq P \cup right(P)$ we can take $Q = P'$. Otherwise, i.e. if $P' \cap left(P) \neq \emptyset$, then as the graph is planar this means that P and P' share internal vertices. In this case we can construct Q from $P \cup P'$ where for subpaths of P and P' between two vertices in $P \cap P'$ we always take the subpath to the right. $\qquad\square$

Corollary 4.11. *Let $s, t \in V(G)$ be vertices in G such that there is a directed path from s to t. Then there is a unique right-most s-t-path in G.*

The corollary states that between any two s and t, if there is an s-t path then there is a rightmost one. The proof of Lemma 4.10 also indicates how such a path can be computed. This is formalised in the next lemma.

Lemma 4.12. *There is a linear time algorithm which, given an upward planar drawing of a graph G and two vertices $s, t \in V(G)$ computes the right-most s-t-path in G, if such a path exists.*

Proof. We first use a depth-first search starting at s to compute the set of vertices $U \subseteq V(G)$ reachable from s. Clearly, if $t \notin U$ then there is no s-t-path and we can stop. Otherwise we use a second, inverse depth-first search to compute the set $U' \subseteq U$ of vertices from which t can be reached. Finally, we compute the right-most s-t path inductively by starting at $s \in U'$ and always choosing the right-most successor of the current vertex until we reach t. The right-most successor is determined by the planar embedding of G. As G is acyclic, this procedure produces the right-most path and can clearly be implemented in linear time. $\qquad\square$

We show next that in any solution $\mathcal{P}$ to the disjoint paths problem in an upward planar digraph, if $P \in \mathcal{P}$ is a maximal element with respect to $\prec^*_\mathcal{P}$, we can replace P by the right-most s-t path and still get a valid solution, where s and t are the endpoints of P.

Lemma 4.13. *Let G be an upward planar graph with a fixed upward planar embedding and let $(s_1, t_1), \ldots, (s_k, t_k)$ be pairs of vertices. Let $\mathcal{P}$ be a set of pairwise disjoint paths connecting (s_i, t_i) for all i. Let $P \in \mathcal{P}$ be path connecting s_i and t_i, for some i, which is maximal with respect to $\prec^*_\mathcal{P}$. Let P' be the right-most $s_i - t_i$-path in G. Then $\mathcal{P} \setminus \{P\} \cup \{P'\}$ is also a valid solution to the disjoint paths problem on G and $(s_1, t_1), \ldots, (s_k, t_k)$.*

Proof. All we have to show is that P' is disjoint from all $Q \in \mathcal{P} \setminus \{P\}$. Clearly, as P and P' are both upward s_i-t_i paths, we have $P' \subseteq P \cup left(P) \cup right(P)$.

By the remark above, if P and Q are incomparable with respect to $\prec^*_\mathcal{P}$, then one is above the other and therefore Q and P' must be disjoint. Now suppose Q and P are comparable and therefore $Q \prec^*_\mathcal{P} P$. This implies that $(P \cup right(P)) \cap Q = \emptyset$ and therefore $Q \cap P' = \emptyset$. $\qquad\square$

The previous lemma yields the key to the proof of Theorem 4.1 :

Proof of Theorem 4.1. Let G with an upward planar drawing of G and k pairs $(s_1, t_1), \ldots, (s_k, t_k)$ be given. To decide whether there is a solution to the disjoint paths problem on this instance we proceed as follows. In the first step we compute for each s_i the set of vertices reachable from s_i. If for some i this does not include t_i we reject the input as obviously there cannot be any solution.

In the second step, for every possible permutation π of $\{1, \ldots, k\}$ we proceed as follows. Let $i_1 := \pi(k), \ldots, i_k := \pi(1)$ be the numbers 1 to k ordered as indicated by π and let $u_j := s_{i_j}$ and $v_j := t_{i_j}$, for all $j \in [k]$. We can view π as a linear order on $1, \ldots, k$ and for every such π we will search for a solution $\mathcal{P}$ of the disjoint paths problem for which $\prec_{\mathcal{P}}^*$ is consistent with π.

For a given π as above we inductively construct a sequence $\mathcal{P}_0, \ldots, \mathcal{P}_k$ of sets of pairwise vertex disjoint paths such that for all i, $\mathcal{P}_i$ contains a set of i paths $P_1, \ldots, P_i$ such that for all $j \in [i]$ P_j links u_j to v_j. We set $\mathcal{P}_0 := \emptyset$ which obviously satisfies the condition. Suppose for some $0 \leq i < k$, $\mathcal{P}_i$ has already been constructed. To obtain $\mathcal{P}_{i+1}$ we compute the right-most path linking u_{i+1} to v_{i+1} in the graph $G \backslash \bigcup \mathcal{P}_i$. By Lemma 4.12, this can be done in linear time for each such pair (s_{i+1}, t_{i+1}). If there is such a path P we define $\mathcal{P}_{i+1} := \mathcal{P}_i \cup \{P\}$. Otherwise we reject the input. Once we reach $\mathcal{P}_k$ we stop and output $\mathcal{P}_k$ as solution.

Clearly, for every permutation π the algorithm can be implemented to run in time $\mathcal{O}(k \cdot n)$, using Lemma 4.12, so that the total running time is $\mathcal{O}(k! \cdot k \cdot n)$ as required.

Obviously, if the algorithm outputs a set $\mathcal{P}$ of disjoint paths then $\mathcal{P}$ is indeed a solution to the problem. What is left to show is that whenever there is a solution to the disjoint path problem, then the algorithm will find one.

So let $\mathcal{P}$ be a solution, i.e. a set of k paths $P_1, \ldots, P_k$ so that P_i links s_i to t_i. Let $\leq$ be a linear order on $\{1, \ldots, k\}$ that extends $\prec_{\mathcal{P}}^*$ and let π be the corresponding permutation such that $(u_1, v_1), \ldots, (u_k, v_k)$ is the ordering of $(s_1, t_1), \ldots, (s_k, t_k)$ according to $\leq$. We claim that for this permutation π the algorithm will find a solution. Let P be the right-most u_k-v_k-path in G as computed by the algorithm. By Lemma 4.13, $\mathcal{P} \backslash \{P_k\} \cup P$ is also a valid solution so we can assume that $P_k = P$. Hence, $P_1, \ldots, P_{k-1}$ form a solution of the disjoint paths problem for $(u_1, v_1), \ldots, (u_{k-1}, v_{k-1})$ in $G \backslash P$. By repeating this argument we get a solution $\mathcal{P}' := \{P_1', \ldots, P_k'\}$ such that each P_i' links u_i to v_i and is the right-most u_i-v_i-path in $G \backslash \bigcup_{j > i} P_j'$. But this is exactly the solution found by the algorithm. This prove the correctness of the algorithm and concludes the proof of the theorem. $\qquad \square$

We remark that we can easily extend this result to "almost upward planar" graphs, i.e., to graphs such that the deletion of at most h edges yields an upward planar graph. As finding an upward planar drawing of an upward planar graph is NP-complete, this might be of use if we have an approximation algorithm that produces almost upward planar embeddings.

5 Conclusion

In this paper we showed that the k-vertex disjoint paths problem is NP-complete on a restricted and yet very interesting class of planar digraphs. On the other hand, we provided a fast algorithm to approach this hard problem by finding good partial order. It is an interesting question to investigate whether the $k!$ factor in the running time of our algorithm can be improved. Another direction of research is to extend our result to more general but still restricted graph classes, such as to digraphs embedded on a torus such that all edges are monotonically increasing in the z-direction or to acyclic planar graphs.

References

1. Bang-Jensen, J., Gutin, G.Z.: Digraphs - Theory, Algorithms and Applications, 2nd edn. Springer (2010)
2. Battista, G.D., Eades, P., Tamassia, R., Tollis, I.G.: Graph Drawing: Algorithms for the Visualization of Graphs. Prentice-Hall (1999)
3. Bertolazzi, P., Di Battista, G., Liotta, G., Mannino, C.: Upward drawings of tri-connected digraphs. Algorithmica 12(6), 476–497 (1994)
4. Bertolazzi, P., Di Battista, G., Mannino, C., Tamassia, R.: Optimal upward planarity testing of single-source digraphs. SIAM J. Comput. 27(1), 132–169 (1998)
5. Cygan, M., Marx, D., Pilipczuk, M., Pilipczuk, M.: The planar directed k-vertex-disjoint paths problem is fixed-parameter tractable. In: 2013 IEEE 54th Annual Symposium on Foundations of Computer Science, pp. 197–206 (2013)
6. Diestel, R.: Graph Theory, 3rd edn. Springer (2005)
7. Downey, R., Fellows, M.: Parameterized Complexity. Springer (1998)
8. Fortune, S., Hopcroft, J.E., Wyllie, J.: The directed subgraph homeomorphism problem. Theor. Comput. Sci. 10, 111–121 (1980)
9. Ganian, R., Hliněný, P., Kneis, J., Langer, A., Obdržálek, J., Rossmanith, P.: On digraph width measures in parameterized algorithmics. In: Chen, J., Fomin, F.V. (eds.) IWPEC 2009. LNCS, vol. 5917, pp. 185–197. Springer, Heidelberg (2009)
10. Garey, M.R., Johnson, D.S.: Computers and Intractability: A Guide to the Theory of NP-Completeness. W.H. Freeman (1979)
11. Garg, A., Tamassia, R.: On the computational complexity of upward and rectilinear planarity testing. SIAM J. Comput. 31(2), 601–625 (2001)
12. Johnson, T., Robertson, N., Seymour, P.D., Thomas, R.: Directed tree-width. J. Comb. Theory, Ser. B 82(1), 138–154 (2001)
13. Lynch, J.F.: The equivalence of theorem proving and the interconnection problem. ACM SIGDA Newsletter 5(3), 31–36 (1975)
14. Ohtsuki, T.: The two disjoint path problem and wire routing design. In: Saito, N., Nishizeki, T. (eds.) Graph Theory and Algorithms. LNCS, vol. 108, pp. 207–216. Springer, Heidelberg (1981)
15. Papakostas, A.: Upward planarity testing of outerplanar dags (extended abstract) (1995), 10.1007/3-540-58950-3_385
16. Robertson, N., Seymour, P.D.: Graph minors XIII. The disjoint paths problem. Journal of Combinatorial Theory, Series B 63, 65–110 (1995)
17. Schrijver, A.: Finding k disjoint paths in a directed planar graph. SIAM Jornal on Computing 23(4), 780–788 (1994)
18. Seymour, P.D.: Disjoint paths in graphs. Discrete Math. 29, 293–309 (1980)
19. Shiloach, Y.: A polynomial solution to the undirected two paths problem. J. ACM 27, 445–456 (1980)
20. Slivkins, A.: Parameterized tractability of edge-disjoint paths on directed acyclic graphs. In: European Symposium on Algorithms, pp. 482–493 (2003)
21. Thomassen, C.: 2-linked graphs. European Journal of Combinatorics 1, 371–378 (1980)

On Lower Bounds for Multiplicative Circuits and Linear Circuits in Noncommutative Domains

V. Arvind[1], S. Raja[1], and A.V. Sreejith[2]

[1] The Institute of Mathematical Sciences (IMSc), Chennai, India
{arvind,rajas}@imsc.res.in
[2] Tata Institute of Fundamental Research (TIFR), Mumbai, India
sreejith@imsc.res.in

Abstract. In this paper we show some lower bounds for the size of multiplicative circuits computing multi-output functions in some *noncommutative* domains such as monoids and finite groups. We also introduce and study a generalization of linear circuits in which the goal is to compute MY where Y is a vector of indeterminates and M is a matrix whose entries come from *noncommutative* rings. We show some lower bounds in this setting as well.

1 Introduction

Let $(S, \circ)$ be a semigroup, i.e., S is a set closed under the binary operation $\circ$ which is associative. A natural multi-output computational model is a circuit over $(S, \circ)$. The circuit is given by a directed acyclic graph with input nodes labeled $x_1, ..., x_n$ of indegree 0 and output nodes $y_1, ..., y_m$ of outdegree 0.

The gates of the circuit all compute the monoid product. We assume that all gates have fanin 2. The size of the circuit is the number of nodes in it and it computes a function $f : S^n \to S^m$.

This provides a general setting to some well studied problems in circuit complexity. For example:

(1) If $S = \mathbb{F}_2$ and $\circ$ is addition in $\mathbb{F}_2$, the problem is one of computing $A\mathbf{x}$ for an $m \times n$ matrix over $\mathbb{F}_2$. The problem of giving an explicit A such that the size of any circuit for it is superlinear is a longstanding open problem. By means of counting arguments, we know that there exist such matrices A [11].

This problem has a rich literature with many interesting developments. Morgenstern [7] showed an $\Omega(n \log n)$ lower bound for the Hadamard matrix in the bounded coefficient model when $\mathbb{F} = \mathbb{C}$. Valiant [11] developed matrix rigidity as a means to attack the problem in the case of logarithmic depth circuits. In spite of many interesting results and developments, superlinear size lower bounds remain elusive over any field $\mathbb{F}$ even for the special case of log-depth circuits (Lokam's monograph [6] contains most of the recent results).

(2) When $S = \{0, 1\}$ and $\circ$ is the boolean OR, this problem is also well studied and due to its monotone nature it has explicit lower bounds of circuit size $n^{2-o(1)}$ (e.g. see section 3.4 in [3]).

E.A. Hirsch et al. (Eds.): CSR 2014, LNCS 8476, pp. 65–76, 2014.

A more restricted form is $S = (\mathbb{N}, +)$ called SUM circuits also well studied e.g. [3]. While for monotone settings (OR,SUM circuits) there are nontrivial lower bounds, in the commutative case for S we do not have strong lower bounds results. In this paper, we explore the case when $(S, \circ)$ is noncommutative and manage to prove strong lower bounds in some cases.

An interesting aspect is that the number of inputs can be restricted to just two: x_0, x_1. The explicit functions y_i, $1 \leq i \leq m$ are defined as words $y_i = y_{i1}y_{i2}...y_{in}$ where $y_{ij} \in \{x_0, x_1\}$ and $\{y_1, y_2, ..., y_m\}$ are explicitly defined. We show that any circuit $C : \{x_0, x_1\} \to \{y_1, y_2, ..., y_m\}$ is of size $\Omega(\frac{mn}{\log^2 n})$ in the following four settings:

1. When $(S, \circ)$ is the free monoid X^* for X such that $|X| \geq 2$.
2. When $(S, \circ)$ is the finite matrix semigroup over the boolean ring and matrices are of dimension $n^c \times n^c$ for some constant $c > 0$.
3. When $(S, \circ)$ is the free group G_X generated by $X = \{x_1, x_2, x_1^{-1}, x_2^{-1}\}$.
4. When $(S, \circ)$ is the permutation group where $S = S_N$ for $N = n^d$ for some constant $d > 0$.

In Section 6, we show lower bounds for a generalization of linear circuits model. In this model we allow coefficients to come from *noncommutative rings*.

2   Circuits over Free Monoids

We consider the free monoid X^* where X is a finite alphabet and the monoid operation is concatenation with the empty string ϵ as identity. The notion of a multiplicative circuits over a free monoid is also known in th area of data compression as a straight line program [5].

Notice that when X is a singleton set $X = \{1\}$ then $(1^*, \circ)$ is essentially the semigroup $(\mathbb{N}, +)$. We consider the simplest noncommutative setting with $X = \{0, 1\}$. In the problem, we consider circuits that take the "generating set" X as input and the m outputs $y_1, y_2, ..., y_m \in X^n$ (where n is the "input" parameter).

Since each y_i is of length n, clearly n gates are sufficient to compute each y_i and hence $O(mn)$ is an obvious upper bound for the circuit size. We will give an explicit set $y_1, y_2, ..., y_m \in \{0, 1\}^n$ so that $\Omega(\frac{mn}{\log^2 n})$ is the circuit size lower bound. We will let $m = n$ in the construction and it can be suitably generalized to larger values of m. We now explain the construction of the set $S = \{y_1, y_2, ..., y_m\} \subseteq \{0, 1\}^n$.

Construction of S

Consider the set $[n^2]$ of the first n^2 natural numbers. Each $i \in [n^2]$ requires $2 \log_2 n$ bits to represent in binary. Initially let $D = [n^2]$.

for $i = 1, ..., n$ do

pick the first $\lceil \frac{n}{2 \log n} \rceil$ numbers from the current D, concatenate their binary representations to obtain y_i and remove these numbers from D.
end for

This defines the set $S = \{y_1, y_2, ..., y_n\}$. Each y_i constructed has the property that y_i has $\geq \frac{n}{2 \log n}$ distinct substrings of length $2 \log n$. We show the following two result about these strings:

- For each $y_i \in S$ any concatenation circuit that generates y_i from input $X = \{0, 1\}$ requires size $\Omega(\frac{n}{\log^2 n})$.
- Any concatenation circuit that takes $X = \{0, 1\}$ as input and outputs $S = \{y_1, y_2, ..., y_n\}$ at n output gates requires size $\Omega(\frac{n^2}{\log^2 n})$.

Lemma 1. *Let $s \in X^n$ be any string where $|X| \geq 2$, such that the number of distinct substrings of s of length l is N. Then any concatenation circuit for s will require $\Omega(\frac{N}{l})$ gates.*

Proof. Let C be any circuit that computes the string s. Now each gate g of C computes some string s_g. Suppose $g = g_1 \circ g_2$ is a gate whose inputs are gates g_1, g_2.

Suppose s_{g_1} has k_1 distinct substrings of length l and s_{g_2} has k_2 distinct substrings of length l. Now, in s_g notice that the *new* substrings of length l (not occurring in s_{g_1} or s_{g_2}) could only arise as a concatenation of some suffix of s_{g_1} and prefix of s_{g_2} such that neither of them is the empty string. The number of such substrings is at most $l - 1$.

Hence, s_g can have at most $k_1 + k_2 + l - 1$ distinct substrings of length l. Thus, each new gate of C can generate at most $l - 1$ new substrings of length l. Since the output string s has N distinct length l substrings, it follows that number of gates in C is $\Omega(\frac{N}{l})$. □

Note the case not covered by the lemma: $|X| = 1$. In that case we know that every string of length n (for every n) has a concatenation circuit of size $\leq 2 \log_2 n$ and the circuit exploits the fact that for each length l there is a unique string. Similar to Lemma 1 is known earlier (e.g. see Lemma 3 in [2]).

Theorem 1. *Let $S \subseteq \{0, 1\}^n$ be the explicit set of n strings defined above. Any concatenation circuit that takes $X = \{0, 1\}$ as input and outputs S at its n output gates will require size $\Omega(\frac{n^2}{\log^2 n})$.*

Proof. Let $S = \{y_1, y_2, ..., y_n\}$ as defined above. Notice that, each y_i can be generated by size n circuit. Let C be any concatenation circuit that takes $X = \{0, 1\}$ as inputs and at its n output gates generates $y_1, y_2, ..., y_n$ respectively. Let C' be a concatenation circuit obtained from C by adding $n - 1$ new gates such that C' outputs the concatenation $y = y_1 y_2 ... y_n$. By construction $size(C') = size(C) + n - 1$. The number of distinct length $2 \log n$ strings in the string y is, by construction, $\geq \frac{n^2}{2 \log n}$. This is because each y_i has $\geq \frac{n}{2 \log n}$ distinct substrings and these are disjoint for different y_i. Hence by Lemma 1, $size(C') = \Omega(\frac{n^2}{\log^2 n})$ which implies $size(C) = \Omega(\frac{n^2}{\log^2 n})$. □

3   Circuits over Matrix Semigroups

The setting now is that of a finite monoid $(M, \circ)$ where M consisting of $p(n) \times p(n)$ matrices whose entries come from the Boolean semiring $\{0, 1, \vee, \wedge\}$. We will modify the lower bound of the previous section to make it work over $(M, \circ)$ which is a finite monoid.

Recall we constructed $S = \{y_1, y_2, ..., y_n\} \subseteq \{0, 1\}^n$. Let D_l be the set of all length l substrings of each $y_i \in S$. Let $D = \bigcup_{l=0}^{n} D_l$. Clearly $|D| = \sum_{l=0}^{n} |D_l| \leq n^3$. The matrices in M are $|D| \times |D|$. We now define two functions $f_0, f_1 : D \to D$ corresponding to the generating set $X = \{0, 1\}$ of the free monoid. For $b \in \{0, 1\}$, define

$$f_b(s) = \begin{cases} s \circ b & s \circ b \in D \\ \epsilon & \text{otherwise} \end{cases}$$

where $s \in \{0, 1\}^*$. These give rise to two matrices M_b, $b \in \{0, 1\}$. The rows and columns of M_b are indexed by elements of D and $M_b(s, s \circ b) = 1$ if $s \circ b \in D$ and $M_b(s, s') = 0$ if $s \circ b \neq s'$. If $s \circ b \notin D$ then $M_b(s, \epsilon) = 1$ and $\forall s' \neq \epsilon$, $M_b(s, s') = 0$.

Thus, we have defined a morphism, $\Phi : (X^*, \circ) \to (M, \circ)$ which maps $b \to M_b$, $b \in \{0, 1\}$ and by natural extension maps a string $s \in X^*$ to M_s. In particular, the set $S = \{y_1, y_2, ..., y_n\}$ defined in section 2 is mapped to $\hat{S} = \{M_{y_1}, M_{y_2}, ..., M_{y_n}\}$.

Theorem 2. *Any circuit over* $(M, \circ)$ *that takes* M_0, M_1 *as input and computes* $\{M_{y_i} | y_i \in S\}$ *at its n output gates is of size* $\Omega(\frac{n^2}{\log^2 n})$.

Proof. Let C be a circuit over $(M, \circ)$ computing M_{y_i}, $1 \leq i \leq n$ at the n output gates and input M_0, M_1. Consider the corresponding circuit C' over the free monoid X^* with input $X = \{0, 1\}$. Let g_i be the output gate of C computing M_{y_i}, $1 \leq i \leq n$. In C' let $w_i \in X^*$ be the word computed at g_i. We know that $M_{w_i} = M_{y_i}$ for $1 \leq i \leq n$. That means $M_{w_i}(\epsilon, y_i) = 1$. By definition of the matrices M_b, the only way this can happen is when $w_i = y_i \circ z_i$ for some $z_i \in X^*$ for each i. Now, let C'' be a new circuit obtained from C' that outputs the concatenation of $w_1, w_2, ..., w_n$ in that order. Then $size(C'') \leq size(C') + n - 1$. The output string by C'' is of the form $y_1 \circ z_1 \circ y_2 \circ z_2 \circ ... \circ y_n \circ z_n$. Since the number of distinct substrings of length $2 \log n$ in $\{y_1, y_2, ..., y_n\}$ we know is $\geq \frac{n^2}{\log^2 n}$, it follows by Lemma 1 that $size(C'') = \Omega(\frac{n^2}{\log^2 n})$. Consequently, $size(C) = size(C') = \Omega(\frac{n^2}{\log^2 n})$. This completes the proof. □

4   Circuits over Free Groups

We consider the free group G_X generated by the set $X = \{x_1, x_2, x_1^{-1}, x_2^{-1}\}$ consisting of x_1, x_2 and their inverses. The group operation is concatenation with the empty string ϵ as identity and the only cancellation rules we can repeatedly use are $x_i x_i^{-1} = x_i^{-1} x_i = \epsilon$ for $i \in \{1, 2\}$. Given a word $w \in X^*$ we can repeatedly apply these rules and obtain a *normal form* $w' \in G_X$ from it which cannot be simplified further. This normal form, by Church-Rosser property, is unique and independent of how we apply the rules.

Recall the set of binary strings we constructed in Section 2. Replacing 0 by x_1 and 1 by x_2 we obtain $S = \{y_1, y_2, ..., y_n\} \subseteq \{x_1, x_2\}^n \subseteq G_X$. Each word y_i constructed has the property that y_i has $\geq \frac{n}{2 \log n}$ distinct subwords of length $2 \log n$. These words are already in their normal forms.

Lemma 2. *Let* $w \in G_X$ *be any word where* $X = \{x_1, x_2, x_1^{-1}, x_2^{-1}\}$, *such that the number of distinct subwords of length l in its normal form w' is N. Then any concatenation circuit for w will require size $\Omega(\frac{N}{l})$ gates.*

Proof. Let C be any circuit that computes the word w. Now each gate g of C computes some word w_g and, as above, w'_g denotes its normal form.

Suppose $g = g_1 \circ g_2$ is a gate whose inputs are gates g_1, g_2. Then, by the Church-Rosser property of cancellations, the normal form for w_g satisfies

$$w'_g = (w'_{g_1} \circ w'_{g_2})'.$$

Suppose w'_{g_1} has k_1 distinct subwords of length l and w'_{g_2} has k_2 distinct subwords of length l. Now, in w'_g notice that the *new* subwords of length l (not occurring in w'_{g_1} or w'_{g_2}) could only arise as a concatenation of some suffix of word w'_{g_1} and prefix of word w'_{g_2} such that neither of them is the empty string. The number of such new subwords is at most l. Hence, w'_g can have at most $k_1 + k_2 + l$ distinct subwords of length l.

Now, since the normal form w' for the output word w has N distinct length l subwords, it follows that number of gates in C is $\Omega(\frac{N}{l})$. $\qquad \square$

Theorem 3. *Let* $S \subseteq \{x_1, x_2\}^n \subseteq G_X$ *be the explicit set of n words defined above. Any concatenation circuit that takes $X = \{x_1, x_2, x_1^{-1}, x_2^{-1}\}$ as input and outputs S at its n output gates will require size $\Omega(\frac{n^2}{\log^2 n})$.*

Proof. Let $S = \{y_1, y_2, ..., y_n\}$ as defined above and let C be any concatenation circuit that takes $X = \{x_1, x_2, x_1^{-1}, x_2^{-1}\}$ as inputs and at its n output gates generates $y_1, y_2, ..., y_n$ respectively. Let C' be a concatenation circuit obtained from C by adding $n - 1$ new gates such that C' outputs the concatenation $y = y_1 y_2 ... y_n$. By construction $size(C') = size(C) + n - 1$. The number of distinct length $2 \log n$ words in the words y is, by construction, $\geq \frac{n^2}{2 \log n}$. This is because each y_i has $\geq \frac{n}{2 \log n}$ distinct subwords and these are disjoint for different y_i. Hence by Lemma 2, $size(C') = \Omega(\frac{n^2}{\log^2 n})$ which implies $size(C) = \Omega(\frac{n^2}{\log^2 n})$. $\qquad \square$

Remark 1. Let $\mathbf{M_0} = \begin{pmatrix} 1 & 2 \\ 0 & 1 \end{pmatrix}$, $\mathbf{M_1} = \begin{pmatrix} 1 & 0 \\ 2 & 1 \end{pmatrix}$ be 2×2 matrices. Consider the infinite group G generated by these elements and their inverses over the field of rationals $\mathbb{Q}$. It is well known (e.g. see [4] for a nice complexity theoretic application) that the group G is isomorphic to the free group G_X, where the isomorphism is defined by $x_1 \to M_0$ and $x_2 \to M_1$. It follows that Theorem 3 also applies to the group G by setting $x_1 = M_0$ and $x_2 = M_1$.

5   Circuits over Permutation Groups

We now present a lower bound in the setting of finite groups. We will transform our free monoid construction to this setting. Recall the set of binary strings S we constructed in Section 2. To this end, we will define two permutations $\pi_0, \pi_1 \in S_N$ (where $N = poly(n)$ will be defined later). These permutations correspond to $X = \{0, 1\}$ and by multiplication the target output permutations are defined:

$$G_S = \{\pi_{y_i} = \Pi_{j=1}^n \pi_{y_i[j]} | y_i \in S\}, \text{ where } y_i[j] \text{ is the } j\text{-th bit of string } y_i.$$

Definition of π_0, π_1:

We pick r primes $p_1, p_2, ..., p_r$ where $r = n^2$ such that $n < p_1 < p_2 < ... < p_r < n^4$. The permutation π_0 is defined as the product of $r + 1$ disjoint cycles, $\pi_0 = C_0.C_1...C_r$ where C_0, C_1 are of length p_1 and for $i \geq 2$, C_i is of length p_i. Similarly, $\pi_1 = C_0'.C_1'...C_r'$ is a product of $r + 1$ disjoint cycles with C_0' and C_1' of length p_1 and for $i \geq 2$, C_i' is of length p_i. Let $\text{supp}(C)$ denote the set of points moved by C for a cycle C (i.e., if we write $C = (i_1 i_2...i_p)$ it means C maps i_1 to i_2 and so on i_p to i_1 and moves no other element of the domain. Hence, $\text{supp}(C) = \{i_1, i_2, ..., i_p\}$). In the construction above we pick the cycles C_i and C_i', $0 \leq i \leq r$ such that $\text{supp}(C_0) \cap \text{supp}(C_0') = \{1\}$ and $\forall (i, j) \neq (0, 0)$ $\text{supp}(C_i) \cap \text{supp}(C_j') = \phi$. The domain $[N]$ on which these permutations are defined is $\bigcup_{i=0}^r (\text{supp}(C_i) \cup \text{supp}(C_i'))$. Note that $N \leq 4p_1 + 2\sum_{i=2}^r p_i = O(n^6)$. Thus, the problem we consider is that of designing a circuit over S_N that takes as input x_0, x_1 and outputs at the n output gates $\pi_{y_i} = \Pi_{j=1}^n x_{y_i[j]}$ where $y_i[j]$ is the j-th bit of string y_i for each $y_i \in S$.

Theorem 4. *Any circuit over the group* $(S_N, \circ)$ *that takes as input* π_0, π_1 *and computes* $G_S = \{\pi_{y_i} | y_i \in S\}$ *as output is of size* $\Omega(\frac{n^2}{\log^2 n})$.

Proof. Let C be the circuit that solves this problem of computing G_S from x_0, x_1. We fix the input as $x_0 = \pi_0$ and $x_1 = \pi_1$. Now, consider the corresponding concatenation circuit C' with input $x_0, x_1 \in X$. At each output gate g_i, $1 \leq i \leq m$, circuit C' computes some word $w_i \in X^*$ such that $\forall i$, $\pi_{w_i} = \pi_{y_i}$ where π_{w_i} is the permutation in S_N obtained by putting $x_0 = \pi_0$ and $x_1 = \pi_1$ in w_i. If $w_i = y_i$ for all i, then in fact C' as a concatenation circuit computes the set S at its output gates. This implies by Theorem 1 that $size(C') = \Omega(\frac{n^2}{\log^2 n})$ and $size(C) = \Omega(\frac{n^2}{\log^2 n})$.

Suppose $w_i \neq y_i$ at some output gate g_i. We can write $w_i = u \circ b_2 \circ s$ and $y_i = v \circ b_1 \circ s$ where $b_1 \neq b_2$. Assume, without loss of generality, that $b_1 = 0$ and $b_2 = 1$. Since $\pi_{w_i} = \pi_{y_i}$, we know $\pi_u \pi_{b_2} \pi_s = \pi_v \pi_{b_1} \pi_s$ (i.e., $\pi_u \pi_1 \pi_s = \pi_v \pi_0 \pi_s$). Let $\alpha \in [N]$ such that $\pi_s(\alpha) = 1$. In $\pi_{y_i} = \pi_v \pi_0 \pi_s$, the permutation π_0 will map 1 to $\beta \in C_0 \backslash \{1\}$, whereas in $\pi_{w_i} = \pi_u \pi_1 \pi_s$ the permutation π_1 maps 1 to $\gamma \in C_0' \backslash \{1\}$. Since $|v| < n$ the point β cannot be moved back to 1 and subsequently to $C_0' \backslash \{1\}$. This is because $p_1 > n$ and the length of cycle C_0' is p_1. Therefore by π_{y_i} the point α is mapped to some point in $C_0 \backslash \{1\}$. Since π_{w_i} must map α to the same point and $\pi_1 \pi_s$ has mapped α to a point in $\gamma \in C_0' \backslash \{1\}$, π_u must have at least $p_1 > n$ occurrences of π_1 in it to move γ to 1 and subsequently to the final point in $C_0 \backslash \{1\}$ (using some π_0 applications). We will now argue that this forces w_i to be a long string.

Pick any tuple of points $(\alpha_1, \alpha_2, ..., \alpha_r)$ where $\alpha_i \in C_i'$, $1 \le i \le r$. Notice that only π_1 moves this tuple because α_i, $1 \le i \le r$ do not belong to $\text{supp}(\pi_0)$. Since $p_1, ..., p_r$ are distinct primes, the permutation π_1 maps $(\alpha_1, \alpha_2, ..., \alpha_r)$ to a set of $\Pi_{i=1}^r p_i - 1$ distinct r-tuples before returning to $(\alpha_1, \alpha_2, ..., \alpha_r)$. Suppose there are l occurrences of π_1 in π_{y_i}, $l < n$. Thus, if $\pi_{y_i}(\alpha_1, \alpha_2, ..., \alpha_r) = (\beta_1, \beta_2, ..., \beta_r)$ then $\pi_1^l(\alpha_1, \alpha_2, ..., \alpha_r) = (\beta_1, \beta_2, ..., \beta_r)$. Then $\pi_{w_i}(\alpha_1, \alpha_2, ..., \alpha_r) = (\beta_1, \beta_2, ..., \beta_r)$. However we know number of occurrences of π_1 in π_{w_i} is some $k \ge n$ which means $\pi_{w_i}(\alpha_1, \alpha_2, ..., \alpha_r) = (\beta_1, \beta_2, ..., \beta_r) = \pi_1^k(\alpha_1, \alpha_2, ..., \alpha_r)$.

It follows that $\pi_1^{k-l}(\alpha_1, \alpha_2, ..., \alpha_r) = (\alpha_1, \alpha_2, ..., \alpha_r)$ which implies $k - l$ is a multiple of $\Pi_{i=1}^r p_i$. Hence $|w_i| \ge \Pi_{i=1}^r p_i$. This implies that the circuit needs at least $\log \Pi_{i=1}^r p_i$ multiplication gates to compute w_i. This gives, $size(C) \ge \log \Pi_{i=1}^r p_i \ge \log 2^{n^2} = n^2$.

Putting it together $size(C) = \Omega(\frac{n^2}{\log^2 n})$ in any case. This completes the proof.

$\square$

6 Linear Circuits over Rings

In this section we consider a generalization of the linear circuits model. In this generalization we allow the coefficients come from *noncommutative rings*. In principle, we can expect lower bounds could be easier to prove in this model. The circuits are more constrained when coefficients come from a noncommutative ring as fewer cancellations can take place. This is in the same spirit as Nisan's [8] work on lower bounds for noncommutative algebraic branching programs. However, in this paper we succeed in showing only some limited lower bounds. We leave open problems that might be more accessible than the notorious problems for linear circuits over fields.

Let $(R, +, \cdot)$ be an arbitrary ring (possibly noncommutative). A *linear circuit* over R takes n inputs $y_1, y_2, ..., y_n$ labeling the indegree 0 nodes of a directed acyclic graph. The circuit has m output nodes. Each edge of the graph is labeled by some element of the ring R. The indegree of each non-input node is two. Each node of the circuit computes a linear form $\sum_{i=1}^n \alpha_i y_i$ for $\alpha_i \in R$ as follows: the input node labeled y_i computes y_i. Suppose g is a node with incoming edges from nodes g_1 and g_2, and the edges (g_1, g) and (g_2, g) are labeled by α and β respectively. If g_1 and g_2 computes the linear forms ℓ_1 and ℓ_2 respectively, then g computes $\alpha \ell_1 + \beta \ell_2$. Thus, for an $m \times n$ matrix A over the ring R, the circuit computes $A\mathbf{y}$ at the m output gates.

When R is a field we get the well-studied linear circuits model [7,11,6]. However, no explicit superlinear size lower bounds are known for this model over fields (except for some special cases like the bounded coefficient model [7] or in the cancellation free case [1]).

When the coefficients to come from a *noncommutative* ring R, we prove lower bounds for certain restricted linear circuits. Suppose the coefficient ring is $R = \mathbb{F}\langle x_0, x_1 \rangle$ consisting of polynomials over the field $\mathbb{F}$ in noncommuting variables x_0 and x_1.

Let $M \in \mathbb{F}^{n \times n} \langle x_0, x_1 \rangle$ where x_0, x_1 are noncommuting variables and $Y = (y_1, y_2, \ldots, y_n)^T$ is a column vector of input variables. The first restriction we consider are *homogeneous* linear circuits over the ring $\mathbb{F}\langle x_0, x_1 \rangle$ for computing MY. The restriction is that for every gate g in the circuit, if g has its two incoming edges from nodes g_1 and g_2, then the edges (g_1, g) and (g_2, g) are labeled by α and β respectively, where $\alpha, \beta \in \mathbb{F}\langle x_0, x_1 \rangle$ are restricted to be *homogeneous polynomials* of same degree in the variables x_0 and x_1. It follows, as a consequence of this restriction, that each gate g of the circuit computes a linear form $\sum_{i=1}^{n} \alpha_i y_i$, where the $\alpha_i \in \mathbb{F}\langle x_0, x_1 \rangle$ are all homogeneous polynomials of the same degree. Our goal is to construct an explicit matrix $M \in \mathbb{F}^{n \times n} \langle x_0, x_1 \rangle$ such that MY can not be computed by any circuit C with size $O(n)$ and depth $O(\log n)$. We prove this by suitably generalizing Valiant's matrix rigidity method [11] as explained below.

Consider $n \times n$ matrices $\mathbb{F}^{n \times n}$ over field $\mathbb{F}$. The *support* of a matrix $A \in \mathbb{F}^{n \times n}$ is the set of locations $\mathrm{supp}(A) = \{(i, j) \mid A_{ij} \neq 0\}$.

Definition 1. *Let $\mathbb{F}$ be any field. The rigidity $\rho_r(\mathcal{A})$ of a deck of matrices $\mathcal{A} = \{A_1, A_2, \ldots, A_N\} \subseteq \mathbb{F}^{n \times n}$ is the smallest number t for which there are a set of t positions $S \subseteq [n] \times [n]$ and a deck of matrices $\mathcal{B} = \{B_1, B_2, \ldots, B_N\}$ such that for all i: $\mathrm{supp}(B_i) \subseteq S$ and the rank of $A_i + B_i$ is bounded by r. A collection $\mathcal{A} = \{A_1, A_2, \ldots, A_N\} \subseteq \mathbb{F}^{n \times n}$ is a rigid deck if $\rho_{\epsilon \cdot n}(\mathcal{A}) = \Omega(n^{2-o(1)})$, where $\epsilon > 0$ is a constant.*

Notice that for $N = 1$ this is precisely the notion of rigid matrices. We are interested in constructing *explicit* rigid decks: I.e. a deck $\mathcal{A}$ such that for each $k \in [N]$ and each $1 \leq i, j \leq n$ there is a polynomial (in n) time algorithm that outputs the $(i, j)^{th}$ entry of A_k. We describe an explicit deck of size $N = 2^{n^2}$ over any field $\mathbb{F}$ and use it to prove our first lower bound result. It is convenient to write the deck as $\mathcal{A} = \{A_m \mid m \in \{x_0, x_1\}^{n^2}\}$ with matrices A_m indexed by monomials m of degree n^2 in the noncommuting variables x_0 and x_1. The matrix A_m is defined as follows:

$$A_m[i, j] = \begin{cases} 1 \text{ if } m_{ij} = x_1 \\ 0 \text{ if } m_{ij} = x_0 \end{cases}$$

Note that all the matrices A_m in the deck $\mathcal{A}$ are in $\mathbb{F}^{n \times n}$. Clearly, $\mathcal{A}$ is an explicit deck. We prove that it is a rigid deck.

Lemma 3. *The deck $\mathcal{A} = \{A_m \mid m \in \{x_0, x_1\}^{n^2}\}$ is an explicit rigid deck for any field $\mathbb{F}$.*

Proof. Valiant [11] showed that almost all $n \times n$ 0-1 matrices over any field $\mathbb{F}$ have rigidity $\Omega(\frac{(n-r)^2}{\log n})$ for target rank r. In particular, for $r = \epsilon \cdot n$, over any field $\mathbb{F}$, there is a 0-1 matrix R for which we have $\rho_r(R) \geq \frac{\delta \cdot n^2}{\log n}$ for some constant $\delta > 0$ depending on ϵ.

We claim that for the deck $\mathcal{A}$ we have $\rho_{\epsilon n}(\mathcal{A}) \geq \frac{\delta \cdot n^2}{\log n}$. To see this, let $E = \{E_m \in \mathbb{F}^{n \times n} | m \in \{x_0, x_1\}^{n^2}\}$ be any collection of matrices such that $|\mathrm{supp}(E_m)| \leq \frac{\delta n^2}{\log n}$ for each m. Since the deck $\mathcal{A}$ contains all 0-1 matrices, in particular $R \in \mathcal{A}$ and $R = A_m$ for some monomial m. From the rigidity of R we know that the rank of $R + E_m$ is at least ϵn. This proves the claim and the lemma follows. $\qquad\square$

We now turn to the lower bound result for homogeneous linear circuits where the coefficient ring is $\mathbb{F}\langle x_0, x_1 \rangle$. We define an explicit $n \times n$ matrix M as

$$M_{ij} = (x_0 + x_1)^{(i-1)n+j-1} \cdot x_1 \cdot (x_0 + x_1)^{n^2 - ((i-1)n+j)}. \qquad (1)$$

It is easy to see that we can express the matrix M as $M = \sum_{m \in \{x_0, x_1\}^{n^2}} A_m m$, where $\mathcal{A} = \{A_m \mid m \in \{x_0, x_1\}^{n^2}\}$ is the deck defined above.

Theorem 5. *Any homogeneous linear circuit C over the coefficient ring $\mathbb{F}\langle x_0, x_1 \rangle$ computing MY, for M defined above, requires either size $\omega(n)$ or depth $\omega(\log n)$.*

Proof. Assume to the contrary that C is a homogeneous linear circuit of size $O(n)$ and depth $O(\log n)$ computing MY. We know that by Valiant's graph-theoretic argument (see e.g. [6]) that in the circuit C there is a set of gates V of cardinality $s = \frac{c_1 n}{\log \log n} = o(n)$ such that at least $n^2 - n^{1+\delta}$, for $\delta < 1$, input-output pairs have all their paths going through V. Thus, we can write $M = B_1 B_2 + E$ where $B_1 \in \mathbb{F}^{n \times s} \langle x_0, x_1 \rangle$ and $B_2 \in \mathbb{F}^{s \times n} \langle x_0, x_1 \rangle$ and $E \in \mathbb{F}^{n \times n} \langle x_0, x_1 \rangle$, and $|\mathrm{supp}(E)| \leq n^{1+\delta}$. By collecting the matrix coefficient of each monomial we can express M and E as

$$M = \sum_{m \in \{x_0, x_1\}^{n^2}} A_m m, \text{ and } E = \sum_{m \in \{x_0, x_1\}^{n^2}} E_m m,$$

where A_m are already defined and $|\cup_{m \in \{x_0, x_1\}^{n^2}} \mathrm{supp}(E_m)| \leq n^{1+\delta}$. Now consider the matrix $B_1 B_2$. By collecting matrix coefficients of monomials we can write $B_1 B_2 = \sum_{m \in \{x_0, x_1\}^{n^2}} B_m m$.

We now analyze the matrices B_m. Crucially, by the homogeneity condition on the circuit C, we can partition $V = V_1 \cup V_2 \cup \ldots V_\ell$, where each gate g in V_i computes a linear form $\sum_{j=1}^n \gamma_j y_j$ and $\gamma_j \in \mathbb{F}\langle x_0, x_1 \rangle$ is a homogeneous degree d_i polynomial. Let $s_i = |V_i|, 1 \leq i \leq \ell$. Then we have $s = s_1 + s_2 + \ldots s_\ell$. Every monomial m has a unique prefix of length d_i for each degree d_i associated with the gates in V. Thus, we can write $B_m = \sum_{j=1}^\ell B_{m,j,1} B_{m,j,2}$, where $B_{m,j,1}$ is the $n \times s_j$ matrix corresponding to the d_j-prefix of m and $B_{m,j,2}$ is the $s_j \times n$ matrix corresponding to the $n^2 - d_j$-suffix of m. It follows that for each monomial m

the rank of B_m is bounded by s. Putting it together, for each monomial m we have $A_m = B_m + E_m$, where B_m is rank s and $|\cup_{m\in\{x_0,x_1\}^{n^2}} \mathrm{supp}(E_m)| \leq n^{1+\delta}$. This contradicts the fact that $\mathcal{A}$ is a rigid deck.   □

Remark 2. For the matrix $M = (m_{ij})$, as defined above, it does not seem that Shoup-Smolensky dimension method [10] can be used to prove a similar lower bound. To see this, suppose $\Gamma_M(n)$ is the set of all monomials of degree n in $\{m_{ij}\}$ and let $D_M(n)$ be the dimension of the vector space over $\mathbb{F}$ spanned by the set $\Gamma_M(n)$. The upper bound for $D_M(n)$ that we can show for a depth d and size $O(n)$ linear circuit over the ring $\mathbb{F}\langle x_0, x_1\rangle$ is as large as $(\frac{O(n)}{d})^{dn}$. This bound, unfortunately, is much larger than the bounds obtainable for the commutative case [10]. On the other hand, the lower bound for $D_M(n)$ is only $n^{\Theta(n)}$. Thus, we do not get a superlinear size lower bound for the size using Shoup-Smolensky dimensions when the coefficient ring is $\mathbb{F}\langle x_0, x_1\rangle$.

We next consider homogeneous depth 2 linear circuits. These are linear circuits of depth 2, where each addition gate can have unbounded fanin. More precisely, if g is an addition gate with inputs from $g_1, g_2, \ldots, g_t$ then the gate g computes $\sum_{i=1}^{t} \alpha_i g_i$, where each edge (g_i, g) is labeled by $\alpha_i \in \mathbb{F}\langle x_0, x_1\rangle$ such that $\alpha_i, 1 \leq i \leq t$ are all homogeneous polynomials of the same degree. We again consider the problem of computing MY for $M \in \mathbb{F}^{n\times n}\langle x_0, x_1\rangle$. The goal is to lower bound the number of wires in the linear circuit. This problem is also well studied for linear circuits over fields and only an explicit $\Omega(n \log^2 n/\log\log n)$ lower bound is known for it [6,9], although for random matrices the lower bound is $\Omega(n^2/\log n)$.

We show that for the explicit matrix M as defined above, computing MY by a depth 2 homogeneous linear circuit (with unbounded fanin) requires $\Omega(\frac{n^2}{\log n})$ wires.

Theorem 6. *Let $M \in \mathbb{F}_2^{n\times n}\langle x_0, x_1\rangle$ as defined in Equation 1. Any homogeneous linear circuit C of depth 2 computing MY requires $\Omega(\frac{n^2}{\log n})$ wires.*

Proof. Let C be a homogeneous linear circuit of depth 2 computing MY. Let $w(C)$ denote the number of wires in C. Let s be the number of gates in the middle layer of C. We can assume without loss of generality that, all input to output paths in C are of length 2 and hence pass through the middle layer. A *level 1* edge connects an input gate to a middle-layer gate and a *level 2* edge is from middle layer to output. Thus, we can factorize $M = M' * M''$ where the matrix M' is in $\mathbb{F}^{n\times s}\langle x_0, x_1\rangle$ and M'' is in $\mathbb{F}^{s\times n}\langle x_0, x_1\rangle$, and the complexity of C is equivalent to total number of nonzero entries in M' and M''. As before, write $M = \sum_{m\in\{x_0,x_1\}^{n^2}} A_m m$.

Given A_m for $m \in \{x_0, x_1\}^{n^2}$, we show how to extract from C a depth-2 linear circuit over the *field* $\mathbb{F}$, call it $C^{(m)}$, that computes A_m such that the number of wires in $C^{(m)}$ is at most the number of wires in C. Indeed, we do not add any new gate or wires in obtaining $C^{(m)}$ from C.

For each gate g in the middle layer, there are at most n incoming edges and n outgoing edges. As C is a homogeneous linear circuit we can associate a degree d_g

to gate g. Each edge (i, g) to g is labeled by a homogeneous degree-d_g polynomial $\alpha_{i,g}$ in $\mathbb{F}\langle x_0, x_1 \rangle$. Likewise, each edge (g, j) from g to the output layer is labeled by a degree $(n^2 - d_g)$ homogeneous polynomial $\beta_{g,j}$. Let $m = m_1 m_2$, where m_1 is of degree d_g and m_2 of degree $n^2 - d_g$. For each incoming edge (i, g) to g we keep as label the coefficient of the monomial m_1 in $\alpha_{i,g}$ and for outgoing edge (g, j) from g we keep as label the coefficient of the monomial m_2 in $\beta_{g,j}$. We do this transformation for each gate g in the middle layer of C. This completes the description of the depth-2 circuit $C^{(m)}$. By construction it is clear that $C^{(m)}$ computes A_m and the number of wires $w(C^{(m)})$ in $C^{(m)}$ is bounded by $w(C)$ for each monomial $m \in \{x_0, x_1\}^{n^2}$. However, $\{A_m \mid m \in \{x_0, x_1\}^{n^2}\}$ is the set of all 0-1 matrices over $\mathbb{F}$ and it is known that there are $n \times n$ 0-1 matrices A_m such that any depth-2 linear circuit for it requires $\Omega(\frac{n^2}{\log n})$ wires (e.g. see [6]). Hence, the number of wires in C is $\Omega(\frac{n^2}{\log n})$. □

If we restrict the edge labels in the linear circuit computing MY to only *constant-degree polynomials*, then we can obtain much stronger lower bounds using Nisan's lower bound technique for noncommutative algebraic branching programs. We can define the matrix M as follows. Let $M_{ij} = w_{ij} w_{ij}^R$, where $w_{ij} \in \{x_0, x_1\}^{2 \log n}$ and $1 \le i, j \le n$ are all distinct monomials of degree $2 \log n$. We refer to M as a *palindrome matrix*. All entries of M are distinct and note that each entry of MY can be computed using $O(n \log n)$ gates.

Theorem 7. *Any linear circuit over $\mathbb{F}\langle x_0, x_1 \rangle$ computing MY, where edge labels are restricted to be constant degree polynomials, requires size $\Omega(\frac{n^2}{\log n})$.*

Proof. Let C be such a linear circuit computing MY. Since edges can be labeled by constant-degree polynomials, we can first obtain a linear circuit C' computing MY such that each edge is labeled by a *homogeneous linear form*. The size $\text{size}(C') = O(\text{size}(C) \log n)$. From C', we can obtain a noncommutative algebraic branching program $\hat{C}$ that computes the palindrome polynomial $\sum_{w \in \{x_0, x_1\}^{2 \log n}} w w^R$ such that $\text{size}(\hat{C}) = O(\text{size}(C'))$. By Nisan's lower bound [8] $\text{size}(\hat{C}) = \Omega(n^2)$, which implies $\text{size}(C) = \Omega(\frac{n^2}{\log n})$. □

Theorem 8. *Any linear circuit, whose edge labels are restricted to be either a homogeneous degree $4 \log n$ polynomial __or__ a scalar, computing MY requires $\Omega(n^2)$ size, where M is the palindrome matrix. Moreover, there is a matching upper bound.*

Proof. Let C be any linear circuit computing MY. Each entry m_{ij} of the matrix M can be written as sum of products of polynomials $m_{ij} = \sum_{\rho_{ij}} \prod_{e \in \rho_{ij}} l(e)$ where ρ_{ij} is a path from input y_j to output gate i in C and $l(e)$ is the label of edge e in C. Let S be set of all edge labels in C with degree $4 \log n$ polynomial. Thus, each m_{ij} is a linear combinations of elements in the set S over $\mathbb{F}$. This implies that $m_{ij} \in \text{Span}(S)$ where $i \le i, j \le n$. Since all m_{ij} are distinct, $|S| \ge n^2$. Since fan in is 2, $\text{size}(C) \ge \frac{n^2}{2} = \Omega(n^2)$.

For upper bound, we use n^2 edges (n edges starting from each input y_i) each labeled by a corresponding monomial in M (of degree $4 \log n$) and then we add relevant edges to get the output gates. Thus, upper bound is $O(n^2)$ for computing MY.   □

Note that, since we have not used noncommutativity in the proof, Theorem 8 also holds in the commutative settings (we require $\Omega(n^2)$ entries of M to be distinct).

7   Concluding Remarks

For multiplicative circuits we could prove lower bounds only for large monoids and large groups. The main question here is whether we can prove lower bounds for an explicit function $f : S^n \to S^m$, for some constant size nonabelian group or monoid S.

We introduced the notion of rigidity for decks of matrices, but the only explicit example we gave was the trivial one with a deck of size 2^{n^2}. A natural question is to give explicit constructions for smaller rigid decks of $n \times n$ matrices, say of size $n!$ or less. Or is the construction of rigid decks of smaller size equivalent to the original matrix rigidity problem?

Acknowledgments. We are grateful to the referees for their detailed comments and useful suggestions.

References

1. Boyar, J., Find, M.G.: Cancellation-free circuits in unbounded and bounded depth. In: Gąsieniec, L., Wolter, F. (eds.) FCT 2013. LNCS, vol. 8070, pp. 159–170. Springer, Heidelberg (2013)
2. Charikar, M., Lehman, E., Liu, D., Panigrahy, R., Prabhakaran, M., Sahai, A., Shelat, A.: The smallest grammar problem. IEEE Transactions on Information Theory 51(7), 2554–2576 (2005)
3. Jukna, S., Sergeev, I.: Complexity of linear boolean operators. Foundations and Trends in Theoretical Computer Science 9(1), 1–123 (2013)
4. Lipton, R.J., Zalcstein, Y.: Word problems solvable in logspace. Journal of the ACM (JACM) 24(3), 522–526 (1977)
5. Lohrey, M.: Algorithmics on slp-compressed strings: A survey. Groups Complexity Cryptology 4(2), 241–299 (2012)
6. Lokam, S.V.: Complexity lower bounds using linear algebra. Foundations and Trends in Theoretical Computer Science 4(1-2), 1–155 (2009)
7. Morgenstern, J.: Note on a lower bound on the linear complexity of the fast fourier transform. Journal of the ACM (JACM) 20(2), 305–306 (1973)
8. Nisan, N.: Lower bounds for non-commutative computation (extended abstract). In: STOC, pp. 410–418 (1991)
9. Pudlak, P.: Large communication in constant depth circuits. Combinatorica 14(2), 203–216 (1994)
10. Shoup, V., Smolensky, R.: Lower bounds for polynomial evaluation and interpolation problems. Computational Complexity 6(4), 301–311 (1996)
11. Valiant, L.G.: Graph-theoretic arguments in low-level complexity. In: Gruska, J. (ed.) MFCS 1977. LNCS, vol. 53, pp. 162–176. Springer, Heidelberg (1977)

Testing Low Degree Trigonometric Polynomials

Martijn Baartse and Klaus Meer*

Computer Science Institute, BTU Cottbus-Senftenberg
Platz der Deutschen Einheit 1
D-03046 Cottbus, Germany
baartse@tu-cottbus.de, meer@informatik.tu-cottbus.de

Abstract. We design a probabilistic test verifying whether a given table
of real function values corresponds to a trigonometric polynomial $f :
F^k \mapsto \mathbb{R}$ of certain (low) degree. Here, F is a finite field. The problem
is studied in the framework of real number complexity as introduced by
Blum, Shub, and Smale. Our main result is at least of a twofold interest.
First, it provides one of two major lacking ingredients for proving a real
PCP theorem along the lines of the proof of the original PCP theorem
in the Turing model. Secondly, beside the PCP framework it adds to the
still small list of properties that can be tested in the BSS model over $\mathbb{R}$.

1 Introduction

Probabilistically checkable proofs and property testing represent some of the
most important areas in theoretical computer science within the last two decades.
Among the many deep results obtained one highlight is the proof of the PCP
theorem [2,3] giving an alternative characterization of complexity class NP in
the Turing model. An alternative proof of the theorem has been given by Dinur
more recently [10].

A branch of computability and complexity theory alternative to the Turing
approach and dealing with real and complex number computations has been
developed by Blum, Shub, and Smale in [8], see also [7]. It presents a model of
uniform algorithms in an algebraic context following the tradition in algebraic
complexity theory [9]. As major problem both for real and complex number
complexity theory the analogue of the classical P versus NP question remains
unsolved in the BSS model as well. We assume the reader to be familiar with
the basic definitions of complexity classes in this model, see [7].

Given the tremendous importance probabilistically checkable proofs and the
PCP theorem exhibit in classical complexity theory, for example, with respect
to the areas of property testing and approximation algorithms it seems natural
to analyze such verification procedures and the corresponding classes as well
in the BSS model. This refers both to the question which kind of interesting
properties can be verified with high probability in the algebraic framework and
to the validity of PCP theorems.

---

* Both authors were supported under projects ME 1424/7-1 and ME 1424/7-2 by the
Deutsche Forschungsgemeinschaft DFG. We gratefully acknowledge the support.

E.A. Hirsch et al. (Eds.): CSR 2014, LNCS 8476, pp. 77–96, 2014.

Starting point of this paper is the original proof of the classical PCP theorem for NP [2,3]. It consists basically of three major steps. The easiest is the construction of so called long transparent proofs for problems in NP. It relies on testing linear functions. Here, given a table of function values a verifier checks with high probability whether the table in a certain sense is close to a linear function. Next, another verification procedure is designed, this time based on testing low-degree (algebraic) polynomials over finite fields instead of linear functions. It is combined with a so called sum-check procedure to obtain a different verifier for problems in NP. The two verification procedures then in a third step are combined applying a clever new technique of composing verifiers.

Our main result in the present work is that the second step above, i.e., a structurally similar verification procedure for certain low degree trigonometric polynomials can be set up in the real number model as well.

1.1 Previous Work

The first result dealing with probabilistically checkable proofs in the framework of real number computations established the existence of long transparent proofs used as one part in both proofs of the classical PCP theorem. Given a table of real function values on a large (i.e., superpolynomial in n) set $X \subset \mathbb{R}^n$ it can be verified probabilistically whether the represented table on a certain subset of X with high probability represents a linear function. The same holds for the complex number model, see [12,6] for those results which need considerable more efforts than the discrete counterpart.

The real and complex PCP theorems were recently proven along the lines of Dinur's proof in [5]. Independently of this result we consider it an interesting question whether as well in the BSS model a proof can be carried out that follows the orginial one by Arora and co-authors. First, since the two classical proofs are quite different we hope for a better insight what kind of results and techniques can (or cannot) be transferred to the real or complex numbers as computational structures. This is a prominent general theme in BSS complexity theory. Secondly, such investigations might open as well new research questions. Testing algorithmically meaningful mathematical objects like it is done for trigonometric polynomials in this paper in our eyes provides a promising example.

An attempt to extend the original proof, before the results of [5] were obtained, was reported in [13]. There it is studied in how far low-degree algebraic polynomials are useful as coding objects of real vectors in $\mathbb{R}^n$. Using such polynomials the main result in [13] shows $NP_{\mathbb{R}} = PCP_{\mathbb{R}}(O(\log n), poly \log n)$, saying that for each problem in $NP_{\mathbb{R}}$ there is a verification procedure that generates a logarithmic number of random bits and then inspects a polylogarithmic number of components of a potential verification proof to make a correct decision with high probability. A main ingredient in the proof is a test for algebraic low degree polynomials based on work by Friedl et al. [11] and adapted to the real number model. Basically, it says that given $d, k \in \mathbb{N}$, a large enough finite set $F \subset \mathbb{R}$ and a function value table for an $f : F^k \mapsto \mathbb{R}$, there is a verification procedure accomplishing the following task: If f corresponds on F^k to a multivariate polynomial of degree d in each of its variables the verifier accepts; if f in a precise sense is not close to such a polynomial the verifier rejects with high probability; as its resources the verifier uses $O(k \cdot \log |F|)$ many random bits and inspects $O(k \cdot d)$ components of the table. Choosing appropriate values for k, d, and $|F|$ depending on the size n of a problem instance from $NP_{\mathbb{R}}$ this test is sufficient to derive with some additional work the above mentioned characterization of $NP_{\mathbb{R}}$. However, it seems not enough in order to obtain the full real PCP theorem $NP_{\mathbb{R}} = PCP_{\mathbb{R}}(O(\log n), O(1))$. The reason is hidden behind the structure of the low degree test, more precisely the way the verifier accesses the components of a verification proof. In the original approach [2,3] the verifiers testing linear functions and low degree polynomials are cleverly composed in order to obtain a third verifier sharing the better parameters of the two. This verifier composition crucially relies on the low-degree verifier reading the proof components in a very structured manner, namely in constantly many blocks of polylogarithmic many components. The above low-degree test does not have this structure. It is in a certain sense comparable to the low-degree verifier designed in [3]. A major amount of work towards the full PCP theorem was done in [2] in order to optimize such a low-degree verifier to make it suitable for verifier composition.

1.2 Alternative Idea for a Test Scenario

The crucial progress made in [2] with respect to a low degree test is that given a function value table for $f : F^k \mapsto F$ for a finite field F the test is designed to perform certain checks along arbitrary straight lines through F^k. In order to make this idea working the structure of a finite field is essential with respect to both the resources needed and the probability arguments. There seem to arise major difficulties if one tries to generalize those ideas to the real numbers, i.e., when $F \subset \mathbb{R}$ is not any longer structured. The test performed in [13] uses paraxial lines and, as mentioned before, does not obey the structure required in order to use it as part of a verifier composition. It is unclear how to generalize it. One reason for this is the lacking structure of lines. Seeing F^k as subset of $\mathbb{R}^k$ will imply that a real line through F^k will leave the set. So one likely would have to enlarge the domain on which the initial function value table is given. But direct approaches seem to require much too large domains then. Note that a similar effect occured when testing linear functions over real domains in [12,6]. However, there they are of no major concern due to the way the result is used in the proof of the full real PCP theorem [5].

A major goal we try to follow in our approach is to still make use of the properties of lines through an F^k, where F is a finite field. The solution we follow is to use other coding objects than algebraic polynomials for vectors in $\mathbb{R}^n$, namely multivariate trigonometric polynomials: The latter should, for a finite field F, map the set F^k to $\mathbb{R}$. The period of such trigonometric polynomials is taken as the field's cardinality q. This has the huge advantage that in the domain it does not matter whether we consider arguments as real numbers or as finite field elements. As nice consequence, straight lines through F^k correspond to lines in $\mathbb{R}^n$ modulo that period. Though this gives back at least some of the advantages dealing with finite fields new difficulties arise that way. The major drawback is the following. All the above mentioned tests rely on restricting the given function table to one-dimensional subsets, thus working with univariate polynomials during the test. However, in contrast to algebraic polynomials the degree of a univariate restriction of such a trigonometric polynomial to an arbitrary line in F^k is not bounded by its original multivariate degree. Depending on the line chosen the degree of such a restriction can grow too much. This implies that not all lines are appropriate as restrictions to work with. As consequence, the design of a suitable set $H \subset F^k$ of directions of test-lines and the analysis of a corresponding test require considerable additional technical efforts. The latter are twofold. First, using the theory of expander graphs one has to establish that the set H is small, but still rich enough to cover in a reasonable sense all F^k. Secondly, it must be shown that a function table which does not give errors on H with high probability is close to a trigonometric polynomial on F^k.

As main result we obtain a verification procedure for trigonometric polynomials that inspects a constant number of relatively small blocks of proof components, thus giving a low degree test which respects the structural requirements necessary for verifier composition. Independently of this aspect, we extend the still small list of interesting real number properties for which a probabilistic verification is possible. In particular, as far as we know trigonometric polynomials have not yet been used in the realm of real number complexity theory. Given the huge importance of Fourier analysis this might be interesting to be studied further.

The paper is organized as follows. Section 2 introduces trigonometric polynomials that map elements from a k-dimensional vector space over a finite field into the reals. The main task of testing whether a table of real values arises from such a polynomial is described precisely and a test to figure this out is given. The rest of the paper then is devoted to prove that the test has the required properties. Towards this aim two major theorems have to be shown; this is done in Section 3. The concluding section will discuss the importance of the result in view of (re-)proving the real PCP theorem; we explain how to choose the parameters in our statements to use it for PCPs and outline what has to be done to obtain the full PCP theorem over $\mathbb{R}$.

A final remark: Some of the proofs necessary to establish our main result are quite technical. Given the page restriction in this extended abstract we thus focus on presenting the main proof structure. Full proofs have to be postponed in most cases to the full paper.

2   Trigonometric Polynomials, Problem Task, and Main Result

In this section we will describe a probabilistic test performed by a real verifier to check whether a given multivariate function is close to a trigonometric polynomial of low degree. In the following sections we show the main result which specifies query complexity and success probability of the verifier.

Let us start with defining the main objects of this paper, namely trigonometric polynomials. Let F be a finite field with $q := |F|$ being prime. As usual, we identify F with $\{0, \ldots, |F| - 1\}$. We want to consider particular real valued functions defined on some F^k.

Definition 1. *a) Let F be a finite field as above with cardinality q. For $d \in \mathbb{N}$ a univariate trigonometric polynomial of degree d from F to $\mathbb{R}$ is a function f of form*

$$f(x) = a_0 + \sum_{m=1}^{d} a_m \cdot \cos(2\pi m \frac{x}{q}) + b_m \cdot \sin(2\pi m \frac{x}{q}),$$

where $a_0, \ldots, a_d$ and $b_1, \ldots, b_d$ are elements in $\mathbb{R}$.

b) For $k \in \mathbb{N}$ a multivariate trigonometric polynomial $f : F^k \mapsto \mathbb{R}$ of max-degree d is defined recursively via

$$f(x_1, \ldots, x_k) = a_0(x_1, \ldots, x_{k-1}) + \sum_{m=1}^{d} a_m(x_1, \ldots, x_{k-1}) \cdot \cos(2\pi m \frac{x_k}{q}) +$$
$$+ \sum_{m=1}^{d} b_m(x_1, \ldots, x_{k-1}) \cdot \sin(2\pi m \frac{x_k}{q}),$$

where the a_i, b_j are trigonometric polynomials of max-degree d in $k - 1$ variables. Alternatively, one can write

$$f(x_1, \ldots, x_k) = \sum_{t} c_t \cdot exp(2\pi i \sum_{j=1}^{k} x_j t_j),$$

where the sum is taken over all $t := (t_1, \ldots, t_k) \in \mathbb{Z}^k$ with $|t_1| \leq d, \ldots, |t_k| \leq d$ and $c_t \in \mathbb{C}$ satisfy $c_t = \overline{c_{-t}}$ for all such t.

In all situations below the potential degrees we work with will be much less than the field's cardinality.

Since we shall mainly deal with trigonometric polynomials in this paper we drop most of the times the term 'trigonometric'. Whenever we refer to usual algebraic polynomials we state it explicitly.

The ultimate goal of this paper is to design a verifier which performs a test whether a given table of function values is generated with high probability by a multivariate polynomial. More precisely, the following is our main result.

Theorem 1 (Main Theorem). *Let $d \in \mathbb{N}$, $h = 10^{15}$, $k \geq \frac{3}{2}(2h + 1)$ and let F be a finite field with $q := |F|$ being a prime number larger than $10^4(2hkd + 1)^3$. We fix these values for the entire paper. There exists a probabilistic verification algorithm in the BSS-model of computation over the reals with the following properties:*

1. *The verifier gets as input a function value table of a multivariate function $f : F^k \to \mathbb{R}$ and a proof string π consisting of at most $|F|^{2k}$ segments (blocks). Each segment consists of $2hkd + k + 1$ real components.*

The verifier first uniformly generates $O(k \cdot \log q)$ random bits; next, it uses the random bits to determine a point $x \in F^k$ together with one segment in the proof string it wants to read. Finally, using the values of $f(x)$ and those of the chosen segment it performs a test (to be described below). According to the outcome of the test the verifier either accepts or rejects the input.

The running time of the verifier is polynomially bounded in the quantity $k \cdot \log q$, i.e., polylogarithmic in the input size $O(k \cdot q^{2k})$.

2. *For every function value table representing a trigonometric max-degree d polynomial there exists a proof string such that the verifier accepts with probability 1.*

3. *For any $0 < \epsilon < 10^{-19}$ and for every function value table whose distance to a closest max-degree $2hkd$ polynomial is at least 2ϵ the probability that the verifier rejects is at least ϵ, no matter what proof string is given. Here, for two functions $f, g : F^k \mapsto \mathbb{R}$ their distance is defined as $\mathrm{dist}(f,g) := \frac{1}{|F^k|} \cdot |\{x \in F^k | f(x) \neq g(x)\}|$.*

The first and the second property in the theorem will follow directly from the description of the test. Proving the last property - as usual with such statements - is the main task. Repeating the verifier's computation constantly many times decreases the error probability below any given fixed positive constant.

Performing (one round of) the test the verifier reads $2hkd+k+1$ real numbers. Thus, it can only test for a local property of low degree polynomials $f : F^k \mapsto \mathbb{R}$. A major amount of work will be to figure out what this local property should look like. The starting idea is common for low degree tests, namely to consider univariate restrictions along certain lines of F^k. The segments of a proof string mentioned above precisely present the coefficients of such a univariate restriction. An advantage using a finite field as domain is that such lines only contain $|F|$ many points. So we do not have to deal with the problem of splitting the domains into a large test domain and a small safe domain as it is, for example, the case with the real linearity test from [12]. On the other hand, it will turn out not to be a good idea to allow any arbitrary line in F^k for such a test as it is done in the classical approach [2]. A fair amount of work will be necessary to figure out a suitable subset $H \subset F^k$ of lines for which the test shall be performed.

2.1 Appropriate Domain of Test Directions; Structure of a Verification Proof

As mentioned above, the verifier expects each segment of the proof to specify a univariate polynomial of *appropriate* degree on a line. Since univariate restrictions of trigonometric polynomials along a line behave a bit differently than of algebraic polynomials some care is necessary. Let $f : F^k \mapsto \mathbb{R}$ be a (trigonometric) max-degree d polynomial and let $\ell := \{x+tv | t \in F\}, x, v \in F^k$ be a line. For determining an upper bound of the degree of the univariate restriction of f on ℓ it turns out to be helpful to define a kind of absolute value for the elements of F. The definition is inspired by the fact that if we later on restrict a trigonometric

polynomial to lines with small components in absolute value the resulting uni-
variate polynomials have a relatively small degree. For $t \in F = \{0, \ldots, |F| - 1\}$
put
$$|t| = \begin{cases} t & \text{if } t < |F|/2 \\ |F| - t & \text{if } t > |F|/2 \end{cases}.$$

If a univariate polynomial $p : F \to \mathbb{R}$ has degree d, then for $a, b \in F$ the
polynomial $t \mapsto p(a + bt)$ has degree at most $d \cdot |b|$ and thus $t \mapsto f(x + tv)$ has
degree at most $d \cdot \sum_{i=1}^{k} |v_i|$, where $v = (v_1, \ldots, v_k)$. This is an easy consequence
of Definition 1.

For the test performed by the verifier we want to specify a suitable set $H \subset F^k$
of lines along which univariate restrictions are considered. Suitable here refers
to the maximal degree such a restriction could have given a max-degree d multivariate polynomial. This maximal degree in a certain sense should be small. The
constant parameter h in Theorem 1 determines what we mean by small. Though
$h = 10^{15}$ of course is a large constant the decisive point is its independence of
d, k, q.

Definition 2. *Let F be a finite field, $k \in \mathbb{N}$ and h be as in Theorem 1. The set
H is defined to be any subset of $F^k \setminus \{0\}$ satisfying the following two conditions:*

i) *For every $0 \neq v := (v_1, \ldots, v_k) \in H$ we have $|v| := \max\{|v_1|, \ldots, |v_k|\} \leq h$
and*

ii) *if for a fixed $v \subset F^k$ several points in the set $\{tv | t \in F\}$ satisfy condition i)
only one of them is included in H.*

Condition i) requires the direction of lines that we consider to have small components, whereas condition ii) just guarantees that each line (as point set) is
included at most once. We abstain from specifying which v we include in such a
case and just fix H as one such set.

If a k-variate polynomial of max-degree d is restricted to a line $\{x + tv | t \in F\}$
whose direction v belongs to H, then the resulting univariate polynomial has
degree at most hkd. Note that for the values chosen hkd is much smaller than
$\frac{|F|}{2}$. In later arguments the cardinality of H will be important, so let us say
something about it already now. For h sufficiently smaller than $|F|$ there are
$(2h + 1)^k - 1$ elements $v \in F^k - \{0\}^k$ such that $|v| \leq h$. For every such v it is
$|-v| \leq h$, therefore $\frac{1}{2}((2h + 1)^k - 1)$ is an upper bound for $|H|$. It can be shown
that for increasing k the fraction $\frac{|H|}{\frac{1}{2}((2h+1)^k - 1)}$ approaches 1.

Given a table of function values for an $f : F^k \mapsto \mathbb{R}$ the verifier now expects
the following information from a potential proof that f is a trigonometric polynomial of max-degree d. For every line ℓ in F^k which has a direction $v \in H$
the proof should provide a segment of real numbers which represent a univariate
polynomial as follows. The segment consists of a point $x \in F^k$ on the line as well
as reals $a_0, \ldots, a_{hkd}, b_1, \ldots, b_{hkd}$. The verifier will interpret this information as
the univariate polynomial with coefficients a_i, b_j that ideally, i.e., for a trigonometric polynomial, represents f's restriction to ℓ parametrized as $t \mapsto f(x + tv)$.
Obviously, there are several different parametrizations depending on the point x.

Any such point $x \in F^k$ and direction $v \in H$ define a line $\ell_{v,x} = \{x + tv : t \in F\}$. Actually, for all $y \in \ell_{v,x}$, $\ell_{v,y} = \ell_{v,x}$. Thus, there are $|F|$ possible parametrizations of a same line. We only code one of them, that is we arbitrarily fix an origin. To be able to discuss about points on these lines, without referring to the parametrization that is used, we define for all $y \in \ell_{v,x}$ the value $\tau_y \in F$ such that $x + \tau_y v = y$, that is the abscissa of y on $\ell_{v,x}$. In this way, we can completely forget the origins of the lines we consider.

The total length of a proof as described above is easily calculated. The point $x \in F^k$ requires k reals, thus each segment in the proof needs $2hkd + k + 1$ reals. Each of the $|H|$ directions contributes $|F|^{k-1}$ different (parallel) lines[1], so the total number of segments is $|H| \cdot |F|^{k-1}$ and the length of the proof is upper bounded by $(2hkd + k + 1) \cdot \frac{1}{2}((2h+1)^k - 1) \cdot |F|^{k-1}$. For our choices of parameters $k, d, |F|$ this is smaller than $|F|^{2k}$.

We are now prepared to describe the test in detail:

Low degree test: Let F and H be as above.

Input: A function $f : F^k \to \mathbb{R}$, given by a table of its values; a list of univariate trigonometric polynomials defined for each line in F^k with direction in H and specified by its coefficients as described above.

1. Pick uniformly at random a direction $v \in H$ and a random point $x \in F^k$;
2. compute deterministically the unique segment in the proof that specifies the univariate polynomial $p_{v,x}(t)$ which is defined on the line through x in direction v; [2]
3. if $f(x) = p_{v,x}(\tau_x)$ accept, otherwise reject.

Since F is discrete the objects picked in Step 1 can be addressed using $O(k \cdot \log|F|)$ random bits. Note that this first step is the same as saying we pick a random direction $v \in H$ together with a random line among those having direction v. There are $|F|$ many points on each such line, i.e., $|F|$ choices for x result in the same line.

3   Proof of Main Theorem

The proof of Theorem 1 relies on two major theorems together with several propositions and lemmas. Let us outline the main ideas behind those theorems. The first one is Theorem 2. It states that the rejection probability of the low-degree test is about proportional to the distance δ of the given function f to a closest max-degree $2hkd$ polynomial. However, this will hold only in case that this distance δ is not too big. For too large distances the lower bound which the

---

[1] Here we need that no element in H is a multiple of another one and that each line contains $|F|$ points.

[2] In order to evaluate such a polynomial the verifier could use $\cos\frac{2\pi}{q}$ and $\sin\frac{2\pi}{q}$ as machine constants. For varying input sizes we address this point again in the final section.

theorem gives for the rejection probability gets very small (and even can become negative). So intuitively the theorem states that if the rejection probability is small, then f is either very close or very far away from a max-degree $2hkd$ polynomial.

Theorem 2. *Let δ denote the distance of a function $f : F^k \to \mathbb{R}$ to a closest max-degree $2hkd$ polynomial. Then for any proof the probability that the low-degree test rejects is at least*

$$\left(\frac{2h}{2h+1}(1-\delta) - \frac{4h^2 k^2 d}{|F|-1} \right) \delta.$$

Thus we have to deal with the case that though the rejection probability might be small given the above lower bound f is far away from such a polynomial. Theorem 3 basically shows that this case cannot occur using the following idea: if for a function f and a proof string π the probability of rejection is small, then f and π can be changed in a number of small steps such that these changes do not increase the rejection probability too much and in the end a max-degree $2hkd$ polynomial f_s is obtained. Since by Theorem 2 such a transformation process would not be possible if f would be far away from any max-degree $2hkd$ polynomial (the process would have to cross functions for which the test rejects with higher probability), it follows that a reasonably small rejection probability only occurs for functions f that were already close to a max-degree $2hkd$ polynomial.

Theorem 3. *Let $0 < \epsilon \le 10^{-19}$ and let a function f_0 together with a proof π_0 be given. If the low-degree test rejects with probability at most ϵ, then there is a sequence $(f_0, \pi_0), (f_1, \pi_1), \ldots, (f_s, \pi_s)$ such that*

1. *for every $i \le s$ the probability that the test rejects input (f_i, π_i) is at most 2ϵ,*
2. *for every $i < s$ the functions f_i and f_{i+1} differ in at most $|F|$ arguments and*
3. *the function f_s is a max-degree $2hkd$ polynomial.*

Note that it is only the existence of such a sequence that we need for the proof of Theorem 1, it does not describe anything that the test does. So it does not have to be (efficiently) computable.

Assuming validity of both theorems the Main Theorem can be proven as follows:

Proof. (of Main Theorem 1) Statements 1. and 2. of the theorem being obvious from the foregoing explanations let us assume we are given a function value table for $f : F^k \mapsto \mathbb{R}$ together with a verification proof π such that the low-degree test rejects with a probability at most $\epsilon \le 10^{-19}$. We will show that the distance δ from f to the set of max-degree $2hkd$ polynomials is at most 2ϵ. In order to avoid tedious calculations the arguments are given on a slightly more abstract level based on continuity, however it is no problem to fill in the precise values for δ in all cases.

Since h is large and $|F| \gg 4h^2k^2d$ Theorem 2 reads $(c_1(1-\delta)-c_2) \cdot \delta \le \epsilon$, where constant c_1 is close to 1 and c_2 is close to 0. This implies that either δ is at most slightly larger than ϵ or δ is close to 1. We want to exclude the second case. Assume for a contradiction that δ is close to 1. By Theorem 3 there exists a function f' and a verification proof π' such that the probability that the test rejects the pair (f', π') is at most 2ϵ and the distance δ' from f' to the closest max-degree $2hkd$ polynomial is close to $\frac{1}{2}$. This is true because each new element (f_i, π_i) in the sequence reduces the number of errors by at most $|F|$. Now choose the number of reducing steps such that the number of errors is reduced from $\delta|F|^k$ to approximately $\frac{1}{2}|F|^k$. This must happen because finally a polynomial is obtained. Again using Theorem 2 it follows that the test must reject (f', π') with a probability which is at least about $\frac{1}{4}$. This contradicts the fact that (f', π') is rejected with probability at most 2ϵ and we have reached the desired contradiction. □

3.1 Proof of Theorem 2

We shall now turn to the proofs of the two major theorems. Especially that for Theorem 3 needs considerable additional technical efforts. As said in the introduction due to the page limit some of those additional results will only be stated in the present section. Lacking proofs are postponed to a full version.

Let us first prove Theorem 2. Suppose we are given a function $f : F^k \to \mathbb{R}$ with (exact) distance $\delta > 0$ to a closest max-degree $2hkd$ polynomial $\tilde{f}$. Let π denote an arbitrary verification proof specifying univariate polynomials $p_{v,x}$ as described above. Define the set $U \subset F^k$ as those points where f and $\tilde{f}$ disagree. Thus $|U| = \delta|F|^k$. The idea behind our analysis is to guarantee the existence of relatively (with respect to δ) many pairs $(x, y) \in U \times \bar{U}$ that are located on a line with direction $v \in H$. More precisely, we shall consider the set C of triples (x, v, y) with $x \in U, v \in H$, and $y \in \bar{U} = F^k \setminus U$ on the line $\ell_{v,x}$. The goal is to show that C contains many triples (x, v, y) for which $p_{v,x}(\tau_x) \ne f(x)$ or $p_{v,x}(\tau_y) \ne f(y)$. [3]

Due to $x \in U$ and $y \in \bar{U}$ for any $(x, v, y) \in C$ one of the following two alternatives holds:

1. The polynomial $p_{v,x}$ is different from $\tilde{f}$ on $\ell_{v,x}$ but agrees with $\tilde{f}$ in y.
2. The polynomial $p_{v,x}$ disagrees with f in x or in y.

Claim: Alternative 1. is satisfied by at most $\delta|F|^k \cdot |H| \cdot 4h^2k^2d$ triples from C.

Proof of Claim: Since $\tilde{f}$ is a max-degree $2hkd$ polynomial, the restriction of $\tilde{f}$ to any line yields a univariate degree $2h^2k^2d$ polynomial. By usual interpolation for trigonometric polynomials different univariate polynomials of degree at most $2h^2k^2d$ agree in at most $4h^2k^2d$ points. Applying this argument to each of the

---

[3] To avoid misunderstandings we point out once more that here we mean the value of the univariate polynomial $\tau \mapsto p_{v,x}(\tau)$ in the respective arguments τ_x and τ_y that on the corresponding line through x and y in direction v result in points x and y, respectively.

$|H|$ lines and taking into account that there are $|U| = \delta|F|^k$ choices for x there can be at most

$$\delta|F|^k \cdot |H| \cdot 4h^2k^2d$$

triples in C satisfying the first alternative.

Next we aim for a lower bound on the total number of triples in C. This easily implies a lower bound on the number of triples satisfying the second alternative; from that a lower bound for the probability that the test rejects can be obtained.

A lower bound on $|C|$ is given in the following proposition. It basically says that the set H of test directions is sufficiently large. Its proof requires considerable technical efforts relying on the theory of expander graphs, so we just give the statement here.

Proposition 1. *Let $U \subseteq F^k$ as above (or any other arbitrary set with cardinality $\delta|F|^k$) with $0 \le \delta \le 1$. Then there are at least*

$$\frac{2h}{2h+1}\delta(1-\delta)|H|(|F|-1)|F|^k$$

pairs $(x,y) \in U \times (F^k - U)$ such that the line through x and y has direction in H.

This proposition together with the above claim implies that the number of triples in C satisfying the second alternative is at least

$$\frac{2h}{2h+1}\delta(1-\delta)|H|(|F|-1)|F|^k - \delta|F|^k \cdot |H| \cdot 4h^2k^2d.$$

In order to finish the proof of Theorem 2 an alternative view of the low-degree test helps. The test can as well be seen as first choosing randomly two points x, y such that they determine a direction $v \in H$. Since there are $|H|$ directions and $|F|$ points on each line there are $|H|(|F|-1)|F|^k$ such triples (x,v,y) in total. Then, with probability $\frac{1}{2}$ the test decided whether to check if $p_{v,x}(\tau_x) = f(x)$ or if $p_{v,x}(\tau_y) = f(y)$. Since triples in C that satisfy alternative 2 result in an error for the low-degree test if the appropriate point for evaluation is chosen, its probability for rejection is at least

$$\left(\frac{2h}{2h+1}(1-\delta) - \frac{4h^2k^2d}{|F|-1}\right)\delta.$$

Half of this value is contributed by triples $(x,v,y) \in C$ for which $p_{v,x}(\tau_x) \ne f(x)$ or $p_{v,x}(\tau_y) \ne f(y)$. The other half arises from triples (y,v,x) for which $(x,v,y) \in C$ and $p_{v,x}(\tau_x) \ne f(x)$ or $p_{v,x}(\tau_y) \ne f(y)$. □

3.2   Proof of Theorem 3

In order to prove Theorem 3 reducing the inconsistency of a proof string plays an important role. This inconsistency is defined as follows.

Definition 3. *The* inconsistency *of a proof string π is the fraction of triples $(x, v, v') \in F^k \times H \times H$ for which $p_{v,x}(\tau_x) \neq p_{v',x}(\tau_x)$.*

The attentive reader might wonder that in the definition *ordered* triples are counted and that $v = v'$ is not excluded, so the inconsistency can never equal 1. These are only technical issues to make some calculations below easier.

In proving the existence of a sequence $\{(f_i, \pi_i)\}_i$ as in the theorem's statement the main idea is to find for a current pair (f, π) a segment in π that can be changed in a way which decreases the inconsistency. After that, also f is changed to a function f' which fits the best to the new proof string π'. The best fit is defined using majority decisions.

Definition 4. *Let π be a proof string for verifying a low-degree polynomial as explained above. As usual, denote by $\{p_{v,x}|v \in H, x \in F^k\}$ the univariate restrictions given with π. A majority function for π is a function $f : F^k \to \mathbb{R}$ such that for all $x \in F^k$ the value $f(x)$ maximizes $|\{v \in H|p_{v,x}(\tau_x) = f(x)\}|$.*

Ties can be broken arbitrarily in the definition of a majority function, so it does not necessarily have to be unique.

When proving Theorem 3 the sequence $(f_0, \pi_0), \ldots, (f_s, \pi_s)$ is defined by first changing π_i to π_{i+1} on a single proof segment and then taking f_{i+1} as a majority function for π_{i+1}. During this process there should occur no pair (f_i, π_i) for which the rejection probability is more than twice as large as the rejection probability of the initial (f_0, π_0). The following easy lemma will be useful in this context. It relates the rejection probability of the low-degree test for a pair (f, π) with inconsistency of π.

Lemma 1. *If f is a majority function for π, then the inconsistency of π is at least as large as the rejection probability of the low-degree test for (f, π) and at most twice as large.*

Proof. For a fixed $x \in F^k$ let m_x be the the number of $v \in H$ such that the value $p_{v,x}(\tau_x)$ specified by π satisfies $p_{v,x}(\tau_x) = f(x)$. The rejection probability is $1 - \frac{\sum_x m_x}{|F^k| \cdot |H|}$. Now for any $v \in H$ there cannot be more than m_x many $v' \in H$ such that (x, v, v') satisfies $p_{v,x}(\tau_x) = p_{v',x}(\tau_x)$. Applying this to the $|H|$ many directions v there are at most $|H| \cdot \sum_x m_x$ triples that do not contribute to π's inconsistency. Rearranging shows that the inconsistency is at least as large as the rejection probability.

Vice versa, for fixed x there must be an $H' \subset H$ with $|H'| = m_x$ such that for all $v, v' \in H'$ the m_x^2 many equations $p_{v,x}(\tau_x) = p_{v',x}(\tau_x)$ are satisfied. Thus the inconsistency is upper bounded by $1 - \frac{\sum_x m_x^2}{|F^k| \cdot |H|^2}$. The latter is easily shown to be at most twice the rejection probability, i.e., $2 \cdot \left(1 - \frac{\sum_x m_x}{|F^k| \cdot |H|}\right)$. □

Without loss of generality we may assume that for the inputted pair (f_0, π_0) function f_0 is already a majority function for π_0. Else we could just define the first pair in the sequence of (f_i, π_i) by changing stepwise one value of f_0 while leaving proof string π_0 unchanged until we have reached a majority function of π_0. Clearly, during this process the rejection probability will not increase.

Now in each further step the inconsistency of a current proof string π_i is strictly reduced for π_{i+1}. This is achieved by changing only one of π_i's segments. Furthermore, the new function f_{i+1} is obtained from f_i in such a way that it becomes a majority function for π_{i+1}. Since inconsistency is reduced step by step Lemma 1 implies that for every $i \leq s$ the rejection probability of (f_i, π_i) can be at most twice as large as the rejection probability of (f_0, π_0). Of course, we have to guarantee that the way the f_i's are changed finally turn them into a trigonometric polynomial.

The following proposition is a key stone to make the above idea precise.

Proposition 2. *Let π be a proof string obeying the general structure required for the low-degree test and having inconsistency at most 2ϵ, where $\epsilon \leq 10^{-19}$. Let f be a majority function of π which is not already a max-degree $2hkd$ polynomial.*

Then there exists a proof string π' such that π' differs from π in only one segment and its inconsistency is strictly less than that of π.

The proof needs several additional technical results. Let us first collect them and then prove the proposition. The following definition specifies certain point sets important in the further analysis.

Definition 5. *Let a pair (f, π) as above be given. Let $\alpha := 10^{-2}$ for the rest of the paper.*

a) *Define $S \subseteq F^k$ to consist of those points x for which the fraction of directions $v \in H$ satisfying $p_{v,x}(\tau_x) = f(x)$ is less than $1 - \alpha$.*

b) *For $v \in H$ define $\widehat{S}(v) \subseteq F^k$ as $\widehat{S}(v) := \{x \in F^k | x \notin S \text{ and } p_{v,x}(\tau_x) \neq f(x)\}$.*

The set S contains those points for which there are relatively many, namely at least $\alpha|H|$, inconsistencies between different line polynomials through x and the value $f(x)$. The set $\widehat{S}(v)$ on the other hand contains the points for which most of the line polynomials agree with f on x, but the particular $p_{v,x}$ does not. As consequence, the latter disagrees with most of the others with respect to point x.

The main purpose using the following proposition is to pick out a line along which the given proof can be changed in such a way that its inconsistency reduces. For obtaining this line ℓ_{v^*,x^*} the objects x^* and v^* are determined by the following crucial proposition. Due to its length and technical feature the proof has to be postponed to the full version. We also note that at this point the significance of Proposition 3 may be hard to see. Its meaning will become clear in the proof of Proposition 2.

Proposition 3. *Let π be a proof string as in Proposition 2. There exist $x^* \in F^k$, $v^* \in H$ and a set $H' \subseteq H$ such that*

1. $x^* \in \widehat{S}(v^*)$;
2. at most $\frac{1}{40}\alpha \cdot |F|$ points on ℓ_{v^*,x^*} belong to S;
3. $|H'| \geq (1 - 4\alpha)|H|$ and
4. for all $v \in H'$
    i) the fraction of pairs $(t,s) \in F^2$ for which $p_{v^*,x^*+sv}(\tau_{x^*}+tv^*+sv) \neq$
       $p_{v,x^*+tv^*}(\tau_{x^*}+tv^*+sv)$ is at most $\frac{1}{4}$ and
    ii) the fraction of $s \in F$ for which $p_{v^*,x^*+sv}(\tau_{x^*}+sv) \neq p_{v,x^*}(\tau_{x^*}+sv)$ is at
        most $\frac{1}{2}$.

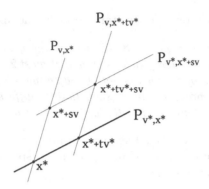

Fig. 1. The figure shows those values that are compared in the fourth item of Proposition 3

The second technical result that we need is a direct adaption of a similar lemma by Arora and Safra [3] to trigonometric polynomials. It says that if the entries of an $|F| \times |F|$ matrix both row-wise and column-wise arise to a large extent from univariate polynomials, then the majority of values of the entire matrix arise from a bivariate polynomial.

Lemma 2. *(see [2], adapted for trigonometric polynomials) Let* $\tilde{d} \in \mathbb{N}, |F| \geq$
$10^4(2\tilde{d}+1)^3$. *Suppose there are two sets of univariate trigonometric degree* $\tilde{d}$
polynomials $\{r_s\}_{s \in F}$ *and* $\{c_t\}_{t \in F}$ *such that the fraction of pairs* $(s,t) \in F^2$ *for which there is a disagreement, i.e.,* $r_s(t) \neq c_t(s)$, *is at most* $\frac{1}{4}$. *Then there exists a bivariate trigonometric max degree* $\tilde{d}$ *polynomial* $Q(s,t)$ *such that for at least a* $\frac{2}{3}$*-fraction of rows* s *it holds that* $r_s(t) \equiv Q(s,t)$; *similarly for at least a* $\frac{2}{3}$*-fraction of columns* t *it holds that* $c_t(s) \equiv Q(s,t)$.

Having all technical auxiliary material at hand we can now prove Proposition 2 and Theorem 3.

Proof. (of Proposition 2) Let x^*, v^* and H' be fixed according to Proposition 3. The segment we are going to change in π is the segment claiming a univariate polynomial on line ℓ_{v^*,x^*}. We need to show that this can be done in a way that decreases inconsistency.

Consider a plane given by two directions v^*, v' in H; note that the definition of H implies that two different points are linearly independent and thus constitute a plane. We want to apply Lemma 2 to every such (v^*, v')- plane through x^*, where $v' \in H'$.

For such a v' define $r_s^{(v')}(t) := p_{v^*, x^*+sv'}(\tau_{x^*}+tv^*+sv')$ and $c_t^{(v')}(s) := p_{v', x^*+tv^*}(\tau_{x^*}+tv^*+sv')$. Proposition 3, item 4,i) implies that the assumptions of Lemma 2 are satisfied for each v'. Let $Q^{(v')}$ denote the corresponding bivariate polynomial of max-degree $\tilde{d} := hkd$.

Note that every (v^*, v')-plane through x^* contains the line ℓ_{v^*, x^*} and that for every $v' \in H'$ it holds $r_0^{(v')}(t) = p_{v^*, x^*}(\tau_{x^*}+tv^*)$ independently of v'. Thus we abbreviate $r_0 := r_0^{(v')}$. The idea now is to show that there exists a degree hkd polynomial $R : F \to \mathbb{R}$ such that R is different from r_0 and for most $v' \in H'$ the function $t \mapsto c_t^{(v')}(0)$ is close to R. From the precise version of this statement it will then follow that changing in π the segment containing r_0 to R will decrease the inconsistency.

For $v' \in H'$ let $R^{(v')}(t) := Q^{(v')}(0, t)$. We want to show that for many $v', v'' \in H', R^{(v')} \equiv R^{(v'')}$. The majority of these polynomials then defines R.

Claim 1: Let $v', v'' \in H'$. If $R^{(v')} \not\equiv R^{(v'')}$, then the distance between $t \mapsto c_t^{(v')}(0)$ and $t \mapsto c_t^{(v'')}(0)$ is at least

$$\frac{1}{3} - \frac{2hkd}{|F|}.$$

Proof of Claim 1: By Lemma 2 $R^{(v')}$ is the unique polynomial to which the function $t \mapsto c_t^{(v')}(0)$ is close with a distance of at most $\frac{1}{3}$. If $R^{(v')} \not\equiv R^{(v'')}$, then as polynomials of degree hkd they differ in at least $|F| - 2hkd$ points, thus $t \mapsto c_t^{(v')}(0)$ and $t \mapsto c_t^{(v'')}(0)$ have at least the claimed distance.

Next consider the number of inconsistencies on ℓ_{v^*, x^*}, i.e., the number of triples $(y, v, w) \in \ell_{v^*, x^*} \times H^2$ for which $p_{v,y}(\tau_y) \neq p_{w,y}(\tau_y)$. Proposition 3 intuitively implies that the number of inconsistencies cannot be too large. On the other hand, Claim 1 above implies that any two v', v'' for which $R^{(v')} \not\equiv R^{(v'')}$ will lead to many inconsistencies on ℓ_{v^*, x^*}. Hence, for most $v', v'' \in H'$ it will be the case that $R^{(v')} \equiv R^{(v'')}$. More precisely:

Claim 2: The number of pairs $(v', v'') \in (H')^2$ for which $R^{(v')} \equiv R^{(v'')}$ is at least

$$\left((1 - 4\alpha)^2 - \frac{2\alpha}{\frac{1}{3} - \frac{1}{40}\alpha - \frac{2hkd}{|F|}} \right) \cdot |H|^2. \tag{1}$$

Proof of Claim 2: The lower bound is obtained by comparing the number of inconsistencies caused by triples $(y, v, w) \in \ell_{v^*, x^*} \times H^2$ on the one hand side and those caused by triples $(y, v', v'') \in \ell_{v^*, x^*} \times (H')^2$ where $R^{(v')} \not\equiv R^{(v'')}$ on the other. We restrict y to belong to $\ell_{v^*, x^*} \cap F^k \setminus S$ (recall Definition 5) and give an upper bound on the first quantity and a lower bound on the second that allows to conclude the claim.

For any $y \notin S$ there are at least $(1 - \alpha)|H|$ directions $w \in H$ such that the values $p_{w,y}(\tau_y)$ coincide with $f(y)$ and thus with each other; so for such a fixed y at least $(1 - \alpha)^2|H|^2$ triples will not result in an inconsistency. Vice versa, at most $(1 - (1 - \alpha)^2)|H|^2 \leq 2\alpha|H|^2$ inconsistencies can occur. Since there are at most $|F|$ choices for $y \in \ell_{v^*,x^*}$ we have the following upper bound:

$$|\{(y, v, w) \in \ell_{v^*,x^*} \times H^2 \,|\, y \notin S \text{ and } p_{v,y}(\tau_y) \neq p_{w,y}(\tau_y)\}| \leq 2\alpha|H|^2|F|. \quad (2)$$

Next consider inconsistencies (y, v', v'') caused by $(v', v'') \in H'$ such that $R^{(v')} \not\equiv R^{(v'')}$. According to Claim 1 each such pair (v', v'') implies the existence of at least $\frac{1}{3}|F| - 2hkd$ points $y \in \ell_{v^*,x^*}$ such that (y, v', v'') is an inconsistency for π. Requiring in addition $y \notin S$ according to Proposition 3 will still give at least $\frac{1}{3}|F| - 2hkd - \frac{1}{40}\alpha|F|$ many such y, i.e., for each $(v', v'') \in H', R^{(v')} \not\equiv R^{(v'')}$ it holds

$$|\{y \in \ell_{v',v''} \,|\, y \notin S \text{ and } p_{v',y}(\tau_y) \neq p_{v'',y}(\tau_y)\}| \geq \frac{1}{3}|F| - 2hkd - \frac{1}{40}\alpha|F|. \quad (3)$$

Combining (2) and (3) it follows that the number of pairs $(v', v'') \in (H')^2$ for which $R^{(v')} \not\equiv R^{(v'')}$ is upper bounded by

$$\frac{2\alpha|H|^2|F|}{\frac{1}{3}|F| - 2hkd - \frac{1}{40}\alpha|F|} = \frac{2\alpha}{\frac{1}{3} - \frac{1}{40}\alpha - \frac{2hkd}{|F|}} \cdot |H|^2.$$

Since $|H'| \geq (1 - 4\alpha)|H|$ this in turn means that the number of pairs $(v', v'') \in (H')^2$ for which $R^{(v')} \equiv R^{(v'')}$ must be at least

$$\left((1 - 4\alpha)^2 - \frac{2\alpha}{\frac{1}{3} - \frac{1}{40}\alpha - \frac{2hkd}{|F|}} \right) \cdot |H|^2. \quad (4)$$

This yields Claim 2.

Next, define the univariate polynomial R as the majority polynomial among all $R^{(v')}$, i.e., the polynomial which maximizes $|\{v' \in H' | R^{(v')} \equiv R\}|$.

Claim 3: The number of choices $v' \in H'$ such that $R \equiv R^{(v')}$ is at least $\beta|H|$, where

$$\beta \geq (1 - 4\alpha)^2 - \frac{2\alpha}{\frac{1}{3} - \frac{1}{40}\alpha - \frac{2hkd}{|F|}} > 0.84.$$

Proof of Claim 3: Let β be the fraction in H (not in H'!) of directions v' which belong to H' and satisfy $R^{(v')} \equiv R$, i.e., $\beta|H| = |\{v' \in H' | R^{(v')} \equiv R\}|$. Clearly, for each $v'' \in H$ there can be at most $\beta|H|$ directions $v' \in H'$ for which $R^{(v'')} \equiv R^{(v')}$. Hence, by Claim 2 it is

$$|H'| \cdot \beta \cdot |H| \geq \left((1 - 4\alpha)^2 - \frac{2\alpha}{\frac{1}{3} - \frac{1}{40}\alpha - \frac{2hkd}{|F|}} \right) \cdot |H|^2$$

and thus, using $|H'| \leq |H|$, we obtain

$$\beta \geq (1 - 4\alpha)^2 - \frac{2\alpha}{\frac{1}{3} - \frac{1}{40}\alpha - \frac{2hkd}{|F|}}. \quad (5)$$

From $\alpha := 10^{-2}$ in Definition 5 and our assumption that $|F| \geq 10^4(2hkd + 1)^3$ it follows that $\beta > 0.84$.

Claim 4: The majority polynomial R and r_0 are different: $R \not\equiv r_0$.

Proof of Claim 4: Recall that by definition $r_0(t)$ equals $r_0^{(v')}(t)$ for each $v' \in H'$ and is the polynomial which is claimed by π on ℓ_{v^*,x^*}. Similarly, for the majority of $v' \in H'$ polynomial $R(t)$ equals $R^{(v')}(t)$. We prove Claim 4 by showing that the particular value $R(0)$ is attained for more choices of $v' \in H'$ than $r_0(0)$.

First note that item 4,ii) of Proposition 3 for all $v' \in H'$ implies $c_0^{(v')}(s) := p_{v^*,x^*+sv'}(\tau_{x^*+sv'}) = p_{v',x^*}(\tau_{x^*+sv'})$ for at least $\frac{1}{2}|F|$ values of s. Next, Lemma 2 implies for each $v' \in H'$ that for at least $\frac{2}{3}|F|$ values of s it holds $r_s^{(v')}(t) = Q^{(v')}(s,t)$ as polynomials in t. For those s it follows in particular that $Q^{(v')}(s,0) = r_s^{(v')}(0) = p_{v^*,x^*+sv'}(\tau_{x^*+sv'})$.

Combining the two equations results for each v' in at least $(\frac{2}{3} - \frac{1}{2})|F|$ many values for s for which $c_0^{(v')}(s) = Q^{(v')}(s,0)$. Now since both functions are univariate polynomials of degree at most khd they are equal as long as $|F|$ is large enough.

Next, it follows that $p_{v',x^*}(\tau_{x^*}) = c_0^{(v')}(0) = Q^{(v')}(0,0) = R^{(v')}(0)$; the latter by definition of $R^{(v')}$ and $p_{v',x^*}(\tau_{x^*})$ equals the value $R(0)$ for at least $\beta|H|$ choices of $v' \in H'$.

On the other hand it is $x^* \in \widehat{S}(v^*)$, thus for at most $\alpha|H|$ many $w \in H$ the value $r_0(0) = p_{v^*,x^*}(\tau_{x^*})$ conincides with $p_{w,x}(\tau_{x^*})$. But $\beta > \alpha$, therefore the claim $R \not\equiv r_0$ follows.

What remains to be done is to show that using R instead of r_0 in the corresponding segment of π strictly reduces its inconsistency.

Claim 5: The number of pairs (y,w) with $y \in \ell_{v^*,x^*}$ and $w \in H$ for which $p_{y,w}$ agrees with R on y is larger than the number of such pairs for which $p_{y,w}$ agrees with r_0 on y.

Proof of Claim 5: Since $R \neq r_0$ they agree in at most $2hkd$ points on ℓ_{v^*,x^*}. By the inclusion-exclusion principle it thus suffices to show that among the $|F||H|$ triples of form $(y,v^*,w), y \in \ell_{v^*,x^*}, w \in H$ (note that v^* is fixed) there are more than $(\frac{1}{2}|F| + 2hkd)|H|$ many for which $p_{w,y}$ agrees with R on x^*. By Lemma 2 and Claim 3 there exist $\beta|H|$ directions $v' \in H'$ for which the distance from $t \mapsto c_t^{(v')}(0)$ to R is at most $\frac{1}{3}$. It follows that $|\{(y,w) \in \ell_{v^*,x^*} \times H | p_{w,y}$ agrees with R on $y\}| \geq \beta|H| \cdot \frac{2}{3}|F|$.

Plugging in the bounds for β and $|F|$ gives $\beta|H| \cdot \frac{2}{3}|F| > (\frac{1}{2}|F| + 2hkd)|H|$. This finishes the proof of Claim 5 and thus also the one of Proposition 2.   □

Proof. (of Theorem 3) We have shown that given a verification proof π and a function f which is a majority function of π and not a max-degree $2hkd$ polynomial we can construct a verification proof π' with a majority function f' such that the following holds.

- The univariate polynomials that π and π' claim differ on one line (i.e. π and π' differ in one segment) and f and f' disagree in at most $|F|$ places.
- The inconsistency of π' is strictly less than the inconsistency of π.

If we apply this construction iteratively it must come to an end after finitely many steps because the inconsistency cannot be reduced an unbounded number of times. Hence, at some point we must obtain a function f_s which is a max-degree $2hkd$ polynomial. Lemma 1 implies that for each (f_i, π_i) in the sequence the rejection probability is at most 2ϵ and this finishes the proof. $\square$

4   Conclusion and Future Problems

The main result of the present paper is the design and analysis of a verifier performing a local test to check whether a given function has small distance to the set a k-variate trigonometric polynomials of low degree d. The latter are defined over a finite field and are real-valued. The verifier accesses a proof that consists of two parts, one giving the function values of the potential polynomial, the other giving (the coefficient vectors of) certain univariate restrictions along a particular set of directions. The entire proof has size $O(kd(h|F|)^{2k})$. The main features of the verifier are: it generates $O(k \cdot \log |F|)$ random bits; it then probes a constant number of blocks in the proof each of algebraic size $O(hkd)$, where h is a constant. After a number of steps polynomial in $k \cdot \log |F|$ the verifier makes its decision whether to accept the given proof or not. Our result shows that by repeating the test a constant number of times with high probability the verifier rejects tables that are far away from trigonometric polynomials.

The result is interesting with respect to the following questions. In the framework of showing the real PCP theorem along [3,2] our low-degree test can be used as follows. Take an $NP_{\mathbb{R}}$-complete problem, for example, the Quadratic Polynomial System problem QPS. Its instances are finitely many real polynomials of degree 2 in n variables; the question is to decide whether these polynomials have a common real zero $x^* \in \mathbb{R}^n$. Using the low-degree approach one can code such a potential zero as k-variate trigonometric polynomial from $F^k \to \mathbb{R}$. This is possible choosing $k := \lceil \frac{\log n}{\log \log n} \rceil$, $d := O(\log n)$ and $|F| := O(\log^6 n)$. Then our result shows that there exists a verifier for checking whether a function value table is close to such a polynomial using $O(\log n)$ many random bits and reading a constant number of segments in the proof each of size $poly \log n$. [4] The important advantage in comparison to the low-degree test for algebraic polynomials from [13] is the fact that the verifier only reads a *constant* number of blocks. This is a crucial requirement in the proof of the classical PCP theorem by Arora et al. in order to apply the technique of verifier composition.

---

[4] Note a technical detail here: if our verifier is used as BSS algorithm for inputs of varying size, then for different cardinalities q of the finite field F it needs to work with different constants $\cos \frac{2\pi}{q}, \sin \frac{2\pi}{q}$. Given q one could add two real numbers to the verification proof which in the ideal case represent a complex primitive q-th root of unity. The verifier deterministically checks in polynomial time whether this is the case and then continues to use these constants for performing its test as described.

Here is a brief outline of how to finish a proof of the real PCP theorem along these lines. The next step is to use the low-degree test of the present paper to design a sum-check procedure establishing (once again) the characterization $NP_\mathbb{R} = PCP_\mathbb{R}(O(\log n), poly \log n)$. This in principle is done using the ideas from [13]. As in the original proof the verifier resulting from combining the low-degree test and the sum-checking procedure lacks the necessary segmentation properties for applying verifier composition; it reads too many blocks. To repair this a second test is developed which together with using the low-degree test allows to restructure sum-checking in an appropriate way so that a properly segmented version is obtained. Though this procedure is not as general as the original ongoing by Arora et al., which gives quite a general segmentation procedure, it turns out to be sufficient for our purposes. In a final step, composition of real verifiers has to be worked out in detail and applied to the long transparent verifier from [12,6] and the verifier obtained from the above considerations. Filling all details requires a significant amount of work and space. Therefore we postpone it to the full version as explained in the introduction.

Let us mention briefly another outcome once a proof of the PCP theorem as indicated above is at hand. In our main result there seemingly is a discrepancy with respect to the test accepting degree d polynomials on one side and rejecting functions far away from degree $2hkd$ polynomials on the other. This lacking sharpness is of no concern for our results but seems a bit unusual in comparison to similar results in testing. However, the full proof can be used to close this degree gap and make the result sharp from that point of view.

Another line of future research addresses the area of locally testable codes. The paper shows that trigonometric polynomials can be used as such codes over the real numbers. So far, not many results into this direction have been obtained in the DSS model. For example, what about tests for algebraic polynomials? Using our test for trigonometric polynomials we expect it is possible to design a test for algebraic polynomials which uses a logarithmic number of random bits and makes a constant number of queries.

Acknowledgement. Thanks go to the anonymous referees for very helpful remarks.

References

1. Arora, S., Barak, B.: Computational Complexity: A Modern Approach. Cambridge University Press (2009)
2. Arora, S., Lund, C., Motwani, R., Sudan, M., Szegedy, M.: Proof verification and hardness of approximation problems. Journal of the ACM 45(3), 501–555 (1998)
3. Arora, S., Safra, S.: Probabilistic checking proofs: A new characterization of NP. Journal of the ACM 45(1), 70–122 (1998)
4. Ausiello, G., Crescenzi, P., Gambosi, G., Kann, V., Marchetti-Spaccamela, A., Protasi, M.: Complexity and Approximation: Combinatorial Optimization Problems and Their Approximability Properties. Springer (1999)
5. Baartse, M., Meer, K.: The PCP theorem for NP over the reals. To appear in Foundations of Computational Mathematics. Springer

6. Baartse, M., Meer, K.: Topics in real and complex number complexity theory. In: Montana, J.L., Pardo, L.M. (eds.) Recent Advances in Real Complexity and Computation, Contemporary Mathematics, vol. 604, pp. 1–53. American Mathematical Society (2013)
7. Blum, L., Cucker, F., Shub, M., Smale, S.: Complexity and Real Computation. Springer (1998)
8. Blum, L., Shub, M., Smale, S.: On a theory of computation and complexity over the real numbers: NP-completeness, recursive functions and universal machines. Bull. Amer. Math. Soc. 21, 1–46 (1989)
9. Bürgisser, P., Clausen, M., Shokrollahi, M.A.: Algebraic Complexity Theory. Springer (1997)
10. Dinur, I.: The PCP theorem by gap amplification. Journal of the ACM 54(3) (2007)
11. Friedl, K., Hátsági, Z., Shen, A.: Low-degree tests. In: Proc. SODA, pp. 57–64 (1994)
12. Meer, K.: Transparent long proofs: A first PCP theorem for $NP_{\mathbb{R}}$.. Foundations of Computational Mathematics 5(3), 231–255 (2005)
13. Meer, K.: Almost transparent short proofs for $NP_{\mathbb{R}}$. In: Owe, O., Steffen, M., Telle, J.A. (eds.) FCT 2011. LNCS, vol. 6914, pp. 41–52. Springer, Heidelberg (2011)

Property Testing Bounds for Linear and Quadratic Functions via Parity Decision Trees

Abhishek Bhrushundi[1], Sourav Chakraborty[1], and Raghav Kulkarni[2]

[1] Chennai Mathematical Institute, India
{abhishek_bhr,sourav}@cmi.ac.in
[2] Center for Quantum Technologies, Singapore
kulraghav@gmail.com

Abstract. In this paper, we study linear and quadratic Boolean functions in the context of property testing. We do this by observing that the query complexity of testing properties of linear and quadratic functions can be characterized in terms of complexity in another model of computation called *parity decision trees*.

The observation allows us to characterize testable properties of linear functions in terms of the approximate l_1 norm of the Fourier spectrum of an associated function. It also allows us to reprove the $\Omega(k)$ lower bound for testing k-linearity due to Blais et al [8]. More interestingly, it rekindles the hope of closing the gap of $\Omega(k)$ vs $O(k \log k)$ for testing k-linearity by analyzing the randomized parity decision tree complexity of a fairly simple function called E_k that evaluates to 1 if and only if the number of 1s in the input is exactly k. The approach of Blais et al. using communication complexity is unlikely to give anything better than $\Omega(k)$ as a lower bound.

In the case of quadratic functions, we prove an adaptive two-sided $\Omega(n^2)$ lower bound for testing affine isomorphism to the inner product function. We remark that this bound is tight and furnishes an example of a function for which the trivial algorithm for testing affine isomorphism is the best possible. As a corollary, we obtain an $\Omega(n^2)$ lower bound for testing the class of *Bent* functions.

We believe that our techniques might be of independent interest and may be useful in proving other testing bounds.

1 Introduction

The field of property testing broadly deals with determining whether a given object satisfies a property $\mathcal{P}$ or is very different from all the objects that satisfy $\mathcal{P}$. In this paper, the objects of interest are Boolean functions on n variables, i.e. functions of the form

$$f : \{0,1\}^n \to \{0,1\}.$$

A Boolean function property $\mathcal{P}$ is a collection of Boolean functions. Given a function g and a parameter ϵ, the goal of a tester is to distinguish between the following two cases:

E.A. Hirsch et al. (Eds.): CSR 2014, LNCS 8476, pp. 97–110, 2014.
© Springer International Publishing Switzerland 2014

- $g \in \mathcal{P}$
- g differs from every function in $\mathcal{P}$ in at least ϵ fraction of points in $\{0,1\}^n$.

The query complexity for testing $\mathcal{P}$ is the number of queries (of the form "what is the value of g at $x \in \{0,1\}^n$?") made by the best tester that distinguishes between the above two cases. If the queries made by the tester depend on the answers to the previous queries, the tester is called *adaptive*. Also, if the tester accepts whenever $g \in \mathcal{P}$, it is called *one-sided*.

Testing of Boolean function properties has been extensively studied over the last couple of decades (See [16,28]). Examples of problems that have been studied are linearity testing [11], k-junta testing [17,7], monotonicity testing [18,13], k-linearity testing [19,8,12] etc. An important problem in the area is to characterize Boolean function properties whose query complexity is *constant* (i.e., independent of n, though it can depend on ϵ). For example, such a characterization is known in the case of graph properties [1]. Though a general characterization for function properties is not yet known, there has been progress for some special classes of properties. In this paper, we attempt characterizing one such class: properties which only consist of linear functions. More specifically, we try to characterize all properties $\mathcal{P}$ of linear Boolean functions which can be tested using constant number of queries.

An example of a property of linear functions is one that contains all parities on k variables. The problem of testing this property is known as k-linearity testing. While this problem had been studied earlier [19], recently Blais et al. [8] used communication complexity to obtain a lower bound of $\Omega(k)$ on the query complexity of adaptive testers for k-linearity. The best known upper bound in the case of adaptive testers is $O(k \log k)$. Whereas a tight bound of $\Theta(k \log k)$ is known for the non-adaptive case [12], a gap still exists for adaptive testing: $\Omega(k)$ vs $O(k \log k)$. In this paper we give another approach to obtain the $\Omega(k)$ lower bound for the adaptive query complexity. While the lower bound technique of Blais et al.[8] is unlikely to give a bound beyond $\Omega(k)$, our technique has the potential of proving a better lower bound. We remark that other proof techniques for the lower bound have also been studied [9].

A rich class of properties for which characterizing constant query testability has been studied are properties that are invariant under natural transformations of the domain. For example, [22,4,3] study invariance under affine/linear transformations in this context. Properties that consist of functions isomorphic to a given function also form an important subclass. The testing of such properties is commonly referred to as *isomorphism testing*, and has seen two directions of study: testing if a function is equivalent to a given function up to permutation of coordinates [14,10], and testing affine/linear isomorphism.

Our second result[1] concerns testing affine isomorphism. A function f is affine isomorphic to g if there is an invertible affine transformation T such that $f \circ T = g$. Recently, Wimmer and Yoshida [30] characterized the query complexity of testing affine/linear isomorphism to a function in terms of its *Fourier norm*. We complement their work by providing the first example of a function for which

---

[1] This result appears in the preprint [5].

the query complexity of testing affine isomorphism is the largest possible. As a corollary, we also prove an adaptive two-sided $\Omega(n^2)$ lower bound for testing the class of *Bent* functions which are an important and well-studied class of Boolean functions in cryptography (See [24,25]).

Grigorescu et al. concurrently and independently obtained these results in [20] using a different proof technique. In fact, they prove an $2^{\Omega(n)}$ lower bound for testing Bent functions. We believe that our proof is arguably simpler and more modular, and is also amenable to generalizations (for example, to the quantum setting), even though the bound we obtain for Bent functions is weaker.

The main technique used in proving all our results is a connection between testing properties of linear and quadratic functions, and parity decision trees. Connections between linear functions and parity decision trees have been both implicitly [9] and explicitly [12] observed in earlier papers. Another connection that we exploit for proving some of our results is the one between parity decision tree depth and communication complexity. Similar connections were known earlier, see for example [31]. We remark that, to the best of our knowledge, our result is the first that combines the two connections, giving yet another way of relating property testing lower bounds to communication complexity (Blais et al. [8] observe such a connection in much more generality). Thus, we believe that our techniques might be of independent interest.

1.1   Our Results and Techniques

Property Testing and Parity Decision Trees. We give a connection between testing properties of linear functions and parity decision trees. The following is an informal statement of the connection:

Theorem 1. *For every property $\mathcal{P}$ of linear functions on n variables, one can associate a Boolean function $E_{\mathcal{P}}$ on n variables such that there is an adaptive q-query tester for determining if a given f is in $\mathcal{P}$ or $1/2$-far from $\mathcal{P}$ only if there is a randomized parity decision tree that makes q queries to compute $E_{\mathcal{P}}$.*

A similar connection holds in the case of quadratic functions.

Theorem 2. *For every property $\mathcal{P}$ of quadratic functions on n variables, one can associate a Boolean function $E_{\mathcal{P}}$ on n^2 variables such that there is an adaptive q query tester for determining if a given f is in $\mathcal{P}$ or $1/4$-far from $\mathcal{P}$ only if there is a randomized parity decision tree that makes q queries to compute $E_{\mathcal{P}}$.*

All the results that follow use the above connections crucially. Another important ingredient for some of the results is a connection between parity decision trees and the communication complexity of XOR functions. We discuss this in detail in Section 3.

Characterization of Testable Properties of Linear Functions. Theorem 1 allows us to characterize the constant query testability of a property $\mathcal{P}$ of linear functions in terms of the approximate L_1 norm of $E_{\mathcal{P}}$.

Theorem 3. *A property $\mathcal{P}$ of linear functions is constant query testable if and only if $\|\widehat{E_{\mathcal{P}}}\|_1^{1/4}$ is constant.*

This is the first such characterization of linear function properties and we hope our result is a small step towards our understanding of function properties testable in constant number of queries.

Testing k-linearity. We also obtain an alternate proof of the lower bound for testing k-linearity due to Blais et al. [8].

Theorem 4. *Any adaptive two-sided tester for testing k-linearity requires $\Omega(k)$ queries.*

The idea behind the proof is as follows. Applying Theorem 1 in the case of k-linearity, $E_{\mathcal{P}}$ turns out to be equal to the function E_k that outputs 1 if and only if there are exactly k 1s in the input string. Thus, to prove Theorem 4 we lower bound the randomized parity decision tree complexity of E_k by $\Omega(k)$.

Note that this leaves open the possibility of proving a tight $\Omega(k \log k)$ bound for testing k-linearity by improving our lower bound on the randomized parity decision tree complexity of E_k.

Lower Bound for Testing Affine Isomorphism. Let $\mathbf{IP}_n(x)$ denote the inner product function $\sum_{i=1}^{n/2} x_i x_{n/2+i}$. We consider the problem of testing affine isomorphism to $\mathbf{IP}_n(x)$ and prove a tight lower bound.

Theorem 5. *Any adaptive two-sided tester for testing affine isomorphism to $\mathbf{IP}_n(x)$ requires $\Omega(n^2)$ queries.*

The proof of Theorem 5 is similar to that of Theorem 4, though in this case, $E_{\mathcal{P}}$ turns out to be E_n, a function that maps graphs on n vertices to $\{0, 1\}$, and outputs 1 if and only if the input graph's adjacency matrix is nonsingular over $\mathbb{F}_2$. As mentioned before, this is the first example of a function for which testing affine isomorphism requires $\Omega(n^2)$ queries ($O(n^2)$ is a trivial upper bound for any function and follows from a folklore result).

It can be show that testing the set of quadratic Bent functions reduces to testing affine isomorphism to $\mathbf{IP}_n(x)$. Thus, Theorem 5 gives a lower bound for testing the set of quadratic Bent functions. Furthermore, using a result from [15], the following corollary can be obtained.

Corollary 1. *Any adaptive two-sided tester for testing the set of Bent functions requires $\Omega(n^2)$ queries.*

2   Preliminaries

2.1   Boolean Functions

Recall that functions mapping $\{0, 1\}^n$ to $\{0, 1\}$ are called Boolean functions. A Boolean function is linear if it is expressible as $\sum_{i \in S} x_i$ for $S \subseteq [n]$ over $\mathbb{F}_2$. The set of linear functions will be denoted by $\mathcal{L}$.

A Boolean function is quadratic if it can be expressed as a polynomial of degree at most two over $\mathbb{F}_2$. We shall denote the set of quadratic functions by $\mathcal{Q}$, and the set of homogeneous quadratic functions (no linear terms) by $\mathcal{Q}_0$. By a property of linear or quadratic functions, we shall always mean a subset of $\mathcal{L}$ or $\mathcal{Q}$.

For Boolean functions f and g, $dist(f, g) = \Pr_x[f(x) \neq g(x)]$. The notion can be extended to sets of Boolean functions S and T in a natural way: $dist(S, T) = \min_{f \in S, g \in T} dist(f, g)$. We state a simple but useful observation:

Observation 1. *If f and g are linear (quadratic) functions then either $f = g$ or $dist(f, g) \geq 1/2$ ($dist(f, g) \geq 1/4$).*

2.2 Property Testing

Let $\mathcal{P}$ be a property of Boolean functions on n variables. We say a randomized algorithm $\mathcal{A}$ ϵ-tests $\mathcal{P}$, if given oracle access to the truth table of an input function f, $\mathcal{A}$ determines with probability at least $2/3$ whether $f \in \mathcal{P}$, or $dist(f, \mathcal{P}) \geq \epsilon$. The number of queries made by the best tester for ϵ-testing $\mathcal{P}$ is known as the query complexity of $\mathcal{P}$. It is denoted by $Q^\epsilon(\mathcal{P})$ and may be a function of n.

Remark. When testing properties of linear functions, it is common to assume that the input function is promised to be a linear function. For a property $\mathcal{P}$ of linear functions, we denote the query complexity of testing $\mathcal{P}$ under such a promise by $Q_1(\mathcal{P})$.

For technical reasons, it will be useful to consider such a notion for quadratic functions. For a property $\mathcal{P} \subseteq \mathcal{Q}$ of quadratic functions, we shall denote by $Q_2(\mathcal{P})$ the query complexity of testing $\mathcal{P}$ under the promise that the input is always a function in $\mathcal{Q}_0$. Observation 1 implies the following statement.

Observation 2. *Let $\mathcal{P}$ be a property of linear functions. Then, $Q^{1/2}(\mathcal{P}) \geq Q_1(\mathcal{P})$. Similarly, in the case of quadratic functions, $Q^{1/4}(\mathcal{P}) \geq Q_2(\mathcal{P})$*

It can also be shown that:

Observation 3. *If $Q_1(\mathcal{P}) = Q$ then $\forall \epsilon \in (0, 1/4)$, $Q^\epsilon(\mathcal{P}) \leq O_\epsilon(Q \log Q)$*

A proof appears in the appendix of the full version [6].

Let G be a group that acts on $\{0, 1\}^n$. A function f is G-isomorphic to another function g if there is a $\phi \in G$ such that $f \circ \phi = g$. For a fixed function g, the problem of testing G-isomorphism to g is to test if an input function f is G-isomorphic to g, or ϵ-far from all functions that are G-isomorphic to g. A folklore result gives a trivial upper bound for the problem:

Lemma 1. *Testing G-isomorphism to a function g can be done in $O(\log |G|)$ queries.*

When G is the group of invertible affine transformations, the problem is known as affine isomorphism testing. The above lemma gives us the following corollary:

Corollary 2. *$O(n^2)$ queries suffice to test affine isomorphism.*

2.3   Parity Decision Trees

Parity decision trees extends the model of ordinary decision trees such that one may query the parity of a subset of input bits, i.e. the queries are of form "is $\sum_{i \in S} x_i \equiv 1 \pmod{2}$? " for an arbitrary subset $S \subseteq [n]$. We call such queries parity queries.

Let f be a Boolean function. For a parity decision tree P_f for f, let $C(P_f, x)$ denote the number of parity queries made by P_f on input x. The parity decision tree complexity of f is $D_{\oplus}(f) = \min_{P_f} \max_x C(P_f, x)$.

Note that $D_{\oplus}(f) \leq D(f)$, where $D(f)$ is the deterministic decision tree complexity of f, as the queries made by a usual decision tree, "is $x_i = 1$?", are also valid parity queries.

A bounded error randomized parity decision tree $R_{\oplus}^f$ is a probability distribution over all deterministic decision trees such that for every input the expected error of the algorithm is bounded by $1/3$. The cost $C(R_{\oplus}^f, x)$ is the highest possible number of queries made by $R_{\oplus}^f$ on x, and the bounded error randomized decision tree complexity of f is $R_{\oplus}(f) = \min_{R_{\oplus}^f} \max_x C(R_{\oplus}^f, x)$

For a Boolean function f, it turns out that $R_{\oplus}(f)$ can be lower bounded by the randomized communication complexity of the so-called XOR function $f(x \oplus y)$ (See [23] for the definition of randomized communication complexity and XOR functions). So we have the following lemma.

Lemma 2. $R_{\oplus}(f) \geq \frac{1}{2} RCC(f(x \oplus y))$

Proof. Given a Boolean function $f : \{0,1\}^n \to \{0,1\}$ on n consider the communication game where x is with Alice and y is with Bob and they want to compute $f(x \oplus y)$ with error bounded by $1/3$. Let $RCC(f(x \oplus y))$ denote the randomized communication complexity of this communication game.

Given a randomized parity decision tree $R_{\oplus}^f$, Alice and Bob can convert it into a protocol by simulating the parity queries made by $R_{\oplus}^f$ by two bits of communication, and thus the inequality follows.

3   Property Testing and Parity Trees

In this section we describe a relation between the testing complexity of a property of linear/quadratic functions and the parity decision tree complexity of an associated function. We remark that such connections have been observed before in the case of linear functions, though, to the best of our knowledge, such an observation had not been made for quadratic functions before our work.

3.1   Parity Trees and Linear Functions

Let $e_i \in \{0,1\}^n$ denote the Boolean string whose i^{th} bit is 1 and all other bits are 0. For any linear function f let us define a string $B(f) \in \{0,1\}^n$ such that the ith bit of $B(f)$ is 1 iff $f(e_i) = 1$. The following lemma is easy to prove:

Lemma 3. *The map $B : \mathcal{L} \to \{0,1\}^n$ gives a bijection between the set $\mathcal{L}$ and strings of length n.*

Now let $\mathcal{P} \subseteq \mathcal{L}$ be a set of linear functions. Given a linear function f we want a tester $\mathcal{T}$ that makes queries to the truth table of f and determines whether f is in $\mathcal{P}$ or is ϵ-far from $\mathcal{P}$. Let us define a set $S_{\mathcal{P}} \subseteq \{0,1\}^n$ as $S_{\mathcal{P}} := \{B(f) \mid f \in \mathcal{P}\}$.

Lemma 4. *For any $\mathcal{P} \subseteq \mathcal{L}$ and any $f \in \mathcal{L}$ we have:*

- *$f \in \mathcal{P}$ if and only if $B(f) \in S_{\mathcal{P}}$ and*
- *f is $1/2$-far from $\mathcal{P}$ if and only if $B(f) \notin S_{\mathcal{P}}$*

We omit the proof of Lemma 4 as it follows directly from Lemma 3 and Observation 1.

Now, by Lemma 4, testing where f is in $\mathcal{P}$ or is $1/2$-far from $\mathcal{P}$ is exactly same as deciding if $B(f) \in S_{\mathcal{P}}$. Furthermore, we can translate the queries made by the tester $\mathcal{T}$ to the truth table of f into parity queries to the string $B(f)$, and vice-versa. Since f is linear, we have $f(x) = \bigoplus_i x_i \cdot f(e_i)$. Now, if $S_x := \{i \mid x_i = 1\}$ then, whenever $\mathcal{T}$ queries f at x, it can be equivalently viewed as the query $\bigoplus_{i \in S_x} (B(f))_i$ made to $B(f)$.

Consider the Boolean function $E_{\mathcal{P}} : \{0,1\}^n \to \{0,1\}$, where $E_{\mathcal{P}}(x) = 1$ iff $B^{-1}(x) \in \mathcal{P}$. From the above discussion, deciding "is $x \in S_{\mathcal{P}}$?" is same as deciding "is $E_{\mathcal{P}}(x) = 1$?" Thus we have:

Theorem 6. *There is a tester that makes q queries for determining if a linear function f satisfies the property $\mathcal{P}$ or is $1/2$-far from satisfying $\mathcal{P}$ if and only if there is a randomized parity decision that makes q queries for computing $E_{\mathcal{P}}$. Equivalently, $Q_1(\mathcal{P}) = R_\oplus(E_{\mathcal{P}})$.*

Notice that Theorem 1 follows from this by using $Q^{1/2}(\mathcal{P}) \geq Q_1(\mathcal{P})$ from Observation 2.

3.2 Parity Trees and Quadratic Functions

Let $\mathcal{G}_n \subseteq \{0,1\}^{n^2}$ denote the set of graphs on n vertices. For any homogeneous quadratic function $f \in \mathcal{Q}_0$ let us define a graph $G(f)$ with vertex set $[n]$ such that the edge $\{i,j\}$ is present in $G(f)$ iff $x_i x_j$ occurs as a monomial when f is expressed as a polynomial over $\mathbb{F}_2$. The following observation follows from the way we constructed $G(f)$:

Observation 4. *The map $G : \mathcal{Q}_0 \to \mathcal{G}_n$ is a bijection.*

Let $\mathcal{P} \subseteq \mathcal{Q}$ be a property of quadratic functions, and let $S_{\mathcal{P}} = \{G(f) \mid f \in \mathcal{P} \cap \mathcal{Q}_0\}$. It now easily follows from Observation 4 and 1 that:

Lemma 5. *For any $\mathcal{P} \subseteq \mathcal{Q}$ and any $f \in \mathcal{Q}_0$ we have:*

- *$f \in \mathcal{P}$ if and only if $G(f) \in S_{\mathcal{P}}$ and*
- *f is $1/4$-far from $\mathcal{P}$ if and only if $G(f) \notin S_{\mathcal{P}}$*

Thus, the above lemma says that testing whether a given $f \in \mathcal{Q}_0$ is in $\mathcal{P}$ or $1/4$-far from $\mathcal{P}$ is exactly the same as deciding if $G(f)$ is in $S_{\mathcal{P}}$.

Let $\mathcal{A}$ be an algorithm that tests if a $f \in \mathcal{Q}_0$ is in $\mathcal{P}$ or $1/4$-far from it. We now describe how to translate queries made by $\mathcal{A}$ to the truth table of f to parity queries to the adjacency matrix of the graph $G(f)$. Given $y \in \{0, 1\}^n$ and a graph G on the vertex set $[n]$, we denote by $G[y]$ the induced graph on the vertex set $\{i | \ y_i = 1\}$. It is not hard to see that the value $f(y)$ is exactly the parity of the number of edges in $G(f)[y]$. Thus, any query to the truth table of f can be translated to a parity query to the adjacency matrix of $G(f)$, but unlike in the case of linear functions, the translation works only in one direction. To be more precise, an arbitrary parity query to the adjacency matrix of $G(f)$ cannot be translated into a query to the truth table of f.

Consider the Boolean function $E_{\mathcal{P}} : \mathcal{G}_n \rightarrow \{0, 1\}$, where $E_{\mathcal{P}}(H) = 1$ iff $G^{-1}(H) \in \mathcal{P}$, and observe that deciding "is $H \in S_{\mathcal{P}}$?" is same as deciding "is $E_{\mathcal{P}}(H) = 1$?". Combining the observations made above, we get:

Lemma 6. *There is an adaptive tester that makes q queries for determining if a given $f \in \mathcal{Q}_0$ satisfies the property $\mathcal{P}$ or is $1/4$-far from satisfying $\mathcal{P}$ only if there is a randomized parity decision that makes q queries for computing $E_{\mathcal{P}}$. Equivalently, $Q_2(\mathcal{P}) \geq R_{\oplus}(E_{\mathcal{P}})$.*

Combining Lemma 6 and Observation 2, we get a more general result:

Theorem 7 (Formal statement of Theorem 2). *There is an adaptive tester that makes q queries for determining if a given f satisfies the property $\mathcal{P}$ or is $1/4$-far from satisfying $\mathcal{P}$ only if there is a randomized parity decision tree that makes q queries for computing $E_{\mathcal{P}}$. Equivalently, $Q^{1/4}(\mathcal{P}) \geq R_{\oplus}(E_{\mathcal{P}})$.*

4   Characterizing Testable Properties of Linear Functions

In this section we give a characterization of properties of linear functions that are testable using constant number of queries. We will use some standard concepts from Fourier analysis of Boolean functions and the reader is referred to [26] for an introduction to the same.

Recall that for a Boolean[2] function f, $||\widehat{f}||_1^{\epsilon}$ denotes the minimum possible $||\widehat{g}||_1$ over all g such that $|f(x) - g(x)| \leq \epsilon$ for all x. We begin by proving the following lemma:

Lemma 7. *There are constants $c_1, c_2 > 0$ such that for sufficiently large n, if $f : \{0, 1\}^n \rightarrow \{-1, +1\}$ is a Boolean function, then $c_1 \cdot \log ||\widehat{f}||_1^{1/4} \leq R_{\oplus}(f) \leq c_2 \cdot (||\widehat{f}||_1^{1/4})^2$*

---

[2] For the purpose of this section, it will be convenient to assume that the range of a Boolean function is $\{-1, +1\}$.

Proof. For the first inequality, we obtain from Lemma 2 that $RCC(f(x \oplus y)) \leq 2R_\oplus(f)$. Now, it is well known that $RCC(f(x \oplus y)) = \Omega(\log ||\widehat{f}||_1^{1/4})$ (see for instance [23]) and thus we have

$$R_\oplus(f) \geq 1/2 \cdot RCC(f(x \oplus y)) = \Omega(\log ||\widehat{f}||_1^{1/4}) \tag{1}$$

To see the second inequality, we will construct a randomized parity decision tree[3] $\mathcal{T}$ with query complexity $O((||\widehat{f}||_1^{1/4})^2)$ that computes f. Let $g : \{0,1\}^n \to \mathbb{R}$ be a function that point-wise 1/4-approximates f (i.e. for all x, $|f(x) - g(x)| \leq 1/4$) such that $||\widehat{g}||_1$ is the minimum among all functions that 1/4-approximate f. Let $\mathcal{D}_g$ denote a distribution on subsets of $[n]$ such that a set S has probability $|\hat{g}(S)|/||\widehat{g}||_1$.

We define the randomized parity decision tree $\mathcal{T}$ as follows. $\mathcal{T}$ makes d (the parameter will be fixed later) random parity queries $S_1, S_2 \ldots S_d$, such that each S_i is distributed according to $\mathcal{D}_g$. Let $X_1, X_2, \ldots X_d$ be random variables such that

$$X_i = \frac{\text{sign}(\hat{g}(S_i))(-1)^{\sum_{j \in S_i} x_j}}{||\widehat{g}||_1}$$

Here the sign function $\text{sign}(x)$ outputs -1 is $x < 0$, and 1 otherwise. Finally, the tree outputs $\text{sign}(\sum_{i=1}^d X_i)$.

The first thing to note is that

$$\mathbb{E}[X_i] = \sum_{S \subseteq [n]} \frac{\text{sign}(\hat{g}(S_i))(-1)^{\sum_{j \in S_i} x_j}}{||\widehat{g}||_1} \frac{|\hat{g}(S)|}{||\widehat{g}||_1} = \frac{g(x)}{(||\widehat{g}||_1)^2}$$

Let $X = \sum_{i=1}^d X_i$. Then, $\mathbb{E}[X] = d \cdot g(x)/(||\widehat{g}||_1)^2$. Setting $d = 100 \cdot (||\widehat{g}||_1)^2$, we get $\mathbb{E}[X] = 100 \cdot g(x)$.

Now each X_i is bounded and lies in $[-1/||\widehat{g}||_1, +1/||\widehat{g}||_1]$. Thus by Hoeffding's inequality we have

$$\Pr[|X - \mathbb{E}[X]| \geq 50] \leq \exp\left(\frac{-2 \cdot (50)^2}{400}\right) = \exp\left(\frac{-25}{2}\right). \tag{2}$$

Since g point-wise 1/4-approximates f, $\text{sign}(g(x)) = \text{sign}(f(x)) = f(x)$. Also, it is easy to see that, if $|X - \mathbb{E}[X]| \leq 50$, $\text{sign}(X) = \text{sign}(\mathbb{E}[X]) = \text{sign}(g(x))$. Thus, by Equation 2, $\text{sign}(X) = f(x)$ with very high probability.

The above argument shows that $\mathcal{T}$ is a randomized decision tree that computes f with high probability and makes $O((||\widehat{g}||_1)^2) = O((||\widehat{f}||_1^{1/4})^2)$ queries. This proves that

$$R_\oplus(f) = O((||\widehat{f}||_1^{1/4})^2) \tag{3}$$

Combining Equations 1 and 3 we can obtain the statement of the Lemma.

---

[3] We shall assume that $\mathcal{T}$'s range is $\{-1, +1\}$.

Let $\mathcal{P}$ be a property of linear functions, and $Q_1(\mathcal{P})$ denote the query complexity of testing $\mathcal{P}$ when the input function is promised to be linear. Then, from the above lemma and Theorem 6, there exist constants $c_1, c_2 > 0$ such that for large enough n,

$$c_1 \cdot \log ||\widehat{E_\mathcal{P}}||_1^{1/4} \leq Q_1(\mathcal{P}) \leq c_2 \cdot (||\widehat{E_\mathcal{P}}||_1^{1/4})^2 \tag{4}$$

Using Observation 2 and 3 and Equation 4, we get, for $\epsilon \in (0, 1/4)$, there exists a constant c_2' that depends on ϵ such that for large enough n:

$$c_1 \cdot \log ||\widehat{E_\mathcal{P}}||_1^{1/4} \leq Q^{1/4}(\mathcal{P}) \leq Q^\epsilon(\mathcal{P}) \leq c_2'(||\widehat{E_\mathcal{P}}||_1^{1/4})^2 \log \left(||\widehat{E_\mathcal{P}}||_1^{1/4} \right)$$

Thus, we can conclude the Theorem 3 from the discussion: a property $\mathcal{P}$ of linear functions is testable using constant number of queries if and only if $||\widehat{E_\mathcal{P}}||_1^{1/4}$ is constant.

5   Testing k-linearity

In this section we apply the result from Section 3 to prove a lower bound for testing k-linearity. We shall use $\mathcal{P}$ to denote the set of k-linear functions on n variables.

Let $E_k : \{0,1\}^n \to \{0,1\}$ denote the Boolean function that outputs 1 if and only if the number of 1s is *exactly* k. Recall a notation from Section 3: for any linear function f we can define a string $B(f) \in \{0,1\}^n$ such that $B(f)_i = 1$ iff $f(e_i) = 1$. We observe the following:

Observation 5. *A Boolean function f is k-linear if and only if $B(f)$ has exactly k 1s.*

Thus, $E_\mathcal{P}$ is exactly the function E_k. Using Theorem 6 we have the following:

$$Q_1(\mathcal{P}) = R_\oplus(E_k) \tag{5}$$

Thus, if we can obtain a lower bound of $\Omega(k \log k)$ on the randomized parity decision tree complexity of E_k then we would obtain a tight bound for adaptive k-linearity testing (This would follow from Observation 2: $Q^{1/2}(\mathcal{P}) \geq Q_1(\mathcal{P})$). Unfortunately we are unable to obtain such a lower bound yet. Instead we can obtain a lower bound of $\Omega(k)$ that matches the previous known lower bound for k-linearity testing [8].

Using Lemma 2, we have that $R_\oplus(E_k) \geq \frac{1}{2}RCC(E_k(x \oplus y))$. Furthermore, Huang et al. [21] show that[4]:

Lemma 8. $RCC(E_k(x \oplus y)) = \Omega(k)$

Using Equation 5 and Lemma 8, we have $Q_1(\mathcal{P}) = \Omega(k)$. Finally, Observation 2 gives us $Q^{1/2}(\mathcal{P}) = \Omega(k)$:

---

[4] Actually, Huang et al. show that $RCC(E_{>k}(x \oplus y)) = \Omega(k)$, but their proof can be used to obtain the same lower bound for $RCC(E_k(x \oplus y))$. Alternatively, the lower bound may be obtained via a reduction to k-DISJ, a problem considered in [8].

Theorem 8 (Formal statement of Theorem 4). *Any adaptive two-sided tester for $1/2$-testing k-linearity must make $\Omega(k)$ queries.*

Thus we obtain a lower bound of $\Omega(k)$ using the lower bound for the randomized communication complexity of the XOR function $E_k(x \oplus y)$. Note that using this method we cannot expect to obtain a better lower bound as there is an upper bound of $O(k)$ on the communication complexity. But there is hope that one may be able to obtain a better lower bound for the parity decision tree complexity of E_k directly.

On the other hand, if one is able to construct a randomized parity decision tree of depth $O(k)$ for deciding E_k, Lemma 5 immediately implies a tester for k-linearity that makes $O(k)$ queries under the promise that the input function is linear. Notice that the exact complexity for even the promise problem is not known and the best upper bound is $O(k \log k)$. (while, naturally, the lower bound is $\Omega(k)$.)

6   Testing Affine Isomorphism to the Inner Product Function

The main result of this section is that $1/4$-testing affine isomorphism to the inner product function $\mathbf{IP}_n(x)$[5] requires $\Omega(n^2)$ queries. As a corollary, we show that testing the set of Bent functions requires $\Omega(n^2)$ queries.

Let $\mathcal{B}$ denote the set of Bent functions (See [27] for a definition). The following is a consequence of Dickson's lemma (We omit the proof here, but it appears in the full version [6])

Lemma 9. *Let $Q(n)$ denote the the query complexity of $1/4$-testing affine isomorphism to the inner product function. Then $Q^{1/4}(\mathcal{B} \cap \mathcal{Q}) = O(Q(n))$.*

Thus, it is sufficient to lower bound $Q^{1/4}(\mathcal{B} \cap \mathcal{Q})$. In fact, by Observation 2, $Q^{1/4}(\mathcal{B} \cap \mathcal{Q}) \geq Q_2(\mathcal{B} \cap \mathcal{Q})$, and thus we can restrict out attention to lower bounding $Q_2(\mathcal{B} \cap \mathcal{Q})$.

Recall from Section 3 that we can associate a graph $G(f)$ with every function $f \in \mathcal{Q}_0$. We now state a criterion for a quadratic function to be Bent that follows from a result due to Rothaus [27].

Lemma 10. *A function $f \in \mathcal{Q}_0$ is Bent iff the adjacency matrix of $G(f)$ is nonsingular.*

We omit the proof due to space constraints and give a proof in the appendix of the full version [6].

Recall from Section 3 that $\mathcal{G}_n \subseteq \{0,1\}^{n^2}$ is the set of graphs on the vertex set $[n]$. Let $\mathcal{P} := \mathcal{B} \cap \mathcal{Q}$, and let $E_n : \mathcal{G}_n \to \{0,1\}$ be a Boolean function such that $E_n(G) = 1$ iff the adjacency matrix of G is nonsingular. Due to Lemma 10, $E_{\mathcal{P}}$ turns out to be exactly equal to E_n. Combining with Theorem 5, we have

$$Q_2(\mathcal{P}) \geq R_\oplus(E_n) \tag{6}$$

---

[5] For the rest of the section we shall assume that the number of variables n is even.

As in the case of E_k, analyzing the decision tree complexity of E_n directly is hard, and we turn to communication complexity. Lemma 2 tells us that $R_\oplus(E_n) \geq \frac{1}{2}RCC(E_n(x \oplus y))$.

Let $M_n(\mathbb{F}_2)$ denote the set of $n \times n$ matrices over $\mathbb{F}_2$, and $Det_n : M_n(\mathbb{F}_2) \to \{0, 1\}$ be the function such that $Det_n(A) = 1$ iff $A \in M_n(\mathbb{F}_2)$ is nonsingular. The following result from [29] analyzes the communication complexity of Det_n.

Lemma 11. $RCC(Det_n(x \oplus y)) = \Omega(n^2)$

It turns out that the communication complexity of Det_n relates to that of E_n.

Lemma 12. $= RCC(Det_n(x \oplus y)) \leq RCC(E_{2n}(x \oplus y))$

Proof. Let $A \in M_n(\mathbb{F}_2)$. Consider the $2n \times 2n$ matrix A' given by

$$A' := \begin{pmatrix} 0 & A^t \\ A & 0 \end{pmatrix}$$

$A' \in \mathcal{G}_{2n}$ by construction and it can be easily verified that A' is nonsingular iff A is nonsingular.

Now, let the inputs to Alice and Bob be A and B respectively. Since $(A \oplus B)' = A' \oplus B'$, $Det_n(A \oplus B) = 1$ iff $E_{2n}((A \oplus B)') = 1$ iff $E_{2n}(A' \oplus B') = 1$. Thus, to determine if $Det_n(A \oplus B)$ is 1, Alice and Bob can construct A' and B' from A and B respectively, and run the protocol for E_{2n} on A' and B'. This completes the proof.

Thus, using Lemma 11, we have $RCC(E_n(x \oplus y)) = \Omega(n^2)$. Using Lemma 2 and Equation 6, we have that $Q_2(\mathcal{P}) = \Omega(n^2)$.

Thus, based on earlier observations, we can conclude:

Theorem 9 (Formal statement of Theorem 5). *Any adaptive two-sided tester for $1/4$-testing affine isomorphism to the inner product function $\mathbf{IP}_n(x)$ requires $\Omega(n^2)$ queries.*

Corollary 2 tells us that our result is tight. Thus, $\mathbf{IP}_n(x)$ is an example of a function for which the trivial bound for testing affine isomorphism is the best possible.

We have shown that $Q^{1/4}(\mathcal{B} \cap \mathcal{Q}) = \Omega(n^2)$. We now state a result due to Chen et al.(Lemma 2 in [15]) in a form that is suitable for application in our setting:

Lemma 13. *Let $\mathcal{P}_1$ and $\mathcal{P}_2$ be two properties of Boolean functions that have testers (possibly two-sided) T_1 and T_2 respectively. Let the query complexity of tester T_i be $q_i(\epsilon, n)$. Suppose $dist(\mathcal{P}_1 \backslash \mathcal{P}_2, \mathcal{P}_2 \backslash \mathcal{P}_1) \geq \epsilon_0$ for some absolute constant ϵ_0. Then, $\mathcal{P}_1 \cap \mathcal{P}_2$ is ϵ-testable with query complexity*

$$O(max\{q_1(\epsilon, n), q_1(\frac{\epsilon_0}{2}, n)\} + max\{q_2(\epsilon, n), q_2(\frac{\epsilon_0}{2}, n)\})$$

In its original form, the lemma has been proven for the case when T_1, T_2 are one-sided, and q_1, q_2 are independent of n, but the proof can be easily adapted to this more general setting.

Another consequence of Dickson's lemma is the following (A proof appears in the full version [6]):

Lemma 14. *Let f, g be Boolean functions. If $f \in \mathcal{B} \setminus \mathcal{Q}$ and $g \in \mathcal{Q} \setminus \mathcal{B}$, then $dist(f, g) \geq 1/4$.*

We are now ready to prove a lower bound for testing Bent functions.

Theorem 10 (Formal statement of Corollary 1). *Any adaptive two-sided tester that $1/8$-tests the set of Bent functions requires $\Omega(n^2)$ queries.*

Proof. It is well known via [2] that $\mathcal{Q}$ is testable with constant number of queries (say $q_1(\epsilon)$). Suppose there is a tester that tests $\mathcal{B}$ using $q_2(\epsilon, n)$ queries. From Lemma 14, we know that $dist(\mathcal{B} \setminus \mathcal{Q}, \mathcal{Q} \setminus \mathcal{B}) \geq \frac{1}{4}$. Thus, by Lemma 13, we have that there is a tester that makes $O(max\{q_1(\epsilon), q_1(\frac{1}{8})\} + max\{q_2(\epsilon, n), q_2(\frac{1}{8}, n)\})$ queries to ϵ-test $\mathcal{B} \cap \mathcal{Q}$.

Setting $\epsilon - \frac{1}{4}$, we have a tester that makes $O(q_1(\frac{1}{8}) + q_2(\frac{1}{8}, n))$ queries to test if a given f is in $\mathcal{B} \cap \mathcal{Q}$, or $1/4$-far from it. Since $Q^{1/4}(\mathcal{B} \cap \mathcal{Q}) = \Omega(n^2)$ and $q_1(\frac{1}{8})$ is a constant, we get $q_2(\frac{1}{8}, n) = \Omega(n^2)$, which completes the proof. $\square$

References

1. Alon, N., Fischer, E., Newman, I., Shapira, A.: A combinatorial characterization of the testable graph properties: It's all about regularity. SIAM J. Comput. 39(1), 143–167 (2009)
2. Alon, N., Kaufman, T., Krivelevich, M., Litsyn, S.N., Ron, D.: Testing low-degree polynomials over $GF(2)$. In: Arora, S., Jansen, K., Rolim, J.D.P., Sahai, A. (eds.) APPROX 2003 + RANDOM 2003. LNCS, vol. 2764, pp. 188–199. Springer, Heidelberg (2003)
3. Bhattacharyya, A., Fischer, E., Hatami, H., Hatami, P., Lovett, S.: Every locally characterized affine-invariant property is testable. In: Proceedings of the 45th Annual ACM Symposium on Symposium on Theory of Computing, STOC 2013, pp. 429–436. ACM Press, New York (2013), http://doi.acm.org/10.1145/2488608.2488662
4. Bhattacharyya, A., Grigorescu, E., Shapira, A.: A unified framework for testing linear-invariant properties. In: Proceedings of the 51st Annual IEEE Symposium on Foundations of Computer Science, pp. 478–487 (2010)
5. Bhrushundi, A.: On testing bent functions. Electronic Colloquium on Computational Complexity (ECCC) 20, 89 (2013)
6. Bhrushundi, A., Chakraborty, S., Kulkarni, R.: Property testing bounds for linear and quadratic functions via parity decision trees. Electronic Colloquium on Computational Complexity (ECCC) 20, 142 (2013)
7. Blais, E.: Testing juntas nearly optimally. In: Proc. ACM Symposium on the Theory of Computing, pp. 151–158. ACM, New York (2009)
8. Blais, E., Brody, J., Matulef, K.: Property testing via communication complexity. In: Proc. CCC (2011)
9. Blais, E., Kane, D.: Tight bounds for testing k-linearity. In: Gupta, A., Jansen, K., Rolim, J., Servedio, R. (eds.) APPROX/RANDOM 2012. LNCS, vol. 7408, pp. 435–446. Springer, Heidelberg (2012)
10. Blais, E., Weinstein, A., Yoshida, Y.: Partially symmetric functions are efficiently isomorphism-testable. In: FOCS, pp. 551–560 (2012)
11. Blum, M., Luby, M., Rubinfeld, R.: Self-testing/correcting with applications to numerical problems. In: STOC, pp. 73–83 (1990)

12. Buhrman, H., García-Soriano, D., Matsliah, A., de Wolf, R.: The non-adaptive query complexity of testing k-parities. CoRR abs/1209.3849 (2012)
13. Chakrabarty, D., Seshadhri, C.: A o(n) monotonicity tester for boolean functions over the hypercube. CoRR abs/1302.4536 (2013)
14. Chakraborty, S., Fischer, E., García-Soriano, D., Matsliah, A.: Junto-symmetric functions, hypergraph isomorphism and crunching. In: IEEE Conference on Computational Complexity, pp. 148–158 (2012)
15. Chen, V., Sudan, M., Xie, N.: Property testing via set-theoretic operations. In: ICS, pp. 211–222 (2011)
16. Fischer, E.: The art of uninformed decisions: A primer to property testing. Science 75, 97–126 (2001)
17. Fischer, E., Kindler, G., Ron, D., Safra, S., Samorodnitsky, A.: Testing juntas. Journal of Computer and System Sciences 68(4), 753–787 (2004), Special Issue on FOCS 2002
18. Fischer, E., Lehman, E., Newman, I., Raskhodnikova, S., Rubinfeld, R., Samorodnitsky, A.: Monotonicity testing over general poset domains. In: STOC, pp. 474–483 (2002)
19. Goldreich, O.: On testing computability by small width obdds. In: Serna, M., Shaltiel, R., Jansen, K., Rolim, J. (eds.) APPROX 2010. LNCS, vol. 6302, pp. 574–587. Springer, Heidelberg (2010)
20. Grigorescu, E., Wimmer, K., Xie, N.: Tight lower bounds for testing linear isomorphism. In: Raghavendra, P., Raskhodnikova, S., Jansen, K., Rolim, J.D.P. (eds.) RANDOM 2013 and APPROX 2013. LNCS, vol. 8096, pp. 559–574. Springer, Heidelberg (2013)
21. Huang, W., Shi, Y., Zhang, S., Zhu, Y.: The communication complexity of the hamming distance problem. Inf. Process. Lett. 99(4), 149–153 (2006)
22. Kaufman, T., Sudan, M.: Algebraic property testing: the role of invariance. In: STOC, pp. 403–412 (2008)
23. Lee, T., Shraibman, A.: Lower bounds in communication complexity. Foundations and Trends in Theoretical Computer Science 3(4), 263–398 (2009)
24. MacWilliams, F.J., Sloane, N.J.A.: The Theory of Error-Correcting Codes (North-Holland Mathematical Library). North Holland Publishing Co. (June 1988), http://www.worldcat.org/isbn/0444851933
25. Neumann, T.: Bent functions, Master's thesis (2006)
26. O'Donnell, R.: Analysis of boolean functions (2012), http://www.analysisofbooleanfunctions.org
27. Rothaus, O.: On bent functions. Journal of Combinatorial Theory, Series A 20(3), 300–305 (1976), http://www.sciencedirect.com/science/article/pii/0097316576900248
28. Rubinfeld, R., Shapira, A.: Sublinear time algorithms. Electronic Colloquium on Computational Complexity (ECCC) 11(013) (2011)
29. Sun, X., Wang, C.: Randomized communication complexity for linear algebra problems over finite fields. In: STACS, pp. 477–488 (2012)
30. Wimmer, K., Yoshida, Y.: Testing linear-invariant function isomorphism. In: Fomin, F.V., Freivalds, R., Kwiatkowska, M., Peleg, D. (eds.) ICALP 2013, Part I. LNCS, vol. 7965, pp. 840–850. Springer, Heidelberg (2013)
31. Zhang, Z., Shi, Y.: On the parity complexity measures of boolean functions. Theor. Comput. Sci. 411(26-28), 2612–2618 (2010)

A Fast Branching Algorithm
for Cluster Vertex Deletion*

Anudhyan Boral[1], Marek Cygan[2], Tomasz Kociumaka[2], and Marcin Pilipczuk[3]

[1] Chennai Mathematical Institute, Chennai, India
anudhyan@cmi.ac.in
[2] Institute of Informatics, University of Warsaw, Poland
{cygan,kociumaka}@mimuw.edu.pl
[3] Department of Informatics, University of Bergen, Norway
Marcin.Pilipczuk@ii.uib.no

Abstract. In the family of clustering problems we are given a set of objects (vertices of the graph), together with some observed pairwise similarities (edges). The goal is to identify clusters of similar objects by slightly modifying the graph to obtain a cluster graph (disjoint union of cliques).

Hüffner et al. [LATIN 2008, Theory Comput. Syst. 2010] initiated the parameterized study of CLUSTER VERTEX DELETION, where the allowed modification is vertex deletion, and presented an elegant $\mathcal{O}(2^k k^9 + nm)$-time fixed-parameter algorithm, parameterized by the solution size. In the last 5 years, this algorithm remained the fastest known algorithm for CLUSTER VERTEX DELETION and, thanks to its simplicity, became one of the textbook examples of an application of the iterative compression principle. In our work we break the 2^k-barrier for CLUSTER VERTEX DELETION and present an $\mathcal{O}(1.9102^k(n+m))$-time branching algorithm.

1 Introduction

The problem to cluster objects based on their pairwise similarities has arisen from applications both in computational biology [6] and machine learning [5]. In the language of graph theory, as an input we are given a graph where vertices correspond to objects, and two objects are connected by an edge if they are observed to be similar. The goal is to transform the graph into a cluster graph (a disjoint union of cliques) using a minimum number of modifications.

The set of allowed modifications depends on the particular problem variant and an application considered. Probably the most studied variant is the CLUSTER EDITING problem, known also as CORRELATION CLUSTERING, where we seek for a minimal number of edge edits to obtain a cluster graph. The study of CLUSTER EDITING includes [3, 4, 14, 20, 31] and, from the parameterized perspective, [7–11, 15, 16, 19, 22–24, 27–29].

The main principle of parameterized complexity is that we seek algorithms that are efficient if the considered parameter is small. However, the distance

---

* Partially supported by NCN grant N206567140 and Foundation for Polish Science.

E.A. Hirsch et al. (Eds.): CSR 2014, LNCS 8476, pp. 111–124, 2014.

measure in CLUSTER EDITING, the number of edge edits, may be quite large in practical instances, and, in the light of recent lower bounds refuting the existence of subexponential FPT algorithms for CLUSTER EDITING [19, 27], it seems reasonable to look for other distance measures (see e.g. Komusiewicz's PhD thesis [27]) and/or different problem formulations.

In 2008, Hüffner et al. [25, 26] initiated the parameterized study of the CLUSTER VERTEX DELETION problem (CLUSTERVD for short). Here, the allowed modifications are vertex deletions.

CLUSTER VERTEX DELETION (CLUSTERVD)   **Parameter:** k
Input: An undirected graph G and an integer k.
Question: Does there exist a set S of at most k vertices of G such that $G \setminus S$ is a cluster graph, i.e., a disjoint union of cliques?

In terms of motivation, we want to refute as few objects as possible to make the set of observations completely consistent. Since a vertex deletion removes as well all its incident edges, we may expect that this new editing measure may be significantly smaller in practical applications than the edge-editing distance.

As CLUSTERVD can be equivalently stated as the problem of hitting, with minimum number of vertices, all induced P_3s (paths on 3 vertices) in the input graph, CLUSTERVD can be solved in $\mathcal{O}(3^k(n + m))$ time by a straightforward branching algorithm [13], where n and m denote the number of vertices and edges of G, respectively. The dependency on k can be improved by considering more elaborate case distinction in the branching algorithm, either directly [21], or via a general algorithm for 3-HITTING SET [17]. Hüffner et al. [26] provided an elegant $\mathcal{O}(2^k k^9 + nm)$-time algorithm, using the iterative compression principle [30] and a reduction to the weighted maximum matching problem. This algorithm, presented at LATIN 2008 [25], quickly became one of the textbook examples of an application of the iterative compression technique.

In our work we pick up this line of research and obtain the fastest algorithm for (unweighted) CLUSTERVD.

Theorem 1. CLUSTER VERTEX DELETION *can be solved in* $\mathcal{O}(1.9102^k(n+m))$ *time and polynomial space on an input* (G, k) *with* $|V(G)| = n$ *and* $|E(G)| = m$.

The source of the exponential 2^k factor in the time complexity of the algorithm of [26] comes from enumeration of all possible intersections of the solution we are looking for with the previous solution of size $(k + 1)$. As the next step in each subcase is a reduction to the weighted maximum matching problem (with a definitely nontrivial polynomial-time algorithm), it seems hard to break the 2^k-barrier using the approach of [26]. Hence, in the proof of Theorem 1 we go back to the bounded search tree approach. However, to achieve the promised time bound, and at the same time avoiding very extensive case analysis, we do not follow the general 3-HITTING SET approach. Instead, our methodology is to carefully investigate the structure of the graph and an optimum solution around a vertex already guessed to be not included in the solution. We note that a somehow similar approach has been used in [26] to cope with a variant of CLUSTERVD where we restrict the number of clusters in the resulting graph.

More precisely, the main observation in the proof of Theorem 1 is that, if for some vertex v we know that there exists a minimum solution S not containing v, in the neighbourhood of v the CLUSTERVD problem reduces to VERTEX COVER. Let us define N_1 and N_2 to be the vertices at distance 1 and 2 from v, respectively, and define the auxiliary graph H_v to be a graph on $N_1 \cup N_2$ having an edge for each edge of G between N_1 and N_2 and for each non-edge in $G[N_1]$. In other words, two vertices are connected by an edge in H_v if, together with v, they form a P_3 in G. We observe that a minimum solution S not containing v needs to contain a vertex cover of H_v. Moreover, one can show that we may greedily choose a vertex cover with inclusion-wise maximal intersection with N_2, as deleting vertices from N_2 helps us resolve the remaining part of the graph.

Branching to find the 'correct' vertex cover of H_v is very efficient, with worst-case $(1, 2)$ (i.e., golden-ratio) branching vector. However, we do not have the vertex v beforehand, and branching to obtain such a vertex is costly. Our approach is to get as much gain as possible from the vertex cover-style branching on the graph H_v, to be able to balance the loss from some inefficient branches used to obtain the vertex v to start with. Consequently, we employ involved analysis of properties and branching algorithms for the auxiliary graph H_v.

Note that the algorithm of Theorem 1 can be pipelined with the kernelization algorithm of 3-HITTING SET [1], yielding the following corollary.

Corollary 2. CLUSTER VERTEX DELETION *can be solved in* $\mathcal{O}(1.9102^k k^4 + nm)$ *time and polynomial space on an input* (G, k) *with* $|V(G)| = n$ *and* $|E(G)| = m$.

However, due to the $\mathcal{O}(nm)$ summand in the complexity of Corollary 2, for a wide range of input instances the running time bound of Theorem 1 is better than the one of Corollary 2. In fact, the advantage of our branching approach is that the obtained dependency on the graph size in the running time is linear, whereas with the approach of [26], one needs to spend at least quadratic time either on computing weighted maximum matching or on kernelizing the instance.

In the full version [12] we also analyse the co-cluster setting, where one aims at obtaining a co-cluster graph instead of a cluster one, and show that the linear dependency on the size of the input can be maintained also in this case.

The paper is organised as follows. We give some preliminary definitions and notation in Section 2. In Section 3 we analyse the structural properties of the auxiliary graph H_v. Then, in Section 4 we prove Theorem 1, with the main tool being a subroutine branching algorithm finding all relevant vertex covers of H_v.

2   Preliminaries

We use standard graph notation. All our graphs are undirected and simple. For a graph G, by $V(G)$ and $E(G)$ we denote its vertex- and edge-set, respectively. For $v \in V(G)$, the set $N_G(v) = \{u \mid uv \in E(G)\}$ is the neighbourhood of v in G and $N_G[v] = N_G(v) \cup \{v\}$ is the closed neighbourhood. We extend these notions to sets of vertices $X \subseteq V(G)$ by $N_G[X] = \bigcup_{v \in X} N_G[v]$ and $N_G(X) = N_G[X] \setminus X$. We omit the subscript if it is clear from the context. For a set $X \subseteq V(G)$ we also define $G[X]$ to be the subgraph induced by X and $G \setminus X$ is a shorthand for $G[V(G) \setminus X]$. An even cycle is a cycle with an even number of edges, and an even path is a path with an even number of edges. A set $X \subseteq V(G)$ is called a *vertex cover* of G if $G \setminus X$ is edgeless. By $\text{MinV}(G)$ we denote the size of the minimum vertex cover of G.

In all further sections, we assume we are given an instance (G, k) of CLUSTER VERTEX DELETION, where $G = (V, E)$. That is, we use V and E to denote the vertex- and edge-set of the input instance G.

A P_3 is an ordered set of 3 vertices (u, v, w) such that $uv, vw \in E$ and $uw \notin E$. A graph is a cluster graph iff it does not contain any P_3; hence, in CLUSTERVD we seek for a set of at most k vertices that hits all P_3s. We note also the following.

Lemma 3. *Let G be a connected graph which is not a clique. Then, for every $v \in V(G)$, there is a P_3 containing v.*

Proof. Consider $N(v)$. If there exist vertices $u, w \in N(v)$ such that $uw \notin E(G)$ then we have a P_3 (u, v, w). Otherwise, since $N[v]$ induces a clique, we must have $w \in N(N[v])$ such that $uw \in E(G)$ for some $u \in N(v)$. Thus we have a P_3, (v, u, w) involving v.   □

If at some point a vertex v is fixed in the graph G, we define sets $N_1 = N_1(v)$ and $N_2 = N_2(v)$ as follows: $N_1 = N_G(v)$ and $N_2 = N_G(N_G[v])$. That is, N_1 and N_2 are sets of vertices at distance 1 and 2 from v, respectively. For a fixed $v \in V$, we define an auxiliary graph H_v with $V(H_v) = N_1 \cup N_2$ and

$$E(H_v) = \{uw \mid u, w \in N_1, uw \notin E\} \cup \{uw \mid u \in N_1, w \in N_2, uw \in E\}.$$

Thus, H_v consists of the vertices in N_1 and N_2 along with non-edges among vertices of N_1 and edges between N_1 and N_2. Note that N_2 is an independent set in H_v. Observe the following.

Lemma 4. *For $u, w \in N_1 \cup N_2$, we have $uw \in E(H_v)$ iff u, w and v form a P_3 in G.*

Proof. For every $uw \in E(H_v)$ with $u, w \in N_1$, (u, v, w) is a P_3 in G. For $uw \in E(H_v)$ with $u \in N_1$ and $w \in N_2$, (v, u, w) forms a P_3 in G. In the other direction, for any P_3 in G of the form (u, v, w) we have $u, w \in N_1$ and $uw \notin E$, thus $uw \in E(H_v)$. Finally, for any P_3 in G of the form (v, u, w) we have $u \in N_1$, $w \in N_2$ and $uw \in E$, hence $uw \in E(H_v)$.   □

We call a subset $S \subseteq V$ a *solution* when $G \setminus S$ is a cluster graph, that is, a collection of disjoint cliques. A solution with minimal cardinality is called a *minimum solution*.

Our algorithm is a typical branching algorithm, that is, it consists of a number of *branching steps*. In a step $(A_1, A_2, \ldots, A_r)$, $A_1, A_2, \ldots, A_r \subseteq V$, we independently consider r subcases. In the i-th subcase we look for a minimum solution S containing A_i: we delete A_i from the graph and decrease the parameter k

by $|A_i|$. If k becomes negative, we terminate the current branch and return a negative answer from the current subcase.

The *branching vector* for a step $(A_1, A_2, \ldots, A_r)$ is $(|A_1|, |A_2|, \ldots, |A_r|)$. It is well-known (see e.g. [18]) that the number of final subcases of a branching algorithm is bounded by $\mathcal{O}(c^k)$, where c is the largest positive root of the equation $1 = \sum_{i=1}^{r} x^{-|A_i|}$ among all branching steps $(A_1, A_2, \ldots, A_r)$ in the algorithm.

At some places, the algorithm makes a greedy (but optimal) choice of including a set $A \subseteq V$ into the constructed solution. We formally treat it as length-one branching step (A) with branching vector $(|A|)$.

3   The Auxiliary Graph H_v

In this section we investigate properties of the auxiliary graph H_v. Hence, we assume that a CLUSTERVD input (G, k) is given with $G = (V, E)$, and a vertex $v \in V$ is fixed.

3.1   Basic Properties

First, note that an immediate consequence of Lemma 4 is the following.

Corollary 5. *Let S be a solution such that $v \notin S$. Then S contains a vertex cover of H_v.*

In the other direction, the following holds.

Lemma 6. *Let X be a vertex cover of H_v. Then, in $G \setminus X$, the connected component of v is a clique.*

Proof. Suppose the connected component of v in $G \setminus X$ is not a clique. Then by Lemma 3, there is a P_3 involving v. Such a P_3 is also present in G. However, by Lemma 4, as X is a vertex cover of H_v, X intersects such a P_3, a contradiction. $\square$

Lemma 7. *Let S be a solution such that $v \notin S$. Denote by X the set $S \cap V(H_v)$. Let Y be a vertex cover of H_v. Suppose that $X \cap N_2 \subseteq Y \cap N_2$. Then $T := (S \setminus X) \cup Y$ is also a solution.*

Proof. Since Y (and hence, $T \cap V(H_v)$) is a vertex cover of H_v and $v \notin T$, we know by Lemma 6 that the connected component of v in $G \setminus T$ is a clique. If T is not a solution, then there must be a P_3 contained in $Z \setminus T$, where $Z = V \setminus (\{v\} \cup N_1)$. But since $S \cap Z \subseteq T \cap Z$, $G \setminus S$ would also contain such a P_3. $\square$

Lemma 7 motivates the following definition. For vertex covers of H_v, X and Y, we say that Y *dominates* X if $|Y| \leq |X|$, $Y \cap N_2 \supseteq X \cap N_2$ and at least one of these inequalities is sharp. Two vertex covers X and Y are said to be *equivalent* if $X \cap N_2 = Y \cap N_2$ and $|X \cap N_1| = |Y \cap N_1|$. We note that the first aforementioned relation is transitive and strongly anti-symmetric, whereas the second is an equivalence relation.

As a corollary of Lemma 7, we have:

Corollary 8. *Let S be a solution such that $v \notin S$. Suppose Y is a vertex cover of H_v which either dominates or is equivalent to the vertex cover $X = S \cap V(H_v)$. Then $T := (S \setminus X) \cup Y$ is also a solution with $|T| \leq |S|$.*

3.2 Special Cases of H_v

We now carefully study the cases where H_v has small vertex cover or has a special structure, and discover some possible greedy decisions that can be made.

Lemma 9. *Suppose X is a vertex cover of H_v. Then there is a minimum solution S such that either $v \notin S$ or $|X \setminus S| \geq 2$.*

Proof. Suppose S is a minimum solution such that $v \in S$ and $|X \setminus S| \leq 1$. We are going to convert S to another minimum solution T that does not contain v.

Consider $T := (S \setminus \{v\}) \cup X$. Clearly, $|T| \leq |S|$. Since T contains X, a vertex cover, by Lemma 6, the connected component of v in $G \setminus T$ is a clique. Thus, there is no P_3 containing v. Since any P_3 in $G \setminus T$ which does not include v must also be contained in $G \setminus S$, contradicting the fact that S is a solution, we obtain that T is also a solution. Hence, T is a minimum solution. □

Corollary 10. *If $MinV(H_v) = 1$ then there is a minimum solution S not containing v.*

Lemma 11. *Let C be the connected component of G containing v, and assume that neither C nor $C \setminus \{v\}$ is a cluster graph. If $X = \{w_1, w_2\}$ is a minimum vertex cover of H_v, then there exists a connected component $\widehat{C}$ of $G \setminus \{v\}$ that is not a clique and $\widehat{C} \cap \{w_1, w_2\} \neq \emptyset$.*

Proof. Assume the contrary. Consider a component $\widehat{C}$ of $C \setminus \{v\}$ which is not a clique. Since v must be adjacent to each connected component of $C \setminus \{v\}$, $\widehat{C} \cap N_1$ must be non-empty. For any $w \in \widehat{C} \cap N_1$, we have that $w_1, w_2 \neq w$ and $ww_1, ww_2 \notin E(G)$, since otherwise the result follows. If $uw \in E(G)$ with $u \in N_2$, then, as $\{w_1, w_2\}$ is a vertex cover of H_v we must have $u = w_1$ or $u = w_2$. We would then have w_1 or w_2 contained in a non-clique $\widehat{C}$, contradicting our assumption. Hence $uw \in E(G) \Rightarrow u \in N_1$. Thus $\widehat{C} \subseteq N_1$. As w_1 and w_2 are not contained in $\widehat{C}$ and they cover all edges in H_v, $\widehat{C}$ must be an independent set in H_v. In $G \setminus \{v\}$, therefore, $\widehat{C}$ must be a clique, a contradiction. □

We now investigate the case when H_v has a very specific structure. The motivation for this analysis will become clear in Section 4.3.

A *seagull* is a connected component of H_v that is isomorphic to a P_3 with middle vertex in N_1 and endpoints in N_2. The graph H_v is called an *s-skein* if it is a disjoint union of s seagulls and some isolated vertices.

Lemma 12. *Let $v \in V$. Suppose that H_v is an s-skein. Then there is a minimum solution S such that $v \notin S$.*

Proof. Let H_v consist of seagulls $(x_1, y_1, z_1), (x_2, y_2, z_2), \ldots, (x_s, y_s, z_s)$. That is, the middle vertices y_i are in N_1, while the endpoints x_i and z_i are in N_2. If $s = 1$, $\{y_1\}$ is a vertex cover of H_v and Corollary 10 yields the result. Henceforth, we assume $s \geq 2$.

Let X be the set N_1 with all the vertices isolated in H_v removed. Clearly, X is a vertex cover of H_v. Thus, we may use X as in Lemma 9 and obtain a minimum solution S. If $v \notin S$ we are done, so let us assume $|X \setminus S| \geq 2$. Take arbitrary i such that $y_i \in X \setminus S$. As $|X \setminus S| \geq 2$, we may pick another $j \neq i$, $y_j \in X \setminus S$. The crucial observation from the definition of H_v is that (y_j, y_i, x_i) and (y_j, y_i, z_i) are P_3s in G. As $y_i, y_j \notin S$, we have $x_i, z_i \in S$. Hence, since the choice of i was arbitrary, we infer that for each $1 \leq i \leq s$ either $y_i \in S$ or $x_i, z_i \in S$, and, consequently, S contains a vertex cover of H_v. By Lemma 6, $S \setminus \{v\}$ is also a solution in G, a contradiction.   □

4   Algorithm

In this section we show our algorithm for CLUSTERVD, proving Theorem 1. The algorithm is a typical branching algorithm, where at each step we choose one branching rule and apply it. In each subcase, a number of vertices is deleted, and the parameter k drops by this number. If k becomes negative, the current subcase is terminated with a negative answer. On the other hand, if k is non-negative and G is a cluster graph, the vertices deleted in this subcase form a solution of size at most k.

4.1   Preprocessing

At each step, we first preprocess simple connected components of G.

Lemma 13. *For each connected component C of G, in linear time, we can:*

1. *conclude that C is a clique; or*
2. *conclude that C is not a clique, but identify a vertex w such that $C \setminus \{w\}$ is a cluster graph; or*
3. *conclude that none of the above holds.*

Proof. On each connected component C, we perform a depth-first search. At every stage, we ensure that the set of already marked vertices induces a clique.

When we enter a new vertex, w, adjacent to a marked vertex v, we attempt to maintain the above invariant. We check if the number of marked vertices is equal to the number neighbours of w which are marked; if so then the new vertex w is marked. Since w is adjacent to every marked vertex, the set of marked vertices remains a clique. Otherwise, there is a marked vertex u such that $uw \notin E(G)$, and we may discover it by iterating once again over edges incident to w. In this case, we have discovered a P_3 (u, v, w) and C is not a clique. At least one of u, v, w must be deleted to make C into a cluster graph. We delete each one of them, and repeat the algorithm (without further recursion) to check if the remaining graph is a cluster graph. If one of the three possibilities returns a cluster graph, then (2) holds. Otherwise, (3) holds.

If we have marked all vertices in a component C while maintaining the invariant that marked vertices form a clique, then the component C is a clique.   □

For each connected component C that is a clique, we disregard C. For each connected component C that is not a clique, but $C \setminus \{w\}$ is a cluster graph for some w, we may greedily delete w from G: we need to delete at least one vertex from C, and w hits all P_3s in C. Thus, henceforth we assume that for each connected component C of G and for each $v \in V(C)$, $C \setminus \{v\}$ is not a cluster graph. In other words, we assume that we need to delete at least two vertices to solve each connected component of G.

4.2   Accessing H_v in Linear Time

Let us now fix a vertex $v \in V$ and let C be its connected component in G. Note that, as H_v contains parts of the complement of G, it may have size superlinear in the size of G. Therefore we now develop a simple oracle access to H_v that allows us to claim linear dependency on the graph size in the time bound.

Lemma 14. *Given a designated vertex $v \in V$, one can in linear time either compute a vertex w of degree at least 3 in H_v, together with its neighbourhood in H_v, or explicitly construct the graph H_v.*

Proof. First, mark vertices of N_1 and N_2. Second, for each vertex of G compute its number of neighbours in N_1 and N_2. This information, together with $|N_1|$, suffices to compute degrees of vertices in H_v. Hence, we may identify a vertex of degree at least 3 in H_v, if it exists. For such a vertex w, computing $N_{H_v}(w)$ takes time linear in the size of G. If no such vertex w exists, the complement of $G[N_1]$ has size linear in $|N_1|$ and we may construct H_v in linear time in a straightforward manner.   □

In the algorithm of Theorem 1, we would like to make a decision depending on the size of the minimum vertex cover of H_v. By the preprocessing step, C is not a clique, and by Lemma 3, H_v contains at least one edge, thus $\mathtt{MinV}(G) \geq 1$. We now note that we can find a small vertex cover of G in linear time.

Lemma 15. *In linear time, we can determine whether H_v has a minimum vertex cover of size 1, of size 2, or of size at least 3. Moreover, in the first two cases we can find the vertex cover in the same time bound.*

Proof. We use Lemma 14 to find, in linear time, a vertex w with degree at least 3, or generate H_v explicitly. In the latter case, H_v has vertices of degree at most 2, and it is straightforward to compute its minimum vertex cover in linear time.

If we find a vertex w of degree at least 3 in H_v, then w must be in any vertex cover of size at most 2. We proceed to delete w and restart the algorithm of Lemma 14 on the remaining graph to check if H_v in $G \setminus w$ has a vertex cover of size 0 or 1. We perform at most 2 such restarts. Finally, if we do not find a vertex cover of size at most 2, it must be the case that the minimum vertex cover contains at least 3 vertices.   □

4.3   Subroutine: Branching on H_v

We are now ready to present a branching algorithm that guesses the 'correct' vertex cover of H_v, for a fixed vertex v. That is, we are now working in the setting where we look for a minimum solution to CLUSTERVD on (G, k) not containing v, thus, by Corollary 5, containing a vertex cover of H_v. Our goal is to branch into a number of subcases, in each subcase picking a vertex cover of H_v. By Corollary 8, our branching algorithm, to be correct, needs only to generate at least one element from each equivalence class of the 'equivalent' relation, among maximal elements in the 'dominate' relation.

The algorithm consists of a number of branching steps; in each subcase of each step we take a number of vertices into the constructed vertex cover of H_v and, consequently, into the constructed minimum solution to CLUSTERVD on G. At any point, the first applicable rule is applied.

First, we disregard isolated vertices in H_v. Second, we take care of large-degree vertices.

Rule 1. *If there is a vertex $u \in V(H_v)$ with degree at least 3 in H_v, include either u or $N_{H_v}(u)$ into the vertex cover. That is, use the branching step $(u, N_{H_v}(u))$.*

Note that Rule 1 yields a branching vector $(1, d)$, where $d \geq 3$ is the degree of u in H_v. Henceforth, we can assume that vertices have degree 1 or 2 in H_v. Assume there exists $u \in N_1$ of degree 1, with $uw \in E(H_v)$. Moreover, assume there exists a minimum solution S containing u. If $w \in S$, then, by Lemma 7, $S \setminus \{u\}$ is also a solution, a contradiction. If $w \in N_2 \setminus S$, then $(S \setminus \{u\}) \cup \{w\}$ dominates S. Finally, if $w \in N_1 \setminus S$, then $(S \setminus \{u\}) \cup \{w\}$ is equivalent to S. Hence, we infer the following greedy rule.

Rule 2. *If there is a vertex $u \in N_1$ of degree 1 in H_v, include $N_{H_v}(u)$ into the vertex cover without branching. (Formally, use the branching step $(N_{H_v}(u))$.)*

Now we assume vertices in N_1 are of degree exactly 2 in H_v. Suppose we have vertices $u, w \in N_1$ with $uw \in E(H_v)$. We would like to branch on u as in Rule 1, including either u or $N_{H_v}(u)$ into the vertex cover. However, note that in the case where u is deleted, Rule 2 is triggered on w and consequently the other neighbour of w is deleted. Hence, we infer the following rule.

Rule 3. *If there are vertices $u, w \in N_1$, $uw \in E(H_v)$ then include either $N_{H_v}(w)$ or $N_{H_v}(u)$ into the vertex cover, that is, use the branching step $(N_{H_v}(w), N_{H_v}(u))$.*

Note that Rule 3 yields the branching vector $(2, 2)$.

We are left with the case where the maximum degree of H_v is 2, there are no edges with both endpoints in N_1, and no vertices of degree one in N_1. Hence H_v must be a collection of even cycles and even paths (recall that N_2 is an independent set in H_v). On each such cycle C, of $2l$ vertices, the vertices of N_1 and N_2 alternate. Note that we must use at least l vertices for the vertex cover of C. By Lemma 7 it is optimal to greedily select the l vertices in $C \cap N_2$.

Rule 4. *If there is an even cycle C in H_v with every second vertex in N_2, include $C \cap N_2$ into the vertex cover without branching. (Formally, use the branching step $(C \cap N_2)$.)*

For an even path P of length $2l$, we have two choices. If we are allowed to use $l+1$ vertices in the vertex cover of P, then, by Lemma 7, we may greedily take $P \cap N_2$. If we may use only l vertices, the minimum possible number, we need to choose $P \cap N_1$, as it is the unique vertex cover of size l of such path. Hence, we have an $(l, l+1)$ branch with our last rule.

Rule 5. *Take the longest possible even path P in H_v and either include $P \cap N_1$ or $P \cap N_2$ into the vertex cover. That is, use the branching step $(P \cap N_1, P \cap N_2)$.*

In Rule 5, we pick the longest possible path to avoid the branching vector $(1, 2)$ as long as possible; this is the worst branching vector in the algorithm of this section. Moreover, note that if we are forced to use the $(1, 2)$ branch, the graph H_v has a very specific structure.

Lemma 16. *If the algorithm of Section 4.3 may only use a branch with the branching vector $(1, 2)$, then H_v is an s-skein for some $s \geq 1$.*

We note that the statement of Lemma 16 is our sole motivation for introducing the notion of skeins and proving their properties in Lemma 12.

We conclude this section with an observation that the oracle access to H_v given by Lemma 14 allows us to execute a single branching step in linear time.

4.4 Main Algorithm

We are now ready to present our algorithm for Theorem 1. We assume the preprocessing (Lemma 13) is done. Pick an arbitrary vertex v. We first run the algorithm of Lemma 15 to determine if H_v has a small minimum vertex cover. Then we run the algorithm of Lemma 14 to check if H_v is an s-skein for some s.

We consider the following cases.

1. $\mathtt{MinV}(H_v) = 1$ **or** H_v **is an s-skein for some s.** Then, by Corollary 10 and Lemma 12, we know there exists a minimum solution not containing v. Hence, we run the algorithm of Section 4.3 on H_v.

2. $\mathtt{MinV}(H_v) = 2$ **and** H_v **is not a 2-skein.**[1] Assume the application of Lemma 15 returned a vertex cover $X = \{w_1, w_2\}$ of H_v. By Lemma 9, we may branch into the following two subcases: in the first we look for minimum solutions containing v and disjoint with X, and in the second, for minimum solutions not containing v.

   In the first case, we first delete v from the graph and decrease k by one. Then we check whether the connected component containing w_1 or w_2 is not a clique. By Lemma 11, for some $w \in \{w_1, w_2\}$, the connected component of $G \setminus \{v\}$ containing w is not a clique; finding such w clearly takes linear time. We invoke the algorithm of Section 4.3 on H_w.

   In the second case, we invoke the algorithm of Section 4.3 on H_v.

---

[1] Note that the size of a minimum vertex cover of an s-skein is exactly s, so this case is equivalent to '$\mathtt{MinV}(H_v) = 2$ and H_v is not an s-skein for any s'.

3. MinV(H_v) ≥ 3 and H_v is not an s-skein for any $s \geq 3$. We branch into
   two cases: we look for a minimum solution containing v or not containing v.
   In the first branch, we simply delete v and decrease k by one. In the second
   branch, we invoke the algorithm of Section 4.3 on H_v.

4.5   Complexity Analysis

In the previous discussion we have argued that invoking each branching step
takes linear time. As in each branch we decrease the parameter k by at least
one, the depth of the recursion is at most k. In this section we analyse branching
vectors occurring in our algorithm. To finish the proof of Theorem 1 we need
to show that the largest positive root of the equation $1 = \sum_{i=1}^{r} x^{-a_i}$ among all
possible branching vectors $(a_1, a_2, \ldots, a_r)$ is strictly less than 1.9102.

As the number of resulting branching vectors in the analysis is rather large,
we use a Python script for automated analysis[2]. The main reason for a large
number of branching vectors is that we need to analyse branchings on the graph
H_v in case when we consider v not to be included in the solution. Let us now
proceed with formal arguments.

Analysis of the Algorithm of Section 4.3. In a few places, the algorithm
of Section 4.3 is invoked on the graph H_v and we know that MinV(H_v) $\geq h$ for
some $h \in \{1, 2, 3\}$. Consider the branching tree $\mathbb{T}$ of this algorithm. For a node
$x \in V(\mathbb{T})$, the *depth* of x is the number of vertices of H_v deleted on the path
from x to the root. We mark some nodes of $\mathbb{T}$. Each node of depth less than h
is marked. Moreover, if a node x is of depth $d < h$ and the branching step at
node w has branching vector $(1, 2)$, we infer that graph H_v at this node is an
s-skein for some $s \geq h - d$, all descendants of x in $V(\mathbb{T})$ are also nodes with
branching steps with vectors $(1, 2)$. In this case, we mark all descendants of x
that are within distance (in $\mathbb{T}$) less than $h - d$. Note that in this way we may
mark some descendants of x of depth equal or larger than h.

We split the analysis of an application of the algorithm of Section 4.3 into two
phases: the first one contains all branching steps performed on marked nodes,
and the second on the remaining nodes. In the second phase, we simply observe
that each branching step has branching vector not worse than $(1, 2)$. In the first
phase, we aim to write a single branching vector summarizing the phase, so that
with its help we can balance the loss from other branches when v is deleted.

We remark that, although in the analysis we aggregate some branching steps
to prove better time bound, we always aggregate only a constant number of
branches (that is, we analyse the branching on marked vertices only for constant
h). Consequently, we maintain a linear dependency on the size of the graph in
the running time bound.

The main property of the marked nodes in $\mathbb{T}$ is that their existence is granted
by the assumption MinV(H_v) $\geq h$. That is, each leaf of $\mathbb{T}$ has depth at least

---

[2] Available at http://www.mimuw.edu.pl/~malcin/research/cvd and in the full version [12].

h, and, if at some node x of depth $d < h$ the graph H_v is an s-skein, we infer that $s \geq h - d$ (as the size of minimum vertex cover of an s-skein is s) and the algorithm performs s independent branching steps with branching vectors $(1, 2)$ in this case. Overall, no leaf of $\mathbb{T}$ is marked.

To analyse such branchings for $h = 2$ and $h = 3$ we employ the Python script. The procedure branch_Hv generates all possible branching vectors for the first phase, assuming the algorithm of Section 4.3 is allowed to pick branching vectors (1), $(1, 3)$, $(2, 2)$ or $(1, 2)$ (option allow_skein enables/disables the use of the $(1, 2)$ vector in the first branch). Note that all other vectors described in Section 4.3 may be simulated by applying a number of vectors (1) after one of the aforementioned branching vectors.

Analysis of the Algorithm of Section 4.4

Case 1. Here the algorithm of Section 4.3 performs branchings with vectors not worse than $(1, 2)$.

Case 2. If v is deleted, we apply the algorithm of Section 4.3 to H_w, yielding at least one branching step (as the connected component with w is not a clique). Hence, in this case the outcoming branching vector is any vector that came out of the algorithm of Section 4.3, with all entries increased by one (for the deletion of v). Recall that in the algorithm of Section 4.3, the worst branching vector is $(1, 2)$, corresponding to the case of H_w being a skein. Consequently, the words branching vector if v is deleted is $(2, 3)$.

If v is not deleted, the algorithm of Section 4.3 is applied to H_v. The script invokes the procedure branch_Hv on $h = 2$ and allow_skein=False to obtain a list of possible branching vectors. For each such vector, we append entries $(2, 3)$ from the subcase when v is deleted.

Case 3. The situation is analogous to the previous case. The script invokes the procedure branch_Hv on $h = 3$ and allow_skein=False to obtain a list of possible branching vectors. For each such vector, we append the entry (1) from the subcase when v is deleted.

Summary. We infer that the largest root of the equation $1 = \sum_{i=1}^{r} x^{-a_i}$ occurs for branching vector $(1, 3, 3, 4, 4, 5)$ and is less than 1.9102. This branching vector corresponds to Case 3 and the algorithm of Section 4.3, invoked on H_v, first performs a branching step with the vector $(1, 3)$ and in the branch with 1 deleted vertex, finds H_v to be a 2-skein and performs two independent branching steps with vectors $(1, 2)$.

This analysis concludes the proof of Theorem 1. We remark that the worst branching vector in Case 2 is $(2, 2, 3, 3, 3)$ (with solution $x < 1.8933$), corresponding to the case with single $(1, 2)$-branch when v is deleted and a 2-skein in the case when v is kept. Obviously, the worst case in Case 1 is the golden-ratio branch $(1, 2)$ with solution $x < 1.6181$.

5   Conclusions and Open Problems

We have presented a new branching algorithm for CLUSTER VERTEX DELETION. We hope our work will trigger a race for faster FPT algorithms for CLUSTERVD, as it was in the case of the famous VERTEX COVER problem.

Repeating after Hüffner et al. [26], we would like to re-pose here the question for a linear vertex-kernel for CLUSTERVD. As CLUSTERVD is a special case of the 3-HITTING SET problem, it admits an $\mathcal{O}(k^2)$-vertex kernel in the unweighted case and an $\mathcal{O}(k^3)$-vertex kernel in the weighted one [1, 2]. However, CLUSTER EDITING is known to admit a much smaller $2k$-vertex kernel, so there is a hope for a similar result for CLUSTERVD.

References

1. Abu-Khzam, F.N.: A kernelization algorithm for d-hitting set. Journal of Computer and System Sciences 76(7), 524–531 (2010)
2. Abu-Khzam, F.N., Fernau, H.: Kernels: Annotated, proper and induced. In: Bodlaender, H.L., Langston, M.A. (eds.) IWPEC 2006. LNCS, vol. 4169, pp. 264–275. Springer, Heidelberg (2006)
3. Ailon, N., Charikar, M., Newman, A.: Aggregating inconsistent information: Ranking and clustering. Journal of the ACM 55(5), 23:1–23:27 (2008)
4. Alon, N., Makarychev, K., Makarychev, Y., Naor, A.: Quadratic forms on graphs. In: Proceedings of STOC 2005, pp. 486–493. ACM (2005)
5. Bansal, N., Blum, A., Chawla, S.: Correlation clustering. Machine Learning 56, 89–113 (2004)
6. Ben-Dor, A., Shamir, R., Yakhini, Z.: Clustering gene expression patterns. Journal of Computational Biology 6(3/4), 281–297 (1999)
7. Böcker, S.: A golden ratio parameterized algorithm for cluster editing. Journal of Discrete Algorithms 16, 79–89 (2012)
8. Böcker, S., Briesemeister, S., Bui, Q.B.A., Truß, A.: Going weighted: Parameterized algorithms for cluster editing. Theoretical Computer Science 410(52), 5467–5480 (2009)
9. Böcker, S., Briesemeister, S., Klau, G.W.: Exact algorithms for cluster editing: Evaluation and experiments. Algorithmica 60(2), 316–334 (2011)
10. Böcker, S., Damaschke, P.: Even faster parameterized cluster deletion and cluster editing. Information Processing Letters 111(14), 717–721 (2011)
11. Bodlaender, H.L., Fellows, M.R., Heggernes, P., Mancini, F., Papadopoulos, C., Rosamond, F.A.: Clustering with partial information. Theoretical Computer Science 411(7-9), 1202–1211 (2010)
12. Boral, A., Cygan, M., Kociumaka, T., Pilipczuk, M.: Fast branching algorithm for cluster vertex deletion. CoRR, abs/1306.3877 (2013)
13. Cai, L.: Fixed-parameter tractability of graph modification problems for hereditary properties. Information Processing Letters 58(4), 171–176 (1996)
14. Charikar, M., Wirth, A.: Maximizing quadratic programs: Extending Grothendieck's inequality. In: Proceedings of FOCS 2004, pp. 54–60. IEEE Computer Society (2004)
15. Damaschke, P.: Fixed-parameter enumerability of cluster editing and related problems. Theory of Computing Systems 46(2), 261–283 (2010)

16. Fellows, M.R., Guo, J., Komusiewicz, C., Niedermeier, R., Uhlmann, J.: Graph-based data clustering with overlaps. Discrete Optimization 8(1), 2–17 (2011)
17. Fernau, H.: A top-down approach to search-trees: Improved algorithmics for 3-hitting set. Algorithmica 57(1), 97–118 (2010)
18. Fomin, F.V., Kratsch, D.: Exact Exponential Algorithms. Texts in theoretical computer science. Springer, Heidelberg (2010)
19. Fomin, F.V., Kratsch, S., Pilipczuk, M., Pilipczuk, M., Villanger, Y.: Tight bounds for parameterized complexity of cluster editing. In: Proceedings of STACS 2013. LIPIcs, vol. 20, pp. 32–43. Schloss Dagstuhl - Leibniz-Zentrum fuer Informatik, Leibniz-Zentrum fuer Informatik (2013)
20. Giotis, I., Guruswami, V.: Correlation clustering with a fixed number of clusters. Theory of Computing 2(1), 249–266 (2006)
21. Gramm, J., Guo, J., Hüffner, F., Niedermeier, R.: Automated generation of search tree algorithms for hard graph modification problems. Algorithmica 39(4), 321–347 (2004)
22. Gramm, J., Guo, J., Hüffner, F., Niedermeier, R.: Graph-modeled data clustering: Exact algorithms for clique generation. Theory of Computing Systems 38(4), 373–392 (2005)
23. Guo, J., Kanj, I.A., Komusiewicz, C., Uhlmann, J.: Editing graphs into disjoint unions of dense clusters. Algorithmica 61(4), 949–970 (2011)
24. Guo, J., Komusiewicz, C., Niedermeier, R., Uhlmann, J.: A more relaxed model for graph-based data clustering: s-plex cluster editing. SIAM Journal of Discrete Mathematics 24(4), 1662–1683 (2010)
25. Hüffner, F., Komusiewicz, C., Moser, H., Niedermeier, R.: Fixed-parameter algorithms for cluster vertex deletion. In: Laber, E.S., Bornstein, C., Nogueira, L.T., Faria, L. (eds.) LATIN 2008. LNCS, vol. 4957, pp. 711–722. Springer, Heidelberg (2008)
26. Hüffner, F., Komusiewicz, C., Moser, H., Niedermeier, R.: Fixed-parameter algorithms for cluster vertex deletion. Theory of Computing Systems 47(1), 196–217 (2010)
27. Komusiewicz, C.: Parameterized Algorithmics for Network Analysis: Clustering & Querying. PhD thesis, Technische Universität Berlin (2011), http://fpt.akt.tu-berlin.de/publications/diss-komusiewicz.pdf
28. Komusiewicz, C., Uhlmann, J.: Alternative parameterizations for cluster editing. In: Černá, I., Gyimóthy, T., Hromkovič, J., Jefferey, K., Královič, R., Vukolić, M., Wolf, S. (eds.) SOFSEM 2011. LNCS, vol. 6543, pp. 344–355. Springer, Heidelberg (2011)
29. Protti, F., da Silva, M.D., Szwarcfiter, J.L.: Applying modular decomposition to parameterized cluster editing problems. Theory of Computing Systems 44(1), 91–104 (2009)
30. Reed, B.A., Smith, K., Vetta, A.: Finding odd cycle transversals. Operations Research Letters 32(4), 299–301 (2004)
31. Shamir, R., Sharan, R., Tsur, D.: Cluster graph modification problems. Discrete Applied Mathematics 144(1-2), 173–182 (2004)

Separation Logic with One Quantified Variable*

Stéphane Demri[1], Didier Galmiche[2],
Dominique Larchey-Wendling[2], and Daniel Méry[2]

[1] New York University & CNRS
[2] LORIA – CNRS – University of Lorraine

Abstract. We investigate first-order separation logic with one record field restricted to a unique quantified variable (1SL1). Undecidability is known when the number of quantified variables is unbounded and the satisfiability problem is PSPACE-complete for the propositional fragment. We show that the satisfiability problem for 1SL1 is PSPACE-complete and we characterize its expressive power by showing that every formula is equivalent to a Boolean combination of atomic properties. This contributes to our understanding of fragments of first-order separation logic that can specify properties about the memory heap of programs with singly-linked lists. When the number of program variables is fixed, the complexity drops to polynomial time. All the fragments we consider contain the magic wand operator and first-order quantification over a single variable.

1 Introduction

Separation Logic for Verifying Programs with Pointers. Separation logic [20] is a well-known logic for analysing programs with pointers stemming from BI logic [14]. Such programs have specific errors to be detected and separation logic is used as an assertion language for Hoare-like proof systems [20] that are dedicated to verify programs manipulating heaps. Any procedure mechanizing the proof search requires subroutines that check the satisfiability or the validity of formulæ from the assertion language. That is why, characterizing the computational complexity of separation logic and its fragments and designing optimal decision procedures remain essential tasks. Separation logic contains a structural separating connective and its adjoint (the separating implication, also known as the magic wand). The main concern of the paper is to study a non-trivial fragment of first-order separation logic with one record field as far as expressive power, decidability and complexity are concerned. Herein, the models of separation logic are pairs made of a variable valuation (store) and a partial function with finite domain (heap), also known as memory states.

Decidability and Complexity. The complexity of satisfiability and model-checking problems for separation logic fragments have been quite studied [6,20,7] (see also new decidability results in [13] or undecidability results in [5,16] in an alternative setting). Separation logic is equivalent to a Boolean propositional logic [17,18] if first-order quantifiers are disabled. Separation logic without first-order quantifiers is decidable, but

---

* Work partially supported by the ANR grant DynRes (project no. ANR-11-BS02-011) and by the EU Seventh Framework Programme under grant agreement No. PIOF-GA-2011-301166 (DATAVERIF).

E.A. Hirsch et al. (Eds.): CSR 2014, LNCS 8476, pp. 125–138, 2014.

it becomes undecidable with first-order quantifiers [6]. For instance, model-checking and satisfiability for propositional separation logic are PSPACE-complete problems [6]. Decidable fragments with first-order quantifiers can be found in [11,4]. However, these known results crucially rely on the memory model addressing cells with two record fields (undecidability of 2SL in [6] is by reduction to the first-order theory of a finite *binary* relation). In order to study decidability or complexity issues for separation logic, two tracks have been observed in the literature. There is the verification approach with decision procedures for fragments of practical use, see e.g. [2,7,12]. Alternatively, fragments, extensions or variants of separation logic are considered from a logical viewpoint, see e.g. [6,5,16].

Our Contributions. In this paper, we study first-order separation logic with one quantified variable, with an unbounded number of program variables and with one record field (herein called 1SL1). We introduce *test formulæ* that state simple properties about the memory states and we show that every formula in 1SL1 is equivalent to a Boolean combination of test formulæ, extending what was done in [18,3] for the propositional case. For instance, separating connectives can be eliminated in a controlled way as well as first-order quantification over the single variable. In that way, we show a quantifier elimination property similar to the one for Presburger arithmetic. This result extends previous ones on propositional separation logic [17,18,3] and this is the first time that this approach is extended to a first-order version of separation logic with the magic wand operator. However, it is the best we can hope for since 1SL with two quantified variables and no program variables (1SL2) has been recently shown undecidable in [9]. Of course, other extensions of 1SL1 could be considered, for instance to add a bit of arithmetical constraints, but herein we focus on 1SL1 that is theoretically nicely designed, even though it is still unclear how much 1SL1 is useful for formal verification. We also establish that the satisfiability problem for Boolean combinations of test formulæ is NP-complete thanks to a saturation algorithm for the theory of memory states with test formulæ, paving the way to use SMT solvers to decide 1SL1 (see e.g. the use of such solvers in [19]).

Even though Boolean combinations of test formulæ and 1SL1 have identical expressive power, we obtain PSPACE-completeness for model-checking and satisfiability in 1SL1. The conciseness of 1SL1 explains the difference between these two complexities. PSPACE-completeness is still a relatively low complexity but this result can be extended with more than one record field (but still with one quantified variable). This is the best we can hope for with one quantified variable and with the magic wand, that is notoriously known to easily increase complexity. We also show that 1SL1 with a bounded number of program variables has a satisfiability problem that can be solved in polynomial time.

Omitted proofs can be found in the technical report [10].

2 Preliminaries

2.1 First-Order Separation Logic with One Selector 1SL

Let PVAR $= \{x_1, x_2, \ldots\}$ be a countably infinite set of *program variables* and FVAR $= \{u_1, u_2, \ldots\}$ be a countably infinite set of *quantified variables*. A *memory state* (also called a *model*) is a pair (s, h) such that s is a variable valuation of the form $s : \text{PVAR} \to$ $\mathbb{N}$ (the *store*) and h is a partial function $h : \mathbb{N} \to \mathbb{N}$ with finite domain (the *heap*) and we write $\text{dom}(h)$ to denote its *domain* and $\text{ran}(h)$ to denote its *range*. Two heaps h_1 and h_2 are said to be *disjoint*, noted $h_1 \perp h_2$, if their domains are disjoint; when this holds, we write $h_1 \uplus h_2$ to denote the heap corresponding to the disjoint union of the graphs of h_1 and h_2, hence $\text{dom}(h_1 \uplus h_2) = \text{dom}(h_1) \uplus \text{dom}(h_2)$. When the domains of h_1 and h_2 are not disjoint, the composition $h_1 \uplus h_2$ is not defined even if h_1 and h_2 have the same values on $\text{dom}(h_1) \cap \text{dom}(h_2)$.

Formulæ of 1SL are built from *expressions* of the form $e ::= x \mid u$ where $x \in \text{PVAR}$ and $u \in \text{FVAR}$, and *atomic formulæ* of the form $\pi ::= e = e' \mid e \hookrightarrow e' \mid \text{emp}$. *Formulæ* are defined by the grammar $\mathcal{A} ::= \perp \mid \pi \mid \mathcal{A} \wedge \mathcal{B} \mid \neg \mathcal{A} \mid \mathcal{A} * \mathcal{B} \mid \mathcal{A} \mathbin{-\!\!*} \mathcal{B} \mid \exists u\, \mathcal{A}$, where $u \in \text{FVAR}$. The connective $*$ is *separating conjunction* and $\mathbin{-\!\!*}$ is *separating implication*, usually called the *magic wand*. The *size* of a formula $\mathcal{A}$, written $|\mathcal{A}|$, is defined as the number of symbols required to write it. An *assignment* is a map $\mathfrak{f} : \text{FVAR} \to$ $\mathbb{N}$. The satisfaction relation $\models$ is parameterized by assignments (clauses for Boolean connectives are omitted):

- $(s, h) \models_{\mathfrak{f}} \text{emp}$ iff $\text{dom}(h) = \varnothing$.
- $(s, h) \models_{\mathfrak{f}} e = e'$ iff $[e] = [e']$, with $[x] \stackrel{\text{def}}{=} s(x)$ and $[u] \stackrel{\text{def}}{=} \mathfrak{f}(u)$.
- $(s, h) \models_{\mathfrak{f}} e \hookrightarrow e'$ iff $[e] \in \text{dom}(h)$ and $h([e]) = [e']$.
- $(s, h) \models_{\mathfrak{f}} \mathcal{A}_1 * \mathcal{A}_2$ iff $h = h_1 \uplus h_2$, $(s, h_1) \models_{\mathfrak{f}} \mathcal{A}_1$, $(s, h_2) \models_{\mathfrak{f}} \mathcal{A}_2$ for some h_1, h_2.
- $(s, h) \models_{\mathfrak{f}} \mathcal{A}_1 \mathbin{-\!\!*} \mathcal{A}_2$ iff for all h', if $h \perp h'$ & $(s, h') \models_{\mathfrak{f}} \mathcal{A}_1$ then $(s, h \uplus h') \models_{\mathfrak{f}} \mathcal{A}_2$.
- $(s, h) \models_{\mathfrak{f}} \exists u\, \mathcal{A}$ iff there is $l \in \mathbb{N}$ such that $(s, h) \models_{\mathfrak{f}[u \mapsto l]} \mathcal{A}$ where $\mathfrak{f}[u \mapsto l]$ is the assignment equal to $\mathfrak{f}$ except that u takes the value l.

Whereas '$\exists$' is clearly a first-order quantifier, the connectives $*$ and $\mathbin{-\!\!*}$ are known to be second-order quantifiers. In the paper, we show how to eliminate the three connectives when only one quantified variable is used.

We write 1SL0 to denote the propositional fragment of 1SL, i.e. without any occurrence of a variable from FVAR. Similarly, we write 1SL1 to denote the fragment of 1SL restricted to a single quantified variable, say u. In that case, the satisfaction relation can be denoted by $\models_l$ where l is understood as the value for the variable under the assignment.

Given $q \geqslant 1$ and $\mathcal{A}$ in 1SL built over $x_1, \ldots, x_q$, we define its *memory threshold* $\text{th}(q, \mathcal{A})$: $\text{th}(q, \mathcal{A}) \stackrel{\text{def}}{=} 1$ for atomic formula $\mathcal{A}$; $\text{th}(q, \mathcal{A}_1 \wedge \mathcal{A}_2) \stackrel{\text{def}}{=} \max(\text{th}(q, \mathcal{A}_1),$ $\text{th}(q, \mathcal{A}_2))$; $\text{th}(q, \neg \mathcal{A}_1) \stackrel{\text{def}}{=} \text{th}(q, \mathcal{A}_1)$; $\text{th}(q, \exists u\, \mathcal{A}_1) \stackrel{\text{def}}{=} \text{th}(q, \mathcal{A}_1)$; $\text{th}(q, \mathcal{A}_1 * \mathcal{A}_2) \stackrel{\text{def}}{=}$ $\text{th}(q, \mathcal{A}_1) + \text{th}(q, \mathcal{A}_2)$; $\text{th}(q, \mathcal{A}_1 \mathbin{-\!\!*} \mathcal{A}_2) \stackrel{\text{def}}{=} q + \max(\text{th}(q, \mathcal{A}_1), \text{th}(q, \mathcal{A}_2))$.

Lemma 1. *Given $q \geqslant 1$ and a formula $\mathcal{A}$ in 1SL, we have $1 \leqslant \text{th}(q, \mathcal{A}) \leqslant q \times |\mathcal{A}|$.*

Let $\mathcal{L}$ be a logic among 1SL, 1SL1 and 1SL0. As usual, the *satisfiability problem* for $\mathcal{L}$ takes as input a formula $\mathcal{A}$ from $\mathcal{L}$ and asks whether there is a memory state (s, h) and an assignment $\mathfrak{f}$ such that $(s, h) \models_{\mathfrak{f}} \mathcal{A}$. The *model-checking problem* for $\mathcal{L}$ takes as input a formula $\mathcal{A}$ from $\mathcal{L}$, a memory state (s, h) and an assignment $\mathfrak{f}$ for free variables from $\mathcal{A}$ and asks whether $(s, h) \models_{\mathfrak{f}} \mathcal{A}$. When checking the satisfiability status of a formula $\mathcal{A}$ in 1SL1, we assume that its program variables are contained in $\{x_1, \ldots, x_q\}$ for some $q \geqslant 1$ and the quantified variable is u. So, PVAR is unbounded but as usual, when dealing with a specific formula, the set of program variables is finite.

Theorem 2. *[6,4,9] Satisfiability and model-checking problems for 1SL0 are* PSPACE-*complete, satisfiability problem for 1SL is undecidable, even restricted to two variables.*

2.2 A Bunch of Properties Stated in 1SL1

The logic 1SL1 allows to express different types of properties on memory states. The examples below indeed illustrate the expressivity of 1SL1, and in the paper we characterize precisely what can be expressed in 1SL1.

- The domain of the heap has at least k elements: $\neg\mathrm{emp} * \cdots * \neg\mathrm{emp}$ (k times).
- The variable x_i is allocated in the heap: $\mathrm{alloc}(x_i) \stackrel{\mathrm{def}}{=} (x_i \hookrightarrow x_i) \rightarrow\!\!* \perp$.
- The variable x_i points to a location that is a loop: $\mathrm{toloop}(x_i) \stackrel{\mathrm{def}}{=} \exists\, u\, (x_i \hookrightarrow u \wedge u \hookrightarrow u)$; the variable x_i points to a location that is allocated: $\mathrm{toalloc}(x_i) \stackrel{\mathrm{def}}{=} \exists\, u\, (x_i \hookrightarrow u \wedge \mathrm{alloc}(u))$.
- Variables x_i and x_j point to a shared location: $\mathrm{conv}(x_i, x_j) \stackrel{\mathrm{def}}{=} \exists\, u\, (x_i \hookrightarrow u \wedge x_j \hookrightarrow u)$; there is a location between x_i and x_j: $\mathrm{inbetween}(x_i, x_j) \stackrel{\mathrm{def}}{=} \exists u\, (x_i \hookrightarrow u \wedge u \hookrightarrow x_j)$.
- Location interpreted by x_i has exactly one predecessor can be expressed in 1SL1: $(\exists\, u\, u \hookrightarrow x_i) \wedge \neg(\exists\, u\, u \hookrightarrow x_i * \exists\, u\, u \hookrightarrow x_i)$.
- Heap has at least 3 self-loops: $(\exists\, u\, u \hookrightarrow u) * (\exists\, u\, u \hookrightarrow u) * (\exists\, u\, u \hookrightarrow u)$.

2.3 At the Heart of Domain Partitions

Given (s, h) and a finite set $\mathcal{V} = \{x_1, \ldots, x_q\} \subseteq \mathrm{PVAR}$, we introduce two partitions of $\mathrm{dom}(h)$ depending on $\mathcal{V}$: one partition takes care of self-loops and predecessors of interpretations of program variables, the other one takes care of locations closely related to interpretations of program variables (to be defined below). This allows us to decompose the domain of heaps in such a way that we can easily identify the properties that can be indeed expressed in 1SL1 restricted to the variables in $\mathcal{V}$. We introduce a first partition of the domain of h by distinguishing the self-loops and the predecessors of variable interpretations on the one hand, and the remaining locations in the domain on the other hand: $\mathrm{pred}(s, h) \stackrel{\mathrm{def}}{=} \bigcup_i \mathrm{pred}(s, h, i)$ with $\mathrm{pred}(s, h, i) \stackrel{\mathrm{def}}{=} \{l' : h(l') = s(x_i)\}$ for every $i \in [1, q]$; $\mathrm{loop}(s, h) \stackrel{\mathrm{def}}{=} \{l \in \mathrm{dom}(h) : h(l) = l\}$; $\mathrm{rem}(s, h) \stackrel{\mathrm{def}}{=} \mathrm{dom}(h)\backslash(\mathrm{pred}(s, h) \cup \mathrm{loop}(s, h))$. So, obviously $\mathrm{dom}(h) = \mathrm{rem}(s, h) \uplus (\mathrm{pred}(s, h) \cup \mathrm{loop}(s, h))$. The sets $\mathrm{pred}(s, h)$ and $\mathrm{loop}(s, h)$ are not necessarily disjoint. As a consequence of h being a partial function, the sets $\mathrm{pred}(s, h, i)$ and $\mathrm{pred}(s, h, j)$ intersect only if $s(x_i) = s(x_j)$, in which case $\mathrm{pred}(s, h, i) = \mathrm{pred}(s, h, j)$.

We introduce a second partition of $\mathrm{dom}(h)$ by distinguishing the locations related to a cell involving a program variable interpretation on the one hand, and the remaining locations in the domain on the other hand. So, the sets below are also implicitly parameterized by $\mathcal{V}$: $\mathrm{ref}(s, h) \stackrel{\mathrm{def}}{=} \mathrm{dom}(h) \cap s(\mathcal{V})$, $\mathrm{acc}(s, h) \stackrel{\mathrm{def}}{=} \mathrm{dom}(h) \cap h(s(\mathcal{V}))$, $\heartsuit(s, h) \stackrel{\mathrm{def}}{=} \mathrm{ref}(s, h) \cup \mathrm{acc}(s, h)$, $\overline{\heartsuit}(s, h) \stackrel{\mathrm{def}}{=} \mathrm{dom}(h)\backslash\heartsuit(s, h)$. The *core* of the memory state, written $\heartsuit(s, h)$, contains the locations l in $\mathrm{dom}(h)$ such that either l is the interpretation of a program variable or it is an image by h of a program variable (that is also in the domain). In the sequel, we need to consider locations that belong to the intersection of sets from different partitions.

Here are their formal definitions:

- $\mathrm{pred}_{\overline{\heartsuit}}(s, h) \stackrel{\mathrm{def}}{=} \mathrm{pred}(s, h) \setminus \heartsuit(s, h)$,
  $\mathrm{pred}_{\overline{\heartsuit}}(s, h, i) \stackrel{\mathrm{def}}{=} \mathrm{pred}(s, h, i) \setminus \heartsuit(s, h)$,
- $\mathrm{loop}_{\overline{\heartsuit}}(s, h) \stackrel{\mathrm{def}}{=} \mathrm{loop}(s, h) \setminus \heartsuit(s, h)$,
  $\mathrm{rem}_{\overline{\heartsuit}}(s, h) \stackrel{\mathrm{def}}{=} \mathrm{rem}(s, h) \setminus \heartsuit(s, h)$.

For instance, $\mathrm{pred}_{\overline{\heartsuit}}(s, h)$ contains the set of locations l from $\mathrm{dom}(h)$, that are predecessors of a variable interpretation but no program variable x in $\{x_1, \ldots, x_q\}$ satisfies $s(x) = l$ or $h(s(x)) = l$ (which means $l \notin \heartsuit(s, h)$).

The above figure presents a memory state (s, h) with the variables $x_1, \ldots, x_4$. Nodes labelled by '$\heartsuit$' [resp. '$\circlearrowleft$', 'p', 'r'] belong to $\heartsuit(s, h)$ [resp. $\mathrm{loop}_{\overline{\heartsuit}}(s, h)$, $\mathrm{pred}_{\overline{\heartsuit}}(s, h)$, $\mathrm{rem}_{\overline{\heartsuit}}(s, h)$]. The introduction of the above sets provides a canonical way to decompose the heap domain, which will be helpful in the sequel.

Lemma 3 (Canonical decomposition). *For all stores s and heaps h, the following identity holds:* $\mathrm{dom}(h) = \heartsuit(s, h) \uplus \mathrm{pred}_{\overline{\heartsuit}}(s, h) \uplus \mathrm{loop}_{\overline{\heartsuit}}(s, h) \uplus \mathrm{rem}_{\overline{\heartsuit}}(s, h)$.

The proof is by easy verification since $\mathrm{pred}(s, h) \cap \mathrm{loop}(s, h) \subseteq \heartsuit(s, h)$.

Proposition 4. $\{\mathrm{pred}_{\overline{\heartsuit}}(s, h, i) \mid i \in [1, q]\}$ *is a partition of* $\mathrm{pred}_{\overline{\heartsuit}}(s, h)$.

Remark that both $\mathrm{pred}_{\overline{\heartsuit}}(s, h, i) = \varnothing$ or $\mathrm{pred}_{\overline{\heartsuit}}(s, h, i) = \mathrm{pred}_{\overline{\heartsuit}}(s, h, j)$ are possible. Below, we present properties about the canonical decomposition.

Proposition 5. *Let s, h, h_1, h_2 be such that $h = h_1 \uplus h_2$. Then,* $\heartsuit(s, h) \cap \mathrm{dom}(h_1) = \heartsuit(s, h_1) \uplus \Delta(s, h_1, h_2)$ *with* $\Delta(s, h_1, h_2) \stackrel{\mathrm{def}}{=} \mathrm{dom}(h_1) \cap h_2(s(V)) \cap \overline{s(V)} \cap \overline{h_1(s(V))}$ *(where $\overline{X} \stackrel{\mathrm{def}}{=} \mathbb{N} \setminus X$).*

The set $\Delta(s, h_1, h_2)$ contains the locations belonging to the core of h and to the domain of h_1, without being in the core of h_1. Its expression in Proposition 5 uses only basic set-theoretical operations. From Proposition 5, we conclude that $\heartsuit(s, h_1 \uplus h_2)$ can be different from $\heartsuit(s, h_1) \uplus \heartsuit(s, h_2)$.

2.4   How to Count in 1SL1

Let us define a formula that states that $\mathrm{loop}_{\overline{\heartsuit}}(s, h)$ has size at least k. First, we consider the following set of formulæ: $T_q = \{\mathtt{alloc}(x_1), \ldots, \mathtt{alloc}(x_q)\} \cup \{\mathtt{toalloc}(x_1), \ldots, \mathtt{toalloc}(x_q)\}$. For any map $\mathfrak{f} : T_q \to \{0, 1\}$, we associate a formula $\mathcal{A}_{\mathfrak{f}}$ defined by $\mathcal{A}_{\mathfrak{f}} \stackrel{\mathrm{def}}{=} \bigwedge \{\mathcal{B} \mid \mathcal{B} \in T_q \text{ and } \mathfrak{f}(\mathcal{B}) = 1\} \wedge \bigwedge \{\neg \mathcal{B} \mid \mathcal{B} \in T_q \text{ and } \mathfrak{f}(\mathcal{B}) = 0\}$. Formula $\mathcal{A}_{\mathfrak{f}}$ is a conjunction made of literals from T_q such that a *positive* literal $\mathcal{B}$ occurs exactly when $\mathfrak{f}(\mathcal{B}) = 1$ and a *negative* literal $\neg \mathcal{B}$ occurs exactly when $\mathfrak{f}(\mathcal{B}) = 0$. We write $\mathcal{A}_{\mathfrak{f}}^{\mathrm{pos}}$ to denote $\bigwedge \{\mathcal{B} \mid \mathcal{B} \in T_q \text{ and } \mathfrak{f}(\mathcal{B}) = 1\}$.

Let us define the formula $\# \mathtt{loop} \geq k$ by $(\exists u\, u \hookrightarrow u) * \cdots * (\exists u\, u \hookrightarrow u)$ (repeated k times). We can express that $\mathtt{loop}_{\overline{\heartsuit}}(s, h)$ has size at least k (where $k \geq 1$) with $\# \mathtt{loop}_{\overline{\heartsuit}} \geq k \stackrel{\text{def}}{=} \bigvee_{\mathfrak{f}} \mathcal{A}_{\mathfrak{f}} \wedge \left(\mathcal{A}_{\mathfrak{f}}^{\mathrm{pos}} * (\# \mathtt{loop} \geq k) \right)$, where $\mathfrak{f}$ spans over the finite set of maps $T_q \rightarrow \{0, 1\}$. So, the idea behind the construction of the formula is to divide the heap into two parts: one subheap contains the full core. Then, any loop in the other subheap is out of the core because of the separation.

Lemma 6. *(I) For any $k \geq 1$, there is a formula $\# \mathtt{loop}_{\overline{\heartsuit}} \geq k$ s.t. for any (s, h), we have $(s, h) \models \# \mathtt{loop}_{\overline{\heartsuit}} \geq k$ iff $\mathrm{card}(\mathtt{loop}_{\overline{\heartsuit}}(s, h)) \geq k$. (II) For any $k \geq 1$ and any $i \in [1, q]$, there is a formula $\# \mathtt{pred}_{\overline{\heartsuit}}^i \geq k$ s.t. for any (s, h), we have $(s, h) \models \# \mathtt{pred}_{\overline{\heartsuit}}^i \geq k$ iff $\mathrm{card}(\mathtt{pred}_{\overline{\heartsuit}}(s, h, i)) \geq k$. (III) For any $k \geq 1$, there is a $\# \mathtt{rem}_{\overline{\heartsuit}} \geq k$ s.t. for any (s, h), we have $(s, h) \models \# \mathtt{rem}_{\overline{\heartsuit}} \geq k$ iff $\mathrm{card}(\mathtt{rem}_{\overline{\heartsuit}}(s, h)) \geq k$.*

All formulae from the above lemma have threshold polynomial in $q + \alpha$.

3 Expressive Completeness

3.1 On Comparing Cardinalities: Equipotence

We introduce the notion of *equipotence* and state a few properties about it. This will be useful in the forthcoming developments. Let $\alpha \in \mathbb{N}$. We say that two finite sets X and Y are α-*equipotent* and we write $X \sim_\alpha Y$ if, either $\mathrm{card}(X) = \mathrm{card}(Y)$ or both $\mathrm{card}(X)$ and $\mathrm{card}(Y)$ are greater that α. The equipotence relation is also decreasing, i.e. $\sim_{\alpha_2} \subseteq \sim_{\alpha_1}$ holds for all $\alpha_1 \leq \alpha_2$. We state below two lemmas that will be helpful in the sequel.

Lemma 7. *Let $\alpha \in \mathbb{N}$ and X, X', Y, Y' be finite sets such that $X \cap X' = \varnothing, Y \cap Y' = \varnothing, X \sim_\alpha Y$ and $\mathrm{card}(X') = \mathrm{card}(Y')$ hold. Then $X \uplus X' \sim_\alpha Y \uplus Y'$ holds.*

Lemma 8. *Let $\alpha_1, \alpha_2 \in \mathbb{N}$ and X, X', Y_0 be finite sets s.t. $X \uplus X' \sim_{\alpha_1 + \alpha_2} Y_0$ holds. Then there are two finite sets Y, Y' s.t. $Y_0 = Y \uplus Y', X \sim_{\alpha_1} Y$ and $X' \sim_{\alpha_2} Y'$ hold.*

3.2 All We Need Is Test formulæ

Test formulæ express simple properties about the memory states; this includes properties about program variables but also global properties about numbers of predecessors or loops, following the decomposition in Section 2.3. These test formulæ allow us to characterize the expressive power of 1SL1, similarly to what has been done in [17,18,3] for 1SL0. Since every formula in 1SL1 is shown equivalent to a Boolean combination of test formulæ (forthcoming Theorem 19), this process can be viewed as a means to eliminate separating connectives in a controlled way; elimination is not total since the test formulæ require such separating connectives. However, this is analogous to quantifier elimination in Presburger arithmetic for which simple modulo constraints need to be introduced in order to eliminate the quantifiers (of course, modulo constraints are defined with quantifiers but in a controlled way too).

Let us introduce the *test formulæ*. We distinguish two types, leading to distinct sets. There are test formulæ that state properties about the direct neighbourhood of program variables whereas others state global properties about the memory states. The test formulæ of the form $\#\mathtt{pred}^i_{\heartsuit} \geq k$ are of these two types but they will be included in Size_α since these are cardinality constraints.

Definition 9 (Test formulæ). *Given $q, \alpha \geq 1$, we define sets of test formulæ:*

$$
\begin{aligned}
\mathsf{Equality} &\stackrel{\text{def}}{=} \{\mathtt{x}_i = \mathtt{x}_j \mid i, j \in [1, q]\} \\
\mathsf{Pattern} &\stackrel{\text{def}}{=} \{\mathtt{x}_i \hookrightarrow \mathtt{x}_j, \mathtt{conv}(\mathtt{x}_i, \mathtt{x}_j), \mathtt{inbetween}(\mathtt{x}_i, \mathtt{x}_j) \mid i, j \in [1, q]\} \\
&\cup \{\mathtt{toalloc}(\mathtt{x}_i), \mathtt{toloop}(\mathtt{x}_i), \mathtt{alloc}(\mathtt{x}_i) \mid i \in [1, q]\} \\
\mathsf{Extra}^{\mathrm{u}} &\stackrel{\text{def}}{=} \{\mathtt{u} \hookrightarrow \mathtt{u}, \mathtt{alloc}(\mathtt{u})\} \cup \{\mathtt{x}_i = \mathtt{u}, \mathtt{x}_i \hookrightarrow \mathtt{u}, \mathtt{u} \hookrightarrow \mathtt{x}_i \mid i \in [1, q]\} \\
\mathsf{Size}_\alpha &\stackrel{\text{def}}{=} \{\#\mathtt{pred}^i_{\heartsuit} \geq k \mid i \in [1, q], k \in [1, \alpha]\} \\
&\cup \{\#\mathtt{loop}_{\heartsuit} \geq k, \#\mathtt{rem}_{\heartsuit} \geq k \mid k \in [1, \alpha]\} \\
\mathsf{Basic} &\stackrel{\text{def}}{=} \mathsf{Equality} \cup \mathsf{Pattern} \qquad \mathsf{Test}_\alpha \stackrel{\text{def}}{=} \mathsf{Basic} \cup \mathsf{Size}_\alpha \cup \{\bot\} \\
\mathsf{Basic}^{\mathrm{u}} &\stackrel{\text{def}}{=} \mathsf{Basic} \cup \mathsf{Extra}^{\mathrm{u}} \qquad \mathsf{Test}^{\mathrm{u}}_\alpha \stackrel{\text{def}}{=} \mathsf{Test}_\alpha \cup \mathsf{Extra}^{\mathrm{u}}
\end{aligned}
$$

Test formulæ express simple properties about the memory states, even though quite large formulæ in 1SL1 may be needed to express them, while being of memory threshold polynomial in $q + \alpha$. An *atom* is a conjunction of test formulæ or their negation (literal) such that each formula from $\mathsf{Test}^{\mathrm{u}}_\alpha$ occurs once (saturated conjunction of literals). Any memory state satisfying an atom containing $\neg\mathtt{alloc}(\mathtt{x}_1) \wedge \neg\#\mathtt{pred}^1_{\heartsuit} \geq 1 \wedge \neg\#\mathtt{loop}_{\heartsuit} \geq 1 \wedge \neg\#\mathtt{rem}_{\heartsuit} \geq 1$ (with $q = 1$) has an empty heap.

Lemma 10. *Satisfiability of conjunctions of test formulæ or their negation can be checked in polynomial time (q and α are not fixed and the bounds k in test formulæ from Size_α are encoded in binary).*

The tedious proof of Lemma 10 is based on a saturation algorithm. The *size* of a Boolean combination of test formulæ is the number of symbols to write it, when integers are encoded in binary (those from Size_α). Lemma 10 entails the following complexity characterization, which indeed makes a contrast with the complexity of the satisfiability problem for 1SL1 (see Theorem 28).

Theorem 11. *Satisfiability problem for Boolean combinations of test formulæ in set $\bigcup_{\alpha \geq 1} \mathsf{Test}^{\mathrm{u}}_\alpha$ (q and α are not fixed, and bounds k are encoded in binary) is* NP-*complete*.

Checking the satisfiability status of a Boolean combination of test formulæ is typically the kind of tasks that could be performed by an SMT solver, see e.g. [8,1]. This is particularly true since no quantification is involved and test formulæ are indeed atomic formulæ about the theory of memory states.

Below, we introduce equivalence relations depending on whether memory states are indistinguishable w.r.t. some set of test formulæ.

Definition 12. *We say that (s, h, l) and (s', h', l') are basically equivalent [resp. extra equivalent, resp. α-equivalent] and we denote $(s, h, l) \simeq_b (s', h', l')$ [resp. $(s, h, l) \simeq_{\mathrm{u}} (s', h', l')$, resp. $(s, h, l) \simeq_\alpha (s', h', l')$] when the condition $(s, h) \models_l B$ iff $(s', h') \models_{l'} B$ is fulfilled for any $B \in \mathsf{Basic}^{\mathrm{u}}$ [resp. $B \in \mathsf{Extra}^{\mathrm{u}}$, resp. $B \in \mathsf{Test}^{\mathrm{u}}_\alpha$].*

Hence (s, h, l) and (s', h', l') are basically equivalent [resp. extra equivalent, resp. α-equivalent] if and only if they cannot be distinguished by the formulæ of Basic^u [resp. Extra^u, resp. Test^u_α]. Since $\text{Extra}^u \subseteq \text{Basic}^u \subseteq \text{Test}^u_\alpha$, it is obvious that the inclusions $\simeq_\alpha \subseteq \simeq_b \subseteq \simeq_u$ hold.

Proposition 13. $(s, h, l) \simeq_\alpha (s', h', l')$ *is equivalent to (1)* $(s, h, l) \simeq_b (s', h', l')$ *and (2)* $\text{pred}_{\overline{\heartsuit}}(s, h, i) \sim_\alpha \text{pred}_{\overline{\heartsuit}}(s', h', i)$ *for any* $i \in [1, q]$, *and (3)* $\text{loop}_{\overline{\heartsuit}}(s, h) \sim_\alpha \text{loop}_{\overline{\heartsuit}}(s', h')$ *and (4)* $\text{rem}_{\overline{\heartsuit}}(s, h) \sim_\alpha \text{rem}_{\overline{\heartsuit}}(s', h')$.

The proof is based on the identity $\text{Basic}^u \cup \text{Size}_\alpha = \text{Test}^u_\alpha$. The *pseudo-core* of (s, h), written $\mathfrak{p}\heartsuit(s, h)$, is defined as $\mathfrak{p}\heartsuit(s, h) = s(\mathcal{V}) \cup h(s(\mathcal{V}))$ and $\heartsuit(s, h)$ is equal to $\mathfrak{p}\heartsuit(s, h) \cap \text{dom}(h)$.

Lemma 14 (Bijection between pseudo-cores). *Let* $l_0, l_1 \in \mathbb{N}$ *and* (s, h) *and* (s', h') *be two memory states s.t.* $(s, h, l_0) \simeq_b (s', h', l_1)$. *Let* $\mathfrak{R}$ *be the binary relation on* $\mathbb{N}$ *defined by:* $l \mathfrak{R} l'$ *iff (a)* $[l = l_0$ *and* $l' = l_1]$ *or (b) there is* $i \in [1, q]$ *s.t.* $[l = s(\mathbf{x}_i)$ *and* $l' = s'(\mathbf{x}_i)]$ *or* $[l = h(s(\mathbf{x}_i))$ *and* $l' = h'(s'(\mathbf{x}_i))]$. *Then* $\mathfrak{R}$ *is a bijective relation between* $\mathfrak{p}\heartsuit(s, h) \cup \{l_0\}$ *and* $\mathfrak{p}\heartsuit(s', h') \cup \{l_1\}$. *Its restriction to* $\heartsuit(s, h)$ *is in bijection with* $\heartsuit(s', h')$ *too if case (a) is dropped out from definition of* $\mathfrak{R}$.

3.3 Expressive Completeness of 1SL1 with Respect to Test Formulæ

Lemmas 15, 16 and 17 below roughly state that the relation $\simeq_\alpha$ behaves properly. Each lemma corresponds to a given quantifier, respectively separating conjunction, magic wand and first-order quantifier. Lemma 15 below states how two equivalent memory states can be split, while loosing a bit of precision.

Lemma 15 (Distributivity). *Let us consider* s, h, h_1, h_2, s', h' *and* $\alpha, \alpha_1, \alpha_2 \geqslant 1$ *such that* $h = h_1 \uplus h_2$ *and* $\alpha = \alpha_1 + \alpha_2$ *and* $(s, h, l) \simeq_\alpha (s', h', l')$. *Then there exists* h'_1 *and* h'_2 *s.t.* $h' = h'_1 \uplus h'_2$ *and* $(s, h_1, l) \simeq_{\alpha_1} (s', h'_1, l')$ *and* $(s, h_2, l) \simeq_{\alpha_2} (s', h'_2, l')$.

Given (s, h), we write $\texttt{maxval}(s, h)$ to denote $\max(s(\mathcal{V}) \cup \text{dom}(h) \cup \text{ran}(h))$. Lemma 16 below states how it is possible to add subheaps while partly preserving precision.

Lemma 16. *Let* $\alpha, q \geqslant 1$ *and* $l_0, l'_0 \in \mathbb{N}$. *Assume that* $(s, h, l_0) \simeq_{q+\alpha} (s', h', l'_0)$ *and* $h_0 \bot h$. *Then there is* $h'_0 \bot h'$ *such that (1)* $(s, h_0, l_0) \simeq_\alpha (s', h'_0, l'_0)$ *(2)* $(s, h \uplus h_0, l_0) \simeq_\alpha (s', h' \uplus h'_0, l'_0)$; *(3)* $\texttt{maxval}(s', h'_0) \leqslant \texttt{maxval}(s', h') + l'_0 + 3(q + 1)\alpha + 1$.

Note the precision lost from $(s, h, l_0) \simeq_{q+\alpha} (s', h', l'_0)$ to $(s, h_0, l_0) \simeq_\alpha (s', h'_0, l'_0)$.

Lemma 17 (Existence). *Let* $\alpha \geqslant 1$ *and let us assume* $(s, h, l_0) \simeq_\alpha (s', h', l_1)$. *We have: (1) for every* $l \in \mathbb{N}$, *there is* $l' \in \mathbb{N}$ *such that* $(s, h, l) \simeq_u (s', h', l')$; *(2) for all* l, l', $(s, h, l) \simeq_u (s', h', l')$ *iff* $(s, h, l) \simeq_\alpha (s', h', l')$.

Now, we state the main property in the section, namely test formulæ provide the proper abstraction.

Lemma 18 (Correctness). *For any* $\mathcal{A}$ *in 1SL1 with at most* $q \geqslant 1$ *program variables, if* $(s, h, l) \simeq_\alpha (s', h', l')$ *and* $\texttt{th}(q, \mathcal{A}) \leqslant \alpha$ *then* $(s, h) \models_l \mathcal{A}$ *iff* $(s', h') \models_{l'} \mathcal{A}$.

The proof is by structural induction on $\mathcal{A}$ using Lemma 15, 16 and 17. Here is one of our main results characterizing the expressive power of 1SL1.

Theorem 19 (Quantifier Admissibility). *Every formula $\mathcal{A}$ in 1SL1 with q program variables is logically equivalent to a Boolean combination of test formulæ in* $\mathsf{Test}^{\mathrm{u}}_{\mathrm{th}(q,\mathcal{A})}$.

The proof of Theorem 19 does not provide a constructive way to eliminate quantifiers, which will be done in Section 4 (see Corollary 30).

Proof. Let $\alpha = \mathrm{th}(q,\mathcal{A})$ and consider the set of literals $\mathcal{S}_\alpha(s,h,l) \stackrel{\mathrm{def}}{=} \{\mathcal{B} \mid \mathcal{B} \in \mathsf{Test}^{\mathrm{u}}_\alpha$ and $(s,h) \models_l \mathcal{B}\} \cup \{\neg\mathcal{B} \mid \mathcal{B} \in \mathsf{Test}^{\mathrm{u}}_\alpha$ and $(s,h) \not\models_l \mathcal{B}\}$. As $\mathsf{Test}^{\mathrm{u}}_\alpha$ is finite, the set $\mathcal{S}_\alpha(s,h,l)$ is finite and let us consider the well-defined atom $\bigwedge \mathcal{S}_\alpha(s,h,l)$. We have $(s',h') \models_{l'} \bigwedge \mathcal{S}_\alpha(s,h,l)$ iff $(s,h,l) \simeq_\alpha (s',h',l')$. The disjunction $\mathcal{T}_{\mathcal{A}} \stackrel{\mathrm{def}}{=} \bigvee\{\bigwedge\mathcal{S}_\alpha(s,h,l) \mid (s,h) \models_l \mathcal{A}\}$ is a (finite) Boolean combination of test formulæ in $\mathsf{Test}^{\mathrm{u}}_\alpha$ because $\bigwedge \mathcal{S}_\alpha(s,h,l)$ ranges over the finite set of atoms built from $\mathsf{Test}^{\mathrm{u}}_\alpha$. By Lemma 18, we get that $\mathcal{A}$ is logically equivalent to $\mathcal{T}_{\mathcal{A}}$. $\square$

When $\mathcal{A}$ in 1SL1 has no free occurrence of u, one can show that $\mathcal{A}$ is equivalent to a Boolean combination of formulæ in $\mathsf{Test}_{\mathrm{th}(q,\mathcal{A})}$. Similarly, when $\mathcal{A}$ in 1SL1 has no occurrence of u at all, $\mathcal{A}$ is equivalent to a Boolean combination of formulæ of the form $\mathrm{x}_i = \mathrm{x}_j$, $\mathrm{x}_i \hookrightarrow \mathrm{x}_j$, $\mathtt{alloc}(\mathrm{x}_i)$ and $\#\,\mathtt{rem}_\heartsuit \geq k$ with the *alternative* definition $\heartsuit(s,h) = \{s(\mathrm{x}_i) : s(\mathrm{x}_i) \in \mathrm{dom}(h), i \in [1,q]\}$ (see also [17,18,3]). Theorem 19 witnesses that the test formulæ we introduced properly abstract memory states when 1SL1 formulæ are involved. Test formulæ from Definition 9 were not given to us and we had to design such formulæ to conclude Theorem 19. Let us see what the test formulæ satisfy. Above all, all the test formulæ can be expressed in 1SL1, see developments in Section 2.2 and Lemma 6. Then, we aim at avoiding redundancy among the test formulæ. Indeed, for any kind of test formulæ from $\mathsf{Test}^{\mathrm{u}}_\alpha$ leading to the subset $X \subseteq \mathsf{Test}^{\mathrm{u}}_\alpha$ (for instance $X = \{\#\,\mathtt{loop}_\heartsuit \geq k \mid k \leq \alpha\}$), there are (s,h), (s',h') and $l,l' \in \mathbb{N}$ such that (1) for every $\mathcal{B} \in \mathsf{Test}^{\mathrm{u}}_\alpha \backslash X$, we have $(s,h) \models_l \mathcal{B}$ iff $(s',h') \models_{l'} \mathcal{B}$ but (2) there is $\mathcal{B} \in X$ such that not $((s,h) \models_l \mathcal{B}$ iff $(s',h') \models_{l'} \mathcal{B})$. When $X = \{\#\,\mathtt{loop}_\heartsuit \geq k \mid k \leq \alpha\}$, clearly, the other test formulæ cannot systematically enforce constraints on the cardinality of the set of loops outside of the core. Last but not least, we need to prove that the set of test formulæ is expressively complete to get Theorem 19. Lemmas 15, 16 and 17 are helpful to obtain Lemma 18 taking care of the different quantifiers. It is in their proofs that the completeness of the set $\mathsf{Test}^{\mathrm{u}}_\alpha$ is best illustrated. Nevertheless, to apply these lemmas in the proof of Lemma 18, we designed the adequate definition for the function $\mathrm{th}(\cdot,\cdot)$ and we arranged different thresholds in their statements. So, there is a real interplay between the definition of $\mathrm{th}(\cdot,\cdot)$ and the lemmas used in the proof of Lemma 18.

A small model property can be also proved as a consequence of Theorem 19 and the proof of Lemma 10, for instance.

Corollary 20 (Small Model Property). *Let $\mathcal{A}$ be a formula in 1SL1 with q program variables. Then, if $\mathcal{A}$ is satisfiable, then there is a memory state (s,h) and $l \in \mathbb{N}$ such that $(s,h) \models_l \mathcal{A}$ and $\max(\mathtt{maxval}(s,h),l) \leq 3(q+1) + (q+3)\mathrm{th}(q,\mathcal{A})$.*

There is no need to count over $\text{th}(q, \mathcal{A})$ (e.g., for the loops outside the core) and the core uses at most $3q$ locations. Theorem 19 provides a characterization of the expressive power of 1SL1, which is now easy to differenciate from 1SL2.

Corollary 21. *1SL2 is strictly more expressive than 1SL1.*

4   Deciding 1SL1 Satisfiability and Model-Checking Problems

4.1   Abstracting Further Memory States

Satisfaction of $\mathcal{A}$ depends only on the satisfaction of formulæ from $\text{Test}^{\text{u}}_{\text{th}(q, \mathcal{A})}$. So, to check satisfiability of $\mathcal{A}$, there is no need to build memory states but rather only abstractions in which only the truth value of test formulæ matters. In this section we introduce *abstract memory states* and we show how it matches indistinguihability with respect to test formulae in $\text{Test}^{\text{u}}_{\alpha}$ (Lemma 24). Then, we use these abstract structures to design a model-checking decision procedure that runs in nondeterministic polynomial space.

Definition 22. *Let $q, \alpha \geqslant 1$. An* abstract memory state $\mathfrak{a}$ *over* (q, α) *is a structure* $((V, E), \mathfrak{l}, \mathfrak{r}, \mathfrak{p}_1, \ldots, \mathfrak{p}_q)$ *such that:*

1. *There is a partition P of $\{x_1, \ldots, x_q\}$ such that $P \subseteq V$. This encodes the store.*
2. *(V, E) is a functional directed graph and a node v in (V, E) is at distance at most two of some set of variables X in P. This allows to encode only the pseudo-core of memory states and nothing else.*
3. *$\mathfrak{l}, \mathfrak{p}_1, \ldots, \mathfrak{p}_q, \mathfrak{r} \in [0, \alpha]$ and this corresponds to the number of self-loops [resp. numbers of predecessors, number of remaining allocated locations] out of the core, possibly truncated over α. We require that if x_i and x_j belong to the same set in the partition P, then $\mathfrak{p}_i = \mathfrak{p}_j$.*

Given $q, \alpha \geqslant 1$, the number of abstract memory states over (q, α) is not only finite but reasonably bounded. Given (s, h), we define its *abstraction* $\text{abs}(s, h)$ over (q, α) as the abstract memory state $((V, E), \mathfrak{l}, \mathfrak{r}, \mathfrak{p}_1, \ldots, \mathfrak{p}_q)$ such that

- $\mathfrak{l} = \min(\text{loop}_{\overline{\heartsuit}}(s, h), \alpha)$, $\mathfrak{r} = \min(\text{rem}_{\overline{\heartsuit}}(s, h), \alpha)$, $\mathfrak{p}_i = \min(\text{pred}_{\overline{\heartsuit}}(s, h, i), \alpha)$ for every $i \in [1, q]$.
- P is a partition of $\{x_1, \ldots, x_q\}$ so that for all x, x', $s(x) = s(x')$ iff x and x' belong to the same set in P.
- V is made of elements from P as well as of locations from the set below:

$$(\{h(s(x_i)) : s(x_i) \in \text{dom}(h), i \in [1, q]\} \cup$$

$$\{h(h(s(x_i))) : h(s(x_i)) \in \text{dom}(h), i \in [1, q]\}) \backslash \{s(x_i) : i \in [1, q]\}$$

- The graph (V, E) is defined as follows:
  1. $(X, X') \in E$ if $X, X' \in P$ and $h(s(x)) = s(x')$ for some $x \in X, x' \in X'$.
  2. $(X, l) \in E$ if $X \in P$ and $h(s(x)) = l$ for some variable x in X and $l \notin \{s(x_i) : i \in [1, q]\}$.

3. $(l, l') \in E$ if there is a set $X \in P$ such that $(X, l) \in E$ and $h(l) = l'$ and $l' \notin \{s(x_i) : i \in [1, q]\}$.

4. $(l, X) \in E$ if there is $X' \in P$ such that $(X', l) \in E$ and $h(l) = s(x)$ for some $x \in X$ and $l \notin \{s(x_i) : i \in [1, q]\}$.

We define abstract memory states to be *isomorphic* if (1) the partition P is identical, (2) the finite digraphs satisfy the same formulæ from Basic when the digraphs are understood as heap graphs restricted to locations at distance at most two from program variables, and (3) all the numerical values are identical. A *pointed abstract memory state* is a pair $(\mathfrak{a}, \mathfrak{u})$ such that $\mathfrak{a} = ((V, E), \mathfrak{l}, \mathfrak{r}, \mathfrak{p}_1, \ldots, \mathfrak{p}_q)$ is an abstract memory state and $\mathfrak{u}$ takes one of the following values: $\mathfrak{u} \in V$ and $\mathfrak{u}$ is at distance at most one from some $X \in P$, or $\mathfrak{u} = L$ but $\mathfrak{l} > 0$ is required, or $\mathfrak{u} = R$ but $\mathfrak{r} > 0$ is required, or $\mathfrak{u} = P(i)$ for some $i \subset [1, q]$ but $\mathfrak{p}_i > 0$ is required, or $\mathfrak{u} = \overline{D}$. Given a memory state (s, h) and $l \in \mathbb{N}$, we define its *abstraction* $\text{abs}(s, h, l)$ with respect to (q, α) as the pointed abstract memory state $(\mathfrak{a}, \mathfrak{u})$ such that $\mathfrak{a} = \text{abs}(s, h)$ and

- $\mathfrak{u} \in V$ if either $l \in V$ and distance is at most one from some $X \in P$, or $\mathfrak{u} = X$ and there is $x \in X \in P$ such that $s(x) = l$,
- or $\mathfrak{u} \stackrel{\text{def}}{=} L$ if $l \in \text{loop}_{\overline{\heartsuit}}(s, h)$, or $\mathfrak{u} \stackrel{\text{def}}{=} R$ if $l \in \text{rem}_{\overline{\heartsuit}}(s, h)$,
- or $\mathfrak{u} \stackrel{\text{def}}{=} P(i)$ if $l \in \text{pred}_{\overline{\heartsuit}}(s, h, i)$ for some $i \in [1, q]$,
- or $\mathfrak{u} \stackrel{\text{def}}{=} \overline{D}$ if none of the above conditions applies (so $l \notin \text{dom}(h)$).

Pointed abstract memory states $(\mathfrak{a}, \mathfrak{u})$ and $(\mathfrak{a}', \mathfrak{u}')$ are *isomorphic* $\stackrel{\text{def}}{\Leftrightarrow}$ $\mathfrak{a}$ and $\mathfrak{a}'$ are isomorphic and, $\mathfrak{u} = \mathfrak{u}'$ or $\mathfrak{u}$ and $\mathfrak{u}'$ are related by the isomorphism.

Lemma 23. *Given a pointed abstract memory state $(\mathfrak{a}, \mathfrak{u})$ over (q, α), there exist a memory state (s, h) and $l \in \mathbb{N}$ such that $\text{abs}(s, h, l)$ and $(\mathfrak{a}, \mathfrak{u})$ are isomorphic*

Abstract memory states is the right way to abstract memory states when the language 1SL1 is involved, which can be formally stated as follows.

Lemma 24. *Let (s, h), (s', h') be memory states and $l, l' \in \mathbb{N}$. The next three propositions are equivalent: (1) $(s, h, l) \simeq_\alpha (s', h', l')$; (2) $\text{abs}(s, h, l)$ and $\text{abs}(s', h', l')$ are isomorphic; (3) there is a unique atom $\mathcal{B}$ from $\text{Test}_\alpha^\mathfrak{u}$ s.t. $(s, h) \models_l \mathcal{B}$ and $(s', h') \models_{l'} \mathcal{B}$.*

Equivalence between (1) and (3) is a consequence of the definition of the relation $\simeq_\alpha$. Hence, a pointed abstract memory state represents an atom of $\text{Test}_\alpha^\mathfrak{u}$, except that it is a bit more concise (only space in $\mathcal{O}(q + \log(\alpha))$ is required whereas an atom requires polynomial space in $q + \alpha$).

Definition 25. *Given pointed abstract memory states $(\mathfrak{a}, \mathfrak{u})$, $(\mathfrak{a}_1, \mathfrak{u}_1)$ and $(\mathfrak{a}_2, \mathfrak{u}_2)$, we write $*_a((\mathfrak{a}, \mathfrak{u}), (\mathfrak{a}_1, \mathfrak{u}_1), (\mathfrak{a}_2, \mathfrak{u}_2))$ if there exist $l \in \mathbb{N}$, a store s and disjoint heaps h_1 and h_2 such that $\text{abs}(s, h_1 \uplus h_2, l) = (\mathfrak{a}, \mathfrak{u})$, $\text{abs}(s, h_1, l) = (\mathfrak{a}_1, \mathfrak{u}_1)$ and $\text{abs}(s, h_2, l) = (\mathfrak{a}_2, \mathfrak{u}_2)$.*

The ternary relation $*_a$ is not difficult to check even though it is necessary to verify that the abstract disjoint union is properly done.

Lemma 26. *Given $q, \alpha \geq 1$, the ternary relation $*_a$ can be decided in polynomial time in $q + \log(\alpha)$ for all the pointed abstract memory states built over (q, α).*

1: **if** $\mathcal{B}$ is atomic **then** return $\mathrm{AMC}((\mathfrak{a}, \mathfrak{u}), \mathcal{B})$;
2: **if** $\mathcal{B} = \neg \mathcal{B}_1$ **then** return not $\mathrm{MC}((\mathfrak{a}, \mathfrak{u}), \mathcal{B}_1)$;
3: **if** $\mathcal{B} = \mathcal{B}_1 \wedge \mathcal{B}_2$ **then** return $(\mathrm{MC}((\mathfrak{a}, \mathfrak{u}), \mathcal{B}_1)$ and $\mathrm{MC}((\mathfrak{a}, \mathfrak{u}), \mathcal{B}_2))$;
4: **if** $\mathcal{B} = \exists\, \mathfrak{u}\, \mathcal{B}_1$ **then** return $\top$ iff there is $\mathfrak{u}'$ such that $\mathrm{MC}((\mathfrak{a}, \mathfrak{u}'), \mathcal{B}_1) = \top$;
5: **if** $\mathcal{B} = \mathcal{B}_1 * \mathcal{B}_2$ **then** return $\top$ iff there are $(\mathfrak{a}_1, \mathfrak{u}_1)$ and $(\mathfrak{a}_2, \mathfrak{u}_2)$ such that $*_a((\mathfrak{a}, \mathfrak{u}), (\mathfrak{a}_1, \mathfrak{u}_1), (\mathfrak{a}_2, \mathfrak{u}_2))$ and $\mathrm{MC}((\mathfrak{a}_1, \mathfrak{u}_1), \mathcal{B}_1) = \mathrm{MC}((\mathfrak{a}_2, \mathfrak{u}_2), \mathcal{B}_2) = \top$;
6: **if** $\mathcal{B} = \mathcal{B}_1 {\twoheadrightarrow} \mathcal{B}_2$ **then** return $\bot$ iff for some $(\mathfrak{a}', \mathfrak{u}')$ and $(\mathfrak{a}'', \mathfrak{u}'')$ such that $*_a((\mathfrak{a}'', \mathfrak{u}''), (\mathfrak{a}', \mathfrak{u}'), (\mathfrak{a}, \mathfrak{u})), \mathrm{MC}((\mathfrak{a}', \mathfrak{u}'), \mathcal{B}_1) = \top$ and $\mathrm{MC}((\mathfrak{a}'', \mathfrak{u}''), \mathcal{B}_2) = \bot$;

Fig. 1. Function $\mathrm{MC}((\mathfrak{a}, \mathfrak{u}), \mathcal{B})$

1: **if** $\mathcal{B}$ is emp **then** return $\top$ iff $E = \varnothing$ and all numerical values are zero;
2: **if** $\mathcal{B}$ is $\mathrm{x}_i = \mathrm{x}_j$ **then** return $\top$ iff $\mathrm{x}_i, \mathrm{x}_j \in X$, for some $X \in P$;
3: **if** $\mathcal{B}$ is $\mathrm{x}_i = \mathfrak{u}$ **then** return $\top$ iff $\mathfrak{u} = X$ for some $X \in P$ such that $\mathrm{x}_i \in X$;
4: **if** $\mathcal{B}$ is $\mathfrak{u} = \mathfrak{u}$ **then** return $\top$;
5: **if** $\mathcal{B}$ is $\mathrm{x}_i \hookrightarrow \mathrm{x}_j$ **then** return $\top$ iff $(X, X') \in E$ where $\mathrm{x}_i \in X \in P$ and $\mathrm{x}_j \in X \in P$;
6: **if** $\mathcal{B}$ is $\mathrm{x}_i \hookrightarrow \mathfrak{u}$ **then** return $\top$ iff $(X, \mathfrak{u}) \in E$ for some $X \in P$ such that $\mathrm{x}_i \in X$;
7: **if** $\mathcal{B}$ is $\mathfrak{u} \hookrightarrow \mathrm{x}_i$ **then** return $\top$ iff either $\mathfrak{u} = P(i)$ or ($\mathfrak{u} \in V$ and there is some $X \in P$ such that $\mathrm{x}_i \in X$ and $(\mathfrak{u}, X) \in E)$;
8: **if** $\mathcal{B}$ is $\mathfrak{u} \hookrightarrow \mathfrak{u}$ **then** return $\top$ iff either $\mathfrak{u} = L$ or $(\mathfrak{u}, \mathfrak{u}) \in E$;

Fig. 2. Function $\mathrm{AMC}((\mathfrak{a}, \mathfrak{u}), \mathcal{B})$

4.2   A Polynomial-Space Decision Procedure

Figure 1 presents a procedure $\mathrm{MC}((\mathfrak{a}, \mathfrak{u}), \mathcal{B})$ returning a Boolean value in $\{\bot, \top\}$ and taking as arguments, a pointed abstract memory state over (q, α) and a formula $\mathcal{B}$ with $\mathrm{th}(q, \mathcal{B}) \leqslant \alpha$. All the quantifications over pointed abstract memory states are done over (q, α). A case analysis is provided depending on the outermost connective. Its structure is standard and mimics the semantics for 1SL1 *except* that we deal with abstract memory states. The auxiliary function $\mathrm{AMC}((\mathfrak{a}, \mathfrak{u}), \mathcal{B})$ also returns a Boolean value in $\{\bot, \top\}$, makes no recursive calls and is dedicated to atomic formulæ (see Figure 2). The design of MC is similar to nondeterministic polynomial space procedures, see e.g. [15,6].

Lemma 27. *Let* $q, \alpha \geqslant 1$, $(\mathfrak{a}, \mathfrak{u})$ *be a pointed abstract memory state over* (q, α) *and* $\mathcal{A}$ *be in 1SL1 built over* $\mathrm{x}_1, \ldots, \mathrm{x}_q$ *s.t.* $\mathrm{th}(q, \mathcal{A}) \leqslant \alpha$. *The propositions below are equivalent: (I)* $\mathrm{MC}((\mathfrak{a}, \mathfrak{u}), \mathcal{A})$ *returns* $\top$; *(II) There exist* (s, h) *and* $l \in \mathbb{N}$ *such that* $\mathrm{abs}(s, h, l) = (\mathfrak{a}, \mathfrak{u})$ *and* $(s, h) \models_l \mathcal{A}$; *(III) For all* (s, h) *and* $l \in \mathbb{N}$ *s.t.* $\mathrm{abs}(s, h, l) = (\mathfrak{a}, \mathfrak{u})$, *we have* $(s, h) \models_l \mathcal{A}$.

Consequently, we get the following complexity characterization.

Theorem 28. *Model-checking and satisfiability pbs. for 1SL1 are* PSPACE-*complete.*

Below, we state two nice by-products of our proof technique.

Corollary 29. *Let* $q \geqslant 1$. *The satisfiability problem for 1SL1 restricted to formulæ with at most q program variables can be solved in polynomial time.*

Corollary 30. *Given a formula $\mathcal{A}$ in 1SL1, computing a Boolean combination of test formulæ in $\mathsf{Test}^{\mathsf{u}}_{\mathsf{th}(q,\mathcal{A})}$ logically equivalent to $\mathcal{A}$ can be done in polynomial space (even though the outcome formula can be of exponential size).*

Here is another by-product of our proof technique. The PSPACE bound is preserved when formulæ are encoded as DAGs instead of trees. The size of a formula is then simply its number of subformulæ. This is similar to machine encoding, provides a better conciseness and complexity upper bounds are more difficult to obtain. With this alternative notion of length, $\mathsf{th}(q, \mathcal{A})$ is only bounded by $q \times 2^{|\mathcal{A}|}$ (compare with Lemma 1). Nevertheless, this is fine to get PSPACE upper bound with this encoding since the algorithm to solve the satisfiability problem runs in logarithmic space in α, as we have shown previously.

5 Conclusion

In [4], the undecidability of 1SL with a unique record field is shown. 1SL0 is also known to be PSPACE-complete [6]. In this paper, we provided an extension with a unique quantified variable and we show that the satisfiability problem for 1SL1 is PSPACE-complete by presenting an original and fine-tuned abstraction of memory states. We proved that in 1SL1 separating connectives can be eliminated in a controlled way as well as first-order quantification over the single variable. In that way, we show a quantifier elimination property. Apart from the complexity results and the new abstraction for memory states, we also show a quite surprising result: when the number of program variables is bounded, the satisfiability problem can be solved in polynomial time. Last but not least, we have established that satisfiability problem for Boolean combinations of test formulæ is NP-complete. This is reminiscent of decision procedures used in SMT solvers and it is a challenging question to take advantage of these features to decide 1SL1 with an SMT solver. Finally, the design of fragments between 1SL1 and undecidable 1SL2 that can be decided with an adaptation of our method is worth being further investigated.

Acknowledgments. We warmly thank the anonymous referees for their numerous and helpful suggestions, improving significantly the quality of the paper and its extended version [10]. Great thanks also to Morgan Deters (New York University) for feedback and discussions about this work.

References

1. Barrett, C., Conway, C.L., Deters, M., Hadarean, L., Jovanović, D., King, T., Reynolds, A., Tinelli, C.: CVC4. In: Gopalakrishnan, G., Qadeer, S. (eds.) CAV 2011. LNCS, vol. 6806, pp. 171–177. Springer, Heidelberg (2011)
2. Berdine, J., Calcagno, C., O'Hearn, P.: Smallfoot: Modular automatic assertion checking with separation logic. In: de Boer, F.S., Bonsangue, M.M., Graf, S., de Roever, W.-P. (eds.) FMCO 2005. LNCS, vol. 4111, pp. 115–137. Springer, Heidelberg (2006)
3. Brochenin, R., Demri, S., Lozes, E.: Reasoning about sequences of memory states. APAL 161(3), 305–323 (2009)
4. Brochenin, R., Demri, S., Lozes, E.: On the almighty wand. IC 211, 106–137 (2012)

5. Brotherston, J., Kanovich, M.: Undecidability of propositional separation logic and its neigh-bours. In: LICS 2010, pp. 130–139. IEEE (2010)
6. Calcagno, C., Yang, H., O'Hearn, P.W.: Computability and complexity results for a spatial as-sertion language for data structures. In: Hariharan, R., Mukund, M., Vinay, V. (eds.) FSTTCS 2001. LNCS, vol. 2245, pp. 108–119. Springer, Heidelberg (2001)
7. Cook, B., Haase, C., Ouaknine, J., Parkinson, M., Worrell, J.: Tractable reasoning in a fragment of separation logic. In: Katoen, J.-P., König, B. (eds.) CONCUR 2011. LNCS, vol. 6901, pp. 235–249. Springer, Heidelberg (2011)
8. de Moura, L., Bjørner, N.: Z3: An efficient SMT solver. In: Ramakrishnan, C.R., Rehof, J. (eds.) TACAS 2008. LNCS, vol. 4963, pp. 337–340. Springer, Heidelberg (2008)
9. Demri, S., Deters, M.: Two-variable separation logic and its inner circle (September 2013) (submitted)
10. Demri, S., Galmiche, D., Larchey-Wendling, D., Mery, D.: Separation logic with one quan-tified variable. arXiv (2014)
11. Galmiche, D., Méry, D.: Tableaux and resource graphs for separation logic. JLC 20(1), 189–231 (2010)
12. Haase, C., Ishtiaq, S., Ouaknine, J., Parkinson, M.J.: SeLoger: A Tool for Graph-Based Rea-soning in Separation Logic. In: Sharygina, N., Veith, H. (eds.) CAV 2013. LNCS, vol. 8044, pp. 790–795. Springer, Heidelberg (2013)
13. Iosif, R., Rogalewicz, A., Simacek, J.: The tree width of separation logic with recursive definitions. In: Bonacina, M.P. (ed.) CADE 2013. LNCS, vol. 7898, pp. 21–38. Springer, Heidelberg (2013)
14. Ishtiaq, S., O'Hearn, P.: BI as an assertion language for mutable data structures. In: POPL 2001, pp. 14–26 (2001)
15. Ladner, R.: The computational complexity of provability in systems of modal propositional logic. SIAM Journal of Computing 6(3), 467–480 (1977)
16. Larchey-Wendling, D., Galmiche, D.: The undecidability of boolean BI through phase se-mantics. In: LICS 2010, pp. 140–149. IEEE (2010)
17. Lozes, E.: Expressivité des logiques spatiales. PhD thesis, LIP, ENS Lyon, France (2004)
18. Lozes, E.: Separation logic preserves the expressive power of classical logic. In: Workshop SPACE 2004 (2004)
19. Piskac, R., Wies, T., Zufferey, D.: Automating separation logic using SMT. In: Sharygina, N., Veith, H. (eds.) CAV 2013. LNCS, vol. 8044, pp. 773–789. Springer, Heidelberg (2013)
20. Reynolds, J.C.: Separation logic: a logic for shared mutable data structures. In: LICS 2002, pp. 55–74. IEEE (2002)

QuickXsort: Efficient Sorting with $n \log n - 1.399n + o(n)$ Comparisons on Average

Stefan Edelkamp[1] and Armin Weiß[2]

[1] TZI, Universität Bremen, Am Fallturm 1, D-28239 Bremen, Germany
[2] FMI, Universität Stuttgart, Universitätsstr. 38, D-70569 Stuttgart, Germany

Abstract. In this paper we generalize the idea of QUICKHEAPSORT leading to the notion of QUICKXSORT. Given some external sorting algorithm X, QUICKXSORT yields an internal sorting algorithm if X satisfies certain natural conditions. We show that up to $o(n)$ terms the average number of comparisons incurred by QUICKXSORT is equal to the average number of comparisons of X.

We also describe a new variant of WEAKHEAPSORT. With QUICKWEAK-HEAPSORT and QUICKMERGESORT we present two examples for the QUICKXSORT construction. Both are efficient algorithms that perform approximately $n \log n - 1.26n + o(n)$ comparisons on average. Moreover, we show that this bound also holds for a slight modification which guarantees an $n \log n + \mathcal{O}(n)$ bound for the worst case number of comparisons.

Finally, we describe an implementation of MERGEINSERTION and analyze its average case behavior. Taking MERGEINSERTION as a base case for QUICKMERGESORT, we establish an efficient internal sorting algorithm calling for at most $n \log n - 1.3999n + o(n)$ comparisons on average. QUICKMERGESORT with constant size base cases shows the best performance on practical inputs and is competitive to STL-INTROSORT.

Keywords: in-place sorting, quicksort, mergesort, analysis of algorithms.

1 Introduction

Sorting a sequence of n elements remains one of the most frequent tasks carried out by computers. A lower bound for sorting by only pairwise comparisons is $\log(n!) \approx n \log n - 1.44n + \mathcal{O}(\log n)$ comparisons for the worst and average case (logarithms denoted by log are always base 2, the average case refers to a uniform distribution of all input permutations assuming all elements are different). Sorting algorithms that are optimal in the leading term are called *constant-factor-optimal*. Tab. 1 lists some milestones in the race for reducing the coefficient in the linear term. One of the most efficient (in terms of number of comparisons) constant-factor-optimal algorithms for solving the sorting problem is Ford and Johnson's MERGEINSERTION algorithm [9]. It requires $n \log n - 1.329n + \mathcal{O}(\log n)$ comparisons in the worst case [12]. MERGEINSERTION has a severe drawback that makes it uninteresting for practical issues: similar to INSERTIONSORT the number of element moves is quadratic in n, i.e., it has quadratic running time.

E.A. Hirsch et al. (Eds.): CSR 2014, LNCS 8476, pp. 139–152, 2014.

With INSERTIONSORT we mean the algorithm that inserts all elements successively into the already ordered sequence finding the position for each element by binary search (*not* by linear search as frequently done). However, MERGEINSERTION and INSERTIONSORT can be used to sort small subarrays such that the quadratic running time for these subarrays is small in comparison to the overall running time. Reinhardt [15] used this technique to design an internal MERGESORT variant that needs in the worst case $n \log n - 1.329n + \mathcal{O}(\log n)$ comparisons. Unfortunately, implementations of this INPLACEMERGESORT algorithm have not been documented. Katajainen et al.'s [11,8] work inspired by Reinhardt is practical, but the number of comparisons is larger.

Throughout the text we avoid the terms *in-place* or *in-situ* and prefer the term *internal* (opposed to *external*). We call an algorithm *internal* if it needs at most $\mathcal{O}(\log n)$ space (computer words) in addition to the array to be sorted. That means we consider QUICKSORT as an internal algorithm whereas standard MERGESORT is external because it needs a linear amount of extra space.

Based on QUICKHEAPSORT [2], we develop the concept of QUICKXSORT in this paper and apply it to MERGESORT and WEAKHEAPSORT, what yields efficient internal sorting algorithms. The idea is very simple: as in QUICKSORT the array is partitioned into the elements greater and less than some pivot element. Then one part of the array is sorted by some algorithm X and the other part is sorted recursively. The advantage of this procedure is that, if X is an external algorithm, then in QUICKXSORT the part of the array which is not currently being sorted may be used as temporary space, what yields an internal variant of X. We give an elementary proof that under natural assumptions QUICKXSORT performs up to $o(n)$ terms on average the same number of comparisons as X. Moreover, we introduce a trick similar to INTROSORT [14] which guarantees $n \log n + \mathcal{O}(n)$ comparisons in the worst case.

The concept of QUICKXSORT (without calling it like that) was first applied in ULTIMATEHEAPSORT by Katajainen [10]. In ULTIMATEHEAPSORT, first the median of the array is determined, and then the array is partitioned into subarrays of equal size. Finding the median means significant additional effort. Cantone and Cincotti [2] weakened the requirement for the pivot and designed QUICKHEAPSORT which uses only a sample of smaller size to select the pivot for partitioning. ULTIMATEHEAPSORT is inferior to QUICKHEAPSORT in terms of average case number of comparisons, although, unlike QUICKHEAPSORT, it allows an $n \log n + \mathcal{O}(n)$ bound for the worst case number of comparisons. Diekert and Weiß [3] analyzed QUICKHEAPSORT more thoroughly and described some improvements requiring less than $n \log n - 0.99n + o(n)$ comparisons on average.

Edelkamp and Stiegeler [5] applied the idea of QUICKXSORT to WEAKHEAPSORT (which was first described by Dutton [4]) introducing QUICKWEAKHEAPSORT. The worst case number of comparisons of WEAKHEAPSORT is $n\lceil \log n \rceil - 2^{\lceil \log n \rceil} + n - 1 \leq n \log n + 0.09n$, and, following Edelkamp and Wegener [6], this bound is tight. In [5] an improved variant with $n \log n - 0.91n$ comparisons in the worst case and requiring extra space is presented. With EXTERNALWEAKHEAPSORT we propose a further refinement with the same worst case bound, but on

Table 1. Constant-factor-optimal sorting with $n \log n + \kappa n + o(n)$ comparisons

	Mem.	Other	κ Worst	κ Avg.	κ Exper.
Lower bound	$\mathcal{O}(1)$	$\mathcal{O}(n \log n)$	-1.44	-1.44	
BOTTOMUPHEAPSORT [16]	$\mathcal{O}(1)$	$\mathcal{O}(n \log n)$	$\omega(1)$	–	[0.35,0.39]
WEAKHEAPSORT [4,6]	$\mathcal{O}(n/w)$	$\mathcal{O}(n \log n)$	0.09	–	[-0.46,-0.42]
RELAXEDWEAKHEAPSORT [5]	$\mathcal{O}(n)$	$\mathcal{O}(n \log n)$	-0.91	-0.91	-0.91
MERGESORT [12]	$\mathcal{O}(n)$	$\mathcal{O}(n \log n)$	-0.91	-1.26	–
EXTERNALWEAKHEAPSORT #	$\mathcal{O}(n)$	$\mathcal{O}(n \log n)$	-0.91	-1.26*	–
INSERTIONSORT [12]	$\mathcal{O}(1)$	$\mathcal{O}(n^2)$	-0.91	-1.38 #	–
MERGEINSERTION [12]	$\mathcal{O}(n)$	$\mathcal{O}(n^2)$	-1.32	-1.3999 #	[-1.43,-1.41]
INPLACEMERGESORT [15]	$\mathcal{O}(1)$	$\mathcal{O}(n \log n)$	-1.32	–	–
QUICKHEAPSORT [2,3]	$\mathcal{O}(1)$	$\mathcal{O}(n \log n)$	$\omega(1)$	-0.03	≈ 0.20
	$\mathcal{O}(n/w)$	$\mathcal{O}(n \log n)$	$\omega(1)$	-0.99	≈ -1.24
QUICKMERGESORT (IS) #	$\mathcal{O}(\log n)$	$\mathcal{O}(n \log n)$	-0.32	-1.38	–
QUICKMERGESORT #	$\mathcal{O}(1)$	$\mathcal{O}(n \log n)$	-0.32	-1.26	[-1.29,-1.27]
QUICKMERGESORT (MI) #	$\mathcal{O}(\log n)$	$\mathcal{O}(n \log n)$	-0.32	-1.3999	[-1.41,-1.40]

Abbreviations: # established in this paper, MI MergeInsertion, – not analyzed, * for $n = 2^k$, w: computer word width in bits; we assume $\log n \in \mathcal{O}(n/w)$.
For QUICKXSORT we assume INPLACEMERGESORT as a worst-case stopper (without $\kappa_{\text{worst}} \in \omega(1)$). The column "Mem." exhibits the amount of computer words of memory needed additionally to the data. "Other" gives the amount of other operations than comparisons performed during sorting.

average requiring approximately $n \log n - 1.26n$ comparisons. Using EXTERNAL-WEAKHEAPSORT as X in QUICKXSORT we obtain an improvement over QUICK-WEAKHEAPSORT of [5].

MERGESORT is another good candidate for applying the QUICKXSORT construction. With QUICKMERGESORT we describe an internal variant of MERGE-SORT which not only in terms of number of comparisons competes with standard MERGESORT, but also in terms of running time. As mentioned before, MERGEIN-SERTION can be used to sort small subarrays. We study MERGEINSERTION and provide an implementation based on weak heaps. Furthermore, we give an average case analysis. When sorting small subarrays with MERGEINSERTION, we can show that the average number of comparisons performed by MERGESORT is bounded by $n \log n - 1.3999n + o(n)$, and, therefore, QUICKMERGESORT uses at most $n \log n - 1.3999n + o(n)$ comparisons in the average case. To our best knowledge this is better than any previously known bound.

The paper is organized as follows: in Sect. 2 the concept of QUICKXSORT is described and our main theorems about the average and worst case number of comparisons are stated. The following sections are devoted to present examples for X in QUICKXSORT: In Sect. 3 we develop EXTERNALWEAKHEAPSORT, analyze it, and show how it can be used for QUICKWEAKHEAPSORT. The next section treats QUICKMERGESORT and the modification that small base cases are sorted with some other algorithm, e.g. MERGEINSERTION, which is then described in Sect. 5. Finally, we present our experimental results in Sect. 6.

Due to space limitations most proofs can be found in the arXiv version [7].

2   QuickXsort

In this section we give a more precise description of QuickXsort and derive some results concerning the number of comparisons performed in the average and worst case. Let X be some sorting algorithm. QuickXsort works as follows: First, choose some pivot element as median of some random sample. Next, partition the array according to this pivot element, i. e., rearrange the array such that all elements left of the pivot are less or equal and all elements on the right are greater or equal than the pivot element. (If the algorithms X outputs the sorted sequence in the extra memory, the partitioning is performed such that the all elements left of the pivot are greater or equal and all elements on the right are less or equal than the pivot element.) Then, choose one part of the array and sort it with algorithm X. (The preferred choice depends on the sorting algorithm X.) After one part of the array has been sorted with X, move the pivot element to its correct position (right after/before the already sorted part) and sort the other part of the array recursively with QuickXsort.

The main advantage of this procedure is that the part of the array that is not being sorted currently can be used as temporary memory for the algorithm X. This yields fast *internal* variants for various *external* sorting algorithms such as MergeSort. The idea is that whenever a data element should be moved to the external storage, instead it is swapped with the data element occupying the respective position in part of the array which is used as temporary memory. Of course, this works only if the algorithm needs additional storage only for data elements. Furthermore, the algorithm has to be able to keep track of the positions of elements which have been swapped. As the specific method depends on the algorithm X, we give some more details when we describe the examples for QuickXsort.

For the number of comparisons we can derive some general results which hold for a wide class of algorithms X. Under natural assumptions the average number of comparisons of X and of QuickXsort differ only by an $o(n)$-term. For the rest of the paper, we assume that the pivot is selected as the median of approximately $\sqrt{n}$ randomly chosen elements. Sample sizes of approximately $\sqrt{n}$ are likely to be optimal as the results in [3,13] suggest.

The following theorem is one of our main results. It can be proved using Chernoff bounds and then solving the linear recurrence.

Theorem 1 (QuickXsort **Average-Case**). *Let X be some sorting algorithm requiring at most $n \log n + cn + o(n)$ comparisons in the average case. Then, QuickXsort implemented with $\Theta(\sqrt{n})$ elements as sample for pivot selection is a sorting algorithm that also needs at most $n \log n + cn + o(n)$ comparisons in the average case.*

Does QuickXsort provide a good bound for the worst case? The obvious answer is "no". If always the $\sqrt{n}$ smallest elements are chosen for pivot selection, $\Theta(n^{3/2})$ comparisons are performed. However, we can prove that such a worst case is very unlikely. Let $R(n)$ be the worst case number of comparisons of the algorithm X.

Proposition 1. *Let $\epsilon > 0$. The probability that* QUICKXSORT *needs more than* $R(n) + 6n$ *comparisons is less than* $(3/4 + \epsilon)^{\sqrt[4]{n}}$ *for n large enough.*

In order to obtain a provable bound for the worst case complexity we apply a simple trick similar to the one used in INTROSORT [14]. We fix some worst case efficient sorting algorithm Y. This might be, e. g., INPLACEMERGESORT. (In order to obtain an efficient internal sorting algorithm, Y has to be internal.) Worst case efficient means that we have a $n \log n + \mathcal{O}(n)$ bound for the worst case number of comparisons. We choose some slowly decreasing function $\delta(n) \in o(1) \cap \Omega(n^{-\frac{1}{4}+\epsilon})$, e. g., $\delta(n) = 1/\log n$. Now, whenever the pivot is more than $n \cdot \delta(n)$ off the median, we stop with QUICKXSORT and continue by sorting both parts of the partitioned array with the algorithm Y. We call this QUICKXYSORT. To achieve a good worst case bound, of course, we also need a good bound for algorithm X. W. l. o. g. we assume the same worst case bounds for X as for Y. Note that QUICKXYSORT only makes sense if one needs a provably good worst case bound. Since QUICKXSORT is always expected to make at most as many comparisons as QUICKXYSORT (under the reasonable assumption that X on average is faster than Y – otherwise one would use simply Y), in every step of the recursion QUICKXSORT is the better choice for the average case.

Theorem 2 (QUICKXYSORT **Worst-Case**). *Let X be a sorting algorithm with at most $n \log n + cn + o(n)$ comparisons in the average case and $R(n) = n \log n + dn + o(n)$ comparisons in the worst case $(d \geq c)$. Let Y be a sorting algorithm with at most $R(n)$ comparisons in the worst case. Then,* QUICKXYSORT *is a sorting algorithm that performs at most $n \log n + cn + o(n)$ comparisons in the average case and $n \log n + (d+1)n + o(n)$ comparisons in the worst case.*

In order to keep the the implementation of QUICKXYSORT simple, we propose the following algorithm Y: Find the median with some linear time algorithm (see e.g. [1]), then apply QUICKXYSORT with this median as first pivot element. Note that this algorithm is well defined because by induction the algorithm Y is already defined for all smaller instances. The proof of Thm. 2 shows that Y, and thus QUICKXYSORT, has a worst case number of comparisons in $n \log n + \mathcal{O}(n)$.

3 QUICKWEAKHEAPSORT

In this section consider QUICKWEAKHEAPSORT as a first example of QUICKXSORT. We start by introducing weak heaps and then continue by describing WEAKHEAPSORT and a novel external version of it. This external version is a good candidate for QUICKXSORT and yields an efficient sorting algorithm that uses approximately $n \log n - 1.2n$ comparisons (this value is only a rough estimate and neither a bound from below nor above). A drawback of WEAKHEAPSORT and its variants is that they require one extra bit per element. The exposition also serves as an intermediate step towards our implementation of MERGEINSERTION, where the weak-heap data structure will be used as a building block. Conceptually, a *weak heap* (see Fig. 1) is a binary tree satisfying the following conditions:

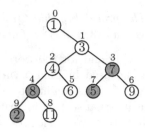

Fig. 1. A weak heap (reverse bits are set for grey nodes, above the nodes are array indices.)

(1) The root of the entire tree has no left child.
(2) Except for the root, the nodes that have at most one child are in the last two levels only. Leaves at the last level can be scattered, i.e., the last level is not necessarily filled from left to right.
(3) Each node stores an element that is smaller than or equal to every element stored in its right subtree.

From the first two properties we deduce that the height of a weak heap that has n elements is $\lceil \log n \rceil + 1$. The third property is called the *weak-heap ordering* or half-tree ordering. In particular, this property enforces no relation between an element in a node and those stored its left subtree. On the other hand, it implies that any node together with its right subtree forms a weak heap on its own. In an array-based implementation, besides the element array s, an array r of *reverse bits* is used, i.e., $r_i \in \{0,1\}$ for $i \in \{0,\ldots,n-1\}$. The root has index 0. The array index of the left child of s_i is $2i + r_i$, the array index of the right child is $2i + 1 - r_i$, and the array index of the parent is $\lfloor i/2 \rfloor$ (assuming that $i \neq 0$). Using the fact that the indices of the left and right children of s_i are exchanged when flipping r_i, subtrees can be reversed in constant time by setting $r_i \leftarrow 1 - r_i$. The *distinguished ancestor* (d-*ancestor*(j)) of s_j for $j \neq 0$, is recursively defined as the parent of s_j if s_j is a right child, and the distinguished ancestor of the parent of s_j if s_j is a left child. The distinguished ancestor of s_j is the first element on the path from s_j to the root which is known to be smaller or equal than s_j by (3). Moreover, any subtree rooted by s_j, together with the distinguished ancestor s_i of s_j, forms again a weak heap with root s_i by considering s_j as right child of s_i.

The basic operation for creating a weak heap is the *join* operation which combines two weak heaps into one. Let $i < j$ be two nodes in a weak heap such that s_i is smaller than or equal to every element in the left subtree of s_j. Conceptually, s_j and its right subtree form a weak heap, while s_i and the left subtree of s_j form another weak heap. (Note that s_i is not part of the subtree with root s_j.) The result of *join* is a weak heap with root at position i. If $s_j < s_i$, the two elements are swapped and r_j is flipped. As a result, the new element s_j will be smaller than or equal to every element in its right subtree, and the new element s_i will be smaller than or equal to every element in the subtree rooted at

s_j. To sum up, *join* requires constant time and involves one element comparison and a possible element swap in order to combine two weak heaps to a new one.

The construction of a weak heap consisting of n elements requires $n - 1$ comparisons. In the standard bottom-up construction of a weak heap the nodes are visited one by one. Starting with the last node in the array and moving to the front, the two weak heaps rooted at a node and its distinguished ancestor are joined. The amortized cost to get from a node to its distinguished ancestor is $\mathcal{O}(1)$ [6].

When using weak heaps for sorting, the minimum is removed and the weak heap condition restored until the weak heap becomes empty. After extracting an element from the root, first the *special path* from the root is traversed top-down, and then, in a bottom-up process the weak-heap property is restored using at most $\lceil \log n \rceil$ join operations. (The special path is established by going once to the right and then to the left as far as it is possible.) Hence, extracting the minimum requires at most $\lceil \log n \rceil$ comparisons.

Now, we introduce a modification to the standard procedure described by Dutton [4], which has a slightly improved performance, but requires extra space. We call this modified algorithm EXTERNALWEAKHEAPSORT. This is because it needs an extra output array, where the elements which are extracted from the weak heap are moved to. On average EXTERNALWEAKHEAPSORT requires less comparisons than RELAXEDWEAKHEAPSORT [5]. Integrated in QUICKXSORT we can implement it without extra space other than the extra bits r and some other extra bits. We introduce an additional array *active* and weaken the requirements of a weak heap: we also allow nodes on other than the last two levels to have less than two children. Nodes where the *active* bit is set to false are considered to have been removed. EXTERNALWEAKHEAPSORT works as follows: First, a usual weak heap is constructed using $n - 1$ comparisons. Then, until the weak heap becomes empty, the root – which is the minimal element – is moved to the output array and the resulting hole has to be filled with the minimum of the remaining elements (so far the only difference to normal WEAKHEAPSORT is that there is a separate output area).

The hole is filled by searching the special path from the root to a node x which has no left child. Note that the nodes on the special path are exactly the nodes having the root as distinguished ancestor. Finding the special path does not need any comparisons since one only has to follow the reverse bits. Next, the element of the node x is moved to the root leaving a hole. If x has a right subtree (i.e., if x is the root of a weak heap with more than one element), this hole is filled by applying the hole-filling algorithm recursively to the weak heap with root x. Otherwise, the *active* bit of x is set to false. Now, the root of the whole weak heap together with the subtree rooted by x forms a weak heap. However, it remains to restore the weak heap condition for the whole weak heap. Except for the root and x, all nodes on the special path together with their right subtrees form weak heaps. Following the special path upwards these weak heaps are joined with their distinguished ancestor as during the weak heap construction (i.e., successively they are joined with the weak heap consisting of the root and the already treated nodes on the special path together with their subtrees). Once, all the weak heaps on the special path are joined, the whole array forms a weak heap again.

Theorem 3. *For* $n = 2^k$ EXTERNALWEAKHEAPSORT *performs exactly the same comparisons as* MERGESORT *applied on a fixed permutation of the same input array.*

By [12, 5.2.4–13] we obtain the following corollary.

Corollary 1 (Average Case EXTERNALWEAKHEAPSORT**).** *For* $n = 2^k$ *the algorithm* EXTERNALWEAKHEAPSORT *uses approximately* $n \log n - 1.26n$ *comparisons in the average case.*

If n is not a power of two, the sizes of left and right parts of WEAKHEAPSORT are less balanced than the left and right parts of ordinary MERGESORT and one can expect a slightly higher number of comparisons. For QUICKWEAKHEAPSORT, the half of the array which is not sorted by EXTERNALWEAKHEAPSORT is used as output area. Whenever the root is moved to the output area, the element that occupied that place before is inserted as a dummy element at the position where the *active* bit is set to false. Applying Thm. 1, we obtain the rough estimate of $n \log n - 1.2n$ comparisons for the average case of QUICKWEAKHEAPSORT.

4   QUICKMERGESORT

As another example for QUICKXSORT we consider QUICKMERGESORT. For the MERGESORT part we use standard (top-down) MERGESORT which can be implemented using m extra spaces to merge two arrays of length m. After the partitioning, one part of the array – we assume the first part – has to be sorted with MERGESORT. In order to do so, the second half of this first part is sorted recursively with MERGESORT while moving the elements to the back of the whole array. The elements from the back of the array are inserted as dummy elements into the first part. Then, the first half the first part is sorted recursively with MERGESORT while being moved to the position of the former second part. Now, at the front of the array, there is enough space (filled with dummy elements) such that the two halves can be merged. The procedure is depicted in Fig. 2. As long as there is at least one third of the whole array as temporary memory left, the larger part of the partitioned array is sorted with MERGESORT, otherwise the smaller part is sorted with MERGESORT. Hence, the part which is not sorted by MERGESORT always provides enough temporary space. Whenever a data element is moved to or from the temporary space, it is swapped with the dummy element occupying the respective position. Since MERGESORT moves through the data from left to right, it is always clear which elements are the dummy elements. Depending on the implementation the extra space needed is $\mathcal{O}(\log n)$ words for the recursion stack of MERGESORT. By avoiding recursion this can be reduced to $\mathcal{O}(1)$. Thm. 1 together with [12, 5.2.4–13] yields the next result.

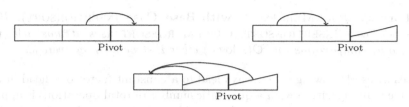

Fig. 2. First the two halves of the left part are sorted moving them from one place to another. Then, they are merged to the original place.

Theorem 4 (Average Case QUICKMERGESORT**).** QUICKMERGESORT *is an internal sorting algorithm that performs at most* $n \log n - 1.26n + o(n)$ *comparisons on average.*

We can do even better if we sort small subarrays with another algorithm Z requiring less comparisons but extra space and more moves, e. g., INSERTION-SORT or MERGEINSERTION. If we use $\mathcal{O}(\log n)$ elements for the base case of MERGESORT, we have to call Z at most $\mathcal{O}(n/\log n)$ times. In this case we can allow additional operations of Z like moves in the order of $\mathcal{O}(n^2)$ given that $\mathcal{O}((n/\log n) \cdot \log^2 n) = \mathcal{O}(n \log n)$. Note that for the next result we only need that the size of the base cases grows as n grows. Nevertheless, when applying an algorithm which uses $\Theta(n^2)$ moves, the size of the base cases has to be in $\mathcal{O}(\log n)$ in order to achieve an $\mathcal{O}(n \log n)$ overall running time.

Theorem 5 (QUICKMERGESORT **with Base Case).** *Let Z be some sorting algorithm with* $n \log n + en + o(n)$ *comparisons on average and other operations taking at most* $\mathcal{O}(n^2)$ *time. If base cases of size* $\mathcal{O}(\log n)$ *are sorted with Z,* QUICKMERGESORT *uses at most* $n \log n + en + o(n)$ *comparisons and* $\mathcal{O}(n \log n)$ *other instructions on average.*

Proof. By Thm. 1 and the preceding remark, the only thing we have to prove is that MERGESORT with base case Z requires on average at most $\leq n \log n + en + o(n)$ comparisons, given that Z needs $\leq U(n) = n \log n + en + o(n)$ comparisons on average. The latter means that for every $\epsilon > 0$ we have $U(n) \leq n \log n + (e+\epsilon) \cdot n$ for n large enough.

Let $S_k(m)$ denote the average case number of comparisons of MERGESORT with base cases of size k sorted with Z and let $\epsilon > 0$. Since $\log n$ grows as n grows, we have that $S_{\log n}(m) = U(m) \leq m \log m + (e + \epsilon) \cdot m$ for n large enough and $(\log n)/2 < m \leq \log n$. For $m > \log n$ we have $S_{\log n}(m) \leq 2 \cdot S_{\log n}(m/2) + m$ and by induction we see that $S_{\log n}(m) \leq m \log m + (e + \epsilon) \cdot m$. Hence, also $S_{\log n}(n) \leq n \log n + (e + \epsilon) \cdot n$ for n large enough. □

Recall that INSERTIONSORT inserts the elements one by one into the already sorted sequence by binary search. Using INSERTIONSORT we obtain the following result. Here, ln denotes the natural logarithm.

Proposition 2 (Average Case of INSERTIONSORT**).** *The sorting algorithm* INSERTIONSORT *needs* $n \log n - 2 \ln 2 \cdot n + c(n) \cdot n + \mathcal{O}(\log n)$ *comparisons on average where* $c(n) \in [-0.005, 0.005]$.

Corollary 2 (QuickMergesort **with Base Case** Insertionsort). *If we use as base case* Insertionsort, QuickMergesort *uses at most* $n \log n - 1.38n + o(n)$ *comparisons and* $\mathcal{O}(n \log n)$ *other instructions on average.*

Bases cases of growing size always lead to a constant factor overhead in running time if an algorithm with a quadratic number of total operations is applied. Therefore, in the experiments we also consider constant size base cases, which offer a slightly worse bound for the number of comparisons, but are faster in practice. We do not analyze them separately since the preferred choice for the size depends on the type of data to be sorted and the system on which the algorithms run.

5   MergeInsertion

MergeInsertion by Ford and Johnson [9] is one of the best sorting algorithms in terms of number of comparisons. Hence, it can be applied for sorting base cases of QuickMergesort what yields even better results than Insertionsort. Therefore, we want to give a brief description of the algorithm and our implementation. Algorithmically, MergeInsertion($s_0, \ldots, s_{n-1}$) can be described as follows (an intuitive example for $n = 21$ can be found in [12]):

1. Arrange the input such that $s_i \geq s_{i+\lfloor n/2 \rfloor}$ for $0 \leq i < \lfloor n/2 \rfloor$ with one comparison per pair. Let $a_i = s_i$ and $b_i = s_{i+\lfloor n/2 \rfloor}$ for $0 \leq i < \lfloor n/2 \rfloor$, and $b_{\lfloor n/2 \rfloor} = s_{n-1}$ if n is odd.
2. Sort the values $a_0, \ldots, a_{\lfloor n/2 \rfloor - 1}$ recursively with MergeInsertion.
3. Rename the solution as follows: $b_0 \leq a_0 \leq a_1 \leq \cdots \leq a_{\lfloor n/2 \rfloor - 1}$ and insert the elements $b_1, \ldots, b_{\lceil n/2 \rceil - 1}$ via binary insertion, following the ordering b_2, $b_1, b_4, b_3, b_{10}, b_9, \ldots, b_5, \ldots, b_{t_{k-1}}, b_{t_{k-1}-1}, \ldots b_{t_{k-2}+1}, b_{t_k}, \ldots$ into the main chain, where $t_k = (2^{k+1} + (-1)^k)/3$.

While the description is simple, MergeInsertion is not easy to implement efficiently because of the different renamings, the recursion, and the change of link structure. Our proposed implementation of MergeInsertion is based on a tournament tree representation with weak heaps as in Sect. 3. It uses $n \log n + n$ extra bits and works as follows: First, step 1 is performed for all recursion levels by constructing a weak heap. (Pseudo-code implementations for all the operations to construct a tournament tree with a weak heap and to access the partners in each round can be found in [7] – note that for simplicity in the above formulation the indices and the order are reversed compared to our implementation.) Then, in a second phase step 3 is executed for all recursion levels, see Fig. 3. One main subroutine of MergeInsertion is binary insertion. The call $binary\text{-}insert(x, y, z)$ inserts the element at position z between position $x - 1$ and $x + y$ by binary insertion. In this routine we do not move the data elements themselves, but we use an additional index array $\phi_0, \ldots, \phi_{n-1}$ to point to the elements contained in the weak heap tournament tree and move these indirect

procedure: $merge(m$: integer)
global: ϕ array of n integers imposed by weak-heap
for $l \leftarrow 0$ **to** $\lfloor m/2 \rfloor - 1$
$\quad | \quad \phi_{m-odd(m)-l-1} \leftarrow d\text{-}child(\phi_l, m - odd(m));$
$k \leftarrow 1; e \leftarrow 2^k; c \leftarrow f \leftarrow 0;$
while $e < m$
$\quad | \quad k \leftarrow k + 1; e \leftarrow 2e;$
$\quad | \quad l \leftarrow \lceil m/2 \rceil + f; f \leftarrow f + (t_k - t_{k-1});$
$\quad | \quad$ **for** $i \leftarrow 0$ **to** $(t_k - t_{k-1}) - 1$
$\quad | \quad | \quad c \leftarrow c + 1;$
$\quad | \quad | \quad$ **if** $c = \lceil m/2 \rceil$ **then**
$\quad | \quad | \quad | \quad$ **return**;
$\quad | \quad | \quad$ **if** $t_k > \lceil m/2 \rceil - 1$ **then**
$\quad | \quad | \quad | \quad binary\text{-}insert(i + 1 - odd(m), l, m - 1);$
$\quad | \quad | \quad$ **else**
$\quad | \quad | \quad | \quad binary\text{-}insert(\lfloor m/2 \rfloor - f + i, e - 1, \lfloor m/2 \rfloor + f);$

Fig. 3. Merging step in MERGEINSERTION with $t_k = (2^{k+1} + (-1)^k)/3$, $odd(m) = m \bmod 2$, and $d\text{-}child(\phi_i, n)$ returns the highest index less than n of a grandchild of ϕ_i in the weak heap (i.e, $d\text{-}child(\phi_i, n) =$ index of the bottommost element in the weak heap which has $d\text{-}ancestor = \phi_i$ and index $< n$)

addresses. This approach has the advantage that the relations stored in the tournament tree are preserved. The most important procedure for MERGEINSERTION is the organization of the calls for $binary\text{-}insert$. After adapting the addresses for the elements b_i (w. r. t. the above description) in the second part of the array, the algorithm calls the binary insertion routine with appropriate indices. Note that we always use k comparisons for all elements of the k-th block (i. e., the elements $b_{t_k}, \ldots, b_{t_{k-1}+1}$) even if there might be the chance to save one comparison. By introducing an additional array, which for each b_i contains the current index of a_i, we can exploit the observation that not always k comparisons are needed to insert an element of the k-th block. In the following we call this the *improved* variant. The pseudo-code of the basic variant is shown in Fig. 3. The last sequence is not complete and is thus tackled in a special case.

Theorem 6 (Average Case of MERGEINSERTION). *The sorting algorithm* MERGEINSERTION *needs* $n \log n - c(n) \cdot n + \mathcal{O}(\log n)$ *comparisons on average, where* $c(n) \geq 1.3999$.

Corollary 3 (QUICKMERGESORT with Base Case MERGEINSERTION). *When using* MERGEINSERTION *as base case,* QUICKMERGESORT *needs at most* $n \log n - 1.3999n + o(n)$ *comparisons and* $\mathcal{O}(n \log n)$ *other instructions on average.*

6   Experiments

Our experiments consist of two parts. First, we compare the different algorithms we use as base cases, i. e., MERGEINSERTION, its improved variant, and INSERTIONSORT. The results can be seen in Fig. 4. Depending on the size of the arrays the displayed numbers are averages over 10-10000 runs[1]. The data elements we sorted were randomly chosen 32-bit integers. The number of comparisons was measured by increasing a counter in every comparison[2].

The outcome in Fig. 4 shows that our improved MERGEINSERTION implementation achieves results for the constant κ of the linear term in the range of $[-1.43, -1.41]$ (for some values of n are even smaller than -1.43). Moreover, the standard implementation with slightly more comparisons is faster than INSERTIONSORT. By the $\mathcal{O}(n^2)$ work, the resulting runtimes for all three implementations raise quickly, so that only moderate values of n can be handled.

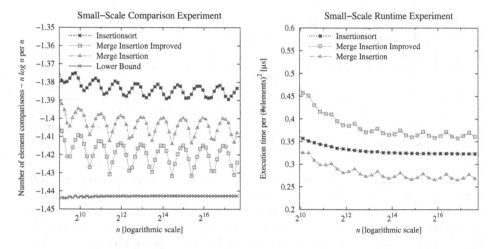

Fig. 4. Comparison of MERGEINSERTION, its improved variant and INSERTIONSORT. For the number of comparisons $n \log n + \kappa n$ the value of κ is displayed.

The second part of our experiments (shown in Fig. 5) consists of the comparison of QUICKMERGESORT (with base cases of constant and growing size) and QUICKWEAKHEAPSORT with state-of-the-art algorithms as STL-INTROSORT (i. e., QUICKSORT), STL-STABLE-SORT (BOTTOMUPMERGESORT) and QUICKSORT with median of $\sqrt{n}$ elements for pivot selection. For QUICKMERGESORT with base cases, the improved variant of MERGEINSERTION is used to sort subarrays of size up to $40 \log_{10} n$. For the normal QUICKMERGESORT we used base cases of size ≤ 9. We also implemented QUICKMERGESORT with median of three for pivot selection, which turns out to be practically efficient, although it needs slightly more comparisons than QUICKMERGESORT with median of $\sqrt{n}$. However,

---

[1] Our experiments were run on one core of an Intel Core i7-3770 CPU (3.40GHz, 8MB Cache) with 32GB RAM; Operating system: Ubuntu Linux 64bit; Compiler: GNU's g++ (version 4.6.3) optimized with flag -O3.

[2] To rely on objects being handled we avoided the flattening of the array structure by the compiler. Hence, for the running time experiments, and in each comparison taken, we left the counter increase operation intact.

since also the larger half of the partitioned array can be sorted with MERGESORT, the difference to the median of $\sqrt{n}$ version is not as big as in QUICKHEAPSORT [3]. As suggested by the theory, we see that our improved QUICKMERGESORT implementation with growing size base cases MERGEINSERTION yields a result for the constant in the linear term that is in the range of $[-1.41, -1.40]$ – close to the lower bound. However, for the running time, normal QUICKMERGESORT as well as the STL-variants INTROSORT (std::sort) and BOTTOMUPMERGE-SORT (std::stable_sort) are slightly better. With about 15% the time gap, however, is not overly big, and may be bridged with additional optimizations. Also, when comparisons are more expensive, QUICKMERGESORT performs faster than INTROSORT and BOTTOMUPMERGESORT, see the arXiv version [7].

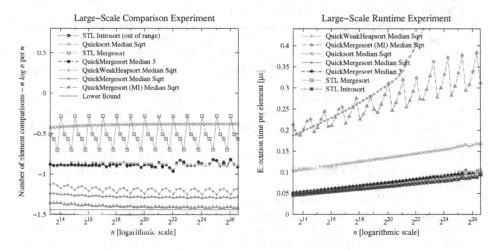

Fig. 5. Comparison of QUICKMERGESORT (with base cases of constant and growing size) and QUICKWEAKHEAPSORT with other sorting algorithms; (MI) is short for including growing size base cases derived from MERGEINSERTION. For the number of comparisons $n \log n + \kappa n$ the value of κ is displayed.

7   Concluding Remarks

Sorting n elements remains a fascinating topic for computer scientists both from a theoretical and from a practical point of view. With QUICKXSORT we have described a procedure how to convert an external sorting algorithm into an internal one introducing only $o(n)$ additional comparisons on average. We presented QUICKWEAKHEAPSORT and QUICKMERGESORT as two examples for this construction. QUICKMERGESORT is close to the lower bound for the average number of comparisons and at the same time is practically efficient, even when the comparisons are fast.

Using MERGEINSERTION to sort base cases of growing size for QUICKMERGE-SORT, we derive an an upper bound of $n \log n - 1.3999n + o(n)$ comparisons for the average case. As far as we know a better result has not been published before. We emphasize that the average of our best implementation has a proven gap of at most $0.05n + o(n)$ comparisons to the lower bound. The value $n \log n - 1.4n$ for $n = 2^k$ matches one side of Reinhardt's conjecture that an optimized in-place algorithm can have $n \log n - 1.4n + \mathcal{O}(\log n)$ comparisons in the average [15]. Moreover, our experimental results validate the theoretical considerations and indicate that the factor -1.43 can be beaten. Of course, there is still room in closing the gap to the lower bound of $n \log n - 1.44n + \mathcal{O}(\log n)$ comparisons.

References

1. Blum, M., Floyd, R.W., Pratt, V., Rivest, R.L., Tarjan, R.E.: Time bounds for selection. J. Comput. Syst. Sci. 7(4), 448–461 (1973)
2. Cantone, D., Cinotti, G.: QuickHeapsort, an efficient mix of classical sorting algorithms. Theoretical Computer Science 285(1), 25–42 (2002)
3. Diekert, V., Weiß, A.: Quickheapsort: Modifications and improved analysis. In: Bulatov, A.A., Shur, A.M. (eds.) CSR 2013. LNCS, vol. 7913, pp. 24–35. Springer, Heidelberg (2013)
4. Dutton, R.D.: Weak-heap sort. BIT 33(3), 372–381 (1993)
5. Edelkamp, S., Stiegeler, P.: Implementing HEAPSORT with $n \log n - 0.9n$ and QUICKSORT with $n \log n + 0.2n$ comparisons. ACM Journal of Experimental Algorithmics 10(5) (2002)
6. Edelkamp, S., Wegener, I.: On the performance of $WEAK - HEAPSORT$. In: Reichel, H., Tison, S. (eds.) STACS 2000. LNCS, vol. 1770, pp. 254–266. Springer, Heidelberg (2000)
7. Edelkamp, S., Weiß, A.: QuickXsort: Efficient Sorting with $n \log n - 1.399n + o(n)$ Comparisons on Average. ArXiv e-prints, abs/1307.3033 (2013)
8. Elmasry, A., Katajainen, J., Stenmark, M.: Branch mispredictions don't affect mergesort. In: Klasing, R. (ed.) SEA 2012. LNCS, vol. 7276, pp. 160–171. Springer, Heidelberg (2012)
9. Ford, J., Lester, R., Johnson, S.M.: A tournament problem. The American Mathematical Monthly 66(5), 387–389 (1959)
10. Katajainen, J.: The Ultimate Heapsort. In: CATS, pp. 87–96 (1998)
11. Katajainen, J., Pasanen, T., Teuhola, J.: Practical in-place mergesort. Nord. J. Comput. 3(1), 27–40 (1996)
12. Knuth, D.E.: Sorting and Searching, 2nd edn. The Art of Computer Programming, vol. 3. Addison Wesley Longman (1998)
13. Martínez, C., Roura, S.: Optimal Sampling Strategies in Quicksort and Quickselect. SIAM J. Comput. 31(3), 683–705 (2001)
14. Musser, D.R.: Introspective sorting and selection algorithms. Software—Practice and Experience 27(8), 983–993 (1997)
15. Reinhardt, K.: Sorting *in-place* with a *worst case* complexity of $n \log n - 1.3n + o(\log n)$ comparisons and $\epsilon n \log n + o(1)$ transports. In: Ibaraki, T., Iwama, K., Yamashita, M., Inagaki, Y., Nishizeki, T. (eds.) ISAAC 1992. LNCS, vol. 650, pp. 489–498. Springer, Heidelberg (1992)
16. Wegener, I.: Bottom-up-Heapsort, a new variant of Heapsort beating, on an average, Quicksort (if n is not very small). Theoretical Computer Science 118(1), 81–98 (1993)

Notions of Metric Dimension of Corona Products: Combinatorial and Computational Results

Henning Fernau[1] and Juan Alberto Rodríguez-Velázquez[2]

[1] FB 4-Abteilung Informatikwissenschaften, Universität Trier, 54286 Trier, Germany
fernau@uni-trier.de
[2] Universitat Rovira i Virgili, Av. Països Catalans 26, 43007 Tarragona, Spain
juanalberto.rodriguez@urv.cat

Abstract. The metric dimension is quite a well-studied graph parameter. Recently, the adjacency metric dimension and the local metric dimension have been introduced. We combine these variants and introduce the local adjacency metric dimension. We show that the (local) metric dimension of the corona product of a graph of order n and some nontrivial graph H equals n times the (local) adjacency metric dimension of H. This strong relation also enables us to infer computational hardness results for computing the (local) metric dimension, based on according hardness results for (local) adjacency metric dimension that we also give.

Keywords: (local) metric dimension, (local) adjacency dimension, NP-hardness.

1 Introduction and Preliminaries

Throughout this paper, we only consider undirected simple loop-free graphs and use standard graph-theoretic terminology. Less known notions are collected at the end of this section.

Let (X, d) be a metric space. The *diameter* of a point set $S \subseteq X$ is defined as $\mathrm{diam}(S) = \sup\{d(x, y) : x, y \in S\}$. A *generator* of (X, d) is a set $S \subseteq X$ such that every point of the space is uniquely determined by the distances from the elements of S. A point $v \in X$ is said to *distinguish* two points x and y of X if $d(v, x) \neq d(v, y)$. Hence, S is a generator if and only if any pair of points of X is distinguished by some element of S.

Four notions of dimension in graphs. Let $\mathbb{N}$ denote the set of non-negative integers. Given a connected graph $G = (V, E)$, we consider the function $d_G : V \times V \to \mathbb{N}$, where $d_G(x, y)$ is the length of a shortest path between u and v. Clearly, (V, d_G) is a metric space. The diameter of a graph is understood in this metric. A vertex set $S \subseteq V$ is said to be a *metric generator* for G if it is a generator of the metric space (V, d_G). A minimum metric generator is called a *metric basis*, and its cardinality the *metric dimension* of G, denoted by $\dim(G)$.

E.A. Hirsch et al. (Eds.): CSR 2014, LNCS 8476, pp. 153–166, 2014.

Motivated by the problem of uniquely determining the location of an intruder in a network, the concept of metric dimension of a graph was introduced by Slater in [33], where the metric generators were called *locating sets*. Independently, Harary and Melter introduced this concept in [16], where metric generators were called *resolving sets*. Applications of this parameter to the navigation of robots in networks are discussed in [25] and applications to chemistry in [22,23]. This graph parameter was studied further in a number of other papers including recent papers like [1,10,13,20,35].

Keeping in mind the robot navigation scenario, where the robot can determine its position by knowing the distances to the vertices in the metric generator, it makes sense to consider local variants of this parameter, assuming that the robot has some idea about its current position. A set S of vertices in a connected graph G is a *local metric generator* for G (also called local metric set for G [29]) if every two adjacent vertices of G are distinguished by some vertex of S. A minimum local metric generator is called a *local metric basis* for G and its cardinality, the *local metric dimension* of G, is denoted by $\dim_l(G)$.

If the distances between vertices are hard to determine, then it might still be the case that the robot can sense whether or not it is within the range of some sender installed on some other vertex. This has motivated the next definition. A set S of vertices in a graph G is an *adjacency generator* for G (also adjacency resolving set for G [21]) if for every $x, y \in V(G) - S$ there exists $s \in S$ such that $|N_G(s) \cap \{x, y\}| = 1$. This concept is very much related to that of a 1-locating dominating set [5]. A minimum adjacency generator is called an *adjacency basis* for G and its cardinality, the *adjacency dimension* of G, is denoted by $\dim_A(G)$. Observe that an adjacency generator of a graph $G = (V, E)$ is also a generator in a suitably chosen metric space, namely by considering $(V, d_{G,2})$, with $d_{G,2}(x, y) = \min\{d_G(x, y), 2\}$, and vice versa.

Now, we combine the two variants of metric dimension defined so far and introduce the local adjacency dimension of a graph. We say that a set S of vertices in a graph G is a *local adjacency generator* for G if for every two adjacent vertices $x, y \in V(G) - S$ there exists $s \in S$ such that $|N_G(s) \cap \{x, y\}| = 1$. A minimum local adjacency generator is called a *local adjacency basis* for G and its cardinality, the *local adjacency dimension* of G, is denoted by $\dim_{A,l}(G)$.

Our main results. In this paper, we study the (local) metric dimension of corona product graphs via the (local) adjacency dimension of a graph. We show that the (local) metric dimension of the corona product of a graph of order n and some non-trivial graph H equals n times the (local) adjacency metric dimension of H. This relation is much stronger and under weaker conditions compared to the results of Jannesari and Omoomi [21] concerning the lexicographic product of graphs. This also enables us to infer NP-hardness results for computing the (local) metric dimension, based on corresponding NP-hardness results for (local) adjacency metric dimension that we also provide. To our knowledge, this is the first time combinatorial results on this particular form of graph product have been used to deduce computational hardness results. The obtained reductions are relatively simple and also allow us to conclude hardness results based on the Exponential Time Hypothesis. We also discuss NP-hardness results for planar graphs, which seem to be of some particular importance to the sketched applications. This also shows the limitations of using corona products to obtain hardness results. Finally, we indicate why computing the (local) adjacency metric dimension is in FPT (under the standard parameterization), contrasting what is known for computing the metric dimension.

Some notions from graph theory. Let $G = (V, E)$ be a graph. A vertex set $D \subseteq V$ is called a *dominating set* if $\bigcup_{v \in D} N_G[v] = V$. The *domination number* of G, denoted by $\gamma(G)$, is the minimum cardinality among all dominating sets in G. A vertex set $C \subseteq V$ is called a *vertex cover* if for each edge $e \in E$, $C \cap e \neq \emptyset$, The *vertex cover number* of G, denoted by $\beta(G)$, is the minimum cardinality among all vertex covers of G.

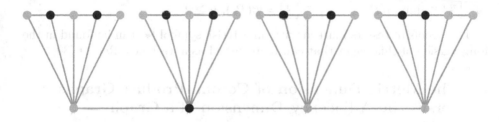

Fig. 1. The bold type forms an adjacency basis for $P_4 \odot P_5$ but not a dominating set

Let G and H be two graphs of order n and n', respectively. The *join (graph)* $G + H$ is defined as the graph obtained from vertex-disjoint graphs G and H by taking one copy of G and one copy of H and joining by an edge each vertex of G with each vertex of H. Graph products is one of the recurring themes in graph theory, see [15]. The *corona product (graph)* $G \odot H$ is defined as the graph obtained from G and H by taking one copy of G and n copies of H and joining by an edge each vertex from the ith copy of H with the ith vertex of G [11]. We will denote by $V = \{v_1, v_2, \ldots, v_n\}$ the set of vertices of G and by $H_i = (V_i, E_i)$ the ith copy of H so that $N_{G \odot H}(v_i) = V_i \cup N_G(v_i)$ and $N_{G \odot H}(x) = \{v_i\} \cup N_{H_i}(x)$ for every $x \in V_i$. Notice that the corona graph $K_1 \odot H$ is isomorphic to the join graph $K_1 + H$. For our computational complexity results, it is important but easy to observe that these graph operations can be performed in polynomial time, given two input graphs. Some of the notions important in this paper are illustrated in Figure 1.

Simple facts. By definition, the following inequalities hold for any graph G:

- $\dim(G) \leq \dim_A(G)$;
- $\dim_l(G) \leq \dim_{A,l}(G)$;
- $\dim_l(G) \leq \dim(G)$;
- $\dim_{A,l}(G) \leq \dim_A(G)$.

Moreover, if S is an adjacency generator, then at most one vertex is not dominated by S, so that

$$\gamma(G) \leq \dim_A(G) + 1.$$

Namely, if x, y are not dominated by S, then no element in S distinguishes them. We also observe that

$$\dim_{A,l}(G) \leq \beta(G),$$

because each vertex cover is a local adjacency generator.

However, all mentioned inequalities could be either equalities or quite weak bounds. Consider the following examples:

1. $\dim_l(P_n) = \dim(P_n) = 1 \leq \lfloor \frac{n}{4} \rfloor \leq \dim_{A,l}(P_n) \leq \lceil \frac{n}{4} \rceil \leq \lfloor \frac{2n+2}{5} \rfloor = \dim_A(P_n)$, $n \geq 7$;
2. $\dim_l(K_{1,n}) = \dim_{A,l}(K_{1,n}) = 1 \leq n-1 = \dim(K_{1,n}) = \dim_A(K_{1,n})$, $n \geq 2$;
3. $\gamma(P_n) = \lceil \frac{n}{3} \rceil \leq \lfloor \frac{2n+2}{5} \rfloor = \dim_A(P_n)$, $n \geq 7$;
4. $\lfloor \frac{n}{4} \rfloor \leq \dim_{A,l}(P_n) \leq \lceil \frac{n}{4} \rceil \leq \lfloor \frac{n}{2} \rfloor = \beta(P_n)$, $n \geq 2$.

The proofs of results marked with an asterisk symbol $(*)$ can be found in the long version of this paper that can be retrieved as a Technical Report [30].

2 The Metric Dimension of Corona Product Graphs versus the Adjacency Dimension of a Graph

The following is the first main combinatorial result of this paper and provides a strong link between the metric dimension of the corona product of two graphs and the adjacency dimension of the second graph involved in the product operation. A seemingly similar formula was derived in [20,35], but there, only the notion of metric dimension was involved (which makes it impossible to use the formula to obtain computational hardness results as we will do), and also, special conditions were placed on the second argument graph of the corona product.

Theorem 1. *For any connected graph G of order $n \geq 2$ and any non-trivial graph H, $\dim(G \odot H) = n \cdot \dim_A(H)$.*

Proof. We first need to prove that $\dim(G \odot H) \leq n \cdot \dim_A(H)$. For any $i \in \{1, \ldots, n\}$, let S_i be an adjacency basis of H_i, the ith-copy of H. In order to show that $X := \bigcup_{i=1}^n S_i$ is a metric generator for $G \odot H$, we differentiate the following four cases for two vertices $x, y \in V(G \odot H) - X$.

1. $x, y \in V_i$. Since S_i is an adjacency basis of H_i, there exists a vertex $u \in S_i$ so that $|N_{H_i}(u) \cap \{x, y\}| = 1$. Hence,

$$d_{G \odot H}(x, u) = d_{\langle v_i \rangle + H_i}(x, u) \neq d_{\langle v_i \rangle + H_i}(y, u) = d_{G \odot H}(y, u).$$

2. $x \in V_i$ and $y \in V$. If $y = v_i$, then for $u \in S_j$, $j \neq i$, we have

$$d_{G \odot H}(x, u) = d_{G \odot H}(x, y) + d_{G \odot H}(y, u) > d_{G \odot H}(y, u).$$

Now, if $y = v_j$, $j \neq i$, then we also take $u \in S_j$ and we proceed as above.

3. $x = v_i$ and $y = v_j$. For $u \in S_j$, we find that

$$d_{G \odot H}(x, u) = d_{G \odot H}(x, y) + d_{G \odot H}(y, u) > d_{G \odot H}(y, u).$$

4. $x \in V_i$ and $y \in V_j$, $j \neq i$. In this case, for $u \in S_i$ we have

$$d_{G \odot H}(x, u) \leq 2 < 3 \leq d_{G \odot H}(u, y).$$

Hence, X is a metric generator for $G \odot H$ and, as a consequence,

$$\dim(G \odot H) \leq \sum_{i=1}^{n} |S_i| = n \cdot \dim_A(H).$$

It remains to prove that $\dim(G \odot H) \geq n \cdot \dim_A(H)$. To do this, let W be a metric basis for $G \odot H$ and, for any $i \in \{1, \ldots, n\}$, let $W_i := V_i \cap W$. Let us show that W_i is an adjacency metric generator for H_i. To do this, consider two different vertices $x, y \in V_i - W_i$. Since no vertex $u \in V(G \odot H) - V_i$ distinguishes the pair x, y, there exists some $u \in W_i$ such that $d_{G \odot H}(x, u) \neq d_{G \odot H}(y, u)$. Now, since $d_{G \odot H}(x, u) \in \{1, 2\}$ and $d_{G \odot H}(y, u) \in \{1, 2\}$, we conclude that $|N_{H_i}(u) \cap \{x, y\}| = 1$ and consequently, W_i must be an adjacency generator for H_i. Hence, for any $i \in \{1, \ldots, n\}$, $|W_i| \geq \dim_A(H_i)$. Therefore,

$$\dim(G \odot H) = |W| \geq \sum_{i=1}^{n} |W_i| \geq \sum_{i=1}^{n} \dim_A(H_i) = n \cdot \dim_A(H).$$

Consequences of Theorem 1 We can now investigate $\dim(G \odot H)$ through the study of $\dim_A(H)$, and vice versa. In particular, results from [3,32,35] allow us to deduce the exact adjacency dimension for several special graphs. For instance, we find that $\dim_A(C_r) = \dim_A(P_r) = \lfloor \frac{2r+2}{5} \rfloor$ for any $r \geq 7$. Other combinatorial results of this type are collected in the long version of this paper [30].

A detailed analysis of the adjacency dimension of the corona product via the adjacency dimension of the second operand. We now analyze the adjacency dimension of the corona product $G \odot H$ in terms of the adjacency dimension of H.

Theorem 2. (*) *Let G be a connected graph of order $n \geq 2$ and let H be a non-trivial graph. If there exists an adjacency basis S for H which is also a dominating set, and if for every $v \in V(H) - S$, it is satisfied that $S \not\subseteq N_H(v)$, then $\dim_A(G \odot H) = n \cdot \dim_A(H)$.*

Corollary 1. (*) *Let $r \geq 7$ with $r \not\equiv 1 \bmod 5$ and $r \not\equiv 3 \bmod 5$. For any connected graph G of order $n \geq 2$, $\dim_A(G \odot C_r) = \dim_A(G \odot P_r) = n \cdot \lfloor \frac{2r+2}{5} \rfloor$.*

Theorem 3. *Let G be a connected graph of order $n \geq 2$ and let H be a non-trivial graph. If there exists an adjacency basis for H which is also a dominating set and if, for any adjacency basis S for H, there exists some $v \in V(H) - S$ such that $S \subseteq N_H(v)$, then $\dim_A(G \odot H) = n \cdot \dim_A(H) + \gamma(G)$.*

Proof. Let W be an adjacency basis for $G \odot H$ and let $W_i = W \cap V_i$ and $U = W \cap V$. Since two vertices belonging to V_i are not distinguished by any $u \in W - V_i$, the set W_i must be an adjacency generator for H_i. Now consider the partition $\{V', V''\}$ of V defined as follows:

$$V' = \{v_i \in V : \ |W_i| = \dim_A(H)\} \text{ and } V'' = \{v_j \in V : \ |W_j| \geq \dim_A(H) + 1\}.$$

Note that, if $v_i \in V'$, then W_i is an adjacency basis for H_i, thus in this case there exists $u_i \in V_i$ such that $W_i \subseteq N_{H_i}(u_i)$. Then the pair u_i, v_i is not distinguished by the elements of W_i and, as a consequence, either $v_i \in U$ or there exists some $v_j \in U$ adjacent to v_i. Hence, $U \cup V''$ must be a dominating set and, as a result, $|U \cup V''| \geq \gamma(G)$. So we obtain the following:

$$\begin{aligned}
\dim_A(G \odot H) = |W| = \bigcup_{v_i \in V'} |W_i| &+ \bigcup_{v_j \in V''} |W_j| + |U| \\
&\geq \sum_{v_i \in V'} \dim_A(H) + \sum_{v_j \in V''} (\dim_A(H) + 1) + |U| \\
&= n \cdot \dim_A(H) + |V''| + |U| \geq n \cdot \dim_A(H) + |V'' \cup U| \\
&\geq n \cdot \dim_A(H) + \gamma(G).
\end{aligned}$$

To conclude the proof, we consider an adjacency basis S for H which is also a dominating set, and we denote by S_i the copy of S corresponding to H_i. We claim that for any dominating set D of G of minimum cardinality $|D| = \gamma(G)$, the set $D \cup (\bigcup_{i=1}^{n} S_i)$ is an adjacency generator for $G \odot H$ and, as a result,

$$\dim_A(G \odot H) \leq \left| D \cup \left(\bigcup_{i=1}^{n} S_i \right) \right| = n \cdot \dim_A(H) + \gamma(G).$$

This can be seen by some case analysis. Let $S' = D \cup \bigcup_{i=1}^{n} S_i$ and let us prove that S' is an adjacency generator for $G \odot H$. We differentiate the following cases for any pair x, y of vertices of $G \odot H$ not belonging to S'.

1. $x, y \in V_i$. Since S_i is an adjacency basis of H_i, there exists $u_i \in S_i$ such that u_i is adjacent to x or to y but not to both.
2. $x \in V_i$, $y \in V_j$, $j \neq i$. As S_i is a dominating set of H_i, there exists $u \in S_i$ such that $u \in N_{H_i}(x)$ and, obviously, $u \notin N_{G \odot H}(y)$.
3. $x \in V_i$, $y = v_i \in V$. As $y = v_i \notin D$, $v_j \in N_G(v_i)$ distinguishes the pair x, y.
4. $x \in V_i \cup \{v_i\}$, $y = v_j \in V$, $i \neq j$. In this case, every $u \in S_j$ is a neighbor of y but not of x.

Corollary 2. *Let $r \geq 2$. Let G be a connected graph of order $n \geq 2$. Then,* $\dim_A(G \odot K_r) = n(r - 1) + \gamma(G)$.

Theorem 4. (*) *Let G be a connected graph of order $n \geq 2$ and let H be a non-trivial graph. If no adjacency basis for H is a dominating set, then we have:* $\dim_A(G \odot H) = n \cdot \dim_A(H) + n - 1.$

It is easy to check that any adjacency basis of a star graph $K_{1,r}$ is composed of $r - 1$ leaves, with the last leaf non-dominated. Thus, Theorem 4 implies:

Corollary 3. *For a connected graph G of order $n \geq 2$, $\dim_A(G \odot K_{1,r}) = n \cdot r - 1$.*

Given a vertex $v \in V$ we denote by $G - v$ the subgraph obtained from G by removing v and the edges incident with it. We define the following auxiliary domination parameter: $\gamma'(G) := \min_{v \in V(G)} \{\gamma(G - v)\}$.

Theorem 5. (∗) *Let H be a non-trivial graph such that some of its adjacency bases are also dominating sets, and some are not. If there exists an adjacency basis S' for H such that for every $v \in V(H) - S'$ it is satisfied that $S' \nsubseteq N_H(v)$, and for any adjacency basis S for H which is also a dominating set, there exists some $v \in V(H) - S$ such that $S \subseteq N_H(v)$, then for any connected graph G of order $n \geq 2$, $\dim_A(G \odot H) = n \cdot \dim_A(H) + \gamma'(G)$.*

As indicated in Figure 1, $H = P_5$ satisfies the premises of Theorem 5, as in particular there are adjacency bases that are also dominating set (see the leftmost copy of a P_5 in Figure 1) as well as adjacency bases that are not dominating sets (see the rightmost copy of a P_5 in that drawing). Hence, we can conclude:

Corollary 4. *For any connected graph G of order $n \geq 2$, $\dim_A(G \odot P_5) = 2n + \gamma'(G)$.*

Since the assumptions of Theorems 2, 3, 4 and 5 are complementary and for any graph G of order $n \geq 3$ it holds that $0 < \gamma'(G) \leq \gamma(G) \leq \frac{n}{2} < n - 1$, we can conclude that in fact, Theorems 2 and 5 are equivalences for $n \geq 3$ (or even $n \geq 2$ in the first case). Therefore, we obtain:

Theorem 6. *Let G be a connected graph of order $n \geq 2$ and let H be a non-trivial graph. The following statements are equivalent:*

(i) *There exists an adjacency basis S for H, which is also a dominating set, such that for every $v \in V(H) - S$ it is satisfied that $S \nsubseteq N_H(v)$.*
(ii) $\dim_A(G \odot H) = n \cdot \dim_A(H)$.
(iii) $\dim_A(G \odot H) = \dim(G \odot H)$.

This should be conferred to the combinatorial results in [20], as it exactly tells when they could possibly apply.

As an example of applying Theorem 6 we can take H as the cycle graphs C_r or the path graphs P_r, where $r \geq 7$, $r \not\equiv 1 \bmod 5$, $r \not\equiv 3 \bmod 5$, see Cor. 1.

Theorem 7. *Let G be a connected graph of order $n \geq 3$ and let H be a non-trivial graph. The following statements are equivalent:*

(i) *No adjacency basis for H is a dominating set.*
(ii) $\dim_A(G \odot H) = n \cdot \dim_A(H) + n - 1$.
(iii) $\dim_A(G \odot H) = \dim(G \odot H) + n - 1$.

3   Locality in Dimensions

First, we consider some straightforward cases. If H is an empty graph, then $K_1 \odot H$ is a star graph and $\dim_l(K_1 \odot H) = 1$. Moreover, if H is a complete graph of order n, then $K_1 \odot H$ is a complete graph of order $n+1$ and $\dim_l(K_1 \odot H) = n$. It was shown in [31] that for any connected nontrivial graph G and any empty graph H, $\dim_l(G \odot H) = \dim_l(G)$. We are going to state results similar to the non-local situation as discussed in the previous section. We omit all proofs as they are along similar lines.

Theorem 8. (∗) *For any connected graph G of order $n \geq 2$ and any non-trivial graph H, $\dim_l(G \odot H) = n \cdot \dim_{A,l}(H)$.*

Based on [31], this allows to deduce quite a number of combinatorial results for the new notion of a local adjacency dimension, as contained in [30].

Fortunately, the comparison of the local adjacency dimension of the corona product with the one of the second argument is much simpler in the local version as in the previously studied non-local version.

Theorem 9. (∗) *Let G be a connected graph of order $n \geq 2$ and let H be a non-trivial graph. If there exists a local adjacency basis S for H such that for every $v \in V(H) - S$ it is satisfied that $S \nsubseteq N_H(v)$, then $\dim_{A,l}(G \odot H) = n \cdot \dim_{A,l}(H)$.*

Theorem 10. (∗) *Let G be a connected graph of order $n \geq 2$ and let H be a non-trivial graph. If for any local adjacency basis for H, there exists some $v \in V(H) - S$ which satisfies that $S \subseteq N_H(v)$, then $\dim_{A,l}(G \odot H) = n \cdot \dim_{A,l}(H) + \gamma(G)$.*

Remark 1. As a concrete example for the previous theorem, consider $H = K_{n'}$. Clearly, $\dim_{A,l}(H) = n' - 1$, and the neighborhood of the only vertex that is not in the local adjacency basis coincides with the local adjacency basis. For any connected graph G of order $n \geq 2$, we can deduce that

$$\dim_{A,l}(G \odot K_{n'}) = n \cdot \dim_{A,l}(K_{n'}) + \gamma(G) = n(n' - 1) + \gamma(G).$$

Since the assumptions of Theorems 9 and 10 are complementary, we obtain the following property for $\dim_{A,l}(G \odot H)$.

Theorem 11. *Let G be a connected graph of order $n \geq 2$ and let H be a non-trivial graph. Then the following assertions are equivalent.*

(i) *There exists a local adjacency basis S for H such that for every $v \in V(H) - S$ it is satisfied that $S \nsubseteq N_H(v)$.*
(ii) *$\dim_{A,l}(G \odot H) = n \cdot \dim_{A,l}(H)$.*
(iii) *$\dim_l(G \odot H) = \dim_{A,l}(G \odot H)$.*

Theorem 12. *Let G be a connected graph of order $n \geq 2$ and let H be a non-trivial graph. Then the following assertions are equivalent.*

(i) *For any local adjacency basis S for H, there exists some $v \in V(H) - S$ which satisfies that $S \subseteq N_H(v)$.*

(ii) $\dim_{A,l}(G \odot H) = n \cdot \dim_{A,l}(H) + \gamma(G)$.
(iii) $\dim_l(G \odot H) = \dim_{A,l}(G \odot H) - \gamma(G)$.

As a concrete example of graph H where we can apply the above result is the star $K_{1,r}$, $r \geq 2$. In this case, for any connected graph G of order $n \geq 2$, we find that $\dim_{A,l}(G \odot K_{1,r}) = n \cdot \dim_{A,l}(K_{1,r}) + \gamma(G) = n + \gamma(G)$.

4   Computational Complexity of the Dimension Variants

In this section, we not only prove NP-hardness of all dimension variants, but also show that the problems (viewed as minimization problems) cannot be solved in time $O(\mathrm{poly}(n + m)2^{o(n)})$ on any graph of order n (and size m). Yet, it is straightforward to see that each of our computational problems can be solved in time $O(\mathrm{poly}(n+m)2^n)$, simply by cycling through all vertex subsets by increasing cardinality and then checking if the considered vertex set forms an appropriate basis. More specifically, based on our reductions we can conclude that these trivial brute-force algorithms are in a sense optimal, assuming the validity of the Exponential Time Hypothesis (ETH). A direct consequence of ETH (using the sparsification lemma) is the hypothesis that 3-SAT instances cannot be solved in time $O(\mathrm{poly}(n + m)2^{o(n+m)})$ on instances with n variables and m clauses; see [19,4].

From a mathematical point of view, the most interesting fact is that most of our computational results are based on the combinatorial results on the dimensional graph parameters on corona products of graphs that are derived above.

Due to the practical motivation of the parameters, we also study their computational complexity on planar graph instances.

We are going to investigate the following problems:

DIM: Given a graph G and an integer k, decide if $\dim(G) \leq k$ or not.

LOCDIM: Given a graph G and an integer k, decide if $\dim_l(G) \leq k$ or not.

ADJDIM: Given a graph G and an integer k, decide if $\dim_A(G) \leq k$ or not.

LOCADJDIM: Given a graph G and an integer k, decide if $\dim_{A,l}(G) \leq k$ or not.

As auxiliary problems, we will also consider:

VC: Given a graph G and an integer k, decide if $\beta(G) \leq k$ or not.

DOM: Given a graph G and an integer k, decide if $\gamma(G) \leq k$ or not.

1-LOCDOM: Given a graph G and an integer k, decide if there exists a 1-locating dominating set of G with at most k vertices or not. (A dominating set $D \subseteq V$ in a graph $G = (V, E)$ is called a *1-locating dominating set* if for every two vertices $u, v \in V \setminus D$, the symmetric difference of $N(u) \cap D$ and $N(v) \cap D$ is non-empty.)

Theorem 13. DIM *is NP-complete, even when restricted to planar graphs.*

Different proofs of this type of hardness result appeared in the literature. While this result is only mentioned in the textbook of Garey and Johnson [12], a proof was first published in [25]. For planar instances, we refer to [9] where this result is stated.

Remark 2. In fact, we can offer a further proof for the NP-hardness of DIM (on general graphs), based upon Theorem 1 and the following reasoning. If there were a polynomial-time algorithm for computing dim(G), then we could compute dim$_A(H)$ for any (non-trivial) graph H by computing dim$(K_2 \odot H)$ with the assumed polynomial-time algorithm, knowing that this is just twice as much as dim$_A(H)$. As every NP-hardness proof adds a bit to the understanding of the nature of the problem, this one does so, as well. It shows that DIM is NP-complete even on the class of graphs that can be written as $G \odot H$, where G is some connected graph of order $n \geq 2$ and H is non-trivial.

Theorem 14. ($*$) 1-LOCDOM *is NP-hard, even when restricted to planar graphs. Moreover, assuming ETH, there is no $O(poly(n + m)2^{o(n)})$ algorithm solving 1-LOCDOM on general graphs of order n and size m.*

Proof. (Sketch) Recall the textbook proof for the NP-hardness of VC (see [12]) that produces from a given 3-SAT instance I with n variables and m clauses a graph G with two adjacent vertices per variable gadget and three vertices per clause gadget forming a C_3 (and $3m$ more edges that interconnect these gadgets to indicate which literals occur in which clauses). So, G has $3m + 2n$ vertices and $3m + n + 3m = 6m + n$ edges. We modify G to obtain G' as follows: Each edge that occurs inside of a variable gadget or of a clause gadget is replaced by a triangle, so that we add $3m + n$ new vertices of degree two. All in all, this means that G' has $(3m+2n)+(3m+n) = 6m+3n$ vertices and $9m+3n+3m = 12m+3n$ edges. Now, assuming (w.l.o.g.) that I contains, for each variable x, at least one clause with x as a literal and another clause with $\bar{x}$ as a literal, we can show that I is satisfiable iff G has a vertex cover of size at most $2m + n$ iff G' has a 1-locating dominating set of size at most $2m + n$.   □

The general case was treated in [7], but that proof (starting out again from 3-SAT) does not preserve planarity, as the variable gadget alone already contains a $K_{2,3}$ subgraph that inhibits non-crossing interconnections with the clause gadgets. However, although not explicitly mentioned, that reduction also yields the non-existence of $O(poly(n + m)2^{o(n)})$ algorithms based on ETH. In [30], we also provide a reduction that works for planar graphs, working on a variant of Lichtenstein's reduction [27] that shows NP-hardness of VC on planar graph instances.

Theorem 15. ADJDIM *is NP-complete, even when restricted to planar graphs. Assuming ETH, there is no $O(poly(n + m)2^{o(n)})$ algorithm solving ADJDIM on graphs of order n and size m.*

Proof. (Sketch) From an instance $G = (V, E)$ and k of 1-LOCDOM, produce an instance (G', k) of ADJDIM by obtaining G' from G by adding a new isolated vertex $x \notin V$ to G. We claim that G has a 1-locating dominating set of size at most k if and only if dim$_A(G') \leq k$.   □

Alternatively, NP-hardness of ADJDIM (and even the ETH-result) can be deduced from the strong relation between the domination number and the adjacency dimension as stated in Cor. 2, based on the NP-hardness of DOM.

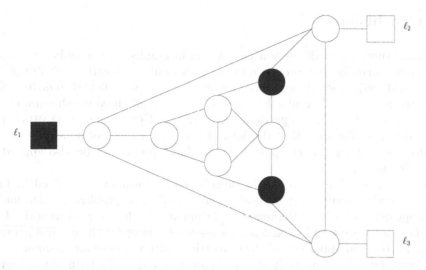

Fig. 2. The clause gadget illustration. The square-shaped vertices do not belong to the gadget, but they are the three literal vertices in variable gadgets that correspond to the three literals in the clause.

As explained in Remark 2, Theorem 1 can be used to deduce furthermore:

Corollary 5. *Assuming ETH, there is no $O(poly(n+m)2^{o(n)})$ algorithm solving* DIM *on graphs of order n and size m.*

Lemma 1. *[28] Assuming ETH, there is no $O(poly(n+m)2^{o(n)})$ algorithm solving* DOM *on graphs of order n and size m.*

From Remark 1 and Lemma 1, we can conclude:

Theorem 16. LOCADJDIM *is NP-complete. Moreover, assuming ETH, there is no $O(poly(n+m)2^{o(n)})$ algorithm solving* LOCADJDIM *on graphs of order n and size m.*

We provide an alternative proof of the previous theorem in [30]. That proof is a direct reduction from 3-SAT and is, in fact, very similar to the textbook proof for the NP-hardness of VC, also see the proof of Theorem 14. This also proves that LOCADJDIM is NP-complete when restricted to planar instances. More precisely, the variable gadgets are paths on four vertices, where the middle two ones interconnect to the clause gadgets in which they occur. The clause gadgets are a bit more involved, as shown in Fig. 2.

As explained in Remark 2, we can (now) use Theorem 8 together with Theorem 16 to conclude the following hitherto unknown complexity result.

Theorem 17. LOCDIM *is NP-complete. Moreover, assuming ETH, there is no $O(poly(n+m)2^{o(n)})$ algorithm solving* LOCDIM *on graphs of order n and size m.*

Notice that the reduction explained in Remark 2 does not help find any hardness results on planar graphs. Hence, we leave it as an open question whether or not LOCDIM is NP-hard also on planar graph instances.

5   Conclusions

We have studied four dimension parameters in graphs. In particular, establishing concise formulae for corona product graphs, linking (local) metric dimension with (local) adjacency dimension of the involved graphs, allowed to deduce NP-hardness results (and similar hardness claims) for all these graph parameters, based on known results, in particular on VERTEX COVER and on DOMINATING SET problems. We hope that the idea of using such types of non-trivial (combinatorial) formulae for computational hardness proofs can be also applied in other situations.

For instance, observe that reductions based on formulae as derived in Theorem 1 clearly preserve the natural parameter of these problems, which makes this approach suitable for Parameterized Complexity. However, let us notice here that DIM is unlikely to be fixed-parameter tractable under the natural parameterization (i.e., an upper bound on the metric dimension) even for subcubic graph instances; see [17]. Conversely, it is not hard to see that the natural parameterization of ADJDIM can be shown to be in FPT by reducing it to TEST COVER. Namely, let $G = (V, E)$ be a graph and k be an integer, defining an instance of ADJDIM. We construct a TEST COVER instance as follows: Let $S = \binom{V}{2}$ be the substances and define the potential test set $T = \{t_v \mid v \in V\}$ by letting

$$t_v(\{x, y\}) = \begin{cases} 1, \text{ if } v \in N[x] \triangle N[y] \\ 0, \text{ otherwise} \end{cases}$$

Now, if D is some adjacency generator, then $T_D = \{t_v \mid v \in D\}$ is some test cover solution, i.e., for any pair of substances, we find a test that differentiates the two. The converse is also true. TEST COVER has received certain interest recently in Parameterized Complexity [8,14]. Does ADJDIM admit a polynomial-size kernel, or does it rather behave like TEST COVER?

From a computational point of view, let us mention (in-)approximability results as obtained in [26,34]. In particular, inapproximability of 1-LOCDOM readily transfers to inapproximability of ADJDIM and this in turn leads to inapproximability results for DIM as in Remark 2; also see [17].

Also, 1-locating dominating sets have been studied (actually, independently introduced) in connection with coding theory [24]. Recall that these sets are basically adjacency bases. Therefore, it might be interesting to try to apply some of the information-theoretic arguments on variants of metric dimension, as well. Conversely, the notion of locality used in this paper connects to the idea of correcting only 1-bit errors in codes. These interconnections deserve further studies.

All these computational hardness results, as well as the various different applications that led to the introduction of these graph dimension parameters, also open up the quest for moderately exponential-time algorithms, i.e., algorithms that should find an optimum solution for any of our dimension problems in time $O(\text{poly}(n + m)c^n)$ on graphs of size m and order n for some $c < 2$, or also to finding polynomial-time algorithms for special graph classes. In this context, we mention results on trees, series-parallel and distance-regular graphs [7,13,18].

In view of the original motivation for introducing these graph parameters, it would be interesting to study their complexity on geometric graphs. Notice that the definition of a metric generator is not exclusively referring to (finite) graphs, which might lead us even back to the common roots of graph theory and topology.

In view of the many different motivations, also the study of computational aspects of other variants of dimension parameters could be of interest. We only mention here the notions of resolving dominating sets [2] and independent resolving sets [6].

References

1. Bailey, R.F., Meagher, K.: On the metric dimension of Grassmann graphs. Discrete Mathematics & Theoretical Computer Science 13, 97–104 (2011)
2. Brigham, R.C., Chartrand, G., Dutton, R.D., Zhang, P.: Resolving domination in graphs. Mathematica Bohemica 128(1), 25–36 (2003)
3. Buczkowski, P.S., Chartrand, G., Poisson, C., Zhang, P.: On k-dimensional graphs and their bases. Periodica Mathematica Hungarica 46(1), 9–15 (2003)
4. Calabro, C., Impagliazzo, R., Paturi, R.: The complexity of satisfiability of small depth circuits. In: Chen, J., Fomin, F.V. (eds.) IWPEC 2009. LNCS, vol. 5917, pp. 75–85. Springer, Heidelberg (2009)
5. Charon, I., Hudry, O., Lobstein, A.: Minimizing the size of an identifying or locating-dominating code in a graph is NP-hard. Theoretical Computer Science 290(3), 2109–2120 (2003)
6. Chartrand, G., Saenpholphat, V., Zhang, P.: The independent resolving number of a graph. Mathematica Bohemica 128(4), 379–393 (2003)
7. Colbourn, C.J., Slater, P.J., Stewart, L.K.: Locating dominating sets in series parallel networks. Congressus Numerantium 56, 135–162 (1987)
8. Crowston, R., Gutin, G., Jones, M., Saurabh, S., Yeo, A.: Parameterized study of the test cover problem. In: Rovan, B., Sassone, V., Widmayer, P. (eds.) MFCS 2012. LNCS, vol. 7464, pp. 283–295. Springer, Heidelberg (2012)
9. Díaz, J., Pottonen, O., Serna, M.J., van Leeuwen, E.J.: On the complexity of metric dimension. In: Epstein, L., Ferragina, P. (eds.) ESA 2012. LNCS, vol. 7501, pp. 419–430. Springer, Heidelberg (2012)
10. Feng, M., Wang, K.: On the metric dimension of bilinear forms graphs. Discrete Mathematics 312(6), 1266–1268 (2012)
11. Frucht, R., Harary, F.: On the corona of two graphs. Aequationes Mathematicae 4, 322–325 (1970)
12. Garey, M.R., Johnson, D.S.: Computers and Intractability: A Guide to the Theory of NP-Completeness. W. H. Freeman & Co., New York (1979)
13. Guo, J., Wang, K., Li, F.: Metric dimension of some distance-regular graphs. Journal of Combinatorial Optimization, 1–8 (2012)
14. Gutin, G., Muciaccia, G., Yeo, A.: (non-)existence of polynomial kernels for the test cover problem. Information Processing Letters 113(4), 123–126 (2013)
15. Hammack, R., Imrich, W., Klavžar, S.: Handbook of product graphs. Discrete Mathematics and its Applications, 2nd edn. CRC Press (2011)

16. Harary, F., Melter, R.A.: On the metric dimension of a graph. Ars Combinatoria 2, 191–195 (1976)
17. Hartung, S., Nichterlein, A.: On the parameterized and approximation hardness of metric dimension. In: Proceedings of the 28th IEEE Conference on Computational Complexity (CCC 2013), pp. 266–276. IEEE (2013)
18. Haynes, T.W., Henning, M.A., Howard, J.: Locating and total dominating sets in trees. Discrete Applied Mathematics 154(8), 1293–1300 (2006)
19. Impagliazzo, R., Paturi, R., Zane, F.: Which problems have strongly exponential complexity? Journal of Computer and System Sciences 63(4), 512–530 (2001)
20. Iswadi, H., Baskoro, E.T., Simanjuntak, R.: On the metric dimension of corona product of graphs. Far East Journal of Mathematical Sciences 52(2), 155–170 (2011)
21. Jannesari, M., Omoomi, B.: The metric dimension of the lexicographic product of graphs. Discrete Mathematics 312(22), 3349–3356 (2012)
22. Johnson, M.: Structure-activity maps for visualizing the graph variables arising in drug design. Journal of Biopharmaceutical Statistics 3(2), 203–236 (1993), pMID: 8220404
23. Johnson, M.A.: Browsable structure-activity datasets. In: Carbó-Dorca, R., Mezey, P. (eds.) Advances in Molecular Similarity, pp. 153–170. JAI Press Inc., Stamford (1998)
24. Karpovsky, M.G., Chakrabarty, K., Levitin, L.B.: On a new class of codes for identifying vertices in graphs. IEEE Transactions on Information Theory 44(2), 599–611 (1998)
25. Khuller, S., Raghavachari, B., Rosenfeld, A.: Landmarks in graphs. Discrete Applied Mathematics 70, 217–229 (1996)
26. Laifenfeld, M., Trachtenberg, A.: Identifying codes and covering problems. IEEE Transactions on Information Theory 54(9), 3929–3950 (2008)
27. Lichtenstein, D.: Planar formulae and their uses. SIAM Journal on Computing 11, 329–343 (1982)
28. Lokshtanov, D., Marx, D., Saurabh, S.: Lower bounds based on the Exponential Time Hypothesis. EATCS Bulletin 105, 41–72 (2011)
29. Okamoto, F., Phinezy, B., Zhang, P.: The local metric dimension of a graph. Mathematica Bohemica 135(3), 239–255 (2010)
30. Rodríguez-Velázquez, J.A., Fernau, H.: On the (adjacency) metric dimension of corona and strong product graphs and their local variants: combinatorial and computational results. Tech. Rep. arXiv:1309.2275 [math.CO], ArXiv.org, Cornell University (2013)
31. Rodríguez-Velázquez, J.A., Barragán-Ramírez, G.A., Gómez, C.G.: On the local metric dimension of corona product graph (2013) (submitted)
32. Saputro, S., Simanjuntak, R., Uttunggadewa, S., Assiyatun, H., Baskoro, E., Salman, A., Bača, M.: The metric dimension of the lexicographic product of graphs. Discrete Mathematics 313(9), 1045–1051 (2013)
33. Slater, P.J.: Leaves of trees. Congressus Numerantium 14, 549–559 (1975)
34. Suomela, J.: Approximability of identifying codes and locating-dominating codes. Information Processing Letters 103(1), 28–33 (2007)
35. Yero, I.G., Kuziak, D., Rodríquez-Velázquez, J.A.: On the metric dimension of corona product graphs. Computers & Mathematics with Applications 61(9), 2793–2798 (2011)

On the Complexity of Computing Two Nonlinearity Measures

Magnus Gausdal Find

Department of Mathematics and Computer Science
University of Southern Denmark

Abstract. We study the computational complexity of two Boolean non-linearity measures: the *nonlinearity* and the *multiplicative complexity*. We show that if one-way functions exist, no algorithm can compute the multiplicative complexity in time $2^{O(n)}$ given the truth table of length 2^n, in fact under the same assumption it is impossible to approximate the multiplicative complexity within a factor of $(2 - \epsilon)^{n/2}$. When given a circuit, the problem of determining the multiplicative complexity is in the second level of the polynomial hierarchy. For nonlinearity, we show that it is #**P** hard to compute given a function represented by a circuit.

1 Introduction

In many cryptographical settings, such as stream ciphers, block ciphers and hashing, functions being used must be deterministic but should somehow "look" random. Since these two desires are contradictory in nature, one might settle with functions satisfying certain *properties* that random Boolean functions possess with high probability. One property is to be somehow different from linear functions. This can be quantitatively delineated using so called "nonlinearity measures". Two examples of nonlinearity measures are the *nonlinearity*, i.e. the Hamming distance to the closest affine function, and the *multiplicative complexity*, i.e. the smallest number of AND gates in a circuit over the basis $(\wedge, \oplus, 1)$ computing the function. For results relating these measures to each other and cryptographic properties we refer to [6,4], and the references therein. The important point for this paper is that there is a fair number of results on the form "if f has low value according to measure μ, f is vulnerable to the following attack ...". Because of this, it was a design criteria in the Advanced Encryption Standard to have parts with high nonlinearity [10]. In a concrete situation, f is an explicit, finite function, so it is natural to ask how hard it is to compute μ given (some representation of) f. In this paper, the measure μ will be either multiplicative complexity or nonlinearity. We consider the two cases where f is being represented by its truth table, or by a circuit computing f.

We should emphasize that multiplicative complexity is an interesting measure for other reasons than alone being a measure of nonlinearity: In many applications it is harder, in some sense, to handle AND gates than XOR gates, so one is interested in a circuit over $(\wedge, \oplus, 1)$ with a small number of AND gates, rather

E.A. Hirsch et al. (Eds.): CSR 2014, LNCS 8476, pp. 167–175, 2014.

than a circuit with the smallest number of *gates*. Examples of this include protocols for secure multiparty computation (see e.g. [8,15]), non-interactive secure proofs of knowledge [3], and fully homomorphic encryption (see for example [20]).

It is a main topic in several papers (see e.g. [5,7,9][1]) to find circuits with few AND gates for specific functions using either exact or heuristic techniques. Despite this and the applications mentioned above, it appears that the computational hardness has not been studied before.

The two measures have very different complexities, depending on the representation of f.

Organization of the Paper and Results. In the following section, we introduce the problems and necessary definitions. All our hardness results will be based on assumptions stronger than $\mathbf{P} \neq \mathbf{NP}$, more precisely the existence of pseudorandom function families and the "Strong Exponential Time Hypothesis". In Section 3 we show that if pseudorandom function families exist, the multiplicative complexity of a function represented by its truth table cannot be computed (or even approximated with a factor $(2 - \epsilon)^{n/2}$) in polynomial time. This should be contrasted to the well known fact that nonlinearity can be computed in almost linear time using the Fast Walsh Transformation. In Section 4, we consider the problems when the function is represented by a circuit. We show that in terms of time complexity, under our assumptions, the situations differ very little from the case where the function is represented by a truth table. However, in terms of complexity classes, the picture looks quite different: Computing the nonlinearity is $\#\mathbf{P}$ hard, and multiplicative complexity is in the second level of the polynomial hierarchy.

2   Preliminaries

In the following, we let $\mathbb{F}_2$ be the finite field of size 2 and $\mathbb{F}_2^n$ the n-dimensional vector space over $\mathbb{F}_2$. We denote by B_n the set of Boolean functions, mapping from $\mathbb{F}_2^n$ into $\mathbb{F}_2$. We say that $f \in B_n$ is *affine* if there exist $\mathbf{a} \in \mathbb{F}_2^n, c \in \mathbb{F}_2$ such that $f(\mathbf{x}) = \mathbf{a} \cdot \mathbf{x} + c$ and *linear* if f is affine with $f(\mathbf{0}) = 0$, with arithmetic over $\mathbb{F}_2$. This gives the symbol "+" an overloaded meaning, since we also use it for addition over the reals. It should be clear from the context, what is meant.

In the following an XOR-AND circuit is a circuit with fanin 2 over the basis $(\wedge, \oplus, 1)$ (arithmetic over $GF(2)$). All circuits from now on are assumed to be XOR-AND circuits. We adopt standard terminology for circuits (see e.g. [21]). If nothing else is specified, for a circuit C we let n be the number of inputs and m be the number of gates, which we refer to as the *size* of C, denoted $|C|$. For a circuit C we let f_C denote the function computed by C, and $c_\wedge(C)$ denote the number of AND gates in C.

For a function $f \in B_n$, the *multiplicative complexity* of f, denoted $c_\wedge(f)$, is the smallest number of AND gates necessary and sufficient in an XOR-AND

---

[1] Here we mean concrete finite functions, as opposed to giving good (asymptotic) upper bounds for an infinite family of functions.

circuit computing f. The *nonlinearity* of a function f, denoted $NL(f)$ is the Hamming distance to its closest affine function, more precisely

$$NL(f) = 2^n - \max_{a \in \mathbb{F}_2^n, c \in \mathbb{F}_2} |\{\mathbf{x} \in \mathbb{F}_2^n | f(\mathbf{x}) = \mathbf{a} \cdot \mathbf{x} + c\}|.$$

We consider four decision problems in this paper: NL_C, NL_{TT}, MC_C and MC_{TT}. For NL_C (resp MC_C) the input is a circuit and a target $s \in \mathbb{N}$ and the goal is to determine whether the nonlinearity (resp. multiplicative complexity) of f_C is at most s. For NL_{TT} (resp. MC_{TT}) the input is a truth table of length 2^n of a function $f \in B_n$ and a target $s \in \mathbb{N}$, with the goal to determine whether the nonlinearity (resp. multiplicative complexity) of f is at most s.

We let $a \in_R D$ denote that a is distributed uniformly at random from D. We will need the following definition:

Definition 1. *A family of Boolean functions* $f = \{f_n\}_{n \in \mathbb{N}}, f_n : \{0,1\}^n \times \{0,1\}^n \to \{0,1\}$, *is a* pseudorandom function family *if f can be computed in polynomial time and for every probabilistic polynomial time oracle Turing machine A,*

$$\left| \Pr_{k \in_R \{0,1\}^n} [A^{f_n(k,\cdot)}(1^n) = 1] - \Pr_{g \in_R B_n} [A^{g(\cdot)}(1^n) = 1] \right| \leq n^{-\omega(1)}.$$

Here A^H denotes that the algorithm A has oracle access to a function H, that might be $f_n(\mathbf{k}, \cdot)$ for some $\mathbf{k} \in \mathbb{F}_2^n$ or a random $g \in B_n$, for more details see [1]. Some of our hardness results will be based on the following assumption.

Assumption 1. *There exist pseudorandom function families.*

It is known that pseudorandom function families exist if one-way functions exist [11,12,1], so we consider Assumption 1 to be very plausible. We will also use the following assumptions on the exponential complexity of SAT, due to Impagliazzo and Paturi.

Assumption 2 (Strong Exponential Time Hypothesis [13]). *For any fixed $c < 1$, no algorithm runs in time 2^{cn} and computes SAT correctly.*

3   Truth Table as Input

It is a well known result that given a function $f \in B_n$ represented by a truth table of length 2^n, the nonlinearity can be computed using $O(n2^n)$ basic arithmetic operations. This is done using the "Fast Walsh Transformation" (See [19] or chapter 1 in [16]).

In this section we show that the situation is different for multiplicative complexity: Under Assumption 1, MC_{TT} cannot be computed in polynomial time.

In [14], Kabanets and Cai showed that if *subexponentially* strong pseudorandom function families exist, the Minimum Circuit Size Problem (MCSP) (the problem of determining the size of a smallest circuit of a function given its truth table) cannot be solved in polynomial time. The proof goes by showing that if MCSP could be solved in polynomial time this would induce a natural combinatorial property (as defined in [17]) useful against circuits of polynomial size. Now by the celebrated result of Razborov and Rudich [17], this implies the nonexistence of subexponential pseudorandom function families.

Our proof below is similar in that we use results from [2] in a way similar to what is done in [14,17] (see also the excellent exposition in [1]). However instead of showing the existence of a natural and useful combinatorial property and appealing to limitations of natural proofs, we give an explicit polynomial time algorithm for breaking any pseudorandom function family, contradicting Assumption 1.

Theorem 1. *Under Assumption 1, on input a truth table of length 2^n, MC_{TT} cannot be computed in time $2^{O(n)}$.*

Proof. Let $\{f_n\}_{n \in \mathbb{N}}$ be a pseudorandom function family. Since f is computable in polynomial time it has circuits of polynomial size (see e.g. [1]), so we can choose $c \geq 2$ such that $c_\wedge(f_n) \leq n^c$ for all $n \geq 2$. Suppose for the sake of contradiction that some algorithm computes MC_{TT} in time $2^{O(n)}$. We now describe an algorithm that breaks the pseudorandom function family. The algorithm has access to an oracle $H \in B_n$, along with the promise either $H(\mathbf{x}) = f_n(\mathbf{k}, \mathbf{x})$ for $\mathbf{k} \in_R \mathbb{F}_2^n$ or $H(\mathbf{x}) = g(\mathbf{x})$ for $g \in_R B_n$. The goal of the algorithm is to distinguish between the two cases. Specifically our algorithm will return 0 if $H(\mathbf{x}) = f(\mathbf{k}, \mathbf{x})$ for some $\mathbf{k} \in \mathbb{F}_2^n$, and if $H(\mathbf{x}) = g(\mathbf{x})$ it will return 1 with high probability, where the probability is only taken over the choice of g.

Let $s = 10c \log n$ and define $h \in B_s$ as $h(\mathbf{x}) = H(\mathbf{x}0^{n-s})$. Obtain the complete truth table of h by querying H on all the $2^s = 2^{10c \log n} = n^{10c}$ points. Now compute $c_\wedge(h)$. By assumption this can be done in time $poly(n^{10c})$. If $c_\wedge(h) > n^c$, output 1, otherwise output 0. We now want to argue that this algorithm correctly distinguishes between the two cases. Suppose first that $H(\mathbf{x}) = f_n(\mathbf{k}, \cdot)$ for some $\mathbf{k} \in \mathbb{F}_2^n$. One can think of h as H where some of the input bits are fixed. But in this case, H can also be thought of as f_n with n of the input bits fixed. Now take the circuit for f_n with the minimal number of AND gates. Fixing the value of some of the input bits clearly cannot increase the number of AND gates, hence $c_\wedge(h) \leq c_\wedge(f_n) \leq n^c$.

Now it remains to argue that if H is a random function, we output 1 with high probability. We do this by using the following lemma.

Lemma 1 (Boyar, Peralta, Pochuev). *For all $s \geq 0$, the number of functions in B_s that can be computed with an XOR-AND circuit using at most k AND gates is at most $2^{k^2+2k+2ks+s+1}$.*

If g is random on B_n, then h is random on $B_{10c \log n}$, so the probability that $c_\wedge(h) \leq n^c$ is at most:

$$\frac{2^{(n^c)^2+2(n^c)+2(n^c)(10c \log n)+10c \log n+1}}{2^{2^{10c \log n}}}.$$

This tends to 0, so if H is a random function the algorithm returns 0 with probability $o(1)$. In total we have

$$\left| \Pr_{k \in_R \{0,1\}^n}[A^{f_n(k,\cdot)}(1^n) = 1] - \Pr_{g \in_R B_n}[A^{g(\cdot)}(1^n) = 1] \right| = |0 - (1 - o(1))|,$$

concluding that if the polynomial time algoritm for deciding MC_{TT} exists, f is not a pseudorandom function family.   □

A common question to ask about a computationally intractable problem is how well it can be *approximated* by a polynomial time algorithm. An algorithm approximates $c_\wedge(f)$ with approximation factor $\rho(n)$ if it always outputs some value in the interval $[c_\wedge(f), \rho(n)c_\wedge(f)]$. By refining the proof above, we see that it is hard to compute $c_\wedge(f)$ within even a modest factor.

Theorem 2. *For every constant $\epsilon > 0$, under Assumption 1, no algorithm takes the 2^n bit truth table of a function f and approximates $c_\wedge(f)$ with $\rho(n) \leq (2 - \epsilon)^{n/2}$ in time $2^{O(n)}$.*

Proof. Assume for the sake of contradiction that the algorithm A violates the theorem. The algorithm breaking any pseudorandom function family works as the one in the previous proof, but instead we return 1 if the value returned by A is at least $T = (n^c + 1) \cdot (2 - \epsilon)^{n/2}$. Now arguments similar to those in the proof above show that if A returns a value larger than T, H must be random, and if H is random, h has multiplicative complexity at most $(n^c + 1) \cdot (2 - \epsilon)^{n/2}$ with probability at most

$$\frac{2\left((n^c+1)\cdot(2-\epsilon)^{(10c\log n)/2}\right)^2 + 2(n^c+1)\cdot(2-\epsilon)^{10c\log n/2}10c\log n + 10c\log n + 1}{2^{2^{10c\log n}}}$$

This tends to zero, implying that under the assumption on A, there is no pseudorandom function family.   □

4   Circuit as Input

From a practical point of view, the theorems 1 and 2 might seem unrealistic. We are allowing the algorithm to be polynomial in the length of the truth table, which is exponential in the number of variables. However most functions used for practical purposes admit small circuits. To look at the entire truth table might (and in some cases should) be infeasible. When working with computational problems on circuits, it is somewhat common to consider the running time in two parameters; the number of inputs to the circuit, denoted by n, and the size of the circuit, denoted by m. In the following we assume that m is polynomial in n. In this section we show that even determining whether a circuit computes an affine function is **coNP**-complete. In addition NL_C can be computed in time $poly(m)2^n$, and is **#P**-hard. Under Assumption 1, MC_C cannot be computed in time $poly(m)2^{O(n)}$, and is contained in the second level of the polynomial hierarchy. In the following, we denote by $AFFINE$ the set of circuits computing affine functions.

Theorem 3. *$AFFINE$ is **coNP** complete.*

Proof. First we show that it actually is in **coNP**. Suppose $C \notin AFFINE$. Then if $f_C(0) = 0$, there exist $\mathbf{x}, \mathbf{y} \in \mathbb{F}_2^n$ such that $f_C(\mathbf{x} + \mathbf{y}) \neq f_C(\mathbf{x}) + f_C(\mathbf{y})$ and if $C(0) = 1$, there exists $\mathbf{x}, \mathbf{y}$ such that $C(\mathbf{x} + \mathbf{y}) + 1 \neq C(\mathbf{x}) + C(\mathbf{y})$. Given $C, \mathbf{x}$ and $\mathbf{y}$ this can clearly be computed in polynomial time. To show hardness, we reduce from $TAUTOLOGY$, which is **coNP**-complete.

Let F be a formula on n variables, $\mathbf{x}_1, \ldots, \mathbf{x}_n$. Consider the following reduction: First compute $c = F(0^n)$, then for every $\mathbf{e}^{(i)}$ (the vector with all coordinates 0 except the ith) compute $F(\mathbf{e}^{(i)})$. If any of these or c are 0, clearly $F \notin TAUTOLOGY$, so we reduce to a circuit trivially not in $AFFINE$. We claim that F computes an affine function if and only if $F \in TAUTOLOGY$. Suppose F computes an affine function, then $F(\mathbf{x}) = \mathbf{a} \cdot \mathbf{x} + c$ for some $\mathbf{a} \in \mathbb{F}_2^n$. Then for every $\mathbf{e}^{(i)}$, we have

$$F(\mathbf{e}^{(i)}) = \mathbf{a}_i + 1 = 1 = F(\mathbf{0}),$$

so we must have that $\mathbf{a} = \mathbf{0}$, and F is constant. Conversely if it is not affine, it is certainly not constant. In particular it is not a tautology. $\square$

So even determining whether the multiplicative complexity or nonlinearity is 0 is **coNP** complete. In the light of the above reduction, any algorithm for $AFFINE$ induces an algorithm for SAT with essentially the same running time, so under Assumption 2, AFFINE needs time essentially 2^n. This should be contrasted with the fact that the seemingly harder problem of computing NL_C can be done in time $poly(m)2^n$ by first computing the entire truth table and then using the Fast Walsh Transformation. Despite the fact that NL_C does not seem to require much more time to compute than $AFFINE$, it is hard for a much larger complexity class.

Theorem 4. NL_C *is* **#P**-*hard.*

Proof. We reduce from $\#SAT$. Let the circuit C on n variables be an instance of $\#SAT$. Consider the circuit C' on $n + 10$ variables, defined by

$$C'(\mathbf{x}_1, \ldots, \mathbf{x}_{n+10}) = C(\mathbf{x}_1, \ldots, \mathbf{x}_n) \wedge \mathbf{x}_{n+1} \wedge \mathbf{x}_{n+2} \wedge \ldots \wedge \mathbf{x}_{n+10}.$$

First we claim that independently of C, the best affine approximation of $f_{C'}$ is always 0. Notice that 0 agrees with $f_{C'}$ whenever at least one of $x_{n+1}, \ldots, x_{n+10}$ is 0, and when they are all 1 it agrees on $|\{\mathbf{x} \in \mathbb{F}_2^n | f_{C'}(\mathbf{x}) = 0\}|$ many points. In total 0 and $f_{C'}$ agree on

$$(2^{10} - 1)2^n + |\{\mathbf{x} \in \mathbb{F}_2^n | f_C(\mathbf{x}) = 0\}|$$

inputs. To see that any other affine function approximates $f_{C'}$ worse than 0, notice that any nonconstant affine function is balanced and thus has to disagree with $f_{C'}$ very often. The nonlinearity of $f_{C'}$ is therefore

$$NL(f_{C'}) = 2^{n+10} - \max_{\mathbf{a} \in \mathbb{F}_2^{n+10}, c \in \mathbb{F}_2} |\{\mathbf{x} \in \mathbb{F}_2^{n+10}| f_{C'}(\mathbf{x}) = \mathbf{a} \cdot \mathbf{x} + c\}|$$
$$= 2^{n+10} - |\{\mathbf{x} \in \mathbb{F}_2^{n+10}| f_{C'}(\mathbf{x}) = 0\}|$$
$$= 2^{n+10} - ((2^{10} - 1)2^n + |\{\mathbf{x} \in \mathbb{F}_2^n| f_C(\mathbf{x}) = 0\}|)$$
$$= 2^n - |\{\mathbf{x} \in \mathbb{F}_2^n| f_C(\mathbf{x}) = 0\}|$$
$$= |\{\mathbf{x} \in \mathbb{F}_2^n| f_C(\mathbf{x}) = 1\}|$$

So the nonlinearity of $f_{C'}$ equals the number satisfying assignments for C.

$\square$

So letting the nonlinearity, s, be a part of the input for NL_C changes the problem from being in level 1 of the polynomial hierarchy to be #**P** hard, but does not seem to change the time complexity much. The situation for MC_C is essentially the opposite, under Assumption 1, the time MC_C needs is strictly more time than $AFFINE$, but is contained in Σ_2^p. By appealing to Theorem 1 and 2, the following theorem follows.

Theorem 5. *Under Assumption 1, no polynomial time algorithm computes MC_C. Furthermore no algorithm with running time $poly(m)2^{O(n)}$ approximates $c_\wedge(f)$ with a factor of $(2 - \epsilon)^{n/2}$ for any constant $\epsilon > 0$.*

We conclude by showing that although MC_C under Assumption 1 requires more time, it is nevertheless contained in the second level of the polynomial hierarchy.

Theorem 6. $MC_C \in \Sigma_2^p$.

Proof. First observe that MC_C written as a language has the right form:

$$MC_C = \{(C, s) | \exists C' \; \forall \mathbf{x} \in \mathbb{F}_2^n \; (C(\mathbf{x}) = C'(\mathbf{x}) \text{ and } c_\wedge(C') \leq s)\}.$$

Now it only remains to show that one can choose the size of C' is polynomial in $n + |C|$. Specifically, for any $f \in B_n$, if C' is the circuit with the smallest number of AND gates computing f, for $n \geq 3$, we can assume that $|C'| \leq 2(c_\wedge(f) + n)^2 + c_\wedge$. For notational convenience let $c_\wedge(f) = M$. C' consists of XOR and AND gates and each of the M AND gates has exactly two inputs and one output. Consider some topological ordering of the AND gates, and call the output of the ith AND gate o_i. Each of the inputs to an AND gate is a sum (in $\mathbb{F}_2$) of x_is, o_is and possibly the constant 1. Thus the $2M$ inputs to the AND gates and the output, can be thought of as $2M + 1$ sums over $\mathbb{F}_2$ over $n + M + 1$ variables (we can think of the constant 1 as a variable with a hard-wired value). This can be computed with at most

$$(2M + 1)(n + M + 1) \leq 2(M + n)^2$$

XOR gates, where the inequality holds for $n \geq 3$. Adding $c_\wedge(f)$ for the AND gates, we get the claim. The theorem now follows, since $c_\wedge(f) \leq |C|$ $\square$

The relation between circuit size and multiplicative complexity given in the proof above is not tight, and we do not need it to be. See [18] for a tight relationship.

Acknowledgements. The author wishes to thank Joan Boyar for helpful discussions.

References

1. Arora, S., Barak, B.: Computational Complexity - A Modern Approach, pp. 1–579. Cambridge University Press (2009)
2. Boyar, J., Peralta, R., Pochuev, D.: On the multiplicative complexity of Boolean functions over the basis $(\wedge,\oplus,1)$. Theoretical Computer Science 235(1), 43–57 (2000)
3. Boyar, J., Damgård, I., Peralta, R.: Short non-interactive cryptographic proofs. J. Cryptology 13(4), 449–472 (2000)
4. Boyar, J., Find, M., Peralta, R.: Four measures of nonlinearity. In: Spirakis, P.G., Serna, M. (eds.) CIAC 2013. LNCS, vol. 7878, pp. 61–72. Springer, Heidelberg (2013)
5. Boyar, J., Matthews, P., Peralta, R.: Logic minimization techniques with applications to cryptology. J. Cryptology 26(2), 280–312 (2013)
6. Carlet, C.: Boolean functions for cryptography and error correcting codes. In: Crama, Y., Hammer, P.L. (eds.) Boolean Models and Methods in Mathematics, Computer Science, and Engineering, ch. 8, pp. 257–397. Cambridge Univ. Press, Cambridge (2010)
7. Cenk, M., Özbudak, F.: On multiplication in finite fields. J. Complexity 26(2), 172–186 (2010)
8. Chaum, D., Crépeau, C., Damgård, I.: Multiparty unconditionally secure protocols (extended abstract). In: Simon, J. (ed.) STOC, pp. 11–19. ACM (1988)
9. Courtois, N., Bard, G.V., Hulme, D.: A new general-purpose method to multiply 3x3 matrices using only 23 multiplications. CoRR abs/1108.2830 (2011)
10. Daemen, J., Rijmen, V.: AES proposal: Rijndael (1999),
    http://csrc.nist.gov/archive/aes/rijndael/Rijndael-ammended.pdf
11. Goldreich, O., Goldwasser, S., Micali, S.: How to construct random functions. J. ACM 33(4), 792–807 (1986)
12. Håstad, J., Impagliazzo, R., Levin, L.A., Luby, M.: A pseudorandom generator from any one-way function. SIAM J. Comput. 28(4), 1364–1396 (1999)
13. Impagliazzo, R., Paturi, R.: On the complexity of k-SAT. J. Comput. Syst. Sci. 62(2), 367–375 (2001)
14. Kabanets, V., yi Cai, J.: Circuit minimization problem. In: Yao, F.F., Luks, E.M. (eds.) STOC, pp. 73–79. ACM (2000)
15. Kolesnikov, V., Schneider, T.: Improved garbled circuit: Free XOR gates and applications. In: Aceto, L., Damgård, I., Goldberg, L.A., Halldórsson, M.M., Ingólfsdóttir, A., Walukiewicz, I. (eds.) ICALP 2008, Part II. LNCS, vol. 5126, pp. 486–498. Springer, Heidelberg (2008)
16. O'Donnell, R.: Analysis of Boolean Functions. Book draft (2012),
    http://www.analysisofbooleanfunctions.org
17. Razborov, A.A., Rudich, S.: Natural proofs. J. Comput. Syst. Sci. 55(1), 24–35 (1997)

18. Sergeev, I.S.: A relation between additive and multiplicative complexity of Boolean functions. CoRR abs/1303.4177 (2013)
19. Sloane, N., MacWilliams, F.J.: The Theory of Error-Correcting Codes. North-Holland Math. Library 16 (1977)
20. Vaikuntanathan, V.: Computing blindfolded: New developments in fully homomorphic encryption. In: Ostrovsky, R. (ed.) FOCS, pp. 5–16. IEEE (2011)
21. Wegener, I.: The Complexity of Boolean Functions. Wiley-Teubner (1987)

Block Products and Nesting Negations in FO²

Lukas Fleischer[1], Manfred Kufleitner[1,2,*], and Alexander Lauser[1,*]

[1] Formale Methoden der Informatik, Universität Stuttgart, Germany
[2] Fakultät für Informatik, Technische Universität München, Germany
fleiscls@studi.informatik.uni-stuttgart.de,
kufleitn@in.tum.de,
lauser@fmi.uni-stuttgart.de

Abstract. The alternation hierarchy in two-variable first-order logic
FO²[<] over words was recently shown to be decidable by Kufleitner
and Weil, and independently by Krebs and Straubing. In this paper we
consider a similar hierarchy, reminiscent of the half levels of the dot-depth
hierarchy or the Straubing-Thérien hierarchy. The fragment Σ_m^2 of FO²
is defined by disallowing universal quantifiers and having at most $m - 1$
nested negations. One can view Σ_m^2 as the formulas in FO² which have
at most m blocks of quantifiers on every path of their parse tree, and the
first block is existential. Thus, the mth level of the FO²-alternation hierarchy is the Boolean closure of Σ_m^2. We give an effective characterization
of Σ_m^2, *i.e.*, for every integer m one can decide whether a given regular
language is definable by a two-variable first-order formula with negation
nesting depth at most m. More precisely, for every m we give ω-terms U_m
and V_m such that an FO²-definable language is in Σ_m^2 if and only if its
ordered syntactic monoid satisfies the identity $U_m \leqslant V_m$. Among other
techniques, the proof relies on an extension of block products to ordered
monoids.

1   Introduction

The study of logical fragments over words has a long tradition in computer
science. The seminal Büchi-Elgot-Trakhtenbrot Theorem from the early 1960s
states that a language is regular if and only if it is definable in monadic second-
order logic [1,5,32]. A decade later, in 1971, McNaughton and Papert showed
that a language is definable in first-order logic if and only if it is star-free [17].
Combining this result with Schützenberger's famous characterization of star-
free languages in terms of finite aperiodic monoids [21] shows that it is decidable
whether a given regular language is first-order definable. Since then, many logical
fragments have been investigated, see *e.g.* [3,25] for overviews.

The motivation for such results is two-fold. First, restricted fragments often
yield more efficient algorithms for computational problems such as satisfiability
or separability. Second, logical fragments give rise to a descriptive complexity:

---

* The last two authors acknowledge the support by the German Research Foundation
(DFG) under grant DI 435/5-1.

E.A. Hirsch et al. (Eds.): CSR 2014, LNCS 8476, pp. 176–189, 2014.

The simpler the fragment to define a language, the simpler the language. This approach can help in understanding the rich structure of regular languages.

Logical fragments are usually defined by restricting some resources in formulas. The three most natural restrictions are the quantifier depth (*i.e.*, the number of nested quantifiers), the alternation depth (*i.e.*, the number alternations between existential and universal quantification), and the number of variables. With respect to decidability questions regarding definability, quantifier depth is not very interesting since for fixed quantifier depth only finitely many languages are definable (which immediately yields decidability), see *e.g.* [4]. The situation with alternation in first-order logic is totally different: Only the very first level (*i.e.*, no alternation) is known to be decidable [8,23]. By a result of Thomas [31] the alternation hierarchy in first-order logic is tightly connected with the dot-depth hierarchy [2] or the Straubing-Thérien hierarchy [24,29], depending on the presence or absence of the successor predicate. Some progress in the study of the dot-depth hierarchy and the Straubing-Thérien hierarchy was achieved by considering the half-levels. For example, the levels ½ and ³⁄₂ in each of the two hierarchies are decidable [6,18,19]. The half levels also have a counterpart in the alternation hierarchy of first-order logic by requiring existential quantifiers in the first block. Another point of view of the same hierarchy is to disallow universal quantifiers and to restrict the number of nested negations.

Regarding the number of variables, Kamp showed that linear temporal logic is expressively complete for first-order logic over words [7]. Since every modality in linear temporal logic can be defined using three variables, first-order logic with only three different names for the variables (denoted by FO^3) defines the same languages as full first-order logic. This result is often stated as $FO^3 = FO$. Allowing only two variable names yields the proper fragment FO^2 of first-order logic. Thérien and Wilke [30] showed that a language is FO^2 definable if and only if its syntactic monoid belongs to the variety **DA** and, since the latter is decidable, one can effectively check whether a given regular language is FO^2-definable. For further information on the numerous characterizations of FO^2 we refer to [3,28].

Inside FO^2, the alternation depth is also a natural restriction. One difference to full first-order logic is that one cannot rely on prenex normal forms as a simple way of defining the alternation depth. Weil and the second author gave an effective algebraic characterization of the mth level FO^2_m of this hierarchy. More precisely, they showed that it is possible to ascend the FO^2-alternation hierarchy using so-called Mal'cev products [15] which in this particular case preserve decidability. There are two main ingredients in the proof. The first one is a combinatorial tool known as *rankers* [33] or *turtle programs* [22], and the second is a relativization property of two-variable first-order logic. These two ingredients are then combined using a proof method introduced in [10]. Krebs and Straubing gave another decidable characterization of FO^2_m in terms of identities of ω-terms using completely different techniques [9,26]; their proof relies on so-called block products.

In this paper we consider the half-levels Σ^2_m of the FO^2-alternation hierarchy. A language is definable in Σ^2_m if and only if it is definable in FO^2 without universal quantifiers and with at most $m-1$ nested negations. It is easy to see that one can avoid negations of atomic predicates. One can think of Σ_m^2 as those FO^2-formulas which on every path of their parse tree have at most m quantifier blocks, and the outermost block is existential. The main contribution of this paper are ω-terms U_m and V_m such that an FO^2-definable language is Σ_m^2-definable if and only if its ordered syntactic monoid satisfies $U_m \leqslant V_m$. For a given regular language it is therefore decidable whether it is definable in Σ_m^2 by first checking whether it is FO^2-definable and if so, then verifying whether $U_m \leqslant V_m$ holds in its ordered syntactic monoid. Moreover, for every FO^2-definable language L one can compute the smallest integer m such that L is definable in Σ_m^2.

The proof step from the identities to logic is a refinement of the approach of Weil and the second author [15] which in turn uses a technique from [10, Section IV]. While the proof method in [10] is quite general and can be applied for solving various other problems [11,12,13,14], it relies on closure under negation. A very specific modification is necessary in order to get the scheme working in the current situation.

The proof for showing that Σ_m^2 satisfies the identity $U_m \leqslant V_m$ is an adaptation of Straubing's proof [26] to ordered monoids. Straubing's proof relies on two-sided semidirect products and the block product principle. We partially extend both tools to ordered monoids. To the best of our knowledge, this extension does not yet appear in the literature. The attribute *partially* is due to the fact that only the first factor in two-sided semidirect products (as used in this paper) is ordered while the second factor is an unordered monoid. As shown by Pin and Weil in the case of one-sided semidirect products [20], one could use ordered alphabets for further extending this approach. We refrain from this in order to focus on the presentation of our main result.

2   Preliminaries

The *free monoid* A^* is the set of finite words over A equipped with concatenation and the empty word ε as neutral element. Let $u = a_1 \cdots a_k$ with $a_i \in A$ be a finite word. The *alphabet* (also known as the *content*) of u is $\mathrm{alph}(u) = \{a_1, \ldots, a_k\}$, its *length* is $|u| = k$, and the *positions* of u are $1, \ldots, k$. We say that i is an *a-position* of u if $a_i = a$. The word u is a *(scattered) subword* of w if $w \in A^* a_1 \cdots A^* a_k A^*$.

First-Order Logic. We consider first-order logic $FO = FO[<]$ over finite words. The syntax of FO-formulas is

$$\varphi ::= \top \mid \bot \mid \lambda(x) = a \mid x = y \mid x < y \mid \neg\varphi \mid \varphi \vee \varphi \mid \varphi \wedge \varphi \mid \exists x\, \varphi$$

where $a \in A$ is a letter, and x and y are variables. We consider universal quantifiers $\forall x\, \varphi$ as an abbreviation of $\neg\exists x\, \neg\varphi$, and $x \leqslant y$ is a shortcut for $(x = y) \vee (x < y)$. The atomic formulas $\top$ and $\bot$ are *true* and *false*, respectively. Variables are interpreted as positions of a word, and $\lambda(x) = a$ is true if x is an a-position. The semantics of the other constructs is as usual; in particular, $\exists x\, \varphi$ means that there exists a position x which makes φ true, and

$x < y$ means that position x is (strictly) smaller than position y. We write $\varphi(x_1, \ldots, x_\ell)$ for a formula φ if at most the variables x_i appear freely in φ; and we write $u, p_1, \ldots, p_\ell \models \varphi(x_1, \ldots, x_\ell)$ if φ is true over u when x_i is interpreted as p_i. A *sentence* is a formula without free variables. A first-order sentence φ defines the language $L(\varphi) = \{u \in A^* \mid u \models \varphi\}$, and a language is *definable* in a first-order fragment $\mathcal{F}$ if it is defined by some sentence in $\mathcal{F}$.

The formulas φ_m in the mth level Σ_m of the *negation nesting* hierarchy in FO are defined as follows:

$$\varphi_m ::= \varphi_{m-1} \mid \neg\varphi_{m-1} \mid \varphi_m \vee \varphi_m \mid \varphi_m \wedge \varphi_m \mid \exists x \, \varphi_m$$
$$\varphi_0 ::= \top \mid \bot \mid \lambda(x) = a \mid x = y \mid x < y \mid \neg\varphi_0 \mid \varphi_0 \vee \varphi_0 \mid \varphi_0 \wedge \varphi_0$$

This means, for $m \geqslant 1$ the formulas in Σ_m have at most $m - 1$ nested negations over quantifier-free formulas φ_0. Using De Morgan's laws and the following equivalences, one can avoid negations in quantifier-free formulas for fixed alphabet A:

$$\lambda(x) \neq a \equiv \bigvee_{b \in A \setminus \{a\}} \lambda(x) = b$$
$$x \neq y \equiv (x < y) \vee (y < x)$$
$$\neg(x < y) \equiv (x = y) \vee (y < x)$$

Also note that, up to logical equivalence, our definition of Σ_m coincides with the more common definition in terms of formulas in prenex normal form with at most m blocks of quantifiers which start with an existential block. This can be seen by the usual procedure of renaming the variables and successively moving quantifiers outwards.

The two-variable fragment FO² of first-order logic uses (and reuses) only two different variables, say x and y. Combining FO² and Σ_m yields the fragment Σ_m^2. That is, we have $\varphi \in \Sigma_m^2$ if both $\varphi \in \Sigma_m$ and $\varphi \in \text{FO}^2$. This also justifies the notation Σ_m^2 which inherits the symbol as well as the subscript from Σ_m and the exponent from FO². The Boolean closure of Σ_m^2 is the mth level FO_m^2 of the *alternation hierarchy* within FO².

Ordered Monoids. *Green's relations* are an important tool in the study of finite monoids. For $x, y \in M$ let $x \leqslant_{\mathcal{R}} y$ if $xM \subseteq yM$, and let $x \leqslant_{\mathcal{L}} y$ if $Mx \subseteq My$. We write $x \mathrel{\mathcal{R}} y$ if both $x \leqslant_{\mathcal{R}} y$ and $y \leqslant_{\mathcal{R}} x$; and we set $x <_{\mathcal{R}} y$ if $x \leqslant_{\mathcal{R}} y$ but not $x \mathrel{\mathcal{R}} y$. The relations $\mathcal{L}$ and $<_{\mathcal{L}}$ are defined similarly. An element $x \in M$ is *idempotent* if $x^2 = x$. For every finite monoid M there exists an integer $\omega_M \geqslant 1$ such that x^{ω_M} is the unique idempotent power generated by $x \in M$. If the reference to M is clear from the context, we simply write ω instead of ω_M.

An *ordered monoid* $(M, \leqslant)$ is a monoid M equipped with a partial order $\leqslant$ which is compatible with multiplication in M; that is, $x \leqslant x'$ and $y \leqslant y'$ implies $xy \leqslant x'y'$. Every monoid can be considered as an ordered monoid by using the identity relation as order. If no ambiguity arises, we subsequently use the notation M without explicitly mentioning the order. An *order ideal* of M is a subset $I \subseteq M$ such that $y \leqslant x$ and $x \in I$ implies $y \in I$.

A *monotone homomorphism* $h\colon M \to N$ is a monoid homomorphism of ordered monoids M and N such that $x \leqslant y$ implies $h(x) \leqslant h(y)$. Submonoids of ordered monoids naturally inherit the order. A monoid N *divides* a monoid M if there exists a surjective homomorphism from a submonoid of M onto N; moreover, if M and N are ordered, then we require the homomorphism to be monotone. The *direct product* of ordered monoids $M_1, \ldots, M_k$ is the usual direct product $M_1 \times \cdots \times M_k$ equipped with the *product order*, i.e., $(x_1, \ldots, x_k) \leqslant (y_1, \ldots, y_k)$ if $x_i \leqslant y_i$ for all $i \in \{1, \ldots, k\}$. The empty direct product is the trivial monoid.

Varieties and Identities. A *variety* (respectively, *positive variety*) is a class of finite monoids (respectively, finite ordered monoids) closed under division and finite direct products. By abuse of notation, we sometimes say that an ordered monoid $(M, \leqslant)$ belongs to a variety $\mathbf{V}$ of unordered monoids if $M \in \mathbf{V}$. Both varieties and positive varieties are often defined by identities of ω-terms. We only describe the formal setting for positive varieties. The *ω-terms* over the variables X are defined inductively: The constant $1 \notin X$ is an ω-term and every variable $x \in X$ is an ω-term. If u and v are ω-terms, then so are uv and u^ω. Here, ω is considered as a formal symbol instead of a fixed integer. Every mapping $h\colon X \to M$ to a finite monoid M uniquely extends to ω-terms by setting $h(1) = 1$, $h(uv) = h(u)h(v)$ and $h(u^\omega) = h(u)^{\omega_M}$. An ordered monoid M *satisfies* the identity $U \leqslant V$ for ω-terms U and V if $h(U) \leqslant h(V)$ for all mappings $h\colon X \to M$. It satisfies $U = V$ if it satisfies both $U \leqslant V$ and $V \leqslant U$. Every class of ordered monoids defined by a set of identities of ω-terms forms a positive variety. In this paper, we need the following varieties:

- The variety $\mathbf{J}$ is the class of all so-called $\mathcal{J}$-trivial finite monoids. There are several well-known characterizations of this class, the most popular being Simon's Theorem on piecewise testable languages [23]. One can define $\mathbf{J}$ by the identities $(xyz)^\omega y = (xyz)^\omega = y(xyz)^\omega$.
- The positive variety $\mathbf{J}^+$ is defined by the identity $x \leqslant 1$. There is a language theoretic characterization similar to Simon's Theorem in terms of so-called shuffle ideals [18].
- The variety $\mathbf{DA}$ is defined by $(xyz)^\omega y(xyz)^\omega = (xyz)^\omega$. Suppose $M \in \mathbf{DA}$ and let $u, v, a \in M$. If $v \,\mathcal{R}\, u \,\mathcal{R}\, ua$, then $v \,\mathcal{R}\, va$; and symmetrically, if $v \,\mathcal{L}\, u \,\mathcal{L}\, au$, then $v \,\mathcal{L}\, av$, see e.g. [12, Lemma 1].

Languages and Syntactic Monoids. A language $L \subseteq A^*$ is *recognized* by a homomorphism $h\colon A^* \to M$ to some ordered monoid M if $L = h^{-1}(I)$ for some order ideal I of M. An ordered monoid M recognizes a language $L \subseteq A^*$ if there exists a homomorphism $h\colon A^* \to M$ which recognizes L. The *syntactic preorder* $\leqslant_L$ on words is defined as follows: We set $u \leqslant_L v$ for $u, v \in A^*$ if $pvq \in L$ implies $puq \in L$ for all $p, q \in A^*$. We write $u \equiv_L v$ if both $u \leqslant_L v$ and $v \leqslant_L u$. The *syntactic monoid* M_L of L is the quotient $A^*/\!\equiv_L$ consisting of the equivalence classes of $\equiv_L$; it is the unique minimal recognizer of L and it is effectively computable from any reasonable presentation of a given regular language. The syntactic preorder induces a partial order on the $\equiv_L$-classes such that M_L becomes an ordered monoid. The *syntactic homomorphism* $h_L\colon A^* \to M_L$ is the natural quotient map.

3 Two-Sided Semidirect Products of Ordered Monoids

The two-sided semidirect product of finite monoids is a useful tool for studying decompositions and hierarchies of varieties, see *e.g.* [25]. In this section, we partially extend the definition to ordered monoids. Let M be an ordered monoid and let N be a monoid. We write the operation in M additively to improve readability, which does not mean that M is commutative. A *left action* of N on M is a mapping $(n, m) \mapsto n \cdot m$ from $N \times M$ to M such that for all $m, m_1, m_2 \in M$ and all $n, n_1, n_2 \in N$ the following axioms hold:

$$n \cdot (m_1 + m_2) = n \cdot m_1 + n \cdot m_2$$
$$(n_1 n_2) \cdot m = n_1 \cdot (n_2 \cdot m)$$
$$1 \cdot m = m$$
$$n \cdot 0 = 0$$
$$n \cdot m_1 \leqslant n \cdot m_2 \quad \text{whenever} \quad m_1 \leqslant m_2$$

To shorten notation, we usually write nm instead of $n \cdot m$. A *right action* of N on M is defined symmetrically. A left and a right action are *compatible* if $(n_1 m)n_2 = n_1(mn_2)$ for all $m \in M$ and all $n_1, n_2 \in N$. For compatible left and right actions of N on M we define the *two-sided semidirect product* $M \ast\ast N$ as the ordered monoid on the set $M \times N$ with the multiplication

$$(m_1, n_1)(m_2, n_2) = (m_1 n_2 + n_1 m_2, n_1 n_2),$$

and the order given by

$$(m_1, n_1) \leqslant (m_2, n_2) \quad \text{if and only if} \quad m_1 \leqslant m_2 \text{ and } n_1 = n_2.$$

It is straightforward to verify that $M \ast\ast N$ indeed is an ordered monoid for each pair of compatible actions. The two-sided semidirect product with left action $(n, m) \mapsto m$ and right action $(m, n) \mapsto m$ yields the direct product of M and N. In this sense the two-sided semidirect product generalizes the usual direct product.

We now define the so-called *block product* as a particular two-sided semidirect product. Let $M^{N \times N}$ be the ordered monoid of all functions from $N \times N$ to the ordered monoid M with componentwise operation. These functions are ordered by $f_1 \leqslant f_2$ if $f_1(n_1, n_2) \leqslant f_2(n_1, n_2)$ for all $n_1, n_2 \in N$. One can view $M^{N \times N}$ as the direct product of $|N|^2$ copies of M. The *block product* $M \square N$ is the two-sided semidirect product $M^{N \times N} \ast\ast N$ induced by the following pair of left and right actions. For $f \in M^{N \times N}$ and $n, n_1, n_2 \in N$ let

$$(nf)(n_1, n_2) = f(n_1, nn_2) \quad \text{and} \quad (fn)(n_1, n_2) = f(n_1 n, n_2).$$

The relationship between two-sided semidirect products and block products is the same as in the unordered case; see *e.g.* [27].

Proposition 1. *Let M, M', N, N' be monoids and suppose that M and M' are ordered. The following properties hold:*

1. *Both M and $(N, =)$ divide every two-sided semidirect product $M \ast\ast N$.*
2. *Every two-sided semidirect product $M \ast\ast N$ divides $M \square N$.*
3. *If M divides M' and N divides N', then $M \square N$ divides $M' \square N'$.*

We now extend the notion of two-sided semidirect products to varieties. For a positive variety $\mathbf{V}$ and a variety $\mathbf{W}$ we let $\mathbf{V} ** \mathbf{W}$ consist of all ordered monoids dividing a two-sided semidirect product $M ** N$ for some $M \in \mathbf{V}$ and $N \in \mathbf{W}$. For two-sided semidirect products $M ** N$ and $M' ** N'$, we define a new two-sided semidirect product $(M \times M') ** (N \times N')$ by the actions

$$(n, n')(m, m') = (nm, n'm')$$
$$(m, m')(n, n') = (mn, m'n')$$

for all $m \in M$, $m' \in M'$, $n \in N$, and $n' \in N'$. An elementary verification shows that this two-sided semidirect product is isomorphic to $(M ** N) \times (M' ** N')$, and $\mathbf{V} ** \mathbf{W}$ forms a positive variety. By Proposition 1 we see that $\mathbf{V} ** \mathbf{W}$ is identical to the positive variety generated by all block products $M \,\square\, N$ with $M \in \mathbf{V}$ and $N \in \mathbf{W}$.

For a homomorphism $h_N : A^* \to N$ we consider the alphabet $A_N = N \times A \times N$ and the length-preserving mapping $\sigma_{h_N} : A^* \to A_N^*$ defined by $\sigma_{h_N}(a_1 \cdots a_n) = b_1 \cdots b_n$, where

$$b_i = (h_N(a_1 \cdots a_{i-1}), a_i, h_N(a_{i+1} \cdots a_n))$$

for all $i \in \{1, \ldots, n\}$. The following proposition uses such mappings to characterize the languages recognized by two-sided semidirect products. It is known as the *block product principle*.

Proposition 2. *Let $\mathbf{V}$ be a positive variety, let $\mathbf{W}$ be a variety, and let $L \subseteq A^*$. The following conditions are equivalent.*

1. *L is recognized by an ordered monoid in $\mathbf{V} ** \mathbf{W}$.*
2. *There exists a homomorphism $h_N : A^* \to N$ with $N \in \mathbf{W}$ such that L is a finite union of languages of the form $\sigma_{h_N}^{-1}(L_K) \cap L_N$ with $L_K \subseteq A_N^*$ being recognized by a monoid in $\mathbf{V}$ and $L_N \subseteq A^*$ being recognized by h_N.*

4   Decidability of Negation Nesting in FO^2

In this section we give two algebraic characterizations of the languages definable in the fragment Σ_m^2 of two-variable first-order logic with a restricted number of nested negations. The first description is in terms of (weakly) iterated two-sided semidirect products with $\mathcal{J}$-trivial monoids. For this we define a sequence of positive varieties by setting $\mathbf{W}_1 = \mathbf{J}^+$ and $\mathbf{W}_m = \mathbf{W}_{m-1} ** \mathbf{J}$. As for the second characterization, we define sequences of ω-terms U_m and V_m by setting

$$U_1 = z, \quad U_m = (U_{m-1} x_m)^\omega U_{m-1} (y_m U_{m-1})^\omega,$$
$$V_1 = 1, \quad V_m = (U_{m-1} x_m)^\omega V_{m-1} (y_m U_{m-1})^\omega,$$

where $x_2, y_2, \ldots, x_m, y_m, z$ are variables.

Theorem 1. *Let $L \subseteq A^*$ and let $m \geqslant 1$. The following conditions are equivalent:*

1. *L is definable in Σ_m^2.*
2. *The ordered syntactic monoid of L is in $\mathbf{W}_m$.*
3. *The ordered syntactic monoid of L is in $\mathbf{DA}$ and satisfies $U_m \leqslant V_m$.*

Since condition 3. in Theorem 1 is decidable for any given regular language L, this immediately yields the following corollary.

Corollary 1. *It is decidable whether a given regular language is definable in Σ_m^2.*

Note that in condition 3. of Theorem 1 one cannot drop requiring that the syntactic monoid is in **DA**. For example, the syntactic monoid of $A^* \setminus A^* aa A^*$ over $A = \{a, b\}$ satisfies the identity $U_m \leqslant V_m$ for all $m \geqslant 2$. It is nonetheless not Σ_m^2-definable, because it is not even FO^2-definable (and thus its syntactic monoid is not in **DA**). The remainder of this section proves Theorem 1. We begin with the direction (1) $\Rightarrow$ (2). The arguments are similar to Straubing's for characterizing FO$_m^2$ in terms of unordered two-sided semidirect products [26].

Lemma 1. *Let $m \geqslant 1$. If L is definable in Σ_m^2, then $M_L \in \mathbf{W}_m$.*

Proof. Let φ be a sentence in Σ_m^2 such that $L = L(\varphi)$. We may assume that quantifier-free subformulas of φ do not contain negations.

The proof proceeds by induction on m. For the base case $m = 1$, the language L is a finite union of languages of the form $A^* a_1 \cdots A^* a_k A^*$ and thus $pq \in L$ implies $puq \in L$ for all $p, u, q \in A^*$. This means that M_L satisfies $x \leqslant 1$ and therefore, $M_L \in \mathbf{J}^+$, see [18].

Let now $m \geqslant 2$. An *innermost block* of φ is a maximal negation-free subformula $\psi(x)$ of φ. As in the unordered case, one can show that each block is equivalent to a disjunction of formulas of the form

$$\lambda(x) = a \wedge \left(\exists y_1 \cdots \exists y_r \bigwedge_{i=1}^{r} (y_i < x \wedge \lambda(y_i) = a_i) \wedge \pi(y_1, \ldots, y_r) \right) \wedge$$
$$\left(\exists z_1 \cdots \exists z_s \bigwedge_{i=1}^{s} (z_i > x \wedge \lambda(z_i) = a_i') \wedge \pi'(z_1, \ldots, z_s) \right),$$

where π and π' are quantifier-free formulas defining an order on their parameters. Hence, each innermost block $\psi(x)$ requires that x is an a-position and that certain subwords appear to the left and to the right of position x. Let k be the maximum of all r and s occurring in these blocks. By Simon's Theorem [23], there exists an unordered monoid $N \in \mathbf{J}$ and a homomorphism $h_N : A^* \to N$ such that $h_N(u) = h_N(v)$ if and only if u and v agree on subwords of length at most k. Now, the aforementioned blocks can be replaced by a disjunction of formulas $\lambda(x) = (n, a, n')$ with $n, n' \in N$ and $a \in A$ to obtain an equivalent formula over the alphabet A_N.

After replacing each innermost block, the resulting formula φ' is in Σ_{m-1}^2. By induction, the corresponding language $L(\varphi')$ is recognized by a monoid $K \in \mathbf{W}_{m-1}$. We have $L = L(\varphi) = \sigma_{h_N}(L(\varphi'))$ by construction. Proposition 2 finally yields $M_L \in \mathbf{W}_{m-1} ** \mathbf{J} = \mathbf{W}_m$. $\qquad\square$

The following lemma can be seen by a similar reasoning as in the unordered case due to Straubing [26].

Lemma 2. *Let $m \geqslant 1$. If $M \in \mathbf{W}_m$, then $M \in \mathbf{DA}$ and M satisfies $U_m \leqslant V_m$.*

We turn to the implication (3) $\Rightarrow$ (1) in Theorem 1, from $U_m \leqslant V_m$ back to logic Σ_m^2. On a high-level perspective, we want to use induction on m, then use the identity $U_{m-1} \leqslant V_{m-1}$ to get to Σ_{m-1}^2, and finally lift this back to Σ_m^2. An important part of this argument is the ability to restrict (or *relativize*) the interpretation of Σ_m^2-formulas to certain factors of the model which are given by first and last occurrences of letters.

In the following we also have to take the *quantifier depth* of a formula into account, *i.e.*, the maximal number of nested quantifiers. For an integer $n \geqslant 0$ let $\Sigma_{m,n}^2$ be the fragment of Σ_m^2 of formulas with quantifier depth at most n.

Lemma 3. *Let $\varphi \in \Sigma_{m,n}^2$ for $m, n \geqslant 0$, and let $a \in A$. There exist formulas $\langle \varphi \rangle_{>\mathsf{X}a} \in \Sigma_{m,n+1}^2$ and $\langle \varphi \rangle_{<\mathsf{X}a} \in \Sigma_{m+1,n+1}^2$ such that for all $u = u_1 a u_2$ with $a \notin \mathrm{alph}(u_1)$ and $i = |u_1 a|$ we have:*

$$u, p, q \models \langle \varphi \rangle_{<\mathsf{X}a} \quad \text{if and only if} \quad u_1, p, q \models \varphi \quad \text{for all } 1 \leqslant p, q < i,$$

$$u, p, q \models \langle \varphi \rangle_{>\mathsf{X}a} \quad \text{if and only if} \quad u_2, p - i, q - i \models \varphi \quad \text{for all } i < p, q \leqslant |u|.$$

Proof. Let $\langle \varphi \rangle_{<\mathsf{X}a} \equiv \varphi$ if φ is an atomic formula. For conjunction and disjunction, and negation we inductively take $\langle \varphi \rangle_{<\mathsf{X}a} \wedge \langle \psi \rangle_{<\mathsf{X}a}$ and $\langle \varphi \rangle_{<\mathsf{X}a} \vee \langle \psi \rangle_{<\mathsf{X}a}$, and $\neg \langle \varphi \rangle_{<\mathsf{X}a}$, respectively. For existential quantification let

$$\langle \exists x \, \varphi \rangle_{<\mathsf{X}a} \equiv \exists x \left(\neg (\exists y \leqslant x \colon \lambda(y) = a) \wedge \langle \varphi \rangle_{<\mathsf{X}a} \right).$$

As usual, swapping the variables x and y yields the corresponding constructions for y. Atomic formulas and Boolean combinations in the construction of $\langle \varphi \rangle_{>\mathsf{X}a}$ are as above. For existential quantification let

$$\langle \exists x \, \varphi \rangle_{>\mathsf{X}a} \equiv \exists x \left((\exists y < x \colon \lambda(y) = a) \wedge \langle \varphi \rangle_{>\mathsf{X}a} \right). \qquad \square$$

The notation in the indices of the formulas mean that we restrict to the positions smaller (respectively, greater) than the first a-position (the neXt a-position, thence X_a). Of course there are dual formulas $\langle \varphi \rangle_{<\mathsf{Y}b} \in \Sigma_{m,n+1}^2$ as well as $\langle \varphi \rangle_{>\mathsf{Y}b} \in \Sigma_{m+1,n+1}^2$ for the last b-position (*i.e.*, the Yesterday b-position). The next lemma handles the case of the first a-position lying beyond the last b-position.

Lemma 4. *Let $\varphi \in \Sigma_{m,n}^2$ for $m, n \geqslant 0$, and let $a, b \in A$. There exists a formula $\langle \varphi \rangle_{(\mathsf{Y}b;\mathsf{X}a)}$ in $\Sigma_{m+1,n+1}^2$ such that for all words $u = u_1 b u_2 a u_3$ with $b \notin \mathrm{alph}(u_2 a u_3)$ and $a \notin \mathrm{alph}(u_1 b u_2)$ and for all $|u_1 b| < p, q \leqslant |u_1 b u_2|$ we have:*

$$u, p, q \models \langle \varphi \rangle_{(\mathsf{Y}b;\mathsf{X}a)} \quad \text{if and only if} \quad u_2, p - |u_1 b|, q - |u_1 b| \models \varphi.$$

Proof. Atomic formulas and Boolean combinations are straightforward. Let the macro $\mathsf{Y}_b < x < \mathsf{X}_a$ stand for $\neg (\exists y \leqslant x \colon \lambda(y) = a) \wedge \neg (\exists y \geqslant x \colon \lambda(y) = b)$. Using this shortcut, we set $\langle \exists x \, \varphi \rangle_{(\mathsf{Y}b;\mathsf{X}a)} \equiv \exists x \left((\mathsf{Y}_b < x < \mathsf{X}_a) \wedge \langle \varphi \rangle_{(\mathsf{Y}b;\mathsf{X}a)} \right). \qquad \square$

Let $h \colon A^* \to M$ be a homomorphism. The $\mathcal{L}$-*factorization* of a word u is the unique factorization $u = s_0 a_1 \cdots s_{\ell-1} a_\ell s_\ell$ with $s_i \in A^*$ and so-called *markers* $a_i \in A$ such that $h(s_\ell) \mathcal{L} 1$ and $h(s_i a_{i+1} \cdots s_{\ell-1} a_\ell s_\ell) >_{\mathcal{L}} h(a_i s_i \cdots a_\ell s_\ell) \mathcal{L}$ $h(s_{i-1} a_i \cdots s_{\ell-1} a_\ell s_\ell)$ for all i. Note that $\ell < |M|$. Furthermore, if $M \in \mathbf{DA}$, then $a_i \notin \mathrm{alph}(s_i)$. Let $D_{\mathcal{L}}(u)$ consist of the positions of the markers, *i.e.*, let

$D_\mathcal{L}(u) = \{|s_0 a_1 \cdots s_{i-1} a_i| \mid 1 \leqslant i \leqslant \ell\}$. The $\mathcal{R}$-factorization is defined left-right symmetrically, and the set $D_\mathcal{R}(u)$ consists of all positions $|pa|$ for prefixes pa of u such that $h(p) >_\mathcal{R} h(pa)$ for some $a \in A$. The following lemma combines the $\mathcal{R}$-factorization with the $\mathcal{L}$-factorization for monoids in **DA** such that, starting with Σ_m^2, one can express Σ_{m-1}^2-properties of the factors. To formulate this feature we set $u \leqslant_{m,n}^{\mathrm{FO}^2} v$ for words $u, v \in A^*$ if $v \models \varphi$ implies $u \models \varphi$ for all $\varphi \in \Sigma_{m,n}^2$.

Lemma 5. *Let* $h \colon A^* \to M$ *be a homomorphism with* $M \in$ **DA**, *let* $m \geqslant 2$ *and* $n \geqslant 0$ *be integers, and let* $u, v \in A^*$ *with* $u \leqslant_{m,2|M|+n}^{\mathrm{FO}^2} v$. *There exist factorizations* $u = s_0 a_1 \cdots s_{\ell-1} a_\ell s_\ell$ *and* $v = t_0 a_1 \cdots t_{\ell-1} a_\ell t_\ell$ *with* $a_i \in A$ *and* $s_i, t_i \in A^*$ *such that the following properties hold for all* $i \in \{1, \ldots, \ell\}$:

1. $s_i \leqslant_{m-1,n}^{\mathrm{FO}^2} t_i$,
2. $h(s_0) \mathcal{R} 1$ *and* $h(t_0 a_1 \cdots t_{i-1} a_i) \mathcal{R} h(t_0 a_1 \cdots t_{i-1} a_i s_i)$,
3. $h(s_\ell) \mathcal{L} 1$ *and* $h(a_i s_i \cdots a_\ell s_\ell) \mathcal{L} h(s_{i-1} a_i \cdots a_\ell s_\ell)$.

Proof. Note that in property 2. the suffix is s_i and not t_i. We want to prove the claim by an induction, for which we have to slightly generalize the claim. Apart from the words u and v from the premises of the lemma we also consider an additional word p which serves as a prefix for v. The proof is by induction on $|D_\mathcal{R}(pv) \setminus D_\mathcal{R}(p)|$. The assumptions are $u \leqslant_{m,n'}^{\mathrm{FO}^2} v$, where $n' = n + |D_\mathcal{R}(pv) \setminus D_\mathcal{R}(p)| + |D_\mathcal{L}(u)| + 1$. We shall construct factorizations $u = s_0 a_1 \cdots s_{\ell-1} a_\ell s_\ell$ and $pv = p t_0 a_1 \cdots t_{\ell-1} a_\ell t_\ell$ such that properties 1. and 3. hold, but instead of 2. we have $h(p t_0 a_1 \cdots t_{i-1} a_i) \mathcal{R} h(p t_0 a_1 \cdots t_{i-1} a_i s_i)$ and $h(p s_0) \mathcal{R}$ $h(p)$. We thus recover the lemma using an empty prefix p.

Let $u = s_0' c_1 \cdots s_{\ell'-1}' c_{\ell'} s_{\ell'}'$ be the $\mathcal{L}$-factorization (in particular $c_i \notin \mathrm{alph}(s_i')$) and let $v = t_0' c_1 \cdots t_{\ell'-1}' c_{\ell'} t_{\ell'}'$ where $c_i \notin \mathrm{alph}(t_i')$ for all i. The factorization of v exists because by assumption u and v agree on subwords of length ℓ'. The dual of Lemma 3 yields $s_0' c_1 \cdots c_{\ell'-i} s_{\ell'-i}' \leqslant_{m-1,n'-i}^{\mathrm{FO}^2} t_0' c_1 \cdots c_{\ell'-i} t_{\ell'-i}'$ as well as $s_i' \leqslant_{m-1,n}^{\mathrm{FO}^2} t_i'$ for all i.

First suppose $D_\mathcal{R}(p) = D_\mathcal{R}(pv)$. In this case $h(p) \mathcal{R} h(pv)$, and therefore, $h(p) \mathcal{R} h(px)$ for all $x \in B^*$, where $B = \mathrm{alph}(v)$. So in particular we have that $h(p t_0' c_1 \cdots t_{i-1}' c_i) \mathcal{R} h(p t_0' c_1 \cdots t_{i-1}' c_i s_i')$ because $\mathrm{alph}(u) = B$. Setting $a_i = c_i$, $s_i = s_i'$, and $t_i = t_i'$ yields a factorization with the desired properties.

Suppose now $D_\mathcal{R}(p) \subsetneq D_\mathcal{R}(pv)$, and let s be the longest prefix of u such that $h(p) \mathcal{R} h(ps) >_\mathcal{R} h(psa)$ for some $a \in A$. Such a prefix exists as $\mathrm{alph}(u) = \mathrm{alph}(v)$. We have $a \notin \mathrm{alph}(s)$ by $M \in$ **DA**. Let t be the longest prefix of v with $a \notin \mathrm{alph}(t)$. Using Lemma 3 we see $\mathrm{alph}(t) \subseteq \mathrm{alph}(s)$. Let k and k' be maximal such that $s_0' c_1 \cdots s_{k-1}' c_k$ is a prefix of s and such that $t_0' c_1 \cdots t_{k'-1}' c_{k'}$ is a prefix of t. We claim $k = k'$. For instance, suppose $k < k'$. Then $a c_{k+1} \cdots c_{\ell'}$ is a subword of u but not of v (since $c_{k+1} t_{k+1}' \cdots c_{\ell'} t_{\ell'}'$ is the shortest suffix of v with the subword $c_{k+1} \cdots c_{\ell'}$ and since there is no a-position in $t_0' c_1 \cdots t_k'$). Let $a_i = c_i$ for $i \in \{1, \ldots, k\}$, let $s_i = s_i'$ and $t_i = t_i'$ for $i \in \{0, \ldots, k-1\}$. Let s_k and t_k such that $s = s_0 c_1 \cdots s_{k-1} c_k s_k$ and $t = t_0 c_1 \cdots t_{k-1} c_k t_k$. Lemma 4 yields $s_k \leqslant_{m-1,n}^{\mathrm{FO}^2} t_k$.

Let $u = sau'$ and $v = tav'$, and let $p' = pta$. For all $i \in \{0, \ldots, k\}$ we have $h(p t_0 a_1 \cdots t_{i-1} a_i) \mathcal{R} h(p t_0 a_1 \cdots t_{i-1} a_i s_i)$ because $\mathrm{alph}(t) \subseteq \mathrm{alph}(s)$. Note that $h(a_{i+1} s_{i+1} \cdots a_k s_k a u') \mathcal{L} h(s_i a_{i+1} s_{i+1} \cdots a_k s_k a u')$. Since $M \in$ **DA** we see

$h(p) >_{\mathcal{R}} h(p')$ and thus $D_{\mathcal{R}}(p) \subsetneq D_{\mathcal{R}}(p')$. Using the formulas $\langle \varphi \rangle_{>\mathsf{X}a}$ from Lemma 3 yields $u' \leqslant^{\text{FO}^2}_{m,n'-1} v'$. As $n' \geqslant |D_{\mathcal{R}}(p'v') \setminus D_{\mathcal{R}}(p')| + |D_{\mathcal{L}}(u')| + 2$ we can apply induction to obtain factorizations $u' = s_{k+1}a_{k+2} \cdots s_{\ell-1}a_\ell s_\ell$ and $v' = t_{k+1}a_{k+2} \cdots t_{\ell-1}a_\ell t_\ell$. Setting $a_{k+1} = a$ yields the desired factorizations. $\square$

The preceding lemma enables induction on the parameter m. We start with a homomorphism onto a monoid satisfying $U_m \leqslant V_m$ and want to show that preimages of $\leqslant$-order ideals are unions of $\leqslant^{\text{FO}^2}_{m,n}$-order ideals for some sufficiently large n. Intuitively, a string rewriting technique yields the largest quotient which satisfies the identity $U_{m-1} \leqslant V_{m-1}$. One rewriting step corresponds to one application of the identity $U_{m-1} \leqslant V_{m-1}$ of level $m-1$. Such rewriting steps can be lifted to the identity $U_m \leqslant V_m$ in the contexts they are applied.

Proposition 3. *Let $m \geqslant 1$ be an integer, let $h\colon A^* \to M$ be a surjective homomorphism onto an ordered monoid $M \in \mathbf{DA}$ satisfying $U_m \leqslant V_m$. There exists a positive integer n such that $u \leqslant^{\text{FO}^2}_{m,n} v$ implies $h(u) \leqslant h(v)$ for all $u, v \in A^*$.*

Proof. We proceed by induction on m. For the base case $m = 1$ a result of Pin [18] shows that, for every $\leqslant$-order ideal I of M, the set $h^{-1}(I)$ is a finite union of languages $A^*a_1 \cdots A^*a_k A^*$ for some $k \geqslant 1$ and $a_i \in A$. Let n be the maximum of all indices k appearing in those unions when considering all order ideals $I \subseteq M$. If $u \leqslant^{\text{FO}^2}_{1,n} v$, then for all languages $P = A^*a_1 \cdots A^*a_k A^*$ with $k \leqslant n$ we have that $v \in P$ implies $u \in P$. Moreover, the preimage L of the order ideal generated by $h(v)$ is a finite union of languages $A^*a_1 \cdots A^*a_k A^*$ with $k \leqslant n$. We have $v \in L$ and thus $u \in L$. This shows $h(u) \leqslant h(v)$.

In the following let $m \geqslant 2$ and fix some integer $\omega \geqslant 1$ such that x^ω is idempotent for all $x \in M$. We introduce a string rewriting system $\to$ on A^* by letting $t \to s$ if $h(s) = h(t)$ or if $t = pv_{m-1}q$ and $s = pu_{m-1}q$ for $p, q \in A^*$, and $v_1 = 1$ and $u_1 = z$, and for $i \geqslant 2$ we have

$$v_i = (u_{i-1}x_i)^\omega v_{i-1}(y_iu_{i-1})^\omega, \quad u_i = (u_{i-1}x_i)^\omega u_{i-1}(y_iu_{i-1})^\omega$$

for $x_i, y_i, z \in A^*$. Note that $t \to s$ implies $p'tq' \to p'sq'$ for all $p', q' \in A^*$. Let $\overset{*}{\to}$ be the transitive closure of $\to$, i.e., let $t \overset{*}{\to} s$ if there exists a chain $t = w_1 \to w_2 \to \cdots \to w_\ell = s$ of rewriting steps for some $\ell \geqslant 1$ and $w_i \in A^*$. We claim that we can lift the rewriting steps of $t \overset{*}{\to} s$ to M within certain contexts in an order respecting way.

Claim. Let $u, v, s, t \in A^*$ with $t \overset{*}{\to} s$. If both $h(u) \, \mathcal{R} \, h(us)$ and $h(v) \, \mathcal{L} \, h(sv)$, then $h(usv) \leqslant h(utv)$.

The proof of the claim is by induction on the length of a minimal $\to$-chain from t to s. The claim is trivial if $h(t) = h(s)$. Suppose $t \overset{*}{\to} t' \to s$ and $t' = pv_{m-1}q$ and $s = pu_{m-1}q$. Since $h(u) \, \mathcal{R} \, h(us)$, there exists $x \in A^*$ such that $h(u) = h(usx)$; and since $h(v) \, \mathcal{L} \, h(sv)$ there exists $y \in A^*$ such that $h(v) = h(ysv)$. Now $h(u) = h(u(pu_{m-1}qx)^\omega)$ and $h(v) = h((ypu_{m-1}q)^\omega v)$. By letting $x_m = qxp$ and $y_m = qyp$, the identity $U_m \leqslant V_m$ of M yields

$$h(usv) = h\big(up(u_{m-1}x_m)^\omega u_{m-1}(y_mu_{m-1})^\omega qv\big)$$
$$\leqslant h\big(up(u_{m-1}x_m)^\omega v_{m-1}(y_mu_{m-1})^\omega qv\big) = h(ut'v).$$

Observe that $(pu_{m-1}qx)^\omega p = p(u_{m-1}qxp)^\omega = p(u_{m-1}x_m)^\omega$. Note that alph$(t') \subseteq$ alph(s). Therefore, $h(u)$ $\mathcal{R}$ $h(us)$ implies $h(u)$ $\mathcal{R}$ $h(ut')$, and symmetrically $h(v)$ $\mathcal{L}$ $h(sv)$ implies $h(v)$ $\mathcal{L}$ $h(t'v)$. Induction yields $h(ut'v) \leqslant h(utv)$ and thus $h(usv) \leqslant h(utv)$. This completes the proof of the claim.

Let $t \sim s$ if $t \stackrel{\cdot}{\to} s$ and $s \stackrel{\cdot}{\to} t$. Let M' be the quotient $A^*/\sim$. The relation $\sim$ is a congruence on A^* and M' is naturally equipped with a monoid structure. Let $h' : A^* \to M'$ be the canonical homomorphism mapping $u \in A^*$ to its equivalence class modulo $\sim$. The preorder $\stackrel{\cdot}{\to}$ on A^* induces a partial order on M' by letting $h'(u) \leqslant h'(v)$ whenever $v \stackrel{\cdot}{\to} u$. Thus M' forms an ordered monoid. Moreover, M' is an unordered quotient of M and, in particular, M' is finite and in **DA**, and x^ω is idempotent for all $x \in M'$.

By construction, M' satisfies the identity $U_{m-1} \leqslant V_{m-1}$ and induction yields an integer n such that $u \leqslant^{\text{FO}^2}_{m-1,n} v$ implies $h'(u) \leqslant h'(v)$. We show that $u \leqslant^{\text{FO}^2}_{m,n'} v$ implies $h(u) \leqslant h(v)$ for $n' = n + 2|M|$. Suppose $u \leqslant^{\text{FO}^2}_{m,n'} v$ and consider the factorizations $u = s_0 a_1 \cdots s_{\ell-1} a_\ell s_\ell$ and $v = t_0 a_1 \cdots t_{\ell-1} a_\ell t_\ell$ from Lemma 5. For all i we have:

- $s_i \leqslant^{\text{FO}^2}_{m-1,n} t_i$ and thus $t_i \stackrel{\cdot}{\to} s_i$ by choice of n,
- $h(t_0 a_1 \cdots t_{i-1} a_i)$ $\mathcal{R}$ $h(t_0 a_1 \cdots t_{i-1} a_i s_i)$, and
- $h(a_{i+1} s_{i+1} \cdots a_\ell s_\ell)$ $\mathcal{L}$ $h(s_i a_{i+1} s_{i+1} \cdots a_\ell s_\ell)$.

For conciseness $t_0 a_1 \cdots t_{i-1} a_i$ is the empty word if $i = 0$, and so is $a_{i+1} s_{i+1} \cdots a_\ell s_\ell$ if $i = \ell$. Applying the above claim repeatedly to substitute s_i with t_i for increasing $i \in \{0, \ldots, \ell\}$ yields the following chain of inequalities:

$$h(u) = h(s_0 a_1 s_1 \cdots s_{\ell-1} a_\ell s_\ell)$$
$$\leqslant h(t_0 a_1 s_1 \cdots s_{\ell-1} a_\ell s_\ell)$$
$$\vdots$$
$$\leqslant h(t_0 a_1 t_1 \cdots t_{\ell-1} a_\ell s_\ell)$$
$$\leqslant h(t_0 a_1 t_1 \cdots t_{\ell-1} a_\ell t_\ell) = h(v). \qquad \square$$

Proof of Theorem 1. The implication 1. $\Rightarrow$ 2. is Lemma 1, and 2. $\Rightarrow$ 3. is Lemma 2. For the implication 3. $\Rightarrow$ 1., let $L \subseteq A^*$ be a language, let $h_L : A^* \to M_L$ be its syntactic homomorphism. Moreover, suppose that M_L is in **DA** and satisfies $U_m \leqslant V_m$. The set $I = h_L(L)$ is an order ideal of M_L. Proposition 3 shows that there exists an integer n such that $L = h_L^{-1}(I)$ is a union of $\leqslant^{\text{FO}^2}_{m,n}$-order ideals. Up to equivalence, there are only finitely many formulas with quantifier depth n. Therefore, $\leqslant^{\text{FO}^2}_{m,n}$-order ideals are $\Sigma^2_{m,n}$-definable. $\qquad \square$

Conclusion

The fragments Σ^2_m of FO$^2[<]$ are defined by restricting the number of nested negations. They can be seen as the half levels of the alternation hierarchy FO^2_m in two-variable first-order logic, and we have $\Sigma^2_m \subseteq \text{FO}^2_m \subseteq \Sigma^2_{m+1}$. It is known that the languages definable in FO^2_m form a strict hierarchy, see *e.g.* [16]. For every $m \geqslant 1$ we have given ω-terms U_m and V_m such that a language L is definable in Σ_m^2 if and only if its ordered syntactic monoid is in the variety **DA** and satisfies the identity $U_m \leqslant V_m$. Using this characterization one can decide whether a given regular language is definable in Σ_m^2. In particular, we have shown decidability for every level of an infinite hierarchy. Note that there is no immediate connection between the decidability of FO_m^2 and the decidability of Σ_m^2.

The block product principle is an important tool in the proof of the direction from Σ_m^2 to identities. In order to be able to apply this tool, we first extended block products to the case where the left factor is an ordered monoid and then stated the block product principle in this context. In order to further extend the block product $M \,\square\, N$ to the case where both M and N are ordered, one has to consider the monotone functions in $N \times N \to M$ instead of $M^{N \times N}$. As in the case of the wreath product principle [20] this leads to ordered alphabets when stating the block product principle. However, one implication in the block product principle fails for ordered alphabets as the universal property does not hold in this setting.

References

1. Büchi, J.R.: Weak second-order arithmetic and finite automata. Z. Math. Logik Grundlagen Math. 6, 66–92 (1960)
2. Cohen, R.S., Brzozowski, J.A.: Dot-depth of star-free events. J. Comput. Syst. Sci. 5(1), 1–16 (1971)
3. Diekert, V., Gastin, P., Kufleitner, M.: A survey on small fragments of first-order logic over finite words. Int. J. Found. Comput. Sci. 19(3), 513–548 (2008)
4. Ebbinghaus, H.-D., Flum, J.: Finite Model Theory. In: Perspectives in Mathematical Logic. Springer (1995)
5. Elgot, C.C.: Decision problems of finite automata design and related arithmetics. Trans. Amer. Math. Soc. 98, 21–51 (1961)
6. Glaßer, C., Schmitz, H.: Languages of dot-depth 3/2. Theory of Computing Systems 42(2), 256–286 (2008)
7. Kamp, J.A.W.: Tense Logic and the Theory of Linear Order. PhD thesis, University of California (1968)
8. Knast, R.: A semigroup characterization of dot-depth one languages. RAIRO, Inf. Théor. 17(4), 321–330 (1983)
9. Krebs, A., Straubing, H.: An effective characterization of the alternation hierarchy in two-variable logic. In: FSTTCS 2012, Proceedings. LIPIcs, vol. 18, pp. 86–98. Dagstuhl Publishing (2012)
10. Kufleitner, M., Lauser, A.: Languages of dot-depth one over infinite words. In: Proceedings of LICS 2011, pp. 23–32. IEEE Computer Society (2011)
11. Kufleitner, M., Lauser, A.: Around dot-depth one. Int. J. Found. Comput. Sci. 23(6), 1323–1339 (2012)
12. Kufleitner, M., Lauser, A.: The join levels of the trotter-weil hierarchy are decidable. In: Rovan, B., Sassone, V., Widmayer, P. (eds.) MFCS 2012. LNCS, vol. 7464, pp. 603–614. Springer, Heidelberg (2012)
13. Kufleitner, M., Lauser, A.: The join of the varieties of R-trivial and L-trivial monoids via combinatorics on words. Discrete Math. & Theor. Comput. Sci. 14(1), 141–146 (2012)

14. Kufleitner, M., Lauser, A.: Quantifier alternation in two-variable first-order logic with successor is decidable. In: Proceedings of STACS 2013. LIPIcs, vol. 20, pp. 305–316. Dagstuhl Publishing (2013)

15. Kufleitner, M., Weil, P.: The FO^2 alternation hierarchy is decidable. In: Proceedings of CSL 2012. LIPIcs, vol. 16, pp. 426–439. Dagstuhl Publishing (2012)

16. Kufleitner, M., Weil, P.: On logical hierarchies within FO^2-definable languages. Log. Methods Comput. Sci. 8, 1–30 (2012)

17. McNaughton, R., Papert, S.: Counter-Free Automata. The MIT Press (1971)

18. Pin, J.-É.: A variety theorem without complementation. Russian Mathematics (Iz. VUZ) 39, 80–90 (1995)

19. Pin, J.-É., Weil, P.: Polynomial closure and unambiguous product. Theory Comput. Syst. 30(4), 383–422 (1997)

20. Pin, J.-É., Weil, P.: The wreath product principle for ordered semigroups. Commun. Algebra 30(12), 5677–5713 (2002)

21. Schützenberger, M.P.: On finite monoids having only trivial subgroups. Inf. Control 8, 190–194 (1965)

22. Schwentick, T., Thérien, D., Vollmer, H.: Partially-ordered two-way automata: A new characterization of DA. In: Kuich, W., Rozenberg, G., Salomaa, A. (eds.) DLT 2001. LNCS, vol. 2295, pp. 239–250. Springer, Heidelberg (2002)

23. Simon, I.: Piecewise testable events. In: Brakhage, H. (ed.) GI-Fachtagung 1975. LNCS, vol. 33, pp. 214–222. Springer, Heidelberg (1975)

24. Straubing, H.: A generalization of the Schützenberger product of finite monoids. Theor. Comput. Sci. 13, 137–150 (1981)

25. Straubing, H.: Finite Automata, Formal Logic, and Circuit Complexity. Birkhäuser (1994)

26. Straubing, H.: Algebraic characterization of the alternation hierarchy in FO^2[<] on finite words. In: Proceedings CSL 2011. LIPIcs, vol. 12, pp. 525–537. Dagstuhl Publishing (2011)

27. Straubing, H., Thérien, D.: Weakly iterated block products of finite monoids. In: Rajsbaum, S. (ed.) LATIN 2002. LNCS, vol. 2286, pp. 91–104. Springer, Heidelberg (2002)

28. Tesson, P., Thérien, D.: Diamonds are forever: The variety DA. In: Proceedings of Semigroups, Algorithms, Automata and Languages, pp. 475–500. World Scientific (2002)

29. Thérien, D.: Classification of finite monoids: The language approach. Theor. Comput. Sci. 14(2), 195–208 (1981)

30. Thérien, D.: Th. Wilke. Over words, two variables are as powerful as one quantifier alternation. In: Proceedings of STOC 1998, pp. 234–240. ACM Press (1998)

31. Thomas, W.: Classifying regular events in symbolic logic. J. Comput. Syst. Sci. 25, 360–376 (1982)

32. Trakhtenbrot, B.A.: Finite automata and logic of monadic predicates (in Russian). Dokl. Akad. Nauk. SSSR 140, 326–329 (1961)

33. Weis, P., Immerman, N.: Structure theorem and strict alternation hierarchy for FO^2 on words. Log. Methods Comput. Sci. 5, 1–23 (2009)

Model Checking for String Problems

Milka Hutagalung and Martin Lange*

School of Electr. Eng. and Computer Science, University of Kassel, Germany

Abstract. Model checking is a successful technique for automatic program verification. We show that it also has the power to yield competitive solutions for other problems. We consider three computation problems on strings and show how the polyadic modal μ-calculus can define their solutions. We use partial evaluation on a model checking algorithm in order to obtain an efficient algorithm for the longest common substring problem. It shows good performance in practice comparable to the well-known suffix tree algorithm. Moreover, it has the conceptual advantage that it can be interrupted at any time and still deliver long common substrings.

1 Introduction

Model checking is the process of automatically evaluating a logical formula on a given interpretation. This logical decision problem has proved to be extremely useful in the area of systems verification where dynamic systems are modelled as transition systems and formulas of temporal logics are being used to formalise behavioural properties [6,20]. Model checking is used to answer the question of whether or not such systems are correct with respect to some specification, namely whether or not they possess the formalised properties. The logics that are typically used in program verification have been designed in order to express typical correctness properties of dynamic systems: LTL [19], CTL [7], CTL* [8], etc. The name "model checking" is derived from the process of checking whether some interpretation in the form of a mathematical structure has the property defined by the formula, i.e. is a model of the formula in logical terms.

The impact that model checking has had for program verification has led to a common understanding of model checking as a program verification method. Still, the applicability of model checking is not limited to that area. Model checking can in principle be used to solve all kinds of decision problems, provided that the used specification language is strong enough to express that problem in the usual sense of a word problem: given a representation of an instance x of a problem, decide whether or not x belongs to some set P. For instance, x could be a directed graph, and P may consist of all graphs having a Hamiltonian path. Take for example Monadic Second-Order Logic (MSO) interpreted over graphs.

---

* The European Research Council has provided financial support under the European Community's Seventh Framework Programme (FP7/2007-2013) / ERC grant agreement no 259267.

E.A. Hirsch et al. (Eds.): CSR 2014, LNCS 8476, pp. 190–203, 2014.

It is not too difficult to construct a formula φ_{ham} which is satisfied by a graph G iff it has a Hamiltonian path. In fact, Fagin's Theorem [10] and the fact that the Hamiltonian path problem is easily seen to belong to NP give such a formula straight-away. Thus, φ_{ham} *expresses* the Hamiltonian path problem, and any model checking algorithm for MSO on graphs can be used to *solve* the Hamiltonian path problem. Likewise, any problem that is definable in a logic with a decidable model checking problem can therefore be solved by model checking.

The logics used in program verification mentioned above are not capable of expressing more complex properties like the Hamiltonian path for instance, unless NLOGSPACE=NP. This is a simple consequence of the fact that the complexity of model checking a fixed formula – known as *data complexity* – in either of these logics is only NLOGSPACE whereas the Hamiltonian path problem is NP-hard.

MSO is not necessarily a good logic for model checking. In order to be useful, such logics must provide a good balance between expressive power on one hand and efficient decision procedures on the other. Clearly, these two goals may be in conflict. It is commonly understood that the use of fixpoint quantifiers provide such a good balance because fixpoints can be computed using iteration methods for instance, and often they increase expressive power.

Fixpoints are implicitly present in the temporal logics mentioned above. Another prominent logic in model checking for program verification is the modal μ-calculus $\mathcal{L}_\mu$ [15] which explicitly adds fixpoint quantifiers to a basic modal logic. Its data complexity is in P, and it is known that it can express more properties or problems than temporal logics like LTL, CTL, etc. Still, its expressive power is limited by other facts, for instance it can only express properties of single nodes in a directed graph. This weakness has been overcome with the introduction of the polyadic or higher-dimensional μ-calculus [1,18] which behaves very much like the ordinary modal μ-calculus when it comes to model checking.

In this paper we show how model checking can be used in order to solve problems in an area that is quite different from program verification: string problems. We consider three of the most prominent examples of such problems: given a set of strings over some finite alphabet, find the *longest common substring*, the *longest common subsequence*, resp. the *shortest common superstring*. Such problems have important applications in bio-informatics as in sequence and genome analysis [12,22], in linguistic information retrieval [24]; plagiarism detection, for instance in publications [11] or source code [17,5]; data compression [21] and so on.

We use the polyadic μ-calculus in order to express these problems on simple graphs encoding string inputs and derive algorithms for these problems from a generic model checking algorithm for this logic. This realises more than a Turing or Cook reduction in two ways. First of all, model checking is a decision problem whereas the string problems mentioned above are computation problems. It is not obvious how model checking could be used to solve them. We do not implement standard tricks like binary search or others to find a solution. Instead, we use fixpoint quantifiers and iteration in the polyadic μ-calculus in order to compute solutions of these strings problems. Second, we want to show that model checking can be used in order to derive competitive algorithms in areas other than program verification. A vital step towards this goal is *partial evaluation*. Model checking takes two inputs: a formula and an interpretation. Algorithms for particular decision problems can be derived from model checking algorithms by fixing the formula input to the one that expresses the problem. Partial evaluation is the process of optimising the generic algorithm to one that operates on a fixed formula and a variable (encoding of) its interpretation.

The fact that highly expressive modal fixpoint logics can be used in order to solve problems that are more complex than the reachability problems arising in automatic program verification has been observed before. For instance, [2] develops a model checking algorithm for an extension of the modal μ-calculus with first-order predicate transformers. It is shown that this can be used to solve problems of higher complexity like NFA universality, QBF, or the shortest common supersequence problem (SCS). The principals are applicable to the other mentioned string problems as well. However, that work only presents the principal applicability of model checking for SCS. The work presented here improves and extends this in the following ways: the use of first-order predicate transformers turned out to be an overkill; here we show that modal fixpoint logics without higher-order features suffice for these string problems. Moreover, while [2] only presents the principal definability of these problems, here we take the work to a further level and show how the generic model checking algorithms can be optimised in order to arrive at practical algorithms for string problems.

The rest of the paper is organised as follows. Sect. 2 recalls the polyadic μ-calculus including the question of how model checking can be done for this logic. Sect. 3 shows how the three aforementioned string problems can be represented as model checking problems in this logic. It also contains a discussion on how model checking logics with fixpoint quantifiers can be used to solve computation problems. In order to evaluate the viability of this approach we concentrate on one particular problem in Sect. 4 where we show how partial evaluation is being used to turn the model checking algorithm for the polyadic μ-calculus on a fixed formula into an algorithm for the longest common substring problem. Sect. 5 compares the obtained algorithm against existing approaches. Sect. 6 finishes the paper with some concluding remarks.

2   The Polyadic μ-Calculus

Syntax and Semantics. The polyadic μ-calculus is interpreted over *transition systems*. Let Σ and $\mathcal{P}$ be finite sets of labels. A transition system is a labeled directed graph $\mathcal{T} = (\mathcal{S}, \rightarrow, \lambda)$ with $\mathcal{S}$ being a set of nodes, $\rightarrow \subseteq \mathcal{S} \times \Sigma \times \mathcal{S}$ the edge relation, and $\lambda : \mathcal{S} \rightarrow 2^{\mathcal{P}}$ a function assigning a set of labels from $\mathcal{P}$ to every node. We write $s \xrightarrow{a} t$ for $(s, a, t) \in \rightarrow$.

Formulas of the polyadic μ-calculus $\mathcal{L}^{\omega}_{\mu}$ are given by

$$\varphi ::= p_i \mid \neg\varphi \mid \varphi \wedge \varphi \mid \langle a \rangle_i \varphi \mid X \mid \mu X.\varphi$$

where $p \in \mathcal{P}$, $a \in \Sigma$, $i \in \mathbb{N}$, and $X \in \mathcal{V}$ for some countably infinite set $\mathcal{V}$ of variables. We additionally assume that any free occurrence of a variable X is under the scope of an even number of negation symbols in $\mu X.\varphi$.

Apart from the usual Boolean operators that can be expressed using $\wedge$ and $\neg$ we introduce the following abbreviations: $[a]_i \varphi := \neg \langle a \rangle_i \neg \varphi$ and $\nu X.\varphi := \neg \mu X.\neg \varphi[\neg X/X]$.

The *dimension* of a formula is the number of different indices occurring in operators of the form p_i, $\langle a \rangle_i$ or $[a]_i$ in it. The fragment $\mathcal{L}_\mu^k$ consists of all formulas of dimension at most k.

A formula of dimension k is interpreted in a k-tuple of nodes in a transition system $\mathcal{T} = (\mathcal{S}, \rightarrow, \lambda)$. The indices an atomic propositions and modal operators refer to a particular dimension, i.e. p_i is to be read as "the i-th component (of the k-tuple) under consideration) satisfies p". Likewise, $\langle a \rangle_i \varphi$ formalises that the tuple can be changed in its i-th component to some successor state such that φ holds.

The semantics assigns to every formula of dimension k the set of all k-tuples that satisfy it as follows. In order to handle free variables we use a variable interpretation $\rho : \mathcal{V} \rightarrow 2^{\mathcal{S}^k}$ assigning to each variable a set of k-tuples of nodes. Then $\rho[X \mapsto S]$ denotes the function that maps X to the set S and agrees with ρ on all other arguments.

$$[\![p_i]\!]_\rho^\mathcal{T} := \{(s_1, \ldots, s_k) \mid p \in \lambda(s_i)\}$$

$$[\![\neg\varphi]\!]_\rho^\mathcal{T} := 2^{\mathcal{S}^k} \setminus [\![\varphi]\!]_\rho^\mathcal{T}$$

$$[\![\varphi \wedge \psi]\!]_\rho^\mathcal{T} := [\![\varphi]\!]_\rho^\mathcal{T} \cap [\![\psi]\!]_\rho^\mathcal{T}$$

$$[\![\langle a \rangle_i \varphi]\!]_\rho^\mathcal{T} := \{(s_1, \ldots, s_k) \mid \exists t \in \mathcal{S} \text{ s.t. } s_i \xrightarrow{a} t \text{ and}$$
$$(s_1, \ldots, s_{i-1}, t, s_{i+1}, \ldots, s_k) \in [\![\varphi]\!]_\rho^\mathcal{T}\}$$

$$[\![X]\!]_\rho^\mathcal{T} := \rho(X)$$

$$[\![\mu X.\varphi]\!]_\rho^\mathcal{T} := \bigcap \{S \subseteq \mathcal{S}^k \mid [\![\varphi]\!]_{\rho[X \mapsto S]}^\mathcal{T} \subseteq S\}$$

Thus, $\mu X.\varphi$ defines the least fixpoint of the function that takes a set of k-tuples of nodes S and returns the set of all k-tuples satisfying φ assuming that X is interpreted as S [14,23]. We write $s \models_\rho \varphi$ if $s \in [\![\varphi]\!]_\rho^\mathcal{T}$ for a k-tuple s of nodes in $\mathcal{T}$, denoting the fact that s satisfies the property formalised by φ. If φ does not contain any free variables we may also drop the interpretation ρ.

A prominent example of a $\mathcal{L}_\mu^2$ formula is

$$\varphi_{\mathsf{bis}} := \nu X.(\bigwedge_{p \in \mathcal{P}} p_1 \leftrightarrow p_2) \wedge \bigwedge_{a \in \Sigma} [a]_1 \langle a \rangle_2 X \wedge [a]_2 \langle a \rangle_1 X .$$

It expresses bisimilarity in the sense that for all pairs of two nodes (s, t) we have $(s, t) \models \varphi_{\mathsf{bis}}$ iff s and t are bisimilar.

Model Checking $\mathcal{L}_\mu^k$. A simple model checking algorithm for $\mathcal{L}_\mu^\omega$ is implicitly given in the semantics of that logic. Given a finite transition system $\mathcal{T}$ and a closed $\mathcal{L}_\mu^k$ formula φ, one can compute the set $[\![\varphi]\!]^{\mathcal{T}}$ by induction on the structure of φ. Fixpoint subformulas of the form $\mu X.\psi$ or $\nu X.\psi$ can be handled using Knaster-Tarski fixpoint iteration: for least fixpoint formulas one binds X to the empty set and computes the value of ψ on $\mathcal{T}$. Then X is bound to this set of k-tuples and so on until a fixpoint is reached. For greatest fixpoints one starts the iteration with $\mathcal{S}^k$ instead.

The model checking problem for the polyadic μ-calculus has been investigated before [1,16]. Essentially, there is no conceptual difference to model checking the ordinary μ-calculus [9] which consists of all formulas of arity 1. In fact, there is a simple reduction from model checking formulas of arity k to formulas of arity 1 on the k-fold product of a transition system. Thus, one of the major parameters in its complexity – besides the formula's arity – is its *alternation depth*. Intuitively, it measures the nesting depth of fixpoints of different type. For formulas with no such nestings we set it to 1. Since alternating fixpoint quantifiers do not play any role in tackling string problems in the next section we omit a formal definition of alternation depth here and refer to the literature instead [4].

The next proposition summarises the findings on the complexity of model checking $\mathcal{L}_\mu^\omega$.

Proposition 1 ([1,16]). *Given a transition system $\mathcal{T}$ with n nodes and a closed $\mathcal{L}_\mu^k$ formula φ of alternation depth d, the set of all k-tuples of nodes in $\mathcal{T}$ satisfying φ can be computed in time $\mathcal{O}((|\varphi| \cdot n^k)^{\lceil d/2 \rceil})$.*

3   Defining String Problems

The three problems under consideration – longest common substring, resp. sequence, and shortest common superstring – all get as input a set $W = \{w_1, \ldots, w_m\}$ of strings over some finite alphabet Σ. For ease of presentation we assume them all to have length n. The theory and procedures to follow are easily adapted to handle strings of varying lengths. First we consider straight-forward representations of such inputs by transition systems. Distinguishing the two cases of finding longest substructures or shortest superstructures turns out to be beneficial.

Longest Common Substring and -Sequence. For these two problems we represent the input strings $W = \{w_1, \ldots, w_m\}$ by a transition system containing a single path for each such string. We also use the symbols $1, \ldots, m$ as propositions on the nodes in order to assess which string a node is in. Let $w_i = a_{i,1} \ldots a_{i,n}$ for $i = 1, \ldots, m$. Then $\mathcal{T}_W$ is given as

Now consider the $\mathcal{L}_\mu^m$-formula $\varphi_{\mathsf{lcst}} := \nu X.(\bigwedge_{i=1}^m i_i) \wedge \bigvee_{a \in \Sigma} \langle a \rangle_1 \ldots \langle a \rangle_m X$ interpreted over transition systems of the form $\mathcal{T}_W$. In order to explain its meaning we consider how the naïve fixpoint iteration algorithm computes $[\![\varphi_{\mathsf{lcst}}]\!]^{\mathcal{T}_W}$. First we observe that $\bigwedge_{i=1}^{m} i_i$ satisfies exactly those m-tuples for which each i-th component belongs to the i-th input string. Thus, fixpoint iteration only yields tuples of positions with exactly one for each input string. Let us call these normal.

The greatest fixpoint iteration starts with the set of all tuples, and we can restrict our attention to all normal tuples only. This set can be seen as a representation of all the position at which the string ε occurs. The next fixpoint iteration forms the union of all sets of normal tuples which represent positions such that some a-edge is possible from all of them, and the resulting tuple represents occurrences of the substring a. Thus, it computes all positions of common substrings of length 1. In general, the j-th fixpoint iteration computes all positions (as normal m-tuples) of a common substring of length j. Clearly, this process is monotonically decreasing and there is some j – at most $n+1$ – such that the j-th iteration returns the empty set.

Indeed we have $\mathcal{T}_W \not\models \varphi_{\mathsf{lcst}}$ for any set W of strings. Nevertheless, model checking via fixpoint iteration computes all common substrings of W before finding out that the formula is not satisfied. This is the basis for an algorithm computing the longest common substring using model checking as described in detail in the next section.

Example 1. Consider the input $W = \{aabab, abaa, babab\}$, represented by the following transition system. For convenience, we have given the nodes names. We also omit the nodes' labels since they are just the same as the names' first components.

A greatest fixpoint iteration for φ_{lcst} on $\mathcal{T}_W$ starts with X^0 as the set of all positions. In order to compute the next iteration, note that $[\![(\bigwedge_{i=1}^{3} i_i)]\!]^{\mathcal{T}_W}$ is the set of all tuples of the form $((1, j_1), (2, j_2), (3, j_3))$ for appropriate j_1, j_2, j_3. Every further iteration intersects some set obtained by evaluating the modal terms with this set. We therefore disregard all other tuples. Under this assumption, $[\![\bigvee_{c \in \{a,b\}} \langle c \rangle_1 \langle c \rangle_2 \langle c \rangle_3 X]\!]^{\mathcal{T}_W}_{[X \mapsto X^0]}$ then evaluates to

$$X^1 := \{(1,0), (1,1), (1,3)\} \times \{(2,0), (2,2), (2,3)\} \times \{(3,1), (3,3)\}$$
$$\cup \ \{(1,2), (1,4)\} \times \{(2,1)\} \times \{(3,0), (3,2), (3,4)\}$$

which is exactly the set of node tuples from which all components can do an a-edge or all components can do a b-edge.

The next iteration for the evaluation of the greatest fixpoint is obtained by evaluating the fixpoint body again, this time under the variable interpretation $[X \mapsto X^1]$, and it yields

$$X^2 := \{(1,1), (1,3)\} \times \{(2,0)\} \times \{(3,1), (3,3)\}$$

$$\cup \ \{(1,2)\} \times \{(2,1)\} \times \{(3,0),(3,2)\}$$

which is the set of positions of ab, resp. ba, in the input strings. Note that aa for instance is no common substring, and this is reflected by the fact that there is no p such that $((1,0),(2,2),p)$ belongs to X^2.

The next iteration yields $X^3 := \{((1,1),(2,0),(3,1))\}$ which denotes the positions of the common substring aba. Finally, we get $X^4 = \emptyset$ at which point the fixpoint is reached, and the solution to this longest common substring instance is obtained as the value of the last iteration beforehand, namely aba at positions 1, 0, and 1 in the three input strings.

$\mathcal{L}_\mu^2$ is also capable of expressing the longest common subsequence problem in the same sense. Let $\langle * \rangle_i \psi$ abbreviate $\mu Y.\psi \vee \bigvee_{a\in\Sigma}\langle a \rangle_i Y$. Informally, it denotes the set of all tuples such that the i-th component can make an arbitrary number of steps along any edge and some resulting tuple satisfies ψ. Now consider $\varphi_{\mathsf{lcsq}} := \nu X.(\bigwedge_{i=1}^m i_i) \wedge \bigvee_{a\in\Sigma}\langle a \rangle_1 \langle * \rangle_1 \dots \langle a \rangle_m \langle * \rangle_m X$. Evaluating this formula on a transition system of the form $\mathcal{T}_W$ will compute the longest common subsequence of the input strings in W in the same way as above. Note that, again, we have $\mathcal{T}_W \not\models \varphi_{\mathsf{lcsq}}$ for any W but all the solutions to this instance are being found in the last iteration of the greatest fixpoint evaluation.

Shortest Common Superstring. In order to model the shortest common superstring problem we change the representation of words by simple paths in a transition system. Given a finite set W of words over Σ, the transition system $\mathcal{T}_W'$ consists of one component for each word $w_i = a_{i,1} \dots a_{i,n}$ which has the form

Two additional atomic propositions s, e are being used to mark the start node and the end node of each component. They, together with the special structure of these transition systems, can also be used to enforce tuples to contain exactly one node from each input string. Thus, propositions $1, \dots, m$ are not needed anymore.

Now consider $\varphi_{\mathsf{scs}} := (\bigwedge_{i=1}^m \mathsf{s}_i) \wedge \mu X.(\bigwedge_{i=1}^m \mathsf{e}_i) \vee \bigvee_{a\in\Sigma}\langle a \rangle_1 \dots \langle a \rangle_m X$. Intuitively, it denotes the set of all tuples such that

1. each component is labeled with s, and
2. there is a sequence of edge labels w such that each component has a path with these labels and the nodes of the tuple at the end of all these paths are all labeled with e.

It should be clear from this description that we have $\mathcal{T}_W' \models \varphi_{\mathsf{scs}}$ for any W. However, evaluating φ_{scs} on $\mathcal{T}_W'$ by a least fixpoint iteration will ultimately construct a shortest common superstring for all the strings in W. This iteration starts with $X^0 := \emptyset$ and – when restricted to tuples with one component in each string – gradually finds tuples $(p_1, \dots, p_m)$ of positions in the i-th iteration such that there is a word w of length i with a path labeled with w from each of their components. By the structure of $\mathcal{T}'_W$, the iteration grows monotonically "to the left", i.e. it only ever adds tuples with positions further left in the input words. Eventually – after no more than $(n+1)m$ iterations in the worst case – the tuple $((1,0),\ldots,(m,0))$ is being found and the least fixpoint is being reached. The number of iterations done to achieve this equals the length of a shortest common superstring, and this string can easily be computed by annotating the found tuples of positions successively.

4   Partial Evaluation and Optimisation

The algorithms sketched above are rather naïve and do not exploit any optimisation potential at all. The descriptions above are only meant to show how model checking with fixpoint logics can in principle be used in order to solve such computation problems. Here we focus on one particular problem, namely finding longest common substrings, and show how partial evaluation of model checking algorithms can be used to obtain an efficient procedure. Also note that a naïve estimation of the worst-case time complexity of these algorithms according to Prop. 1 yields a horrendous overapproximation: in general, model checking $\mathcal{L}^k_\mu$ is exponential in the arity k, here equalling the number of strings m in the input. This, however, ignores the special structure of the transition systems used here and that of the fixed formula.

Consider the algorithm that has been described in Example 1. It basically works on a set X of common substrings, and in each iteration it extends all elements of X to a longer common substring by considering one more letter to the left. For m input strings of length n, the set X is represented by a set of m-tuples, which initially contains n^m tuples that represent the positions of the empty string.

A straightforward optimisation changes the representation of the set X. Instead of using a set of m-tuples, we can represent a single substring w with a set $t(w)$ of pairs such that $(i,j) \in t(w)$ iff w occurs in w_i at position j. By using this representation, initially we have nm positions instead of n^m positions for the empty string. Moreover, it is easy to check whether w is a common substring which is true iff for every $i = 1,\ldots,m$ there is some j with $(i,j) \in t(w)$.

Applying these straight-forward optimisations to the procedure described in Ex. 1 yields Algorithm 1. It collects all non-extendable common substrings in a set Y, and uses that for a return value.

In the following we describe further optimisations for Algorithm 1, so that it can find longest common substrings faster and more efficiently.

Extension Restriction. To extend $w \in X$ in each iteration, it is not necessary to consider all letters from Σ. Suppose $w = w'a$, where $a \in \Sigma$ and $w' \in \Sigma^*$, then to extend w it is enough to consider letters that have successfully extended w', since for every $a' \in \Sigma$ if $a'w'$ is not a common substring, then $a'w$ is not either. We can get the letters that extend w' in constant time by always keeping a pointer from w to w' for each $w \in X$, and from w to all of its extensions

---

Algorithm 1. Finding the longest common substring

1: $X \leftarrow \Sigma, Y \leftarrow \emptyset$
2: **while** $X \neq \emptyset$ **do**   ▷ extend some $w \in X$
3:      take $w \in X$
4:      **if** Ext$(w) \neq \emptyset$ **then**   ▷ Ext$(w) := \{aw \mid a \in \Sigma,\ aw$ common substring $\}$
5:          $X \leftarrow$ Ext$(w) \cup X \backslash w$ ▷ replace w with its extension Ext(w)
6:      **else**
7:          $Y \leftarrow \{w\} \cup Y$ ▷ have w as non-extendable common substring
8:      **end if**
9: **end while**
10: **return** the longest $w \in Y$

---

for each $w \in X$ that have been extended. Moreover, we should always take the shortest $w \in X$ in each iteration, to make sure that the extension of w' is already computed in the previous iteration.

Multiple Substrings Extension. Under some conditions, extending a single $w \in X$ may imply extensions of some other substrings $u \in X$. For any $w \in X$ let $S(w) = \{u \in X \mid u = wv, v \in \Sigma^+\}$. If w is extendable to aw, in general we cannot conclude that $u \in S(w)$ is also extendable to au. However it is the case if $t(aw)$ is equal to $\{(i, j-1)|(i,j) \in t(w)\}$, since this means that all occurrences of w in the input strings are always preceded by a. In this case, we can extend w to aw, and also every $u \in S(w)$ to au. Likewise if w is not extendable to any longer common substring, then every $u \in S(w)$ is also not extendable. In this case we can move w and all $u \in S(w)$ to Y. Extending all $u \in S(w)$ (resp. moving all $u \in S(w)$ to Y) can be done in constant time by exploiting the pointers defined before, i.e. a pointer from $w = w'a$ to w' since every u are successively linked by the pointer to w.

Multiple Letters Extension. It is also possible to extend $w \in X$ with a sequence of letters $a_n a_{n-1} \ldots a_1 \in \Sigma^n$ instead of only one single letter. Suppose $w = w'a$, and the string w' was extended to a common substring $a_n a_{n-1} \ldots a_1 w'$ because of the previous extension policy, i.e. because $t(a_1 w') = \{(i, j-1)|(i,j) \in t(w')\}, \ldots, t(a_n \ldots a_1 w') = \{(i, j-1)|(i,j) \in t(a_{n-1} \ldots a_1 w')\}$. Then if we can extend w to $a_1 w$, we can immediately conclude that w can be extended to $a_n a_{n-1} \ldots a_1 w$.

All of these optimisations will not make the extension of a single substring $w \in X$ harder since we store more information on each common substring, to accommodate the optimisations. The extension policies derived from these optimisations can cut down the number of iterations needed in Algorithm 1.

Example 2. Let $W = \{$cgtacgag, aacgtag, agcgtacg$\}$ be the input strings. We illustrate the computation of the longest common substring using Algorithm 1 in Fig. 1, both with and without optimisation. In each iteration we pick the shortest common substring to be extended. Figure (a) shows the first 9 iterations (of 14 altogether) without any optimisation. Figure (b) shows the

i	X	Y
1	g̲,t,c,a	-
2	ag,cg,t̲,c,a	-
3	ag,cg,gt,c̲,a	-
4	ag,cg,gt,ac,a̲	-
5	ag,cg,gt,ac,ta	-
6	c̲g̲,gt,ac,ta	ag
7	acg,g̲t̲,ac,ta	ag
8	acg,cg,a̲c̲,ta	ag
9	acg,cgt,t̲a̲	ag,ac
⋮	⋮	⋮

(a) without optimisation

i	X	Y
1	g̲,t,c,a	-
2	ag,cg,t̲,c,a	-
3	ag,cg,gt,c̲,a	-
4	ag,cg,gt,ac,a̲	-
5	ag,cg,gt,ac,gta	-
6	c̲g̲,gt,ac,gta	ag
7	acg,g̲t̲,ac,gta	ag
8	acg,cgt,a̲c̲,cgta	ag
9	acg,cgt,cgta	ag,ac
10	c̲g̲t̲,cgta	ag,ac,acg
11	-	ag,ac,acg,cgt,cgta

(b) with optimisation

Fig. 1. Computation for $W = \{$cgtacgag, aacgtag, agcgtacg$\}$

computation with optimisation which finds the longest common substring cgta after 11 iterations.

Note that in Fig.1 (b), we apply the optimisation in the 4th, 7th, and 10th iteration. In the 4th iteration it is found that a can be extended to ta and we have $t(\text{gt}) = \{(i, j - 1) \mid (i, j) \in t(\text{t})\}$ from the previous iteration, so a can be extended directly to gta. In the 7th iteration, by extending gt to cgt we also extend gta to cgta since gta $\in S(\text{gt})$. In the 10th iteration cgt is not extendable so we conclude that cgta are not either.

Theorem 1. *For input strings $w_1, \ldots, w_m$ each of length n, the number of iterations needed by the optimised algorithm is at most $n + n + m(n - 1)$.*

Proof. In each iteration i, we pick the shortest common substring $w \in X^i$ to be extended, which satisfies one of these properties, either:

1. w cannot be extended to the left anymore, or
2. w can be extended to aw and $t(aw) = \{(i, j - 1) \mid (i, j) \in t(w)\}$, i.e. w allows multiple substring extension as described previously, or
3. none of these two conditions apply to w.

Let L_1, L_2, L_3 be the set of common substring of $w_1, \ldots, w_n$, such that $w \in L_i$ iff w satisfies the i-th property.

$|L_1| \leq n$, since we have a one-to-one mapping from L_1 to the set of prefixes of w_1. Note that each $v \in L_1$ is a substring of w_1 (resp. $w_2, \ldots, w_n$), and it can be mapped to a prefix uv of w_1. Every two different $v_1, v_2 \in L_1$ are mapped to two different prefixes of w_1, for otherwise one of them would be a suffix of the other, and thus could be extended to the left which would contradict $v_1, v_2 \in L_1$.

$|L_3| \leq m(n - 1)$, since if $v \in L_3$ then v occurs on all input strings, and there exists an input string w_i such that v occurs more than once on w_i. If $|L_3| = k$, then $|w_1| + \ldots + |w_m|$ is at least $m + k$. However, the added lengths of all input strings is bounded by mn, so $|L_3| \leq mn - m$.

In general, $|L_2| \leq n^2$. But consider $L'_2 \subseteq L_2$ where $v \in L'_2$ iff its longest proper prefix v' does not belong to L_2. Suppose on the i-th iteration, we pick $v \in L_2$ to be extended. If $v \notin L'_2$ then the longest proper prefix v' of v is not yet extended on any j-th iteration, $j < i$. Otherwise either the optimisation rule: multiple substring extension or multiple letters extension have been applied to obtain the extension av of v, and implies $v \notin X^i$ for every $i > j$. So $v \notin L'_2$ contradicts to v being the current shortest common substring found.

$|L'_2| \leq n$, because we have a one-to-one mapping from L'_2 to the set of suffixes of w_1. Each $v \in L'_2$ can be mapped to a suffix vu of w_1. Every two different $v_1, v_2 \in L'_2$ are mapped into two different suffixes of w_1, for otherwise one of them would be a prefix of the other. Suppose w.l.o.g. that v_1 was a prefix of v_2. Then v_1 is also a prefix of the longest proper prefix v'_2 of v_2. Hence $v'_2 \in L_2$ which contradicts that $v_2 \in L'_2$. □

5   A Comparison against the Suffix-Tree Approach

The literature describes two algorithms for solving the longest common substring problem: *dynamic programming* [13] and the *suffix tree algorithm* [12]. However, it is known that the speed and versatility of the suffix tree algorithm is better than dynamic programming. It is therefore the state-of-the-art and standard algorithm used for the longest common substring problem.

A suffix tree of W is a tree storing all suffixes of strings in W. It has many applications in biological sequence data analysis [3], especially for searching patterns in DNA or protein sequences. For a more detailed explanation of suffix trees see [12]. We compare the optimised Algorithm 1 with the suffix tree algorithm empirically on a biological data set, and also conceptually.

Empirical Comparison. An interesting application for the longest common string problem comes from bioinformatics area for identification of common features in genome analysis [12]. We compare the performance of the optimised Algorithm 1 and the suffix tree algorithm on a data set of complete genomes. We consider the 11 species that are named in Fig. 2. All of these complete genomes can be obtained on a public GenBank database of NCBI[1].

We choose mostly virus and bacteria genomes since they usually have only one or two chromosomes, and the size of their complete genome is approximately 3 megabytes, which is still suitable for the experiment. For more complex species such as humans the size of their complete genome is approximately 800 megabytes[1]. The bacteria *Vibrio cholerae* and *Agrobacterium tumefaciens* have two chromosomes, and we treat each chromosome separately. We take two to four species for each experiment and try to find their longest common substring.

Fig. 2 compares the running time of the optimised Algorithm 1 and the suffix tree algorithm on some benchmarks. Both algorithms were implemented in OCAML. The suffix tree algorithm uses Sébastien Ferré's implementation of the

---

[1] See http://www.ncbi.nlm.nih.gov/genbank

Species	Size	Suffix Tree Alg.	Opt. Alg. 1
E. coli, M. tuberculosis	9.8 mb	70 min 33 sec	54 min 26 sec
A. tumefaciens (I), V. cholerae (I)	7.1 mb	42 min 46 sec	38 min 17 sec
S. enterica, B. subtilis (I)	9.2 mb	64 min 2 sec	52 min 13 sec
A. fugildus, N. gonorrhoeae, A. tumefaciens (II)	7.1 mb	35 min 10 sec	33 min 50 sec
C. trachomatis, A. aeolicus, H. influenzae, V. cholerae (II)	5.7 mb	21 min 33 sec	24 min 34 sec
E. coli, V. cholerae (I)	8.2 mb	60 min 53 sec	46 min 21 sec
M. tuberculosis, V. cholerae (I)	7.5 mb	43 min 36 sec	40 min 4 sec

Fig. 2. Comparison with suffix tree algorithm

suffix tree data structure[2]. The experiments have been run on a machine with 16 Intel Xeon cores running at 1.87GHz and 256GB of memory.

The result suggests that the optimised Algorithm 1 is comparable to the suffix tree algorithm. The time needed is often even less than the suffix tree algorithm, except on the data set with 5.7 mb, where the optimised Algorithm 1 was 3min slower than the suffix tree algorithm. However, in general we can conclude that the optimised Algorithm 1 performed well compared to the suffix tree algorithm.

Conceptual Comparison. The suffix tree algorithm builds the tree first and then searches for the deepest node that represents the longest common substring of all input strings. The usual approaches to building the tree are incremental with respect to the number of the input strings [19]. For example, to build a suffix tree of $W = \{w_1, \ldots, w_n\}$, one starts with a suffix tree of w_1, then gradually modifies the tree to include the suffixes of w_2, and so on. This has the disadvantage of not being able to see the common substrings of all $w_1, \ldots, w_m$ during the tree construction. The whole tree has to be constructed first before searching for any common substring of $w_1, \ldots, w_m$. What can be recorded during the tree construction is only the common substring for the first m-input strings.

Now suppose that the input data is large such that despite the linear time complexity an entire run of the algorithm would still take, say, days to terminate. In such a case it would be great if the algorithm was able to report the finding of long common substrings on-the-fly, i.e. incrementally produce longer and longer common substrings. The suffix tree algorithm is not able to do this, because it needs to process all input strings before finding even the shortest common substring. However it is not the case for Algorithm 1. We have seen that the algorithm always maintains the currently longest common substring found in each iteration, and it is able to incrementally report longer and longer common substrings rather than finding them only at the very end of the entire computation.

6   Conclusion and Further Work

We have shown that certain string problems can be expressed as model checking problems for modal fixpoint logics. Fixpoint computation can be used to find

---
[2] See http://www.irisa.fr/LIS/ferre/software.en.html

optimal solutions for these problems. We have assumed straight-forward encoding of input strings as transition systems of disjoint paths. However, the formulas of the polyadic μ-calculus that were used to define these string problems also work on more compact graph encodings when common parts of input strings are being shared.

We have focused on the longest common substring problem and shown that it is possible to derive a new competitive algorithm by partial evaluation of a generic model checking algorithm for the polyadic μ-calculus. It turned out to have the conceptual advantage of being interruptable: roughly speaking, at half of the running time it has computed the half-longest common substrings of all inputs. The suffix tree algorithm, as a standard for this problem, on the other hand has computed, at half of the running time, the longest common substrings of half of the input strings.

Further work on the longest common substring algorithms includes a broader practical evaluation and a thorough study of possibilities to combine features from the suffix tree algorithm and the model checking approach. It also remains to execute the partial evaluation work for the two other string problems considered here, and possibly others as well in order to create hopefully competitive and usable algorithms for these problems, too.

Finally, we believe that model checking technology can contribute to algorithmic solutions for all sorts of other problems as well. This, of course, has to be studied for every possible decision or computation problem of interest separately.

References

1. Andersen, H.R.: A polyadic modal μ-calculus. Technical Report ID-TR: 1994-195, Dept. of Computer Science, Technical University of Denmark, Copenhagen (1994)
2. Axelsson, R., Lange, M.: Model checking the first-order fragment of higher-order fixpoint logic. In: Dershowitz, N., Voronkov, A. (eds.) LPAR 2007. LNCS (LNAI), vol. 4790, pp. 62–76. Springer, Heidelberg (2007)
3. Bieganski, P., Riedl, J., Cartis, J.V., Retzel, E.F.: Generalized suffix trees for biological sequence data: applications and implementation. In: Proc. 27th Hawaii Int. Conf. on System Sciences, vol. 5, pp. 35–44 (January 1994)
4. Bradfield, J., Stirling, C.: Modal mu-calculi. In: Blackburn, P., van Benthem, J., Wolter, F. (eds.) Handbook of Modal Logic: Studies in Logic and Practical Reasoning, vol. 3, pp. 721–756. Elsevier (2007)
5. Campos, R.A.C., Martínez, F.J.Z.: Batch source-code plagiarism detection using an algorithm for the bounded longest common subsequence problem. In: Proc. 9th Int. IEEE Conf. on Electrical Engineering, Computing Science and Automatic Control, CCE 2012, pp. 1–4. IEEE (2012)
6. Emerson, E.A., Clarke, E.M.: Characterizing correctness properties of parallel programs as fixpoints. In: de Bakker, J.W., van Leeuwen, J. (eds.) ICALP 1980. LNCS, vol. 85, pp. 169–181. Springer, Heidelberg (1980)
7. Emerson, E.A., Clarke, E.M.: Using branching time temporal logic to synthesize synchronization skeletons. Science of Computer Programming 2(3), 241–266 (1982)
8. Emerson, E.A., Halpern, J.Y.: "Sometimes" and "not never" revisited: On branching versus linear time temporal logic. Journal of the ACM 33(1), 151–178 (1986)

9. Emerson, E.A., Jutla, C.S., Sistla, A.P.: On model checking for the μ-calculus and its fragments. TCS 258(1-3), 491–522 (2001)
10. Fagin, R.: Generalized first-order spectra and polynomial-time recognizable sets. Complexity and Computation 7, 43–73 (1974)
11. Gipp, B., Meuschke, N.: Citation pattern matching algorithms for citation-based plagiarism detection: greedy citation tiling, citation chunking and longest common citation sequence. In: Proc. 2011 ACM Symp. on Document Engineering, pp. 249–258. ACM (2011)
12. Gusfield, D.: Algorithms on Strings, Trees, and Sequences: Computer Science and Computational Biology. Cambridge University Press (1997)
13. Hirschberg, D.: A linear space algorithm for computing maximal common subsequences. Commun. ACM 18(6), 341–343 (1975)
14. Knaster, B.: Un théorèm sur les fonctions d'ensembles. Annals Soc. Pol. Math. 6, 133–134 (1928)
15. Kozen, D.: Results on the propositional μ-calculus. TCS 27, 333–354 (1983)
16. Lange, M., Lozes, E.: Model checking the higher-dimensional modal μ-calculus. In: Proc. 8th Workshop on Fixpoints in Comp. Science, FICS 2012. Electr. Proc. in Theor. Comp. Sc., vol. 77, pp. 39–46 (2012)
17. Oetsch, J., Pührer, J., Schwengerer, M., Tompits, H.: The system Kato: Detecting cases of plagiarism for answer-set programs. Theory and Practice of Logic Programming 10(4-6), 759–775 (2010)
18. Otto, M.: Bisimulation-invariant PTIME and higher-dimensional μ-calculus. Theor. Comput. Sci. 224(1–2), 237–265 (1999)
19. Pnueli, A.: The temporal logic of programs. In: Proc. 18th Symp. on Foundations of Comp. Science, FOCS 1977, Providence, RI, USA, pp. 46–57. IEEE (1977)
20. Queille, J.P., Sifakis, J.: Specification and verification of concurrent systems in CESAR. In: Dezani-Ciancaglini, M., Montanari, U. (eds.) Programming 1982. LNCS, vol. 137, pp. 337–371. Springer, Heidelberg (1982)
21. Storer, J.A.: Data Compression: Methods and Theory. Comp. Sci. Press (1988)
22. Sung, W.-K.: Algorithms in Bioinformatics: A Practical Approach. CRC Press (2009)
23. Tarski, A.: A lattice-theoretical fixpoint theorem and its application. Pacific Journal of Mathematics 5, 285–309 (1955)
24. Xiao, Y., Luk, R.W.P., Wong, K.F., Kwok, K.L.: Using longest common subsequence matching for chinese information retrieval. Journal of Chinese Language and Computing 15(1), 45–51 (2010)

Semiautomatic Structures

Sanjay Jain[1,*], Bakhadyr Khoussainov[2,**], Frank Stephan[3,***],
Dan Teng[3], and Siyuan Zou[3]

[1] Department of Computer Science, National University of Singapore
13 Computing Drive, COM1, Singapore 117417, Republic of Singapore
sanjay@comp.nus.edu.sg
[2] Department of Computer Science, University of Auckland, New Zealand
Private Bag 92019, Auckland, New Zealand
bmk@cs.auckland.ac.nz
[3] Department of Mathematics, The National University of Singapore
10 Lower Kent Ridge Road, S17, Singapore 119076, Republic of Singapore
fstephan@comp.nus.edu.sg, {tengdanqq930,zousiyuan}@hotmail.com

Abstract. Semiautomatic structures generalise automatic structures in
the sense that for some of the relations and functions in the structure one
only requires the derived relations and structures are automatic when all
but one input are filled with constants. One can also permit that this
applies to equality in the structure so that only the sets of representatives equal to a given element of the structure are regular while equality
itself is not an automatic relation on the domain of representatives. It
is shown that one can find semiautomatic representations for the field of
rationals and also for finite algebraic field extensions of it. Furthermore,
one can show that infinite algebraic extensions of finite fields have semiautomatic representations in which the addition and equality are both
automatic. Further prominent examples of semiautomatic structures are
term algebras, any relational structure over a countable domain with
a countable signature and any permutation algebra with a countable
domain. Furthermore, examples of structures which fail to be semiautomatic are provided.

1 Introduction

General Background. An important topic in computer science and mathematics is concerned with classifying structures that can be presented in a way that
certain operations linked to the structures are computed with low computational
complexity. Automatic functions and relations can, in some sense, be considered

---

* S. Jain was supported in part by NUS grants C252-000-087-001, R146-000-181-112
and R252-000-420-112.
** B. Khoussainov is partially supported by Marsden Fund grant of the Royal Society
of New Zealand. The paper was written while B. Khoussainov was on sabbatical
leave to the National University of Singapore.
*** F. Stephan was supported in part by NUS grants R146-000-181-112 and R252-000-
420-112.

E.A. Hirsch et al. (Eds.): CSR 2014, LNCS 8476, pp. 204–217, 2014.
© Springer International Publishing Switzerland 2014

to have low complexity. The first work in this field centered on the question which sets are regular (that is, recognised by finite automata) and how one can transform the various descriptions of regular sets into each other. Later mathematicians applied the concept also to structures: Thurston automatic groups [3] are one of the pioneering works combining automata theory with structures. Here one has (a) a regular set of representatives A consisting of words over a finite alphabet of generators, (b) an automatic equivalence relation representing equality and (c) for every fixed group member y, an automatic mapping f_y from A to A such that $f_y(x)$ is a representative of the group member $x \circ y$. Here a function is automatic iff its graph can be recognised by a finite automaton or, equivalently, iff it is computed in linear time by a one-tape Turing machine which replaces the input by the output on the tape, starting with the same position [1]. These concepts have been generalised to Cayley automatic groups [7,10] and to automatic structures in general.

For automatic structures, one has to define how to represent the input to functions that have several inputs. This is now explained in more detail. If Σ is the alphabet used in the regular domain $A \subseteq \Sigma^*$ of the structure, one defines the convolution of two strings $a_0 a_1 \ldots a_n$ and $b_0 b_1 \ldots b_m$ to consist of combined characters $c_0 c_1 \ldots c_{\max\{m,n\}}$ where

if $k \leqslant \min\{m,n\}$ then $c_k = \binom{a_k}{b_k}$ else if $m < k \leqslant n$ then $c_k = \binom{a_k}{\#}$ else $c_k = \binom{\#}{b_k}$.

Here $\#$ is a fixed character outside Σ used for padding purposes. Convolution of strings x and y is denoted by $conv(x,y)$. Now the domain of a function $f : A \times A \to A$ is the set $\{conv(x,y) : x, y \in A\}$ which might from now on be identified with $A \times A$. Similarly one can define convolutions of more than two parameters and also define that an automatic relation over A^k is an automatic function from A^k to $\{0,1\}$ taking 1 on those tuples where the relation is true and taking 0 otherwise. A structure $\mathcal{A}$ is automatic iff it is isomorphic to a structure $\mathcal{B}$ such that the domain and all functions and relations in the structure are automatic.

Let $\mathbb{N}$ denote the set of natural numbers, $\mathbb{Z}$ the set of integers and $\mathbb{Q}$ the set of rational numbers. Now $(\mathbb{N}, +, =, <)$ is an automatic structure, as (i) there is a regular set A such that each member of $\mathbb{N}$ is represented by at least one member of A, (ii) there is an automatic function $f : A \times A \to A$ such that for each $x, y \in A$ the value $f(x,y)$ is a representative of the sum of the elements represented by x, y and (iii) the sets $\{conv(x,y) : x, y$ represent the same element of $\mathbb{N}\}$ and $\{conv(x,y) : x$ represents a number n and y represents a number m with $n < m\}$ are both regular. Automatic structures were introduced by Hodgson [5] and later, independently, by Khoussainov and Nerode [8]. Automatic structures have a decidable first-order theory and every function or relation first-order definable in an automatic structure (with quantification over members of the structure, say group elements and using as parameters relations from the structure or other automatic relations introduced into the representation of the structure) are again automatic. These closure properties made automatic structures an interesting field of study; however, a limitation is its expressiveness. For example, the structure $(\mathbb{N}, \cdot, =)$ is not automatic yet its first-order theory is decidable. There is a limited version of multiplication which is automatic in every automatic presentation of $(\mathbb{N}, +)$ or $(\mathbb{Z}, +)$: For every multiplication with a fixed element n, one can find an automatic function which maps every representative of a number m to a representative of the number $m \cdot n$.

Therefore, one would like to overcome the lack of expressivity of automatic structures and address the following questions: (1) Are there general ways to utilise finite automata for the representation of non-automatic structures such as $(Q, +)$ and $(\mathbb{N}, \cdot, =)$? (2) Under such general settings, what properties of automatic structures should be sacrificed and what properties should be preserved to accommodate non-automatic structures as those we mentioned above? (3) What are the limits of finite automata in representations of structures?

The present paper proposes one possible approach to address the questions above. The main concept is motivated by the notion of Thurston automaticity and Cayley automaticity for groups [3,7]. Namely, one says that a function $f : A^k \mapsto A$ is semiautomatic iff whenever one fixes all but one of the inputs of f with some fixed elements of A, then the resulting mapping from A to A is automatic. Similarly a relation $R \subseteq A^k$ is semiautomatic, if it is a semiautomatic function when viewed as a $\{0, 1\}$-valued function (mapping the members of R to 1 and the non-members of R to 0). This permits now to give the general definition using finite automata representing structures. For a structure, say $(\mathbb{N}, +, <, =; \cdot)$, one says that this structure is semiautomatic iff there is a representation $(A, f, B, C; g)$ of this structure such that A is a regular set of representatives of $\mathbb{N}$, f is an automatic function representing $+$, B, C are automatic relations representing $<, =$, respectively, and g is a semiautomatic relation representing the multiplication. Note that the convention here is that the relations and functions before the semicolon have to be automatic while those after the semicolon need only to be semiautomatic. Hence in a structure $(\mathbb{N}, +, <, =; \cdot)$ the operation $+$ and the relations $<$ and $=$ have to be automatic and $\cdot$ is only semiautomatic while in a structure $(\mathbb{Q}; +, \cdot, <, =)$ not only the operations addition and multiplication are semiautomatic but also the relations $<$ and $=$, that is, only the sets which compare to a fixed element (say all representatives of numbers below $1/2$ or all representatives of 5) have to be regular. This difference is crucial, for example, $(\mathbb{Q}, +; =)$ is not semiautomatic [14] and $(\mathbb{Q}, =; +)$ is semiautomatic. It is of course the goal to maintain automaticity for as many operations and relations as possible, therefore one needs to pay attention to these differences. Here are some important comments on the structures.

- The condition that a basic function, say $f : A^2 \to A$, is semiautomatic requires, for all $a \in A$, merely the existence of automata recognising the sets $\{conv(x, y) : y = f(x, a)\}$ and $\{conv(x, y) : y = f(a, x)\}$. This part of the definition is kept as general as possible to accommodate a large class of structures. In particular, this part is needed to address question (3) posed above. Obviously, the requirement that the graph $\{conv(x, y) : y = f(x, a)\}$ is automatic can be made effective; namely, there is an algorithm that given any a from the domain produces a finite automaton recognising the graph $\{conv(x, y) : y = f(x, a)\}$. All the results of the paper, apart from Theorems 6, 7 and 8, satisfy this effectiveness condition. Thus, under this effectiveness condition, semiautomatic structures are still structures with finite presentations.

- For the structures $\mathcal{A}$ with no relation symbols, semiautomaticity is equivalent to saying that all algebraic polynomials with one variable (as defined in the beginning of Section 2) are automatic. Thus, semiautomaticity under the effectiveness condition, is equivalent to saying that the structure $\mathcal{A}' = (A, g_0, g_1, \ldots)$, where $g_0, g_1, \ldots$ is the list of all algebraic polynomials with one variable, is automatic. In particular this implies that the first order theory of this structure derived from $\mathcal{A}$ is decidable. The first order theory of $\mathcal{A}'$, can naturally be embedded into the first order theory of $\mathcal{A}$. Hence, semiautomaticity of $\mathcal{A}$ under the effectiveness condition, implies that a natural fragment of the first order theory of $\mathcal{A}$ is decidable. Moreover, algebraically the structure $\mathcal{A}'$ has exactly the same set of congruences as the original structure $\mathcal{A}$.

- There is a difference between semiautomatic / automatic functions and relations when $=$ is only semiautomatic and not automatic. While a function, for each input, has to find only one representative of an output, a relation must be true for all representatives of a given tuple which is satisfied. Therefore, it can be that a function is automatic while the graph $\{conv(x, y) : y = f(x)\}$ is not automatic. This difference in the effectivity of functions and relations is found in many domains where equality is not fully effective. For example there are many methods to systematically alter computer programs (for example, if p computes $x \mapsto f(x)$ then $F(p)$ computes $x \mapsto f(x) + 1$). For many programming languages, such an F can even be realised by an automatic function transforming the programs. However, it would be impossible to check whether a program q is equal to $F(p)$ in the sense that it has the same input/output behaviour: the relation $\{(p, q)$ such that q computes a function producing outputs one larger than those outputs produced by $p\}$ is indeed an undecidable set, due to Rice's Theorem.

Often one identifies the rationals with the set of all pairs written as a/b with $a \in \mathbb{Z}$ and $b \in \{1, 2, \ldots\}$; so one identifies "one half" with each of $1/2, 2/4, 3/6, \ldots$ and consider all of these to be equal. Similarly, in the case that the distinction is not relevant, the represented structure is often identified with its automatic or semiautomatic presentation and one denotes representatives in the automatic domain by their natural representation or vice versa and denotes the automatic functions realising these operations with the usual notation for operations of the structure represented.

Contributions of the Paper. First, the paper proposes the class of semiautomatic structures that can be defined in terms of finite automata. This class contains all automatic structures. Under the effectiveness condition put on semiautomaticity, (1) these structures have finite presentations, (2) natural fragments of their first order theories are decidable and (3) the class is wide enough to contain structures with undecidable theories. The paper provides many examples of semiautomatic structures, see Section 2.

Second, the paper provides several results of a general character. For example, purely relational structures, countable ordinals and permutation algebras all have semi-automatic presentations. This provides a large class of semiautomatic structures and showcases the power of finite automata in representation of algebraic structures. Note that for these results, no effectivity constraints on the semiautomaticity are made. See Section 3.

Third, the paper proves semiautomaticity for many of the classical algebraic structures which are groups, rings and vector spaces. The main reason for this study is that most of these structures lack automatic presentations (such as $(Q, +)$, $(Z; , +, \cdot, \leqslant)$ and infinite fields). Therefore, it is natural to ask which of these structures admit semiautomaticity. Many of these structures and in particular all concretely given examples are also semiautomatic with the effectivity condition. For instance, the ordered field $(\mathbb{Q}(\sqrt{n}); +, \cdot, <, =)$ is semiautomatic for every natural number n. There are also several counterexamples which are not semiautomatic. These examples and counterexamples are presented in Sections 4, 5 and 6.

A full version with the omitted proofs and results is available as Research Report 457 of the Centre for Discrete Mathematics and Theretical Computer Science (CDMICS), The Univerity of Auckland.

2   Decidability Theorem and Examples

The first result is a simple and general decidability result about semiautomatic structures without relational symbols. So, let $\mathcal{A} = (A, f_1, \ldots, f_n)$ be a semiautomatic structure where each f_i is an operation. An *algebraic polynomial* is a unary operation g of the form $f(a_1, \ldots, a_k, x, a_{k+2}, \ldots, a_n)$, where f is a basic operation of $\mathcal{A}$, and $a_1, \ldots, a_k, a_{k+2}, \ldots, a_n$ are parameters from A. Consider the structure $\mathcal{A}' = (\mathcal{A}; g_0, g_1, \ldots)$, where $g_0, g_1, \ldots$ are a complete list of all algebraic polynomials obtained from $f_1, \ldots, f_n$. There is a close relationship between $\mathcal{A}$ and $\mathcal{A}'$ in terms of congruence relations (that is equivalence relations respected by the basic operations):

Proposition 1. *The set of congruences of the structures $\mathcal{A}$ and $\mathcal{A}'$ coincide.*

The transformation $\mathcal{A} \to \mathcal{A}'$ gives an embedding of the first order theory of $\mathcal{A}'$ (with parameters from A) into the first order theory of $\mathcal{A}$. The embedding is the identity mapping. Assuming the effectivity condition (that is there is an algorithm that given any algebraic polynomial g produces a finite automaton recognising the graph of g), the automatic structure $\mathcal{A}'$ has a decidable first order theory.

Theorem 2 (Decidability Theorem). *If $\mathcal{A}$ is semiautomatic, then under the effectivity condition the first order theory of $\mathcal{A}'$ is decidable.*

The next examples illustrate that there are many semiautomatic structures which are not automatic.

Example 3. $(\{0,1\}^*, =; \circ)$ with $\circ$ being the string concatenation is an example of a structure which is semiautomatic but not automatic. For a fixed string v, the mappings $w \mapsto vw$ and $w \mapsto wv$ are both automatic; however, $conv(v,w) \mapsto vw$ is not an automatic mapping. Indeed, there is no automatic presentation of $(\{0,1\}^*, =, \circ)$.

Furthermore, $(\mathbb{N}, +, <, =; \cdot)$ is an example of a semiautomatic structure which is not automatic, as there is no automatic copy of the multiplicative structure of the natural numbers $(\mathbb{N}, \cdot)$. It is known that multiplication is semiautomatic due to the fact that multiplication with a constant can be implemented as repeated addition. One can augment this example with a semiautomatic function $f : \mathbb{N} \times \mathbb{N} \to \mathbb{N}$ such that $f(x+y, x) = 3^x + 3^{x+1} \cdot y - 1$ and $f(x, x+y+1) = 2 \cdot 3^x + 3^{x+1} \cdot y - 1$. Note that f is a semiautomatic bijection and there exists no infinite regular set A for which there exists an automatic bijection $g : A \times A \to A$.

Every two-sided Cayley automatic group $(G, \circ)$ is an example of a semiautomatic structure $(G, =; \circ)$; the reason is that for finitely generated groups, multiplication with group elements is automatic, as the multiplication with each generator is automatic. Furthermore, the automaticity of $=$ follows from the definition of Cayley automatic groups. In turn, the definition of a semiautomatic structure gives that every semiautomatic group $(G, =; \circ)$ is Cayley automatic. Miasnikov and Sunic [10] provide an example of a group $(G, \circ)$ which is one-sided but not two-sided Cayley automatic, thus one would not have that $(G, =; \circ)$ is semiautomatic but only that $(G, =; \{x \mapsto x \circ y : y \in G\})$ is semiautomatic.

Example 4. Let $\mathbb{S}$ be the set of square numbers. Then $(\mathbb{N}, \mathbb{S}, <, =; +)$ is semiautomatic. This structure is obtained by first using a default automatic representation $(A, +, <, =)$ of the additive monoid of the natural numbers and then to let $B = \{conv(a,b) : a, b \in A \wedge b \leqslant a + a\}$ be the desired structure. Here $conv(a,b)$ represents $a^2 + b$. One has now $conv(a,b) < conv(a',b')$ iff $a < a' \vee (a = a' \wedge b < b')$. Furthermore, $conv(a,b) + 1 = conv(a',b')$ iff $(a = a' \wedge b' = b + 1 \leqslant a + a) \vee (a' = a + 1 \wedge b = a + a \wedge b' = 0)$. Iterated addition with 1 defines the addition with any fixed natural number. Note that $(\mathbb{N}, \mathbb{S}, +, <, =)$ is not automatic.

The term algebra of a binary function f over a constant a consist of the term a and all terms $f(x,y)$ formed from previous terms x and y; for example $f(a,a)$, $f(a, f(a,a))$ and $f(f(a,a), f(a,a))$ are terms. Let T denote the set of all terms formed using the constant a and binary function f.

Theorem 5. *The term algebra* $(T; f, =)$ *is semiautomatic.*

Proof. Let $x_0, x_1, \ldots$ be a one-one enumeration of all terms. One now has to find a representation of the terms in which all mappings $left_k : y \mapsto f(x_k, y)$ and $right_k : y \mapsto f(y, x_k)$ are automatic. The idea is to represent a by 0 and each function $left_k$ by 01^{2k} and each function $right_k$ by 01^{2k+1}. That is, if w

represents a term y, then $01^{2k}w$ denotes $left_k(y)$ and $01^{2k+1}w$ denotes $right_k(y)$. Note that each term starts with a 0 and thus, for each $w \in (01^*)^*0$, there is a unique term represented by w.

For the above representation, the functions $left_k$ and $right_k$ are clearly automatic, as each of them just inserts the prefix 01^{2k} or 01^{2k+1} in front of the input. Thus, f is semiautomatic.

Let $depth(a) = 0$ and $depth(f(x,y)) = 1+max\{depth(x), depth(y)\}$. Now each term y has only finitely many representations, as it can only have representations which have at most $depth(y) + 1$ zeros and each $left_k$ or $right_k$ used in the representation must be a sub-term of y. Thus, $=$ is semiautomatic. □

3   Relational Structures, Permutation Algebras and Ordinals

This section shows that when the signature of the structure is very restricted then the resulting structure is always semiautomatic. Theorem 6 says that if a structure over a countable domain consists only of relations and each of these relations has only to be semiautomatic and not automatic, then one can indeed find a representation for this structure; this first result will then be applied to show that every countable set of ordinals which is closed under $+$ and $<$ has a semiautomatic representation.

Theorem 6. *Every relational structure* $(A; R_1, R_2, \ldots)$, *given by at most countably many relations* $R_1, R_2, \ldots$ *over a countable domain* A, *is semiautomatic.*

Delhommé [2] showed that some automatic ordered set $(A, <, =)$ is isomorphic to the set of ordinals below α iff $\alpha < \omega^\omega$. Furthermore some tree-automatic set $(A, <, =)$ is isomorphic to the set of ordinals below α iff $\alpha < \omega^{\omega^\omega}$. It follows directly from Theorem 6 that every countable set of ordinals is isomorphic to an semiautomatic ordered set, the next result shows that one can combine this result also with a semiautomatic addition of ordinals.

Theorem 7. *Let* α *be a countable ordinal. The structure* $(\{\beta : \beta < \omega^\alpha\}; +, <, =)$ *is semiautomatic.*

Proof. Let $(A; <, =)$ be a semiautomatic representation of the ordinals below α and assume that the symbols $\omega, \char94, (,), +$ are not in the alphabet of A. Now let B be the set of all strings of the form $\omega\char94(a_0) + \omega\char94(a_1) + \ldots + \omega\char94(a_n)$ representing the ordinal $\omega^{a_0} + \omega^{a_1} + \ldots + \omega^{a_n}$ with the empty string representing 0. The set of all possible representations of the ordinals below ω^α is regular. Now one can realise the addition of two non-empty strings v, w by forming $v + w$, so if $v = \omega\char94(5) + \omega\char94(3) + \omega\char94(3)$ and $w = \omega\char94(4) + \omega\char94(1) + \omega\char94(0)$ then $v + w = \omega\char94(5) + \omega\char94(3) + \omega\char94(3) + \omega\char94(4) + \omega\char94(1) + \omega\char94(0)$ representing $\omega^5 + \omega^3 + \omega^3 + \omega^4 + \omega^1 + \omega^0$. Ordinals have the rules that if $a < b$ then $\omega^a + \omega^b = \omega^b$, hence the above ordinal equals $\omega^5 + \omega^4 + \omega^1 + \omega^0$.

Following Cantor's arguments, for every ordinal $\beta < \omega^\alpha$, there is a normal form $\beta = \omega^{b_0} + \omega^{b_1} + \ldots + \omega^{b_n}$ for some n with $b_0, b_1, \ldots, b_n \in A$ and $b_0 \geqslant$

$b_1 \geqslant \ldots \geqslant b_n$. Now, given an ordinal $w = \omega^{a_0} + \omega^{a_1} + \ldots + \omega^{a_m}$, it holds that $w = \beta$ iff there are $i_0, i_1, \ldots, i_n$ such that $i_0 < i_1 < \ldots < i_n = m$ and for all $k \leqslant m$, the least index j with $k \leqslant i_j$ must satisfy $k < i_j \Rightarrow a_k < b_j$ and $k = i_j \Rightarrow a_k = b_j$. Given the automata to check whether some $a \in A$ is below or equal b_j, one can build from these finite automata an automaton which checks whether $w = \beta$. Furthermore, $w < \beta$ iff there are $n' \leqslant n$ and $i_0, i_1, \ldots, i_{n'-1}, i_{n'}$ with $i_0 < i_1 < \ldots < i_{n'} = m + 1$ and for all $k \leqslant m$, the least index j with $k \leqslant i_j$ must satisfy $k < i_j \Rightarrow a_k < b_j$ and $k = i_j \Rightarrow a_k = b_j$. Again the corresponding test can be realised by a finite automaton.   □

The next result shows that permutation algebras are semiautomatic. Here a permutation algebra is a domain A plus a function f such that f is a bijection. Furthermore, the domain A is assumed to be countable and this assumption applies to all structures considered in the present paper.

Theorem 8. *Every permutation algebra $(A, f; =)$ is semiautomatic.*

Proof. Let the orbit of a set z be the set of all z' such that there is an n with $f^n(z') = z$ or $f^n(z) = z'$. The idea is to pick up a set X of elements $x_0, x_1, \ldots$ such that for each z there is exactly one x_k in its orbit and represents X by the set Y of its indices (which is either $\mathbb{N}$ or a finite subset of $\mathbb{N}$). Now the domain is $Y \times \mathbb{Z}$ and one uses the following semiautomatic equivalence relation =: $conv(k, 0)$ represents x_k and $conv(k, h) = conv(k', h')$ iff $k = k'$ and $f^{|h-h'|}(x_k) = x_k$; here $|h - h'|$ is the absolute value of $h - h'$ and if one starts at x_k, then $conv(k, h) = conv(k, h')$ iff $|h - h'|$ times applying f to x_k gives x_k again. Furthermore, $f(k, h) = (k, h + 1)$. It is easy to see that f is automatic on a suitable representation of $Y \times \mathbb{Z}$ and that also every set $\{(k', h') : (k', h') = (k, h)\}$ is regular as either it is the set $\{(k, h)\}$ itself or it is the set of all $\{(k, h + \ell \cdot c) : \ell \in \mathbb{Z}\}$ for some $c \in \{1, 2, \ldots\}$.   □

Theorems 6 and 8 either use relations or a single unary function. If one has both of these concepts, then the result does no longer hold. The next result below stands in contrast with Theorem 6. Note that though relations can be made semi-automatic using the technique of Theorem 6, functions cannot: the intuitive reason is that the graphs of functions over one variable are relations over two variables (one for the input and one for the output), and the semiautomaticity requirement is that the graph of the function (which is two variable relation) is automatic.

Theorem 9. *There is a recursive subset B of $\mathbb{N}$ such that the structure $(\mathbb{N}, B, Succ; =)$ is not semiautomatic.*

Proof. Let B be a recursive subset of $\mathbb{N}$ which is not exponential time computable and let $Succ : \mathbb{N} \mapsto \mathbb{N}$ be the successor function from x to $x + 1$.

If the above structure $(\mathbb{N}, B, Succ; =)$ is semiautomatic, then there exists a regular domain A (representing $\mathbb{N}$), an automaton M accepting $B \subseteq A$ and a linear time computable function $f : A \to A$ representing S, where $\{conv(x, f(x)) : x \in A\}$ is regular. Let $w \in A$ represent 0. Thus, $f^n(w)$ represents n and $f^n(w)$ has length at most $(n + 1) \cdot c$ for some constant c.

Now, $B(n)$ can be decided by first computing $f^n(w)$ and then checking if $M(f^n(w))$ accepts. This can be done in time polynomial in n and thus exponential in the length of the binary representation of n. This is a contradiction, as B was chosen not to be exponential time computable. Thus, the structure $(\mathbb{N}, B, Succ; =)$ cannot be semiautomatic. Note that if the structure contains only one of B and f, then it has to be automatic, as they are a predicate (characteristic function of set) and a function with only one input variable and the proof does not even use whether $=$ is automatic or semiautomatic at all.   □

4   Groups and Order

Khoussainov, Rubin and Stephan [9, Corollary 4.4] showed (in slightly different words) that there is a semiautomatic presentation of $(\mathbb{Z}, =; +)$ in which the order of the integers is not semiautomatic. An important question left open is whether one can improve this result such that the presentation of $(\mathbb{Z}, +, =)$ used is automatic.

Question 10. *Is there an automatic presentation of the integers such that addition and equality are automatic while the set of positive integers is not regular, that is, the ordering of the integers is not semiautomatic?*

Note that a positive answer to this question would be a strengthening of the fact that the order $<$ is not first-order definable in $(\mathbb{Z}, +, =)$. This question motivates to study the connections between automatic and semiautomatic groups and order. For this, recall the definition of ordered groups.

Definition 11. A group $(G, \circ)$ is a structure with a neutral element e such that for all $x, y, z \in G$ there is a $u \in G$ satisfying $x \circ e = e \circ x = x$, $x \circ (y \circ z) = (x \circ y) \circ z$, $u \circ x = e$ and $x \circ u = e$. Such a structure without the last statements on the existence of the inverse is called a monoid. An ordered group $(G, \circ, <)$ satisfies that $<$ is transitive, antisymmetric and that all $x, y, z \in G$ with $x < y$ satisfy $x \circ z < y \circ z$ and $z \circ x < z \circ y$. If the preservation of the order holds only for operations with z from one side, then one calls the corresponding group right-ordered or left-ordered, respectively.

The first result is that in a semiautomatic group $(G, \circ; =)$ and a semiautomatic ordered group $(G, \circ; <, =)$, the relations $=$ and $<$ are indeed automatic.

Proposition 12. *Given a semiautomatic presentation $(G, \circ; =)$ of a group, the equality in this presentation is already automatic; similarly, given any semiautomatic presentation $(G, \circ; <, =)$ of an ordered group, the equality and order in this presentation are both automatic.*

Proof. Note that there are now several members of the presentation G of the group which are equal, for ease of notation one just writes still $x \in G$ in this case.

So let the semiautomatic presentation $(G, \circ; =)$ be given and let e be the neutral element. In particular the set of all representatives of e is regular. Now one can define an automatic function neg which finds for every $x \in G$ an element $neg(x) \in G$ with $x \circ neg(x) = e$. Having this function and using that $\circ$ is automatic, one has that $x = y \Leftrightarrow x \circ neg(y) = e$, hence $=$ is automatic.

Similarly, given an automatic presentation $(G, \circ; <, =)$ of an ordered group, one shows again that $=$ is automatic. Furthermore, as the set $\{u \in G : u < e\}$ is regular, one can use that $x < y$ iff $x \circ neg(y) < e$ in order to show that $<$ is also automatic. $\qquad \square$

There are numerous examples of ordered automatic groups. It is clear that such groups must be torsion-free. Examples would be the additive group of integers, $\mathbb{Z}^n$ with lexicographic order on the components and pointwise addition, subgroups of the rationals generated by elements of the form x^{-k} for some fixed rational x and k ranging over $\mathbb{N}$. So it is natural to look for further examples, in particular noncommutative ones. The next result shows noncommutative automatic ordered groups do not exist; note that the result holds even if in the group below only $(G, \circ, =)$ is automatic and the ordering exists, but is not effective.

Theorem 13. *Every ordered automatic group $(G, \circ, <, =)$ is Abelian.*

Proof. Let an automatic ordered group $(G, \circ, <, =)$ be given, as the equality is automatic, one can without loss of generality assume that every element of the group is given by a unique representative in G. Nies and Thomas [12,13] showed that due to the automaticity every finitely generated subgroup $(F, \circ)$ of G satisfies that it is Abelian by finite. In particular every two elements v, w of F satisfy that there is a power n with $v^n \circ w^n = w^n \circ v^n$. Now, following arguments of Neumann [11] and Fuchs [4, page 38, Proposition 10], one argues that the group is Abelian.

In the case that $v \circ w^n \neq w^n \circ v$, consider the element $w^n \circ v \circ w^{-n} \circ v^{-1}$ which is different from e; without loss of generality $w^n \circ v \circ w^{-n} \circ v^{-1} < e$. By multiplying from both sides inductively with $w^n \circ v \circ w^{-n}$ and v^{-1}, respectively, one gets inductively the relation $(w^n \circ v \circ w^{-n})^{m+1} \circ v^{-(m+1)} < (w^n \circ v \circ w^{-n})^m \circ v^{-m} < e$ for $m = 1, 2, \ldots, n$ and by associativity and cancellation the relation $w^n \circ v^n \circ w^{-n} \circ v^{-n} < e$ can be derived. This contradicts the assumption that v^n and w^n commute and therefore $w^n \circ v^n \circ w^{-n} \circ v^{-n} = e$.

In the case that $v \circ w^n = w^n \circ v$, one again assumes that $v \circ w \circ v^{-1} \circ w^{-1} < e$ and derives that $v \circ w^n \circ v^{-1} \circ w^{-n} < e$ contradicting the assumption that v and w^n commute. Hence one can derive that any two given elements v, w in G commute and $(G, \circ)$ is an Abelian group. $\qquad \square$

Example 14. The Klein bottle group is an example of a noncommutative left-ordered group. This is the group of all $a^i b^j$ with generators a, b and the defining equality $a \circ b = b^{-1} \circ a$. One represents the group as the set of all $conv(i, j)$ with $i, j \in \mathbb{Z}$ using an automatic presentation of $(\mathbb{Z}, +, <)$. Now the group operation $a^i b^j \circ a^{i'} b^{j'}$ is given by the mapping from $conv(i, j), conv(i', j')$ to $conv(i + i', j + j')$ in the case that i' is even and to $conv(i + i', -j + j')$ in the case that i' is odd. Thus the group is automatic. The ordering on the pairs is the lexicographic ordering, that is, $a^i b^j < a^{i'} b^{j'}$ iff $i < i'$ or $i = i' \wedge j < j'$. Using some case distinction, one can show that $a^i b^j < a^{i'} b^{j'}$ iff $a \circ a^i b^j < a \circ a^{i'} b^{j'}$ iff $b \circ a^i b^j < b \circ a^{i'} b^{j'}$ and deduce from these basic relations that the group is left-ordered.

A central motivation of Question 10 is the connection between definability and automaticity of the order in groups. The next example shows that for some semiautomatic groups, the order can be first-order defined from the group operation (which is not the case with the integers). In the example one cannot have that $\circ$ is automatic, as the group is not commutative.

Theorem 15. *There is a semiautomatic noncommutative ordered group $(G, <, =; \circ)$ such that the ordering is first-order definable from the group operation.*

Theorem 16. *The additive ordered subgroup $(\{n \cdot 6^m : n, m \in \mathbb{Z}\}, +, <)$ of the rationals has a presentation in which the addition and equality are automatic while the ordering is not semiautomatic.*

Proof. The idea is to represent group elements as $conv(a, b, c)$ representing $a + b + c$ where $a \in \mathbb{Z}$, $b = b_1 b_2 \ldots b_n \in \{0\} \cup \{0, 1\}^* \cdot \{1\}$ represents $b_1/2 + b_2/4 + \ldots + b_n/2^n$ and $c_1 c_2 \ldots c_m \in \{0\} \cup \{0, 1, 2\}^* \cdot \{1, 2\}$ represents $c_1/3 + c_2/9 + \ldots + c_m/3^m$. The representation of $\mathbb{Z}$ is chosen such that addition is automatic. Furthermore, now one adds $conv(a, b, c)$ and $conv(a', b', c')$ by choosing $conv(a'', b'', c'')$ such that the represented values satisfy $a'' = a + a' + (b + b' - b'') + (c + c' - c'')$ and $b'' \in \{b + b', b + b' - 1\}$ and $c'' \in \{c + c', c + c' - 1\}$ and $0 \leqslant b'' < 1$ and $0 \leqslant c'' < 1$. It can be easily seen that the resulting operation is automatic.

Assume now by way of contradiction that one could compare the fractional parts b and c of a number in order, that is, the relation $\{(b, c) : conv(0, b, 0) < conv(0, 0, c)\}$ would be automatic. Then one could first-order define a function f which maps every ternary string c to the length-lexicographic shortest binary string b satisfying $conv(0, 0, c1) < conv(0, b, 0) < conv(0, 0, c2)$. There are $3^n \cdot 2$ ternary strings c of length $n + 1$ not ending with a 0 representing different values between 0 and 1 and f maps these to $3^n \cdot 2$ different binary strings representing values between 0 and 1; as the resulting strings are binary, some of these values $f(c)$ must have the length at least $n \cdot \log(3) / \log(2)$. However, this contradicts the fact the length of $f(c)$ is at most a constant longer than c for all inputs c from the domain of f (as f is first-order defined from an automatic relation and thus automatic). Thus the function f cannot be automatic and therefore the ordering can also not be automatic. It follows from Proposition 12 that the order is not even semiautomatic.   □

Tsankov [14] showed that the structure $(\mathbb{Q}, +, =)$ is not automatic. However, one can still get the following weaker representation.

Theorem 17. *The ordered group $(\mathbb{Q}, <, =; +)$ of rationals is semiautomatic.*

Theorem 18. *Let G be a Baumslag Solitar group, that is, be a finitely generated group with generators a, b and the defining relation $b^n a = a b^m$ for some $m, n \in \mathbb{Z} - \{0\}$. Then the group $(G; \circ, =)$ is semiautomatic.*

5 Rings

The ring of integers $(\mathbb{Z}, +, <, =; \cdot)$ is semiautomatic, the semiautomaticity of the multiplication stems from the fact that multiplication with fixed constants can be implemented by repeated adding or subtracting the input from 0 a fixed number of times. One can, however, augment the ring of integers with a root of a natural number and still preserve that addition and order are automatic and multiplication is semiautomatic.

Theorem 19. *The ring* $(\mathbb{Z}(\sqrt{n}), +, <, =; \cdot)$ *has for every positive natural number n a semiautomatic presentation.*

The next result deals with noncommutative rings where the multiplication is not commutative and where a 1 does not need to exist.

Theorem 20. *There is a ring* $(R, +, =, \cdot)$ *such that* $(R, +, =)$ *is an automatic group and the family of functions* $\{y \mapsto y \cdot x : x \in R\}$ *is semiautomatic while every function* $y \mapsto x \cdot y$ *with* $x \in R$ *fixed is either constant 0 or not automatic (independent of the automatic representation chosen for the ring).*

6 Fields and Vector Spaces

In the following, let $(A, +, <, =; \cdot)$ be a semiautomatic ordered ring. Note that such a ring is an integral domain, as given two nonzero factors v, w, one can (after multiplication with -1 when needed) assume that $0 < v$ and $0 < w$; then it follows that $0 < v \cdot w$ and therefore $v \cdot w$ differs from 0. Hence the quotient field is always defined.

Theorem 21. *If* $(A, +, <, =; \cdot)$ *is a semiautomatic ordered ring then the unique quotient field F defined by the ring is an ordered semiautomatic field* $(F; +, \cdot, <, =)$.

Proof. The members of F are of the form $\frac{a}{b}$ with $a, b \in A$ and $0 < b$; they are represented by $conv(a, b)$ but for convenience in the following always written as $\frac{a}{b}$. Let $\frac{a'}{b'}$ be a fixed element of F and consider adding, multiplying and comparing with $\frac{a'}{b'}$:

– The addition $\frac{a}{b} \mapsto \frac{a \cdot b' + a' \cdot b}{b \cdot b'}$ is automatic, as multiplication with fixed ring elements a', b' is automatic and adding of ring elements is also automatic;
– The multiplication $\frac{a}{b} \mapsto \frac{a \cdot a'}{b \cdot b'}$ is automatic, for the same reasons as addition;
– The set $\{\frac{a}{b} : a \cdot b' < a' \cdot b\}$ of all representatives of members of F less than $\frac{a'}{b'}$ is regular;
– The set $\{\frac{a}{b} : a \cdot b' = a' \cdot b\}$ of all representatives of $\frac{a'}{b'}$ is regular.

Hence $(F; +, \cdot, <, =)$ is semiautomatic; the verification that the resulting structure is an ordered field follows the verification that the rationals are an ordered field when constructed from the ring of integers, this verification is left to the reader. □

Corollary 22. *The ordered field* $(\mathbb{Q}; +, \cdot, <, =)$ *of the rationals and, for all* $n \in \mathbb{N}$*, the extensions* $(\mathbb{Q}(\sqrt{n}); +, \cdot, <, =)$ *are semiautomatic.*

Theorem 23. *If* $(A, +, <, =; \cdot)$ *is a semiautomatic ordered ring then every finite-dimensional vector space* $(F^n; +, \cdot, =)$ *defined from the quotient field* F *of the ring* A *has a semiautomatic representation and all linear mappings from* F^n *to* F^n *are automatic. In particular, finite algebraic extensions* $(G; +, \cdot, =)$ *of the field* $(F, +, \cdot, =)$ *are semiautomatic.*

Question 24. (a) *Are the structures* $(\mathbb{Q}, <, =; +, \cdot)$ *or* $(\mathbb{Q}, =; +, \cdot)$ *semiautomatic? In other words, is it really needed, as done in the above default representations, that the equality and the order are not automatic?*
(b) *Is the polynomial ring* $(\mathbb{Q}[x]; +, \cdot, =)$ *semiautomatic?*
(c) *Is there a transcendental field extension of the rationals which is semiautomatic?*

The counterpart of Questions 24 (b) and (c) for finite fields has a positive answer.

Theorem 25. *Let* $(F, +, \cdot)$ *be a finite field. Then the following structures are semiautomatic:*

- *Every (possibly infinite) algebraic extension* $(G, +, =; \cdot)$ *of the field;*
- *The polynomial rings* $(F[x], +, =; \cdot)$ *in one variable and* $(F[x, y]; +, \cdot, =)$ *in two or more variables;*
- *The field of fractions* $(\{\frac{a}{b} : a, b \in F[x] \wedge b \neq 0\}; +, \cdot, =)$ *over the polynomial ring with one variable.*

7   Conclusion

The present work gives an overview on initial results on semiautomatic structures and shows that many prominent structures (countable ordinals with addition, the ordered fields of rationals extended perhaps by one root of an integer, algebraic extensions of finite fields) are semiautomatic and investigates to which degree one can still have that some of the involved operators and relations are automatic. Several concrete questions are still open, in particular the following ones: Is there an automatic presentation of the integers such that addition and equality are automatic while the ordering of the integers is not semiautomatic? Are the structures $(\mathbb{Q}, <, =; +, \cdot)$ or $(\mathbb{Q}, =; +, \cdot)$ semiautomatic, that is, can in the semiautomatic field of rationals the order and the equality be made automatic? The corresponding is possible for the additive group of rationals.

Additional questions might relate to the question of effectivity. For example, for a given function f in some given structure, can one effectively find from the parameter y an automaton for $x \mapsto f(x, y)$? While this is impossible for the most general results in Section 3, the concrete structures in Sections 4, 5 and 6 permit that one obtains the automata from the representatives by recursive functions. The complexity of these functions might be investigated in subsequent work for various structures.

Acknowledgements. The authors would like to thank Anil Nerode as well as the participants of the IMS Workshop on Automata Theory and Applications who discussed the topic and initial results with the authors.

References

1. Case, J., Jain, S., Seah, S., Stephan, F.: Automatic Functions, Linear Time and Learning. In: Cooper, S.B., Dawar, A., Löwe, B. (eds.) CiE 2012. LNCS, vol. 7318, pp. 96–106. Springer, Heidelberg (2012)
2. Delhommé, C.: Automaticité des ordinaux et des graphes homogènes. Comptes Rendus Mathematique 339(1), 5–10 (2004)
3. Epstein, D.B.A., Cannon, J.W., Holt, D.F., Levy, S.V.F., Paterson, M.S., Thurston, W.P.: Word Processing in Groups. Jones and Bartlett Publishers, Boston (1992)
4. Fuchs, L.: Partially Ordered Algebraic Systems. Pergamon Press (1963)
5. Hodgson, B.R.: Décidabilité par automate fini. Annales des Sciences Mathématiques du Québec 7(1), 39–57 (1983)
6. Hopcroft, J.E., Motwani, R., Ullman, J.D.: Introduction to Automata Theory, Languages and Computation, 3rd edn. Addison-Wesley (2007)
7. Kharlampovich, O., Khoussainov, B., Miasnikov, A.: From automatic structures to automatic groups. CoRR abs/1107.3645 (2011)
8. Khoussainov, B., Nerode, A.: Automatic presentations of structures. In: Leivant, D. (ed.) LCC 1994. LNCS, vol. 960, pp. 367–392. Springer, Heidelberg (1995)
9. Khoussainov, B., Rubin, S., Stephan, F.: Definability and Regularity in Automatic Structures. In: Diekert, V., Habib, M. (eds.) STACS 2004. LNCS, vol. 2996, pp. 440–451. Springer, Heidelberg (2004)
10. Miasnikov, A., Šunić, Z.: Cayley graph automatic groups are not necessarily Cayley graph biautomatic. In: Dediu, A.-H., Martín-Vide, C. (eds.) LATA 2012. LNCS, vol. 7183, pp. 401–407. Springer, Heidelberg (2012)
11. Neumann, B.H.: On ordered groups. American Journal of Mathematics 71, 1–18 (1949)
12. Nies, A.: Describing Groups. The Bulletin of Symbolic Logic 13(3), 305–339 (2007)
13. Nies, A., Thomas, R.: FA-presentable groups and rings. Journal of Algebra 320, 569–585 (2008)
14. Tsankov, T.: The additive group of the rationals does not have an automatic presentation. The Journal of Symbolic Logic 76(4), 1341–1351 (2011)

The Query Complexity of Witness Finding

Akinori Kawachi[1], Benjamin Rossman[2], and Osamu Watanabe[1]

[1] Dept. of Mathematical and Computing Sciences, Tokyo Institute of Technology
Ookayama 2-12-1, Meguro-ku, Tokyo 152-8552, Japan
[2] National Institute of Informatics
2-1-2 Hitotsubashi, Chiyoda-ku, Tokyo 101-8430, Japan

Abstract. We study the following information-theoretic *witness finding problem*: for a hidden nonempty subset W of $\{0,1\}^n$, how many non-adaptive randomized queries (yes/no questions about W) are needed to guess an element $x \in \{0,1\}^n$ such that $x \in W$ with probability $> 1/2$? Motivated by questions in complexity theory, we prove tight lower bounds with respect to a few different classes of queries:

- We show that the *monotone* query complexity of witness finding is $\Omega(n^2)$. This matches an $O(n^2)$ upper bound from the Valiant-Vazirani Isolation Lemma [8].
- We also prove a tight $\Omega(n^2)$ lower bound for the class of *NP queries* (queries defined by an NP machine with an oracle to W). This shows that the classic search-to-decision reduction of Ben-David, Chor, Goldreich and Luby [3] is optimal in a certain black-box model.
- Finally, we consider the setting where W is an affine subspace of $\{0,1\}^n$ and prove an $\Omega(n^2)$ lower bound for the class of *intersection queries* (queries of the form "$W \cap S \neq \emptyset$?" where S is a fixed subset of $\{0,1\}^n$). Along the way, we show that every monotone property defined by an intersection query has an exponentially sharp threshold in the lattice of affine subspaces of $\{0,1\}^n$.

1 Introduction

We initiate a study of the following information-theoretic search problem, parameterized by a family $\mathcal{W}$ of subsets of $\{0,1\}^n$ and a family $\mathcal{Q}$ of functions $\mathcal{W} \to \{\top, \bot\}$ (i.e. yes/no questions about elements of $\mathcal{W}$, which we refer to as "queries").

Question 1. What is the minimum number of nonadaptive randomized queries from $\mathcal{Q}$ required to guess an element $x \in \{0,1\}^n$ such that $\mathbb{P}[x \in W] > 1/2$ for every nonempty $W \in \mathcal{W}$?

Formally, Question 1 asks for a joint distribution $(\mathbf{Q}_1, \ldots, \mathbf{Q}_m)$ on $\mathcal{Q}^m$ together with a function $f : \{\top, \bot\}^m \to \{0,1\}^n$ such that

$$\mathbb{P}[f(\mathbf{Q}_1(W), \ldots, \mathbf{Q}_m(W)) \in W] > 1/2$$

E.A. Hirsch et al. (Eds.): CSR 2014, LNCS 8476, pp. 218–231, 2014.

for every nonempty $W \in \mathcal{W}$. We emphasize that randomized queries $\mathbf{Q}_1, \ldots, \mathbf{Q}_m$ are non-adaptive, though not necessarily independent.[1]

We refer to Question 1 as the *witness finding problem* and to its answer, $m = m(\mathcal{W}, \mathcal{Q})$, as the *$\mathcal{Q}$-query complexity of $\mathcal{W}$-witness finding*. (We introduce the terminology "witness finding" to distinguish this information-theoretic problem from traditional computational search problems where the solution space is determined by an input, such as a boolean formula φ in the case of the search problem for SAT.) Note that $m(\mathcal{W}, \mathcal{Q})$ is monotone increasing with respect to $\mathcal{W}$ and monotone decreasing with respect to $\mathcal{Q}$. In this paper, we mainly study the setting where $\mathcal{W}$ is the set of all subsets of $\{0,1\}^n$. Here, to simplify notation, we simply write $m(\mathcal{Q})$ and speak of the *$\mathcal{Q}$-query complexity of witness finding*.

Our main results are tight lower bounds on $m(\mathcal{Q})$ for a few specific classes of queries (namely, *intersection queries*, *monotone queries* and *NP queries*). However, before defining these classes and stating our results formally, let us first dispense with the trivial cases where $\mathcal{Q}$ is the class All of all possible queries or the class Direct of *direct queries* of the form "$x \in W$?" where $x \in \{0,1\}^n$. It is easy to see that $m(\mathsf{All}) = n$ and $m(\mathsf{Direct}) = 2^n - 1$. Both lower bounds $m(\mathsf{All}) \geq n$ and $m(\mathsf{Direct}) \geq 2^n - 1$ follow from considering the random singleton witness set $\{\mathbf{x}\}$ where $\mathbf{x}$ is uniform in $\{0,1\}^n$. The upper bound $m(\mathsf{Direct}) \leq 2^n - 1$ is obvious, while the upper bound $m(\mathsf{All}) \leq n$ comes via deterministic queries $Q_1, \ldots, Q_n$ where $Q_i(W)$ asks for the ith coordinate in the lexicographically minimal element of W.

1.1 Intersection Queries and Monotone Queries

The first class $\mathcal{Q}$ that we consider, for which the question of $m(\mathcal{Q})$ is nontrivial, is the class Intersection of *intersection queries* of the form "$S \cap W \neq \emptyset$?" for fixed $S \subseteq \{0,1\}^n$. As we now explain, the Valiant-Vazirani Isolation Lemma [8] gives an elegant upper bound of $m(\mathsf{Intersection}) = O(n^2)$. First, note that if W is a singleton $\{w\}$, then n nonadaptive intersection queries suffice to learn w: for $1 \leq i \leq n$, we ask "$S_i \cap W \neq \emptyset$?" where $S_i = \{x \in \{0,1\}^n : x_i = 0\}$. Moreover, by asking n additional intersection queries "$T_i \cap W \neq \emptyset$?" where $T_i = \{x \in \{0,1\}^n : x_i = 1\}$, we can learn whether or not W is a singleton, in addition to learning w in the event that $W = \{w\}$. The Valiant-Vazirani Isolation Lemma gives a distribution $\mathbf{X}$ on subsets of $\{0,1\}^n$ such that $\mathbb{P}[|W \cap \mathbf{X}| = 1] = \Omega(1/n)$ for every nonempty $W \subseteq \{0,1\}^n$. By taking $s = O(n)$ independent copies of $\mathbf{X}_1, \ldots, \mathbf{X}_s$ of this distribution $\mathbf{X}$, we have $\mathbb{P}[\bigvee_{j=1}^{s} |W \cap \mathbf{X}_j| = 1] > 1/2$ for every nonempty $W \subseteq \{0,1\}^n$. We now get a witness finding procedure which makes $2ns = O(n^2)$ randomized intersection queries for sets $\mathbf{S}_{i,j} := S_i \cap \mathbf{X}_j$ and $\mathbf{T}_{i,j} := T_i \cap \mathbf{X}_j$. (By now the reader will have noticed our convention of designating random variables by bold letters.)

---

[1] That is, $\mathbf{Q}_1$ and $\mathbf{Q}_2$ may be dependent random variables. However, conditioned on $\mathbf{Q}_1 = Q_1$, $\mathbf{Q}_2$ cannot depend on the answer $Q_1(W) \in \{\top, \bot\}$. We remark that Question 1 is trivial for adaptive queries: for any class $\mathcal{Q}$ which includes queries "$\exists x \in W$ such that $x_i = 1$?", n adaptive (deterministic) queries suffice to find an element in every nonempty W.

The present paper started out as an investigation into the question whether $O(n^2)$ is a tight upper bound on $m(\mathsf{Intersection})$. This question arose from work of Dell, Kabanets, van Melkebeek and Watanabe [7], who showed that the Valiant-Vazirani Isolation Lemma is optimal among so-called black-box isolation procedures:

Theorem 1 ([7]). *For every distribution* $\mathbf{X}$ *on subsets of* $\{0,1\}^n$, *there exists nonempty* $W \subseteq \{0,1\}^n$ *such that* $\mathbb{P}[|\mathbf{X} \cap W| = 1] = O(1/n)$.

Borrowing an idea from the proof of Theorem 1 (namely, a particular distribution on subsets of $\{0,1\}^n$), we were able to show $m(\mathsf{Intersection}) = \Omega(n^2)$. (Note that Theorem 1 can be derived from this lower bound, as any black-box isolation procedure with success probability $o(1/n)$ would show that $m(\mathsf{Intersection}) = o(n^2)$ by the argument sketched above.) As a natural next step, we considered the class of *monotone queries*, that is, $Q : \wp(\{0,1\}^n) \to \{\top, \bot\}$ such that $Q(W) = \top \Rightarrow Q(W') = \top$ for all $W \subseteq W' \subseteq \{0,1\}^n$. Note that intersection queries are monotone, hence $n \leq m(\mathsf{Monotone}) \leq m(\mathsf{Intersection}) = \Theta(n^2)$. Generalizing our lower bound for intersection queries, we were able to prove the stronger result:

Theorem 2. *The monotone query complexity of witness finding,* $m(\mathsf{Monotone})$, *is* $\Omega(n^2)$.

We present the proof of Theorem 2 in §2. The proof uses an entropy argument, which hinges on the threshold behavior of monotone queries (in particular, the theorem of Bollobás and Thomason [4]).

1.2   NP Queries

Another motivation for studying Question 1 comes from a question concerning search-to-decision reductions. In the context of SAT, a *search-to-decision reduction* is an algorithm which, given a boolean function $\varphi(x_1, \ldots, x_n)$, constructs a satisfying assignment $x \in \{0,1\}^n$ for φ (if one exists) using an oracle for the SAT decision problem. The standard P^{NP} search-to-decision reduction uses n adaptive deterministic queries. In the setting of nonadaptive randomized queries, Ben-David, Chor, Goldreich and Luby [3] (using the Valiant-Vazirani Isolation Lemma) gave a $\mathrm{BPP}_{||}^{\mathrm{NP}}$ search-to-decision reduction with $O(n^2)$ queries. ($\mathrm{BPP}_{||}^{\mathrm{NP}}$ is the class of BPP algorithms with non-adaptive (parallel) query access to an NP oracle.)

We are interested in lower bounds for the query complexity of search-to-decisions for SAT. Of course, any nontrivial lower bound would separate P from NP. However, we can consider a "black-box" setting where, instead of receiving a boolean formula $\varphi(x_1, \ldots, x_n)$ as input, the $\mathrm{BPP}_{||}^{\mathrm{NP}}$ algorithm (including both the BPP machine and the NP machine) are given input 1^n as well as an oracle to the set $\{x \in \{0,1\}^n : x \text{ is a satisfying assignment for } \varphi\}$. On inspection, it is clear that the reduction of Ben-David et al. (which is indifferent to the syntax of the boolean formula φ) carries over to this black-box setting. Thus, we have the upper bound:

Theorem 3 (follows from [3]). *There is a* $\mathrm{BPP}_{||}^{\mathrm{NP}}$ *algorithm which solves the black-box satisfiability search problem with* $O(n^2)$ *queries.*

Motivated by this connection to complexity theory, we next set our sights on the question whether $O(n^2)$ is tight in Theorem 3. To fit the question into the framework of Question 1, we define the class of *NP queries* as follows.

Definition 1. *Informally, an* NP query *is a query* Q *given by an NP machine* M *with an oracle to* W *where* $Q(W) = M^W(1^n)$ *(i.e.* $Q(W) = \top \Leftrightarrow M^W$ *has an accepting computation on input* 1^n *). Formally, an* NP query *is a sequence* $Q = (Q^1, Q^2, \dots)$ *of queries* $Q^n : \wp(\{0,1\}^n) \to \{\top, \bot\}$ *such that there exists a single NP machine* $M^{()}$ *(with an unspecified oracle) where* $Q^n(W) = M^W(1^n)$ *for every* $W \subseteq \{0,1\}^n$. *An* ensemble of NP queries *is a sequence* $(Q_1, \dots, Q_m)$ *of NP queries given by NP machines* $M_1, \dots, M_m$ *which have a common upper bound* $t(n) = n^{O(1)}$ *on their running time.*

The NP query complexity of witness finding, $m(\mathsf{NP})$, gives a lower bound on the query complexity of $\mathrm{BPP}_{||}^{\mathrm{NP}}$ algorithms solving the black-box satisfiability search problem. Note that NP queries and monotone queries are incomparable: NP queries clearly need not be monotone, while it can be shown that the monotone "majority" query (defined by $Q_{\mathrm{maj}}(W) = \top$ iff $|W| \geq 2^{n-1}$) is not an NP query.[2] Nevertheless, we show that every NP query can be *well-approximated* by a monotone query (Lemma 7). Using this result together with our lower bound for $m(\mathsf{Monotone})$, we show:

Theorem 4. *The NP query complexity of witness finding,* $m(\mathsf{NP})$, *is* $\Omega(n^2)$.

Theorem 4 thus establishes the optimality of the search-to-decision reduction of Ben-David et al. in the black-box setting. The proof is presented in §3.

1.3 Affine Witness Sets

Finally, we consider the setting where $\mathcal{W}$ is the set of affine subspaces of $\{0,1\}^n$. Here, for a class of queries $\mathcal{Q}$, we write $m_{\mathrm{affine}}(\mathcal{Q})$ and speak of the $\mathcal{Q}$-*query complexity of affine witness finding*. While $m_{\mathrm{affine}}(\mathcal{Q}) \leq m(\mathcal{Q})$ by definition, intuitively the affine witness finding problem is easier because there are only $2^{O(n^2)}$ possibilities for W, as opposed to 2^{2^n}. One motivation for studying the affine setting comes from the observation that lower bounds on $m_{\mathrm{affine}}(\mathsf{NP})$ imply lower bounds on the complexity of the black-box satisfiability search problem on *polynomial-size* boolean formulas, since every affine subspace of $\{0,1\}^n$ is the set of satisfying assignments to a polynomial-size boolean formula of n variables. While we were unable to prove any nontrivial lower bounds on $m_{\mathrm{affine}}(\mathsf{Monotone})$ or $m_{\mathrm{affine}}(\mathsf{NP})$, we did get a result for intersection queries:

---

[2] Due to uniformity issues, it does not make sense to compare the classes of NP queries and intersection queries. However, for a natural notion of *non-uniform NP queries*, every intersection query "$S \cap W \neq \emptyset$?" is a non-uniform NP query where the NP machine M hardwires S using 2^n advice bits, non-deterministically guesses $x \in S$ and simply verifies that $x \in W$ using one oracle call to W.

Theorem 5. *The intersection query complexity of affine witness finding,* $m_{\text{affine}}(\textsf{Intersection})$, *is* $\Omega(n^2)$.

The proof is presented in §4. Along the way, we show that every monotone property defined by an intersection query has an *exponentially sharp threshold* in the lattice of affine subspaces of $\{0,1\}^n$ (Theorem 6). This raises the question whether all monotone properties have an exponentially sharp threshold in the affine lattice (Question 2); we note that a positive answer would imply $m_{\text{affine}}(\textsf{Monotone}) = \Omega(n^2)$.

2   Lower Bound for Monotone Queries

In this section, we prove Theorem 2 ($m(\textsf{Monotone}) = \Omega(n^2)$) using an information-theoretic argument. We briefly present the relevant notation. Let $H : [0,1] \to [0,1]$ denote the binary entropy function $H(p) := p\log(1/p) + (1-p)\log(1/(1-p))$. For finite random variables $\mathbf{X}$ and $\mathbf{Y}$, entropy $\mathbb{H}(\mathbf{X})$ and relative entropy $\mathbb{H}(\mathbf{X} \mid \mathbf{Y})$ are defined by

$$\mathbb{H}(\mathbf{X}) := \sum_{x \in \text{Supp}(\mathbf{X})} \mathbb{P}[\mathbf{X} = x] \cdot \log(1/\mathbb{P}[\mathbf{X} = x]),$$

$$\mathbb{H}(\mathbf{X} \mid \mathbf{Y}) := \sum_{y \in \text{Supp}(\mathbf{Y})} \mathbb{P}[\mathbf{Y} = y] \cdot \mathbb{H}(\mathbf{X} \mid \mathbf{Y} = y).$$

(Here $\mathbb{H}(\mathbf{X} \mid \mathbf{Y} = y)$ is the entropy of the marginal distribution of $\mathbf{X}$ conditioned on $\mathbf{Y} = y$.) We assume familiarity with the basic properties of entropy, namely the chain rule $\mathbb{H}(\mathbf{X}, \mathbf{Y}) = \mathbb{H}(\mathbf{X}) + \mathbb{H}(\mathbf{Y} \mid \mathbf{X})$, the fact that $\mathbb{H}(f(\mathbf{X})) \le \mathbb{H}(\mathbf{X})$ for every deterministic function f of $\mathbf{X}$, and the fact that $\mathbb{H}(\mathbf{X}) \le \log|\text{Supp}(\mathbf{X})|$ with equality iff $\mathbf{X}$ is uniform (for more background, see [6]).

Our lower bound uses a standard averaging argument (Yao's principle) to invert the role of randomness in the definition of $m(\mathcal{W}, \mathcal{Q})$. For completeness, the proof is included in Appendix A.

Lemma 1. *Suppose* $\mathbf{W}$ *is a random variable on* $\mathcal{W} \setminus \{\emptyset\}$ *such that for all* $Q_1, \ldots, Q_m \in \mathcal{Q}$ *and every function* $f : \{\top, \bot\}^m \to \{0,1\}^n$,

$$\mathbb{P}[f(Q_1(\mathbf{W}), \ldots, Q_m(\mathbf{W})) \in \mathbf{W}] \le 1/2.$$

Then the $\mathcal{Q}$-*query complexity of* $\mathcal{W}$-*witness finding is* $> m$.

We now define a particular random subset $\mathbf{W}$ of $\{0,1\}^n$. For all $0 \le k \le n$, let $\mathbf{W}_k$ be the random subset of $\{0,1\}^n$ containing each $x \in \{0,1\}^n$ independently with probability n^{k-n}. Let $\mathbf{k}$ be uniformly distributed in $\{1, \ldots, n/2\}$.[3] Finally, let $\mathbf{W} := \mathbf{W}_{\mathbf{k}}$. (A similar distribution was considered by Dell et al. [7] in proving

---

[3] For convenience, we assume $n/2$ is an integer (or an abbreviation for $\lfloor n/2 \rfloor$). For purposes of §2, $\mathbf{k}$ could just as well be monotone in $\{1, \ldots, n\}$. For purposes of §3, we merely require that $\mathbf{k}$ be uniformly distributed in $\{1, \ldots, n'\}$ where $n' \le n - \log^{\omega(1)} n$.

an upper bound of $O(1/n)$ on the success probability of black-box isolation procedures.)

The following lemma is a special case of the Bollobás-Thomason Theorem [4] (informally, "every monotone increasing property of subsets of a fixed set has a threshold function"). For completeness, a simple self-contained proof is included in Appendix B.

Lemma 2. *Let Q be a non-trivial monotone increasing property of subsets of $\{0,1\}^n$. For all $0 \le k \le n$, let $p_k := \mathbb{P}[\mathbf{W}_k$ has property $Q]$. Let θ be the unique index such that $p_\theta \le 1/2 < p_{\theta+1}$. Then*

(1)	$p_{\theta-i} \le 2^{-i} \ln 2$	*for all $0 \le i \le \theta$,*
(2)	$p_{\theta+i+1} \ge 1 - 2^{-2^i}$	*for all $0 \le i \le n - \theta - 1$,*
(3)	$H(p_k) \le (|\theta - k| + 1)/2^{|\theta-k|-1}$	*for all $0 \le k \le n$.*

Using Lemma 2(3), we prove a sharp bound on the relative entropy $Q(\mathbf{W} \mid \mathbf{k})$ all monotone queries Q.

Lemma 3. $\mathbb{H}(Q(\mathbf{W}) \mid \mathbf{k}) = O(1/n)$ *for every monotone query Q.*

Proof. If Q is identically $\bot$ or $\top$, then the statement is trivial (as $\mathbb{H}(Q(\mathbf{W}) \mid \mathbf{k}) = 0$). So assume Q is a non-trivial monotone query and let $p_0, \ldots, p_n$ and θ be as in Lemma 2. Then

$$\mathbb{H}(Q(\mathbf{W}) \mid \mathbf{k}) = \sum_{k=0}^{n/2} \mathbb{P}[\mathbf{k} = k] \cdot \mathbb{H}(Q(\mathbf{W}_k))$$

$$= \frac{2}{n} \sum_{k=1}^{n/2} H(p_k) \le \frac{2}{n} \sum_{k=1}^{n/2} \frac{|\theta - k| + 1}{2^{|\theta-k|-1}} \le \frac{4}{n} \sum_{i=0}^{\infty} \frac{i+1}{2^{i-1}} \le \frac{24}{n}.$$

The next lemma relates the entropy of an arbitrary random variable $\mathbf{z}$ on $\{0,1\}^n$ to the probability that $\mathbf{z} \in \mathbf{W}$.

Lemma 4. *For every random variable $\mathbf{z}$ on $\{0,1\}^n$ (not necessarily independent of $\mathbf{W}$),*

$$\mathbb{P}[\mathbf{z} \in \mathbf{W}] \le \frac{4}{n}\mathbb{H}(\mathbf{z}) + \frac{1}{2^{n/4}}.$$

Proof. Define $S \subseteq \{0,1\}^n$ by $S := \{x \in \{0,1\}^n : \mathbb{P}[\mathbf{z} = x] \ge 2^{-n/4}\}$. Note that

$$\mathbb{P}[\mathbf{z} \in \mathbf{W}] \le \mathbb{P}[\mathbf{z} \notin S] + \mathbb{P}[S \cap \mathbf{W} \ne \emptyset].$$

We bound each these righthand probabilities. First, by definition of S and $\mathbb{H}(\mathbf{z})$,

$$\mathbb{P}[\mathbf{z} \notin S] = \sum_{x \in \{0,1\}^n \setminus S} \mathbb{P}[\mathbf{z} = x] \le \sum_{x \in \{0,1\}^n \setminus S} \mathbb{P}[\mathbf{z} = x]\frac{\log(1/\mathbb{P}[\mathbf{z} = x])}{n/4} \le \frac{4}{n}\mathbb{H}(\mathbf{z}).$$

(Here we used $x \notin S \Rightarrow \mathbb{P}[\mathbf{z} = x] < 2^{-n/4} \Rightarrow \log(1/\mathbb{P}[\mathbf{z} = x]) > n/4$.) Finally, noting that $|S| \leq 2^{n/4}$ and $\mathbb{P}[x \in \mathbf{W}] < 2^{-n/2}$ for all $x \in \{0,1\}^n$, we have

$$\mathbb{P}[\mathbf{W} \cap S \neq \emptyset] \leq \sum_{x \in S} \mathbb{P}[x \in \mathbf{W}] < \frac{1}{2^{n/4}}.$$

Combining Lemmas 3 and 4, we get our main lemma:

Lemma 5. *For all monotone queries $Q_1, \ldots, Q_m$ and every function $f : \{\top, \bot\}^m \to \{0,1\}^n$,*

$$\mathbb{P}[f(Q_1(\mathbf{W}), \ldots, Q_m(\mathbf{W})) \in \mathbf{W}] \leq O(m/n^2) + o(1).$$

Proof. By standard entropy inequalities,

$$
\begin{aligned}
\mathbb{H}(f(Q_1(\mathbf{W}), \ldots, Q_m(\mathbf{W}))) &\leq \mathbb{H}(Q_1(\mathbf{W}), \ldots, Q_m(\mathbf{W})) \\
&\leq \mathbb{H}(Q_1(\mathbf{W}), \ldots, Q_m(\mathbf{W}), \mathbf{k}) \\
&= \mathbb{H}(\mathbf{k}) + \mathbb{H}(Q_1(\mathbf{W}), \ldots, Q_m(\mathbf{W}) \mid \mathbf{k}) \\
&\leq \mathbb{H}(\mathbf{k}) + \mathbb{H}(Q_1(\mathbf{W}) \mid \mathbf{k}) + \cdots + \mathbb{H}(Q_m(\mathbf{W}) \mid \mathbf{k}).
\end{aligned}
$$

Since $\mathbb{H}(\mathbf{k}) = \log(n/2)$ and $\mathbb{H}(Q_i(\mathbf{W}) \mid \mathbf{k}) = O(1/n)$ for all i by Lemma 3, we have

$$\mathbb{H}(f(Q_1(\mathbf{W}), \ldots, Q_m(\mathbf{W}))) \leq O(m/n) + \log n.$$

Since $f(Q_1(\mathbf{W}), \ldots, Q_m(\mathbf{W}))$ is a random variable on $\{0,1\}^n$, we can apply Lemma 4 to get

$$
\begin{aligned}
\mathbb{P}[f(Q_1(\mathbf{W}), \ldots, Q_m(\mathbf{W})) \in \mathbf{W}] &\leq \frac{4}{n} \mathbb{H}(f(Q_1(\mathbf{W}), \ldots, Q_m(\mathbf{W}))) + \frac{1}{2^{n/4}} \\
&\leq O(m/n^2) + \frac{4 \log n}{n} + \frac{1}{2^{n/4}} \\
&= O(m/n^2) + o(1).
\end{aligned}
$$

Finally, we prove the main theorem of this section.

Theorem 2. (restated) *The monotone query complexity of witness finding, $m(\mathsf{Monotone})$, is $\Omega(n^2)$.*

Proof. Let $m = m(\mathsf{Monotone})$. By Lemma 1, there exist monotone queries $Q_1, \ldots, Q_m$ and a function $f : \{\top, \bot\}^m \to \{0,1\}^n$ such that

$$\mathbb{P}[f(Q_1(\mathbf{W}), \ldots, Q_m(\mathbf{W})) \in \mathbf{W} \mid \mathbf{W} \neq \emptyset] > 1/2.$$

By Lemma 5 and the fact that $\mathbb{P}[\mathbf{W} \neq \emptyset] = 1 - o(1)$,

$$
\begin{aligned}
\mathbb{P}[f(Q_1(\mathbf{W}), \ldots, Q_m(\mathbf{W})) \in \mathbf{W} \mid \mathbf{W} \neq \emptyset] &= \frac{\mathbb{P}[f(Q_1(\mathbf{W}), \ldots, Q_m(\mathbf{W})) \in \mathbf{W}]}{\mathbb{P}[\mathbf{W} \neq \emptyset]} \\
&\leq O(m/n^2) + o(1).
\end{aligned}
$$

It follows that $1/2 < O(m/n^2) + o(1)$ and hence $m = \Omega(n^2)$.

3   Lower Bound for NP Queries

In this section, we prove Theorem 4 ($m(\mathsf{NP}) = \Omega(n^2)$). The main idea in the proof involves showing that every NP query is well-approximated by a monotone query. First, we give a normal form for NP queries.

Lemma 6. *For every NP query Q, there exists a sequence $(A_1, B_1), \ldots, (A_s, B_s)$ where $A_i, B_i \subseteq \{0,1\}^n$ and $|A_i|, |B_i| \leq n^{O(1)}$ and $A_i \cap B_i = \emptyset$ such that for all $W \subseteq \{0,1\}^n$,*

$$Q(W) = \top \iff \bigvee_{i=1}^{s} (A_i \subseteq W) \wedge (B_i \cap W = \emptyset).$$

Proof. Let $M^{()}$ be the nondeterministic Turing machine (with an unspecified oracle) which defines Q, that is, $Q(W) = M^W(1^n)$. Let $t = n^{O(1)}$ be the maximum running time of $M^{()}$. For each accepting computation of $M^{()}$ on input 1^n, there is a sequence $\sigma = ((x_1, y_1), \ldots, (x_{t'}, y_{t'})) \in (\{0,1\}^n \times \{\top, \bot\})^{t'}$, $t' \leq t$, such that the computation makes oracle calls $x_1, \ldots, x_{t'}$ and receives answers $y_1, \ldots, y_{t'}$. Let $A_\sigma := \{x_i : y_i = \top\}$ and $B_\sigma := \{x_i : y_i = \bot\}$ and note that $|A_\sigma|, |B_\sigma| \leq t' \leq t$ and $A_\sigma \cap B_\sigma = \emptyset$. Let $(A_1, B_1), \ldots, (A_s, B_s)$ enumerate pairs (A_σ, B_σ) over all σ corresponding to accepting computations of $M^{()}$. This sequence $(A_1, B_1), \ldots, (A_s, B_s)$ satisfies the conditions of the lemma. ∎

The next lemma gives the approximation of NP queries by monotone queries. Let $\mathbf{W}$ continue to denote the random subset of $\{0,1\}^n$ defined in the previous section.

Lemma 7. *For every NP query Q, there is a monotone query Q^+ such that $\mathbb{P}[Q(\mathbf{W}) \neq Q^+(\mathbf{W})] = 2^{-\Omega(n)}$.*

Proof. Let $(A_1, B_1), \ldots, (A_s, B_s)$ be as in Lemma 6. Define Q^+ by

$$Q^+(W) = \top \overset{\text{def}}{\iff} \bigvee_{i=1}^{s} (A_i \subseteq W).$$

Clearly, Q^+ is a monotone query and $Q(W) \Rightarrow Q^+(W)$ (i.e. $Q(W) = \top$ implies $Q^+(W) = \top$). We have

$$\mathbb{P}\Big[Q(\mathbf{W}) \neq Q^+(\mathbf{W})\Big] = \mathbb{P}\Big[\neg Q(\mathbf{W}) \wedge Q^+(\mathbf{W})\Big]$$

$$= \mathbb{P}\Big[\Big(\bigwedge_{i=1}^{s}(A_i \not\subseteq \mathbf{W}) \vee (B_i \cap \mathbf{W} \neq \emptyset)\Big) \wedge \Big(\bigvee_{i=1}^{s}(A_i \subseteq \mathbf{W})\Big)\Big]$$

$$\leq \mathbb{P}\Big[\bigvee_{i=1}^{s}(B_i \cap \mathbf{W} \neq \emptyset) \wedge (A_i \subseteq \mathbf{W}) \wedge \bigwedge_{j=1}^{i-1}(A_i \not\subseteq \mathbf{W})\Big]$$

$$(4) \qquad \leq \max_i \mathbb{P}\Big[B_i \cap \mathbf{W} \neq \emptyset \;\Big|\; (A_i \subseteq \mathbf{W}) \wedge \bigwedge_{j=1}^{i-1}(A_i \not\subseteq \mathbf{W})\Big],$$

where this last inequality is justified by the fact that events $\{(A_i \subseteq \mathbf{W}) \wedge \bigwedge_{j=1}^{i-1}(A_i \not\subseteq \mathbf{W})\}$ are mutually exclusive over $i \in \{1, \ldots, s\}$.

Now fix i which maximizes (4). We claim that

$$(5) \qquad \mathbb{P}\Big[B_i \cap \mathbf{W} \neq \emptyset \,\Big|\, (A_i \subseteq \mathbf{W}) \wedge \bigwedge_{j=1}^{i-1}(A_i \not\subseteq \mathbf{W})\Big] \leq \mathbb{P}[B_i \cap \mathbf{W} \neq \emptyset].$$

This may be seen as follows. For $1 \leq k \leq n/2$, write $\mathbf{X}_k, \mathbf{Y}_k, \mathbf{Z}_k$ for events

$$\mathbf{X}_k := \{B_i \cap \mathbf{W}_k \neq \emptyset\}, \quad \mathbf{Y}_k := \{A_i \subseteq \mathbf{W}_k\}, \quad \mathbf{Z}_k := \{\bigwedge_{j=1}^{i-1} \bigvee_{y \in A_i \setminus A_j}(y \notin \mathbf{W}_k)\}.$$

First, note that $\mathbf{Y}_k \wedge \bigwedge_{j=1}^{i-1}(A_i \not\subseteq \mathbf{W}_k)$ is equivalent to $\mathbf{Y}_k \wedge \mathbf{Z}_k$. Next, note that $(\mathbf{X}_k, \mathbf{Z}_k)$ is independent of $\mathbf{Y}_k$ (by the independence of events $\{x \in \mathbf{W}_k\}$ over $x \in \{0,1\}^n$ and the fact that $A_i \cap B_i = \emptyset$). Therefore, $\mathbb{P}[\mathbf{X}_k \,|\, \mathbf{Y}_k \wedge \mathbf{Z}_k] = \mathbb{P}[\mathbf{X}_k \,|\, \mathbf{Z}_k]$. Next, note that $\mathbf{X}_k$ is monotone increasing and $\mathbf{Z}_k$ is monotone decreasing in the lattice of subsets of $\{0,1\}^n$. By well-known correlation inequalities (the FKG inequality, see Ch. 6 of [1]), it follows that $\mathbb{P}[\mathbf{X}_k \,|\, \mathbf{Z}_k] \leq \mathbb{P}[\mathbf{X}_k]$. Therefore, $\mathbb{P}[\mathbf{X}_k \,|\, \mathbf{Y}_k \wedge \mathbf{Z}_k] \leq \mathbb{P}[\mathbf{X}_k]$ for all $1 \leq k \leq n/2$ and hence $\mathbb{P}[\mathbf{X}_k \,|\, \mathbf{Y}_k \wedge \mathbf{Z}_k] \leq \mathbb{P}[\mathbf{X}_k]$. Finally, note that (5) is equivalent to the statement $\mathbb{P}[\mathbf{X}_k \,|\, \mathbf{Y}_k \wedge \mathbf{Z}_k] \leq \mathbb{P}[\mathbf{X}_k]$.

Picking up from (5), we have

$$(6) \qquad \mathbb{P}[B_i \cap \mathbf{W} \neq \emptyset] \leq \sum_{x \in B_i} \mathbb{P}[x \in \mathbf{W}] \leq \frac{|B_i|}{2^{n/2}} = \frac{n^{O(1)}}{2^{n/2}} = 2^{-\Omega(n)}.$$

Stringing together (4), (5) and (6), we conclude that $\mathbb{P}[Q(\mathbf{W}) \neq Q^+(\mathbf{W})] = 2^{-\Omega(n)}$.

Using this approximation of NP queries by monotone queries, we prove:

Theorem 4. (restated) *The NP query complexity of witness finding, $m(\mathsf{NP})$, is $\Omega(n^2)$.*

Proof. Let $m = m(\mathsf{NP})$. By Lemma 1, there exist NP queries $Q_1, \ldots, Q_m$ and a function $f : \{\top, \bot\}^m \to \{0,1\}^n$ such that

$$\mathbb{P}[f(Q_1(\mathbf{W}), \ldots, Q_m(\mathbf{W})) \in \mathbf{W} \mid \mathbf{W} \neq \emptyset] > 1/2.$$

Let $Q_1^+, \ldots, Q_m^+$ be monotone queries approximating $Q_1, \ldots, Q_m$ as in Lemma 7. We have

$$\mathbb{P}[f(Q_1^+(\mathbf{W}), \ldots, Q_m^+(\mathbf{W})) \in \mathbf{W}]$$

$$\geq \mathbb{P}[f(Q_1(\mathbf{W}), \ldots, Q_m(\mathbf{W})) \in \mathbf{W}] - \sum_{i=1}^{m} \mathbb{P}[Q_i(\mathbf{W}) \neq Q_i^+(\mathbf{W})]$$

$$= \Omega(1) - \frac{m}{2^{\Omega(n)}}.$$

On the other hand, by Lemma 5,

$$\mathbb{P}[f(Q_1^+(\mathbf{W}), \ldots, Q_m^+(\mathbf{W})) \in \mathbf{W}] \leq O(m/n^2) + o(1).$$

It follows that $\Omega(1) - m 2^{-\Omega(n)} \leq O(m/n^2) + o(1)$, which is only possible if $m = \Omega(n^2)$.

4 Affine Witness Sets

At this point, we have shown that $m(\mathsf{Intersection})$, $m(\mathsf{Monotone})$ and $m(\mathsf{NP})$ are all $\Theta(n^2)$ by a combination of our lower bound (Theorems 2 and 4) and the upper bounds mentioned in §1. We now turn our attention to the setting of affine witness sets. We would like to prove lower bounds on $m_{\mathrm{affine}}(\mathsf{Intersection})$, $m_{\mathrm{affine}}(\mathsf{Monotone})$ and $m_{\mathrm{affine}}(\mathsf{NP})$ using similar information-theoretic arguments. We begin by considering the natural affine analogue of the random witness set $\mathbf{W}$. For all $0 \leq k \leq n$, let $\mathbf{A}_k$ be the uniform random k-dimensional subspace of $\{0,1\}^n$. Let $\mathbf{k}$ be uniform in $\{1,\dots,n/2\}$ (as before) and let $\mathbf{A} := \mathbf{A}_{\mathbf{k}}$.

Unfortunately, when we attempt to repeat the argument in §2, we get stuck at Lemma 2 (the Bollobás-Thomason Theorem). In particular, in order to have an appropriate version of Lemma 2(3) in the affine setting, we need a positive answer the following question:

Question 2. Let Q be a non-trivial monotone increasing property of affine subspaces of $\{0,1\}^n$. For all $0 \leq k \leq n$, let $p_k := \mathbb{P}[\mathbf{A}_k$ has property $Q]$. Let θ be the unique index such that $p_\theta \leq 1/2 < p_{\theta+1}$. Is it necessarily true that $\min\{p_k, 1 - p_k\} \leq 2^{-|\theta-k|+O(1)}$ for all k?

In other words, Question 2 asks whether every monotone property has an *exponentially sharp threshold* in the lattice of affine subspaces of $\{0,1\}^n$.

Remark 1. We can ask a similar question with respect to the lattice $\mathcal{L}_n$ of linear subspaces of $\{0,1\}^n$ (we suspect that the answer is the same). Writing $\mathcal{P}_n$ (resp. $\mathcal{P}_{2^n}$) for the lattice of subsets of $[n]$ (resp. $\{0,1\}^n$), note that $\mathcal{L}_n$ has an ambiguous status in relation to $\mathcal{P}_n$ and $\mathcal{P}_{2^n}$: on the one hand, $\mathcal{L}_n$ is the "q-analogue" of $\mathcal{P}_n$; on the other hand, $\mathcal{L}_n$ is a subset (in fact, a sub-meet-semilattice) of $\mathcal{P}_{2^n}$. Using a q-analogue of the Kruskal-Katona Theorem due to Chowdhury and Patkos [5], we can show that $p_k \leq 2^{-\Omega(\theta/k)}$ for all $k < \theta$ and $1 - p_k \leq 2^{-\Omega((n-\theta)/(n-k))}$ for all $k > \theta$. This shows that the threshold behavior of monotone properties in $\mathcal{L}_n$ scales at least like monotone properties in $\mathcal{P}_n$. The linear version of Question 2 asks whether the threshold behavior of monotone properties in $\mathcal{L}_n$ in fact scales like monotone properties in $\mathcal{P}_{2^n}$.

If the answer to Question 2 is "yes", then we get $m_{\mathrm{affine}}(\mathsf{Monotone}) = \Omega(n^2)$ by using the same information-theoretic argument as in our proof of Theorem 2 in §2. While we were unable to answer Question 2 for general monotone queries, the next theorem gives a positive answer in the special case where Q is an intersection query.

Theorem 6. *Let S be any subset of $\{0,1\}^n$. For all $0 \leq k \leq n$, let $p_k := \mathbb{P}[\mathbf{A}_k \cap S \neq \emptyset]$. Let $\tau := n - \log |S|$. Then $\min\{p_k, 1 - p_k\} \leq 2^{-|\tau-k|+O(1)}$ for all k.*

(Note that $|\theta - \tau| = O(1)$ for θ as in Question 2.)

Proof. The case where $k \leq \tau$ follows from a simple union bound. Let $\mathbf{a}_1, \dots, \mathbf{a}_{2^k}$ enumerate the elements of $\mathbf{A}_k$ in any order. Then

$$p_k = \mathbb{P}[\mathbf{A}_k \cap S \neq \emptyset] \leq \sum_{i=1}^{2^k} \mathbb{P}[\mathbf{a}_i \in S] = \sum_{i=1}^{2^k} \frac{|S|}{2^n} = 2^{-(\tau-k)}.$$

The case $k > \tau$ requires a more careful argument. Let $\mathbf{H}$ be a uniform random affine hyperplane (i.e. $(n-1)$-dimensional subspace) in $\{0,1\}^n$. (That is, $\mathbf{H} = \mathbf{A}_{n-1}$.)

Claim 1. *For all $\lambda > 0$, $\mathbb{P}\left[|S \cap \mathbf{H}| \leq (\frac{1}{2} - \lambda)|S|\right] \leq \dfrac{1}{4\lambda^2 |S|}$.*

Proof (Proof of Claim 1). Let $\mathbf{Z} := |S \cap \mathbf{H}|$. We have $\mathbb{E}[\mathbf{Z}] = |S|/2$ and

$$\mathbb{E}[\mathbf{Z}^2] = \sum_{x \in S} \mathbb{P}[x \in \mathbf{H}] + \sum_{x,y \in S \,:\, x \neq y} \mathbb{P}[x, y \in \mathbf{H}]$$

$$= \frac{|S|}{2} + |S|(|S|-1)\frac{2^{n-1}-1}{2(2^n-1)} \leq \frac{1}{4}(|S| + |S|^2).$$

By Chebyshev's inequality,

$$\mathbb{P}\left[\mathbf{Z} \leq (\tfrac{1}{2} - \lambda)|S|\right] \leq \mathbb{P}\left[|\mathbf{Z} - \mathbb{E}[\mathbf{Z}]| \leq \lambda|S|\right] \leq \frac{\mathrm{Var}(\mathbf{Z})}{\lambda^2 |S|^2} = \frac{\mathbb{E}[\mathbf{Z}^2] - \mathbb{E}[\mathbf{Z}]^2}{\lambda^2 |S|^2} \leq \frac{1}{4\lambda^2 |S|}.$$

$\square$Claim

Claim 2. *Let $S \subseteq \{0,1\}^n$, let $\mathbf{B} = \mathbf{A}_{n-j}$ be a uniform random affine subspace of $\{0,1\}^n$ of co-dimension j, and let $b = 2^{-1/4}$. Then*

$$\mathbb{P}[\mathbf{B} \cap S = \emptyset] \leq \frac{2^{j+4(1+b+b^2+\cdots+b^j)}}{|S|}.$$

Proof. We argue by induction on j. In base case $j = 0$ (where $\mathbf{B} = \{0,1\}^n$), the lemma holds since $\mathbb{P}[\mathbf{B} \cap S = \emptyset] = 0$.

For induction step, let $j \geq 1$ and assume the lemma holds for $j - 1$. By the induction hypothesis, for every affine hyperplane H,

$$\mathbb{P}[\mathbf{B} \cap S = \emptyset \mid \mathbf{B} \subseteq H] \leq \frac{2^{j-1+4(1+b+b^2+\cdots+b^{j-1})}}{|S \cap H|}.$$

Let $\mathbf{H}$ be a uniform random affine hyperplane. Note that $\mathbf{H}$ is independent of the event that $\mathbf{B} \subseteq \mathbf{H}$.

Let $\lambda := b^j/4$. We have

$$\mathbb{P}[\mathbf{B} \cap S = \emptyset] = \mathbb{P}[\mathbf{B} \cap S = \emptyset \mid \mathbf{B} \subseteq \mathbf{H}]$$

$$\leq \mathbb{P}\left[\mathbf{B} \cap S = \emptyset \text{ or } |S \cap \mathbf{H}| < (\tfrac{1}{2} - \lambda)|S| \;\Big|\; \mathbf{B} \subseteq \mathbf{H}\right]$$

$$\leq \mathbb{P}\left[|S \cap \mathbf{H}| < (\tfrac{1}{2} - \lambda)|S|\right]$$

$$+ \mathbb{P}\left[\mathbf{B} \cap S = \emptyset \;\Big|\; \mathbf{B} \subseteq \mathbf{H} \text{ and } |S \cap \mathbf{H}| \geq (\tfrac{1}{2} - \lambda)|S|\right]$$

$$\leq \frac{1}{4\lambda^2|S|} + \frac{2^{j-1+4(1+b+b^2+\cdots+b^{j-1})}}{(\tfrac{1}{2} - \lambda)|S|} \quad \text{(Claim 1 and ind. hyp.)}$$

$$= \left(2^{(j+4)/2} + \frac{2^{j+4(1+b+b^2+\cdots+b^{j-1})}}{1 - (b^j/2)}\right)\frac{1}{|S|}.$$

Noting that $1 - (b^j/2) \geq 2^{-b^j}$, we have

$$2^{(j+4)/2} + \frac{2^{j+4(1+b+b^2+\cdots+b^{j-1})}}{1 - (b^j/2)} \leq 2^{(j+4)/2} + 2^{j+4(1+b+b^2+\cdots+b^{j-1})+b^j}$$

$$\leq 2^{j+4(1+b+b^2+\cdots+b^{j-1})+b^j}\left(1 + 2^{-(j+4)/2}\right)$$

$$\leq 2^{j+4(1+b+b^2+\cdots+b^{j-1})+b^j}\, e^{2^{-(j+4)/2}}$$

$$\leq 2^{j+4(1+b+b^2+\cdots+b^{j-1}+b^j)}.$$

The proof is completed by combining the above inequalities. $\qquad\qquad\square$Claim

Returning to the proof of Theorem 6, we now show the case $k > \tau$ using Claim 2 as follows:

$$1 - p_k = \mathbb{P}[\mathbf{A}_k \cap S = \emptyset] \leq \frac{2^{n-k+4(1+b+\cdots+b^{n-k})}}{|S|} \leq 2^{\tau-k+4\sum_{j=0}^{\infty} b^j} \leq 2^{-(k-\tau)+26}.$$

Therefore, $\max\{p_k, 1 - p_k\} \leq 2^{-|\tau-k|+O(1)}$, which completes the proof of the theorem.

As a corollary of Theorem 6, we get:

Theorem 5. (restated) *The intersection query complexity of affine witness finding,* $m_{\text{affine}}(\text{Intersection})$, *is* $\Omega(n^2)$.

Proof. We use the same information-theoretic argument as the proof of Theorem 2 in §2, except $\mathbf{A}$ plays the role of $\mathbf{W}$ and Theorem 6 plays the role of Lemma 2(3) (in particular, we require the bound $H(p_k) \leq (|\tau - k| + O(1))/2^{|\tau-k|-O(1)}$, which follows from Theorem 6).

5   Conclusion

We initiated the study of the information-theoretic witness finding problem. For three natural classes of queries (intersection queries, monotone queries, NP queries), we proved lower bounds of $\Omega(n^2)$ on the query complexity of witness finding over arbitrary subsets of $\{0,1\}^n$. These lower bounds match upper bounds coming from classic results of Valiant and Vazirani [8] and Ben-David et al. [3]. In addition, we considered the setting where witness sets are affine subspaces of $\{0,1\}^n$ and proved a tight lower bound of $\Omega(n^2)$ for intersection queries. (All of our lower bounds hold even under the strong interpretation of Ω, i.e., for all but finitely many n.) Our investigation of affine witness finding led to an interesting and apparently new question about the threshold behavior of monotone properties in the affine lattice (Question 2). Other questions left open by this work are to resolve the monotone and NP query complexity of affine witness finding (i.e. $m_{\text{affine}}(\text{Monotone})$ and $m_{\text{affine}}(\text{NP})$). Finally, we wonder whether the idea in §3 of approximating NP queries by monotone queries might have other applications in complexity theory.

Acknowledgements. We thank Oded Goldreich for feedback on an earlier manuscript. We are also grateful to the anonymous reviewers for their detailed and extremely helpful comments.

References

1. Alon, N., Spencer, J.: The Probablistic Method, 3rd edn. Wiley (2008)
2. Bellare, M., Goldwasser, S.: The complexity of decision versus search. SIAM Journal on Computing 23, 97–119 (1994)
3. Ben-David, S., Chor, B., Goldreich, O., Luby, M.: On the theory of average-case complexity. Journal of Computer and System Sciences 44(2), 193–219 (1992)
4. Bollobás, B., Thomason, A.G.: Threshold functions. Combinatorica 7(1), 35–38 (1987)
5. Chowdhury, A., Patkos, B.: Shadows and intersections in vector spaces. J. of Combinatorial Theory, Ser. A 117, 1095–1106 (2010)
6. Cover, T., Thomas, J.: Elements of Information Theory. Wiley Interscience, New York (1991)
7. Dell, H., Kabanets, V., van Melkebeek, D., Watanabe, O.: Is the Valiant-Vazirani isolation lemma improvable? In: Proc. 27th Conference on Computational Complexity, pp. 10–20 (2012)
8. Valiant, L., Vazirani, V.: NP is as easy as detecting unique solutions. Theoretical Computer Science 47, 85–93 (1986)
9. Yao, A.C.: Probabilistic computations: toward a unified measure of complexity. In: Proc. of the 18th IEEE Sympos. on Foundations of Comput. Sci., pp. 222–227. IEEE (1977)

A   Proof of Lemma 1

In order to apply Yao's minimax principle [9], we express $m(\mathcal{W}, \mathcal{Q})$ in terms of a particular matrix M. Let $\mathcal{F}$ be the set of functions $\{\top, \bot\}^m \to \{0,1\}^n$. Let $\mathcal{A} := \mathcal{Q}^m \times \mathcal{F}$ (representing the set of deterministic witness finding algorithms). Let $\mathcal{W}_0 := \mathcal{W} \setminus \{\emptyset\}$. Finally, let M be the $\mathcal{A} \times \mathcal{W}_0$-matrix defined by

$$M_{(Q_1,\ldots,Q_m;f),W} := \begin{cases} 1 & \text{if } f(Q_1(W),\ldots,Q_m(W)) \in W, \\ 0 & \text{otherwise.} \end{cases}$$

In this context, Yao's minimax principle states that for all random variables $\mathbf{W}$ on $\mathcal{W}_0$ and $(\mathbf{Q}_1,\ldots,\mathbf{Q}_m;\mathbf{f})$ on $\mathcal{A}$,

$$\min_{(Q_1,\ldots,Q_m,f)\in\mathcal{A}} \mathbb{E}[M_{(Q_1,\ldots,Q_m;f),\mathbf{W}}] \leq \max_{W\in\mathcal{W}_0} \mathbb{E}[M_{(\mathbf{Q}_1,\ldots,\mathbf{Q}_m;\mathbf{f}),W}].$$

It follows that, if $\mathbb{P}[f(Q_1(\mathbf{W}),\ldots,Q_m(\mathbf{W})) \in \mathbf{W}] \leq 1/2$ for all $Q_1,\ldots,Q_m \in \mathcal{Q}$ and every function $f : \{\top, \bot\}^m \to \{0,1\}^n$, then for all $(\mathbf{Q}_1,\ldots,\mathbf{Q}_m;\mathbf{f}) \in \mathcal{A}$ (including the special case where $\mathbf{f}$ is deterministic, as in the definition of witness finding procedures), there exists $W \in \mathcal{W}_0$ such that $\mathbb{P}[\mathbf{f}(\mathbf{Q}_1(W),\ldots,\mathbf{Q}_m(W)) \in W] \leq 1/2$. Therefore, the $\mathcal{Q}$-query complexity of $\mathcal{W}$-witness finding is $> m$.

B   Proof of Lemma 2

For inequality (1), let $\mathbf{Y}_1,\ldots,\mathbf{Y}_{2^i}$ be independent copies of $\mathbf{W}_{\theta-i}$. Note that

$$\mathbb{P}[x \in (\mathbf{Y}_1 \cup \cdots \cup \mathbf{Y}_{2^i})] = 1 - (1 - 2^{\theta-i-n})^{2^i} < 2^{\theta-n} = \mathbb{P}[w \in \mathbf{W}_\theta]$$

independently for all $x \in \{0,1\}^n$. Therefore, by monotonicity,

$$\mathbb{P}[Q(\mathbf{Y}_1) \vee \cdots \vee Q(\mathbf{Y}_{2^i})] \leq \mathbb{P}[Q(\mathbf{Y}_1 \cup \cdots \cup \mathbf{Y}_{2^i})] \leq \mathbb{P}[Q(\mathbf{W}_\theta)].$$

Using independence of $\mathbf{Y}_1,\ldots,\mathbf{Y}_{2^i}$, we have

$$1/2 \geq \mathbb{P}[Q(\mathbf{W}_\theta)] \geq \mathbb{P}[\bigvee_{j=1}^{2^i} Q(\mathbf{Y}_j)] = 1 - \mathbb{P}[\neg Q(\mathbf{W}_{\theta-i})]^{2^i} = 1 - (1 - p_{\theta-i})^{2^i}.$$

Therefore, $p_{\theta-i} \leq 1 - (1/2)^{1/2^i} < (\ln 2)/2^i$.

For inequality (2), let $\mathbf{Z}_1,\ldots,\mathbf{Z}_{2^i}$ be independent copies of $\mathbf{W}_{\theta+1}$. By a similar argument, we have

$$p_{\theta+i+1} = \mathbb{P}[Q(\mathbf{W}_{\theta+i+1})] \geq \mathbb{P}[\bigvee_{j=1}^{2^i} Q(\mathbf{Z}_j)] = 1 - \mathbb{P}[\neg Q(\mathbf{W}_{\theta+1})]^{2^i} > 1 - \frac{1}{2^{2^i}}.$$

Finally, for inequality (3), note that for all $p, q \in [0,1]$,

$$0 \leq \min(p, 1-p) \leq q \leq 1/2 \implies H(p) \leq H(q) \leq 2q \log(1/q).$$

By this observation, together with (1) and (2), we have

$$H(p_{\theta-i-1}) \leq 2\frac{\ln 2}{2^{i+1}} \log(\frac{2^{i+1}}{\ln 2}) < \frac{i+2}{2^i}, \quad H(p_{\theta+i+1}) \leq 2\frac{1}{2^{2^i}} \log(2^{2^i}) = \frac{1}{2^{2^i-i-1}}.$$

From these two inequalities, it follows that $H(p_k) \leq (|\theta - k| + 1)/2^{|\theta-k|-1}$.

Primal Implication as Encryption

Vladimir N. Krupski

Faculty of Mechanics and Mathematics, Lomonosov Moscow State University,
Moscow 119992, Russia
krupski@lpcs.math.msu.su

Abstract. We propose a "cryptographic" interpretation for the propositional connectives of primal infon logic introduced by Y. Gurevich and I. Neeman and prove the corresponding soundness and completeness results. Primal implication $\varphi \to_p \psi$ corresponds to the encryption of ψ with a secret key φ, primal disjunction $\varphi \vee_p \psi$ is a group key and $\bot$ reflects some backdoor constructions such as full superuser permissions or a universal decryption key. For the logic of $\bot$ as a universal key (it was never considered before) we prove that the derivability problem has linear time complexity. We also show that the universal key can be emulated using primal disjunction.

1 Introduction

Primal Infon Logic ([1], [2], [3], [4], [5]) formalizes the concept of *infon*, i.e. a message as a piece of information. The corresponding derivability statement $\Gamma \vdash \varphi$ means that the principal can get (by herself, without any communication) the information φ provided she already has all infons $\psi \in \Gamma$.

Primal implication $(\to_p)$ that is used in Primal Infon Logic to represent the conditional information is a restricted form of intuitionistic implication defined by the following inference rules:

$$\frac{\Gamma \vdash \psi}{\Gamma \vdash \varphi \to_p \psi} \, (\to_p I) \quad , \quad \frac{\Gamma \vdash \varphi \quad \Gamma \vdash \varphi \to_p \psi}{\Gamma \vdash \psi} \, (\to_p E) \, .$$

These rules admit cryptographic interpretation of primal implication $\varphi \to_p \psi$ as some kind of digital envelop: it is an infon, containing the information ψ encrypted by a symmetric key (generated from) φ. Indeed, the introduction rule $(\to_p I)$ allows to encrypt any available message by any key. Similarly, the elimination rule $(\to_p E)$ allows to extract the information from the ciphertext provided the key is also available. So the infon logic incorporated into communication protocols ([1], [2]) is a natural tool for manipulating with commitment schemes (see [7]) without detailed analysis of the scheme itself.

Example. (cf. [8]). Alice and Bob live in different places and communicate via a telephone line or by e-mail. They wish to play the following game distantly. Each of them picks a bit, randomly or somehow else. If the bits coincide then Alice wins; otherwise Bob wins. Both of them decide to play fair but don't believe

E.A. Hirsch et al. (Eds.): CSR 2014, LNCS 8476, pp. 232–244, 2014.

in the fairness of the opponent. To play fair means that they honestly declare their choice of a bit, independently of what the other player said. So they use cryptography.

We discuss the symmetric version of the coin flipping protocol from [8] in order to make the policies of both players the same. Consider the policy of one player, say Alice. Her initial state can be represented by the context

$$\Gamma = \{A \ said \ m_a, \ A \ said \ k_a, \ A \ IsTrustedOn \ m_a, \ A \ IsTrustedOn \ k_a\},$$

where infons m_a and k_a represent the chosen bit and the key Alice intends to use for encryption. Her choice is recorded by infons $A \ said \ m_a$ and $A \ said \ k_a$ where $A \ said$ is the quotation modality governed by the modal logic $\mathbf{K}$.[1] Alice simply says, to herself, the infons m_a and k_a.

The remaining two members of Γ reflect the decision to play fair. The infon $X \ IsTrustedOn \ y$ abbreviates $(X \ said \ y) \rightarrow_p y$. It provides the ability to obtain the actual value of y from the declaration $X \ said \ y$, so Alice can deduce the actual m_a and k_a she has spoken about.

The commit phase. Alice derives m_a and $k_a \rightarrow_p m_a$ from her context by rules $(\rightarrow_p E)$, $(\rightarrow_p I)$ and sends the infon $k_a \rightarrow_p m_a$ to Bob. Bob acts similarly, so Alice will receive a message from him and her context will be extended to

$$\Gamma' = \Gamma \cup \{B \ said \ (k_b \rightarrow_p m_b)\}.$$

The reveal phase. After updating the context Alice obtains k_a by rule $(\rightarrow_p E)$ and sends it to Bob. He does the same, so Alice's context will be

$$\Gamma'' = \Gamma' \cup \{B \ said \ k_b\}.$$

Now by reasoning in $\mathbf{K}$ Alice deduces $B \ said \ m_b$. She also has $A \ said \ m_a$, so it is clear to her who wins. Alice simply compares these infons with the patterns $B \ said \ 0$, $B \ said \ 1$ and $A \ said \ 0$, $A \ said \ 1$ respectively.

The standard analysis of the protocol shows that Bob will come to the same conclusion. Moreover, Alice can be sure that she is not cheated provided she successively follows her policy up to the end.[2] The same with Bob.

Note that infon logic is used here as a part of the protocol. It is one of the tools that provide the correctness. But it does not prove the correctness. In order to formalize and prove the correctness of protocols one should use much more powerful formal systems.   ∎

We make our observation precise by defining interpretations of purely propositional part of infon logic in "cryptographic" infon algebras and proving the corresponding soundness and completeness theorems.

---

[1] The only modal inference rule that is used in this paper is $X \ said \ \varphi$, $X \ said \ (\varphi \rightarrow_p \psi) \vdash X \ said \ \psi$. It is admissible in $\mathbf{K}$. For more details about modalities in the infon logic see [4],[5].

[2] Here we suppose that the encryption method is practically strong and unambiguous. It is impossible for a player who does not know the encryption key to restore the plaintext from a ciphertext. It is also impossible for him to generate two key-message pairs with different messages and the same ciphertext.

In Section 2 this is done for the system **P** which is the $\{\top, \wedge, \to_p\}$-fragment of infon logic. We also show that the quasi-boolean semantics for **P** (see [4]) is essentially a special case of our semantics.

In Section 3 we show that $\bot$ can be used to reflect some backdoor constructions. Two variants are considered: system **P**$[\bot]$ from [4] with the usual elimination rule for $\bot$ and a new system **P**$[\bot_w]$ with a weak form of elimination rule for $\bot$. The first one treats $\bot$ as a root password, and the second one — as a universal key for decryption. For almost all propositional primal infon logics the derivability problem has linear time complexity. We prove the same complexity bound for **P**$[\bot_w]$ in Section 4.

Finally we consider a system **P**$[\vee_p]$ which is the modal-free fragment of Basic Propositional Primal Infon Logic **PPIL** from [5]. The primal disjunction $\vee_p$ in **P**$[\vee_p]$ has usual introduction rules and no elimination rules. We treat it as a group-key constructor and provide a linear time reduction of **P**$[\bot_w]$ to **P**$[\vee_p]$. It thus gives another proof of linear time complexity bound for **P**$[\bot_w]$.

2   Semantics for $\{\top, \wedge, \to_p\}$-fragment

Let Σ be a finite alphabet, say $\Sigma = \{0,1\}$. Let us fix a total pairing function $\pi : (\Sigma^*)^2 \to \Sigma^*$ with projections $l, r : \Sigma^* \to \Sigma^*$, where Σ^* is the set of all binary strings,

$$l(\pi(x,y)) = x, \quad r(\pi(x,y)) = y, \tag{1}$$

and two functions $enc, dec : (\Sigma^*)^2 \to \Sigma^*$ such that enc is total and

$$dec(x, enc(x,y)) = y. \tag{2}$$

String $enc(x,y)$ will be treated as a ciphertext containing string y encrypted with key x. Function dec is the decryption method that exploits the same key. In this text we do not restrict ourselves to encryptions that are strong in some sense. For example, $enc(x,y)$ may be the concatenation of strings x and y. Then dec on arguments x, y simply removes the prefix x from y. The totality of functions l, r, dec is not supposed, but the left-hand parts of (1) and (2) must be defined for all $x, y \in \Sigma^*$.

We also fix some set $E \subset \Sigma^*$, $E \neq \emptyset$. It will represent the information known by everyone, for example, facts like $0 < 1$ and $2 \cdot 2 = 4$. The structure $\mathcal{A} = \langle \Sigma^*, \pi, l, r, enc, dec, E \rangle$ will be referred as *an infon algebra*.[3]

Definition 1. A set $M \subseteq \Sigma^*$ will be called *closed* if $E \subseteq M$ and M satisfies the following closure conditions:

1. $a, b \in M \Leftrightarrow \pi(a,b) \in M$,
2. $a, enc(a,b) \in M \Rightarrow b \in M$,
3. $a \in \Sigma^*, b \in M \Rightarrow enc(a,b) \in M$.

---

[3] We use this term differently from [2] where infon algebras are semi-lattices with information order "x is at least as informative as y".

A closed set M represents the information that is potentially available to an agent in a local state, i.e. between two consecutive communication steps of a protocol. The information is represented by texts. M contains all public and some private texts. The agent can combine several texts in a single multi-part document using π function as well as to extract its parts by means of l and r. She has access to the encryption tool enc, so she can convert a plaintext into a ciphertext. The backward conversion (by dec) is also available provided she has the encryption key.

Note that in the closure condition 3 we do not require that $a \in M$. The agent will never need to decrypt the ciphertext $enc(a, b)$ encrypted by herself because she already has the plaintext b. The key a can be generated by some trusted third party and sent to those who really need it. This is the case when the encryption is used to provide secure communications between agents when only the connections to the third party are secure (and the authentication is reliable). On the other hand, some protocols may require the agent to distribute keys by herself. Then she can use a key that is known to her or get it from the third party. In the latter case a will be available in her new local state that will be updated by the communication with the third party.

The natural deduction calculus for primal infon logic $\mathbf{P}$ is considered in [4]. The corresponding *derivability relation* $\Gamma \vdash \varphi$ is defined by the following rules:

$$\frac{}{\vdash \top} \qquad \frac{}{\varphi \vdash \varphi} \qquad \frac{\Gamma \vdash \varphi}{\Gamma, \Delta \vdash \psi} \, (\textit{Weakening}) \qquad \frac{\Gamma \vdash \varphi_1 \quad \Gamma, \varphi_1 \vdash \varphi_2}{\Gamma \vdash \varphi_2} \, (\textit{Cut})$$

$$\frac{\Gamma \vdash \varphi_1 \quad \Gamma \vdash \varphi_2}{\Gamma \vdash \varphi_1 \wedge \varphi_2} \, (\wedge I) \qquad \frac{\Gamma \vdash \varphi_1 \wedge \varphi_2}{\Gamma \vdash \varphi_i} \, (\wedge E_i) \quad (i = 1, 2)$$

$$\frac{\Gamma \vdash \varphi_2}{\Gamma \vdash \varphi_1 \rightarrow_p \varphi_2} \, (\rightarrow_p I) \qquad \frac{\Gamma \vdash \varphi_1 \quad \Gamma \vdash \varphi_1 \rightarrow_p \varphi_2}{\Gamma \vdash \varphi_2} \, (\rightarrow_p E) \, .$$

Here $\varphi, \varphi_1, \varphi_2$ are infons, i.e. the expressions constructed from the set At of atomic infons by the grammar

$$\varphi ::= \top \mid At \mid (\varphi \wedge \varphi) \mid (\varphi \rightarrow_p \varphi),$$

and Γ, Δ are sets of infons.

As usual, *a derivation of* φ *from a set of assumptions* Γ is a sequence of infons $\varphi_1, \ldots, \varphi_n$ where $\varphi_n = \varphi$ and each φ_k is either a member of $\Gamma \cup \{\top\}$ or is obtained from some members of $\{\varphi_j \mid j < k\}$ by one of the rules

$$\frac{\varphi_1 \quad \varphi_2}{\varphi_1 \wedge \varphi_2} \qquad \frac{\varphi_1 \wedge \varphi_2}{\varphi_i} \qquad \frac{\varphi_2}{\varphi_1 \rightarrow_p \varphi_2} \qquad \frac{\varphi_1 \quad \varphi_1 \rightarrow_p \varphi_2}{\varphi_2} \, .$$

It is easy to see that $\Gamma \vdash \varphi$ iff there exists a derivation of φ from Γ. So rules like (*Weakening*) or (*Cut*) from the definition of derivability relation are never used in a derivation itself.

Definition 2. *An interpretation* (of the infon language) is a pair $I = \langle \mathcal{A}, v \rangle$ where $\mathcal{A} = \langle \Sigma^*, \pi, l, r, enc, dec, E \rangle$ is an infon algebra and $v \colon At \cup \{\top\} \to \Sigma^*$ is a total evaluation that assigns binary strings to atomic infons and to constant $\top$, $v(\top) \in E$. We assume that v is extended as follows:

$$v(\varphi_1 \wedge \varphi_2) = \pi(v(\varphi_1), v(\varphi_2)), \quad v(\varphi_1 \to_p \varphi_2) = enc(v(\varphi_1), v(\varphi_2)),$$

$$v(\Gamma) = \{v(\varphi) \mid \varphi \in \Gamma\}.$$

A *model* is a pair $\langle I, M \rangle$ where I is an interpretation and $M \subseteq \Sigma^*$ is a closed set.

In the paper [4] it is established that **P** is sound and complete with respect to quasi-boolean semantics. *A quasi-boolean model* is a validity relation $\models$ that enjoys the following properties:

$$- \models \top,$$
$$- \models \varphi_1 \wedge \varphi_2 \iff \models \varphi_1 \text{ and } \models \varphi_2,$$
$$- \models \varphi_2 \implies \models \varphi_1 \to_p \varphi_2,$$
$$- \models \varphi_1 \to_p \varphi_2 \implies \not\models \varphi_1 \text{ or } \models \varphi_2.$$

An infon φ is derivable in the infon logic **P** from the context Γ iff $\models \Gamma$ implies $\models \varphi$ for all quasi-boolean models $\models$.

It can be seen that the definition of a quasi-boolean model is essentially a special case of Definition 2. Indeed, suppose that atomic infons are distinct words in the unary alphabet $\{|\}$. Then all infons turn out to be words in some finite alphabet Σ_0. Consider a translation $\ulcorner \cdot \urcorner \colon \Sigma_0^* \to \{0,1\}^*$ that maps all elements of Σ_0 into distinct binary strings of the same length, $\ulcorner \Lambda \urcorner = \Lambda$ for the empty word Λ and $\ulcorner a_1 \ldots a_n \urcorner = \ulcorner a_1 \urcorner \ldots \ulcorner a_n \urcorner$ for $a_1 \ldots, a_n \in \Sigma_0$.

The corresponding infon algebra $\mathcal{A}$ and the evaluation v can be defined as follows: $v(a) = \ulcorner a \urcorner$ for $a \in At \cup \{\top\}$,

$$\pi(x, y) = \ulcorner (\urcorner x \ulcorner \wedge \urcorner y \ulcorner) \urcorner, \quad enc(x, y) = \ulcorner (\urcorner x \ulcorner \to_p \urcorner y \ulcorner) \urcorner, \quad E = \{\ulcorner \top \urcorner\}. \quad (3)$$

Projections and the decryption function can be found from (1) and (2). Note that for this interpretation the equality $v(\varphi) = \ulcorner \varphi \urcorner$ holds for every infon φ.

Consider a quasi-boolean model $\models$. Let M be the closure of the set $M_0 = \{\ulcorner \varphi \urcorner \mid\mid \models \varphi\}$, i.e. the least closed extension of M_0.

Lemma 3. $\models \varphi$ *iff* $v(\varphi) \in M$.

Proof. It is sufficient to prove that the set $M \setminus M_0$ does not contain words of the form $v(\varphi)$. Any element $b \in M \setminus M_0$ can be obtained from some elements of M_0 by a finite sequence of steps 1,2,3 that correspond to closure conditions:

1. $x, y \mapsto \ulcorner (\urcorner x \ulcorner \wedge \urcorner y \ulcorner) \urcorner; \quad \ulcorner (\urcorner x \ulcorner \wedge \urcorner y \ulcorner) \urcorner \mapsto x; \quad \ulcorner (\urcorner x \ulcorner \wedge \urcorner y \ulcorner) \urcorner \mapsto y;$
2. $x, \ulcorner (\urcorner x \ulcorner \to_p \urcorner y \ulcorner) \urcorner \mapsto y;$
3. $y \mapsto \ulcorner (\urcorner x \ulcorner \to_p \urcorner y \ulcorner) \urcorner.$

The history of this process is a derivation of b from M_0 with 1,2,3 treated as inference rules. Let $b = v(\varphi)$ and $b_1, \ldots, b_n = b$ be the derivation. Consider the (partial) top-down syntactic analysis of strings $b_1, \ldots, b_n$ using patterns

$$\ulcorner(\urcorner \cdot \ulcorner\wedge\urcorner \cdot \ulcorner)\urcorner, \qquad \ulcorner(\urcorner \cdot \ulcorner\rightarrow_p\urcorner \cdot \ulcorner)\urcorner, \qquad \ulcorner || \ldots |\urcorner.$$

We replace all substrings that remain unparsed by $v(a)$ where $a = || \ldots |$ is some fresh atomic infon. The resulting sequence $c_1, \ldots, c_n$ is also a derivation of b from M_0 because any string of the from $v(\psi)$ has no unparsed substrings. All its members have the form $c_i = v(\varphi_i)$ for some infons φ_i. Moreover, $\varphi_1, \ldots, \varphi_n$ is a derivation of $\varphi = \varphi_n$ in $\mathbf{P}$ from the set of hypotheses $\Gamma = \{\varphi_j \mid c_j \in M_0\}$. But $\models \Gamma$ and $\mathbf{P}$ is sound with respect to quasi-boolean models, so $\models \varphi$ and $b = v(\varphi) \in M_0$. Contradiction. ∎

Theorem 4. $\Gamma \vdash \varphi$ in $\mathbf{P}$ *iff* $v(\varphi) \in M$ *for every model* $\langle I, M \rangle$ *with* $v(\Gamma) \subseteq M$.

Proof. The theorem states that the infon logic $\mathbf{P}$ is sound and complete with respect to the class of models introduced by Definition 2. The soundness can be proven by straightforward induction on the derivation of φ from Γ. The completeness follows from Lemma 3 and the completeness result for quasi-boolean models (see [4]). ∎

A set $\{v(\psi) \mid \psi \in T\} \subseteq \Sigma^*$ will be called *deductively closed* if $T \vdash \psi$ implies $\psi \in T$ for all infons ψ, i.e. T is deductively closed in $\mathbf{P}$. In the proof of Lemma 3 we actually establish that the particular interpretation $\langle \mathcal{A}, v \rangle$ is *conservative* in the following sense: the closure M of any deductively closed set $M_0 \subseteq \Sigma^*$ does not contain "new" strings of the form $v(\psi) \notin M_0$. It is also *injective*: $v(\varphi_1) = v(\varphi_2)$ implies $\varphi_1 = \varphi_2$. An interpretation that enjoys these two properties will be called *plain*.

Lemma 5. *There exists a plain interpretation.*

The completeness part of Theorem 4 can be strengthened.

Theorem 6. *Let the interpretation* $I = \langle \mathcal{A}, v \rangle$ *be plain. For any context* Γ *there exists a model* $\langle I, M \rangle$ *with* $v(\Gamma) \subseteq M$ *such that* $\Gamma \not\vdash \varphi$ *implies* $v(\varphi) \notin M$ *for all infons* φ.

Proof. Let M be the closure of the set $M_0 = \{v(\psi) \mid \Gamma \vdash \psi\}$. Then $v(\Gamma) \subseteq M$. The set M_0 is deductively closed, so $M \setminus M_0$ does not contain strings of the form $v(\psi)$. Suppose $\Gamma \not\vdash \varphi$. Then $v(\varphi) \notin M_0$ because the interpretation is injective. Thus $v(\varphi) \notin M$. ∎

3   Constant $\perp$ and Backdoors

$\perp$ as Superuser Permissions

Infon logic $\mathbf{P}[\perp]$ is the extension of $\mathbf{P}$ by additional constant $\perp$ that satisfies the elimination rule

$$\frac{\Gamma \vdash \perp}{\Gamma \vdash \varphi} \, (\perp E) \, .$$

The corresponding changes in Definition 2 are as follows. We add to the alphabet a new letter $\mathbf{f} \notin \Sigma$ and set $\Sigma_\perp = \Sigma \cup \{\mathbf{f}\}$, $v(\perp) = \mathbf{f}$. Functions π, l, r, enc, dec act on words from $\Sigma_\perp^*$ but still satisfy the conditions (1), (2). We suppose them to preserve Σ^*: the value should be a binary string provided all arguments are. We also suppose that $v(\top) \in E \subseteq \Sigma^*$ and $v(\varphi) \in \Sigma^*$ for $\varphi \in At$ and add new closure condition to Definition 1:

4. $\mathbf{f} \in M, a \in \Sigma_\perp^* \Rightarrow a \in M.$

Models for $\mathbf{P}[\perp]$ are all pairs $\langle I, M \rangle$ where I is an interpretation and M is a closed set, both in the updated sense. The definition of plain interpretation is just the same.

Constant $\perp$ is some kind of root password that grants the superuser permissions to its owner. The owner has the direct access to all the information available in the system without any communication or decryption. At the same time $\perp$ can be incorporated into some messages that will be used in communication.

$\perp$ as Universal Key

The root password provides the direct access to all the information in the system including private information of any agent that was never sent to anybody else. It is also natural to consider a restricted form of superuser permissions that protect the privacy of agents but provide the ability to decrypt any available ciphertext. It can be simulated by infon logic $\mathbf{P}[\perp_w]$ with constant $\perp$ treated as *a universal key*. The corresponding inference rule is a weak form of $(\perp E)$ rule,

$$\frac{\Gamma \vdash \perp \quad \Gamma \vdash \varphi \rightarrow_p \psi}{\Gamma \vdash \psi} \, (\perp E_w),$$

that has an additional premise $\Gamma \vdash \varphi \rightarrow_p \psi$. So the owner of $\perp$ can get an infon only if she already has the same information as a ciphertext. The rule $(\perp E_w)$ is really weaker than $(\perp E)$ because $\psi \rightarrow_p \psi$ is not derivable in $\mathbf{P}$.

All definitions concerning models for $\mathbf{P}[\perp_w]$ are similar to the case of $\mathbf{P}[\perp]$ with closure condition 4 replaced by

4'. $\mathbf{f}, enc(a, b) \in M \Rightarrow b \in M.$

Essentially we extend the signature of infon algebras by additional (partial) operation $crack(x, y)$ that satisfies the equality

$$crack(\mathbf{f}, enc(a, b)) = b \tag{4}$$

and allow any agent to use it, so her local state satisfies the closure condition 4'.

Lemma 7. *There exist plain interpretations for* $\mathbf{P}[\bot]$ *and for* $\mathbf{P}[\bot_w]$.

Proof. We extend the example of plain interpretation for $\{\top, \wedge, \rightarrow_p\}$-fragment from Section 2 (see (3)). Set $\ulcorner \bot \urcorner = \mathbf{f}$ and extend the interpretation in accordance with (3). The resulting interpretation is plain in the sense of $\mathbf{P}[\bot]$. Indeed, it is injective because $\mathbf{f} \notin \Sigma$. It is also conservative. In order to prove this we use the construction from Lemma 3.

Let the set $M_0 = \{v(\psi) \mid \psi \in T\} \subseteq (\Sigma \cup \{\mathbf{f}\})^*$ be deductively closed and M be its closure. Suppose $v(\varphi) \in M \setminus M_0$ for some infon φ. Then $b_n = v(\varphi)$ has a derivation $b_1, \ldots, b_n$ from M_0 in the calculus with closure conditions considered as inference rules:

1. $x, y \mapsto \ulcorner (\urcorner x \ulcorner \wedge \urcorner y \ulcorner) \urcorner$; $\ulcorner (\urcorner x \ulcorner \wedge \urcorner y \ulcorner) \urcorner \mapsto x$; $\ulcorner (\urcorner x \ulcorner \wedge \urcorner y \ulcorner) \urcorner \mapsto y$;
2. x, $\ulcorner (\urcorner x \ulcorner \rightarrow_p \urcorner y \ulcorner) \urcorner \mapsto y$;
3. $y \mapsto \ulcorner (\urcorner x \ulcorner \rightarrow_p \urcorner y \ulcorner) \urcorner$;
4. $\mathbf{f} \mapsto x$.

Consider the (partial) top-down syntactic analysis of strings $b_1, \ldots, b_n$ using patterns

$$\ulcorner (\urcorner \cdot \ulcorner \wedge \urcorner \cdot \ulcorner) \urcorner, \qquad \ulcorner (\urcorner \cdot \ulcorner \rightarrow_p \urcorner \cdot \ulcorner) \urcorner, \qquad \ulcorner || \ldots |\urcorner, \qquad \mathbf{f}.$$

Replace all substrings that remain unparsed by $v(a)$ where $a = || \ldots |$ is some fresh atomic infon. The resulting sequence $c_1, \ldots, c_n$ is also a derivation of $v(\varphi)$ from M_0 because any string of the from $v(\psi)$ has no unparsed substrings. All its members have the form $c_i = v(\varphi_i)$ for some infons φ_i and $\varphi_1, \ldots, \varphi_n$ is a derivation of $\varphi = \varphi_n$ in $\mathbf{P}[\bot]$ from the set of hypotheses T. But T is deductively closed, so $\varphi \in M_0$. Contradiction.

Now set

$$crack(x, y) := \begin{cases} b, & \text{if } x = \mathbf{f} \text{ and } y = \ulcorner (\urcorner a \ulcorner \rightarrow_p \urcorner b \ulcorner) \urcorner, \\ \text{undefined}, & \text{otherwise.} \end{cases}$$

It satisfies the condition (4), so the interpretation for $\mathbf{P}[\bot_w]$ is defined. One can prove in a similar way that the interpretation is plain (w.r.t. $\mathbf{P}[\bot_w]$).   ∎

The completeness results from Section 2 hold for logics $\mathbf{P}[\bot]$ and $\mathbf{P}[\bot_w]$ too. The proofs are essentially the same with one difference: the quasi-boolean semantics from [4] does not cover the case of $\mathbf{P}[\bot_w]$. Let $\mathbf{L}$ be one of the logics $\mathbf{P}[\bot]$ or $\mathbf{P}[\bot_w]$.

Theorem 8. $\Gamma \vdash \varphi$ *in* $\mathbf{L}$ *iff* $v(\varphi) \in M$ *for every model* $\langle I, M \rangle$ *of* $\mathbf{L}$ *with* $v(\Gamma) \subseteq M$.

Proof. The soundness part can be proven by straightforward induction on the derivation of φ from Γ. The completeness follows from Lemma 7 and Theorem 9.   ∎

Theorem 9. *Let I be a plain interpretation of $\mathbf{L}$. For any context Γ there exists a model $\langle I, M \rangle$ of $\mathbf{L}$ with $v(\Gamma) \subseteq M$ such that $\Gamma \not\vdash \varphi$ implies $v(\varphi) \notin M$ for all infons φ.*

Proof. Similar to Theorem 6.   ∎

4   Decision Algorithm for $\mathbf{P}[\perp_w]$

The derivability problems for infon logics $\mathbf{P}$ and $\mathbf{P}[\perp]$ are linear time decidable ([3], [4], [5]). We provide a decision algorithm for $\mathbf{P}[\perp_w]$ with the same complexity bound.

Definition 10. (Positive atoms.) In what follows we assume that the language of $\mathbf{P}$ also contains $\perp$, but it is an ordinary member of At without any specific inference rule for it. Let

$$At^+(\varphi) = \{\varphi\} \text{ for } \varphi \in At \cup \{\top, \perp\},$$
$$At^+(\varphi \wedge \psi) = At^+(\varphi) \cup At^+(\psi),$$
$$At^+(\varphi \rightarrow_p \psi) = At^+(\psi).$$

For a context Γ set $At^+(\Gamma) = \bigcup_{\varphi \in \Gamma} At^+(\varphi)$.

Lemma 11. *Let $\Gamma \vdash \perp$ in $\mathbf{P}[\perp_w]$. Then $\Gamma \vdash \varphi$ in $\mathbf{P}[\perp_w]$ iff $At^+(\varphi) \subseteq At^+(\Gamma)$.*

Proof. Suppose $\Gamma \vdash \varphi$. The inclusion $At^+(\varphi) \subseteq At^+(\Gamma)$ can be proved by straightforward induction on the derivation of φ from Γ.

Now suppose that $\Gamma \vdash \perp$ and $At^+(\varphi) \subseteq At^+(\Gamma)$. By rules $(\wedge E_i)$ and $(\perp E_w)$ we prove that $\Gamma \vdash \psi$ for every infon $\psi \in At^+(\Gamma)$. Then we derive $\Gamma \vdash \varphi$ by rules $(\wedge I)$, $(\rightarrow_p I)$.   ∎

Lemma 12. *If $\Gamma \not\vdash \perp$ in $\mathbf{P}$ and $\Gamma \vdash \varphi$ in $\mathbf{P}[\perp_w]$ then $\Gamma \vdash \varphi$ in $\mathbf{P}$.*

Proof. $\Gamma \not\vdash \perp$ in $\mathbf{P}$ implies that $\Gamma \not\vdash \perp$ in $\mathbf{P}[\perp_w]$ because the shortest derivation of $\perp$ from Γ cannot use the $(\perp E_w)$ rule. So any derivation in $\mathbf{P}[\perp_w]$ from Γ cannot use this rule.   ∎

The decision algorithm for $\mathbf{P}[\perp_w]$ consists of the following three steps:

1. Test whether $\Gamma \vdash \varphi$ in $\mathbf{P}$. If yes, then $\Gamma \vdash \varphi$ in $\mathbf{P}[\perp_w]$ too. Else go to step 2.
2. Test whether $\Gamma \not\vdash \perp$ in $\mathbf{P}$. If yes, then $\Gamma \not\vdash \varphi$ in $\mathbf{P}[\perp_w]$ by Lemma 12. Else go to step 3.
3. We have $\Gamma \vdash \perp$ in $\mathbf{P}$, so it is also true in $\mathbf{P}[\perp_w]$. Test the condition $At^+(\varphi) \subseteq At^+(\Gamma)$. If it is fulfilled then $\Gamma \vdash \varphi$ in $\mathbf{P}[\perp_w]$; otherwise $\Gamma \not\vdash \varphi$ in $\mathbf{P}[\perp_w]$ (Lemma 11).

Linear time complexity bounds for steps 1,2 follow from the linear bound for **P**. In order to prove the same bound for step 3 we use the preprocessing stage of the linear time decision algorithm from [5]. It deals with sequents $\Gamma \vdash \varphi$ in a language that extends the language of $\mathbf{P}[\perp_w]$. The preprocessing stage is purely syntactic, so it does not depend on the logic involved and can be used for $\mathbf{P}[\perp_w]$ as well.

The algorithm constructs the parse tree for the sequent. Two nodes are called homonyms if they represent two occurrences of the same infon. For every homonymy class, the algorithm chooses a single element of it, the homonymy leader, and labels all nodes with pointers that provide a constant time access from a node to its homonymy leader. All this can be done in linear time (see [5]).

Now it takes a single walk through the parse tree to mark by a special flag all homonymy leaders that correspond to infons $\psi \in At^+(\Gamma)$. One more walk is required to test whether all homonymy leaders that correspond to $\psi \in At^+(\varphi)$ already have this flag. Thus we have a linear time test for the inclusion $At^+(\varphi) \subseteq At^+(\Gamma)$.

Theorem 13. *The derivability problem for infon logic $\mathbf{P}[\perp_w]$ is linear time decidable.*

5   Primal Disjunction and Backdoor Emulation

Primal infon logic with disjunction $\mathbf{P}[\vee]$ was studied in [4]. It is defined by all rules of **P** and usual introduction and elimination rules for disjunction. $\mathbf{P}[\vee]$ can emulate the classical propositional logic, so the derivability problem for it is co-NP-complete.

Here we consider the logic $\mathbf{P}[\vee_p]$, an efficient variant of $\mathbf{P}[\vee]$. It was mentioned in [4] and later was incorporated into Basic Propositional Primal Infon Logic **PPIL** [5] as its purely propositional fragment without modalities. In $\mathbf{P}[\vee_p]$ the standard disjunction is replaced by a "primal" disjunction $\vee_p$ with introduction rules

$$\frac{\Gamma \vdash \varphi_i}{\Gamma \vdash \varphi_1 \vee_p \varphi_2} \, (\vee_p I_i) \qquad (i = 1, 2)$$

and without elimination rules. It results in a linear-time complexity bound for $\mathbf{P}[\vee_p]$ (and for **PPIL** too, see [4],[5]).

When the primal implication is treated as encryption, the primal disjunction can be used as a method to construct group keys. An infon of the form

$$(\varphi_1 \vee_p \varphi_2) \rightarrow_p \psi \tag{5}$$

represents a ciphertext that can be decrypted by anyone who has at least one of the keys φ_1 or φ_2. In **P** the same effect can be produced by the infon

$$(\varphi_1 \rightarrow_p \psi) \wedge (\varphi_2 \rightarrow_p \psi), \tag{6}$$

but it requires two copies of ψ to be encrypted. Moreover, a principal A who does not know both keys φ_1 and φ_2 fails to distinguish between (6) and $(\varphi_1 \rightarrow_p \psi_1) \wedge (\varphi_2 \rightarrow_p \psi_2)$. If A receives (6) from some third party and forwards it to some principals B and C, she will never be sure that B and C will get the same plaintext after decryption. Group keys eliminate the length growth and ambiguity.

An infon algebra for $\mathbf{P}[\vee_p]$ has an additional total operation $gr \colon (\Sigma^*)^2 \rightarrow \Sigma^*$ for evaluation of primal disjunction: $v(\varphi \vee_p \psi) = gr(v(\varphi), v(\psi))$. The corresponding closure condition in Definition 1 will be

5. If $a \in M$, $b \in \Sigma^*$ or $b \in M$, $a \in \Sigma^*$ then $gr(a, b) \in M$.

All the results of Section 3 (Lemma 7, Theorems 8, 9) hold for $\mathbf{P}[\vee_p]$ too. The proofs are essentially the same.

$\mathbf{P}[\perp_w]$ *is linear-time reducible to* $\mathbf{P}[\vee_p]$, so $\mathbf{P}[\vee_p]$ and $\mathbf{PPIL}$ can emulate the backdoor based on a universal key. The reduction also gives another proof for Theorem 13.

Remember that in the language of $\mathbf{P}[\vee_p]$ symbol $\perp$ denotes some regular atomic infon. Consider the following translation:

$$q^* = q \ \text{ for } q \in At \cup \{\top, \perp\},$$
$$(\varphi \wedge \psi)^* = \varphi^* \wedge \psi^*,$$
$$(\varphi \rightarrow_p \psi)^* = (\perp \vee_p \varphi^*) \rightarrow_p \psi^*,$$
$$\Gamma^* = \{\varphi^* \mid \varphi \in \Gamma\}.$$

The transformation of Γ, φ into Γ^*, φ^* can be implemented in linear time.

Theorem 14. $\Gamma \vdash \varphi$ *in* $\mathbf{P}[\perp_w]$ *iff* $\Gamma^* \vdash \varphi^*$ *in* $\mathbf{P}[\vee_p]$.

Proof. Part "only if" can be proved by straightforward induction on the derivation of φ from assumptions Γ in $\mathbf{P}[\perp_w]$. For any inference rule of $\mathbf{P}[\perp_w]$, its translation is derivable in $\mathbf{P}[\vee_p]$. For example, consider the elimination rules for $\rightarrow_p$ and $\perp$:

$$\frac{\varphi^* \qquad \perp \vee_p \varphi^* \qquad \perp \vee_p \varphi^* \rightarrow_p \psi^*}{\psi^*}, \qquad \frac{\perp \qquad \perp \vee_p \varphi^* \qquad \perp \vee_p \varphi^* \rightarrow_p \psi^*}{\psi^*}.$$

Part "if". Let $\Gamma^* \vdash \varphi^*$ in $\mathbf{P}[\vee_p]$. Note that $\mathbf{P}[\vee_p]$ is the modal-free fragment of $\mathbf{PPIL}$ and the shortest derivation of φ^* from assumptions Γ^* in $\mathbf{PPIL}$ is also a derivation in $\mathbf{P}[\vee_p]$. Let D be this derivation.

It is proved in [5] that any shortest derivation is local. For the case of $\mathbf{P}[\vee_p]$ it means that all formulas from D are subformulas of Γ^*, φ^*. In particular, $\vee_p$ occurs in D only in subformulas of the form $\perp \vee_p \theta^*$.

Case 1. Suppose that the $(\vee_p I_1)$ rule is never used in D. Remove part "$\perp \vee_p$" from every subformula of the form $\perp \vee_p \psi$ that occurs in D. This transformation eliminates $\lor_p$ and makes all steps correspondent to $(\lor_p I_2)$ rule trivial. The result will be a derivation of φ from assumptions Γ in $\mathbf{P}$. So $\Gamma \vdash \varphi$ in $\mathbf{P}[\bot_w]$ too.

Case 2. Suppose that the $(\lor_p I_1)$ rule is used in D. It has the form

$$\frac{\bot}{\bot \lor_p \theta^*}, \tag{7}$$

so D also contains a derivation of $\bot$. The corresponding subderivation is the shortest one and does not use the $(\lor_p I_1)$ rule. By applying the transformation from Case 1 we prove that $\Gamma \vdash \bot$ in $\mathbf{P}$ and $\bot \in At^+(\Gamma)$.

We extend Definition 10 with new item

$$At^+(\psi_1 \lor_p \psi_2) = At^+(\psi_1) \cup At^+(\psi_2),$$

so $At^+(\psi)$ is defined for every ψ in the language of $\mathbf{P}[\lor_p]$. Moreover, $At^+(\varphi^*) = At^+(\varphi)$ and $At^+(\Gamma^*) = At^+(\Gamma)$. We claim that $At^+(\varphi^*) \subseteq At^+(\Gamma^*)$.

Indeed, consider D as a proof tree and its node ψ with $At^|(\psi) \not\subseteq At^|(\Gamma^*)$ whereas $At^+(\psi') \subseteq At^+(\Gamma^*)$ holds for all predecessors ψ'. The only rule that can produce this effect is (7), so $\psi = \bot \lor_p \theta^*$ for some θ where all occurrences of "new" atoms $q \in At^+(\psi) \setminus At^+(\Gamma^*)$ are inside θ^*.

Consider the path from the node ψ to the root node φ^* and the trace of ψ along it. There is no elimination rule for $\lor_p$, so ψ cannot be broken into pieces. All occurrences of positive atoms in θ^* will be positive in all formulas along the trace. But $\lor_p$ occurs in φ^* only in the premise of primal implication, so the trace does not reach the root node. Thus, at some step the formula containing ψ will be eliminated and "new" atoms from θ^* will never appear in $At^+(\varphi^*)$:

$$\frac{\bot}{\bot \lor_p \theta^*}$$

$$\frac{\eta_1[\bot \lor_p \theta^*] \qquad \eta_1[\bot \lor_p \theta^*] \to_p \eta_2}{\eta_2}$$

We have established that $At^+(\varphi) \subseteq At^+(\Gamma)$. But $\Gamma \vdash \bot$ in $\mathbf{P}$ and in $\mathbf{P}[\bot_w]$, so $\Gamma \vdash \varphi$ in $\mathbf{P}[\bot_w]$ by Lemma 11. ∎

Comment. It is also possible to reduce $\mathbf{P}[\bot_w]$ to $\mathbf{P}$. The corresponding reduction is two-step translation. One should convert φ into φ^* and then replace all subformulas of the form (5) in it with (6). Unfortunately, the second step results in the exponential growth of the length of a formula.

Acknowledgements. I would like to thank Yuri Gurevich, Andreas Blass and Lev Beklemishev for valuable discussion, comments and suggestions.

The research described in this paper was partially supported by Microsoft project DKAL and Russian Foundation for Basic Research (grant 14-01-00127).

References

1. Gurevich, Y., Neeman, I.: DKAL: Distributed-Knowledge Authorization Language. In: Proc. of CSF 2008, pp. 149–162. IEEE Computer Society (2008)
2. Gurevich, Y., Neeman, I.: DKAL 2 — A Simplified and Improved Authorization Language. Technical Report MSR-TR-2009-11, Microsoft Research (February 2009)
3. Gurevich, Y., Neeman, I.: Logic of infons: the propositional case. ACM Transactions on Computational Logic 12(2) (2011)
4. Beklemishev, L., Gurevich, Y.: Propositional primal logic with disjunction. J. of Logic and Computation 22, 26 pages (2012)
5. Cotrini, C., Gurevich, Y.: Basic primal infon logic. Microsoft Research Technical Report MSR-TR-2012-88, Microsoft Research (August 2012)
6. Troelstra, A., Schwichtenberg, H.: Basic proof theory. Cambridge Tracts in Theoretical Computer Science, vol. 43. Cambridge University Press, Cambridge (1996)
7. Goldreich, O.: Foundations of Cryptography: Volume 1, Basic Tools. Cambridge University Press, Cambridge (2001)
8. Blum, M.: Coin Flipping by Telephone. In: Proceedings of CRYPTO, pp. 11–15 (1981)

Processing Succinct Matrices and Vectors*

Markus Lohrey[1] and Manfred Schmidt-Schauß[2]

[1] Universität Siegen, Department für Elektrotechnik und Informatik, Germany
[2] Institut für Informatik, Goethe-Universität, D-60054 Frankfurt, Germany

Abstract. We study the complexity of algorithmic problems for matrices that
are represented by multi-terminal decision diagrams (MTDD). These are a variant of ordered decision diagrams, where the terminal nodes are labeled with arbitrary elements of a semiring (instead of 0 and 1). A simple example shows
that the product of two MTDD-represented matrices cannot be represented by an
MTDD of polynomial size. To overcome this deficiency, we extended MTDDs
to $\mathrm{MTDD_+}$ by allowing componentwise symbolic addition of variables (of the
same dimension) in rules. It is shown that accessing an entry, equality checking,
matrix multiplication, and other basic matrix operations can be solved in polynomial time for $\mathrm{MTDD_+}$-represented matrices. On the other hand, testing whether
the determinant of a MTDD-represented matrix vanishes is PSPACE-complete,
and the same problem is NP-complete for $\mathrm{MTDD_+}$-represented diagonal matrices. Computing a specific entry in a product of MTDD-represented matrices is
#P-complete. Complete proofs can be found in the full version [19] of this paper.

1 Introduction

Algorithms that work on a succinct representation of certain objects can nowadays be
found in many areas of computer science. A paradigmatic example is the use of OBDDs
(ordered binary decision diagrams) in hardware verification [5,21]. OBDDs are a succinct representation of Boolean functions. Consider a boolean function $f(x_1, \ldots, x_n)$
in n input variables. One can represent f by its decision tree, which is a full binary tree
of height n with $\{0, 1\}$-labelled leaves. The leaf that is reached from the root via the
path $(a_1, \ldots, a_n) \in \{0, 1\}^n$ (where $a_i = 0$ means that we descend to the left child
in the i-th step, and $a_i = 1$ means that we descend to the right child in the i-th step)
is labelled with the bit $f(a_1, \ldots, a_n)$. This decision tree can be folded into a directed
acyclic graph by eliminating repeated occurrences of isomorphic subtrees. The result is
the OBDD for f with respect to the variable ordering $x_1, \ldots, x_n$.[1] Bryant was the first
who realized that OBDDs are an adequate tool in order to handle the state explosion
problem in hardware verification [5].

OBDDs can be also used for storing large graphs. A graph G with 2^n nodes and adjacency matrix M_G can be represented by the boolean function $f_G(x_1, y_1, \ldots, x_n, y_n)$,
where $f_G(a_1, b_1, \ldots, a_n, b_n)$ is the entry of M_G at position (a, b); here $a_1 \cdots a_n$ (resp.,

---

* The first (second) author is supported by the DFG grant LO 748/8-2 (SCHM 986/9-2).

[1] Here, we are cheating a bit: In OBDDs a second elimination rule is applied that removes
nodes for which the left and right child are identical. On the other hand, it is known that
asymptotically the compression achieved by this elimination rule is negligible [31].

E.A. Hirsch et al. (Eds.): CSR 2014, LNCS 8476, pp. 245–258, 2014.
© Springer International Publishing Switzerland 2014

$b_1 \cdots b_n$) is the binary representation of the index a (resp. b). Note that we use the so called interleaved variable ordering here, where the bits of the two coordinates a and b are bitwise interleaved. This ordering turned out to be convenient in the context of OBDD-based graph representation, see e.g. [10].

Classical graph problems (like reachability, alternating reachability, existence of a Hamiltonian cycle) have been studied for OBDD-represented graphs in [9,30]. It turned out that these problems are exponentially harder for OBDD-represented graphs than for explicitly given graphs. In [30] an upgrading theorem for OBDD-represented graphs was shown. It roughly states that completeness of a problem A for a complexity class C under quantifier free reductions implies completeness of the OBDD-variant of A for the exponentially harder version of C under polynomial time reductions.

In the same way as OBDDs represent boolean mappings, functions from $\{0,1\}^n$ to any set S can be represented. One simply has to label the leaves of the decision tree with elements from S. This yields multi-terminal decision diagrams (MTDDs) [11]. Of particular interest is the case, where S is a semiring, e.g. $\mathbb{N}$ or $\mathbb{Z}$. In the same way as an adjacency matrix (i.e., a boolean matrix) of dimension 2^n can be represented by an OBDD, a matrix of dimension 2^n over any semiring can be represented by an MTDD. As for OBDDs, we assume that the bits of the two coordinates a and b are interleaved in the order $a_1, b_1, \ldots, a_n, b_n$. This implies that an MTDD can be viewed as a set of rules of the form

$$A \to \begin{pmatrix} A_{1,1} & A_{1,2} \\ A_{2,1} & A_{2,2} \end{pmatrix} \qquad \text{or} \qquad B \to a \text{ with } a \in S. \tag{1}$$

where A, $A_{1,1}$, $A_{1,2}$, $A_{2,1}$, and $A_{2,2}$ are variables that correspond to certain nodes of the MTDD (namely those nodes that have even distance from the root node). Every variable produces a matrix of dimension 2^h for some $h \geq 0$, which we call the height of the variable. The variables $A_{i,j}$ in (1) must have the same height h, and A has height $h+1$. The variable B has height 0. We assume that the additive monoid of the semiring S is finitely generated, hence every $a \in S$ has a finite representation.

MTDDs yield very compact representations of sparse matrices. It was shown that an $(n \times n)$-matrix with m nonzero entries can be represented by an MTDD of size $O(m \log n)$ [11, Thm. 3.2], which is better than standard succinct representations for sparse matrices. Moreover, MTDDs can also yield very compact representations of non-sparse matrices. For instance, the Walsh matrix of dimension 2^n can be represented by an MTDD of size $O(n)$, see [11]. In fact, the usual definition of the n-th Walsh matrix is exactly an MTDD. Matrix algorithms for MTDDs are studied in [11] as well, but no precise complexity analysis is carried out. In fact, the straightforward matrix multiplication algorithm for multi-terminal decision diagrams from [11] has an exponential worst case running time, and this is unavoidable: The smallest MTDD that produces the product of two MTDD-represented matrices may be of exponential size in the two MTDDs, see Thm. 2. The first main contribution of this paper is a generalization of MTDDs that overcomes this deficiency: An MTDD_+ consists of rules of the form (1) together with addition rules of the form $A \to B+C$, where "$+$" refers to matrix addition over the underlying semiring. Here, A, B, and C must have the same height, i.e., produce matrices of the same dimension. We show that an MTDD_+ for the product of two MTDD_+-represented matrices can be computed in polynomial time (Thm. 3). In Sec. 4.1 we also present efficient (polynomial time) algorithms for several other important matrix problems on $MTDD_+$-represented input matrices: computation of a specific matrix entry, computation of the trace, matrix transposition, tensor and Hadamard product. Sec. 5 deals with equality checking. It turns out that equality of $MTDD_+$-represented matrices can be checked in polynomial time, if the additive monoid is cancellative, in all other cases equality checking is coNP-complete.

To the knowledge of the authors, complexity results similar to those from [9,30] for OBDDs do not exist in the literature on MTDDs. Our second main contribution fills this gap. We prove that already for MTDDs over $\mathbb{Z}$ it is PSPACE-complete to check whether the determinant of the generated matrix is zero (Thm. 6). This result is shown by lifting a classical construction of Toda [27] (showing that computing the determinant of an explicitly given integer matrix is complete for the counting class GapL) to configuration graphs of polynomial space bounded Turing machines, which are of exponential size. It turns out that the adjacency matrix of the configuration graph of a polynomial space bounded Turing machine can be produced by a small MTDD. Thm. 6 sharpens a recent result from [14] stating that it is PSPACE-complete to check whether the determinant of a matrix that is represented by a boolean circuit (see Sec. 4.2) vanishes. We also prove several hardness results for counting classes. For instance, computing a specific entry of a matrix power A^n, where A is given by an MTDD over $\mathbb{N}$ is #P-complete (resp. #PSPACE-complete) if n is given unary (resp. binary). Here, #P (resp. #PSPACE) is the class of functions counting the number of accepting computations of a nondeterministic polynomial time Turing machine [29] (resp., a nondeterministic polynomial space Turing machine [15]). An example of a natural #PSPACE-complete counting problem is counting the number of strings not accepted by a given NFA [15].

2 Related Work

Sparse Matrices and Quad-Trees. To the knowledge of the authors, most of the literature on matrix compression deals with sparse matrices, where most of the matrix entries are zero. There are several succinct representations of sparse matrices. One of which are *quad-trees*, used in computer graphics for the representation of large constant areas in 2-dimensional pictures, see for example [24,8]. Actually, an MTDD can be seen as a quad-tree that is folded into a dag by merging identical subtrees.

Two-Dimensional Straight-Line Programs. MTDDs are also a special case of 2-dimensional straight-line programs (SLPs). A (1-dimensional) SLP is a context-free grammar in Chomsky normal form that generates exactly one OBDD. An SLP with n rules can generate a string of length 2^n; therefore an SLP can be seen as a succinct representation of the string it generates. Algorithmic problems that can be solved efficiently (in polynomial time) on SLP-represented strings are for instance equality checking (first shown by Plandowski [23]) and pattern matching, see [18] for a survey.

In [3] a 2-dimensional extension of SLPs (2SLPs in the following) was defined. Here, every variable of the grammar generates a (not necessarily square) matrix (or picture), where every position is labeled with an alphabet symbol. Moreover, there are two (partial) concatenation operations: horizontal composition (which is defined for two pictures if they have the same height) and vertical composition (which is defined for two pictures if they have the same width). This formalism does not share all the nice algorithmic properties of (1-dimensional) SLPs [3]: Testing whether two 2SLPs produce the same picture is only known to be in coRP (co-randomized polynomial time). Moreover, checking whether an explicitly given (resp., 2SLP-represented) picture appears within a 2SLP-represented picture is NP-complete (resp., Σ_2^P-complete). Related hardness results in this direction concern the convolution of two SLP-represented strings of the same length (which can be seen as a picture of height 2). The convolution of strings $u = a_1 \cdots a_n$ and $v = b_1 \cdots b_n$ is the string $(a_1, b_1) \cdots (a_n, b_n)$. By a result from [4] (which is stated in terms of the related operation of literal shuffle), the size of a shortest SLP for the convolution of two strings that are given by SLPs G and H may be exponential in the size of G and H. Moreover, it is PSPACE-complete to check for two SLP-represented strings u and v and an NFA T operating on strings of pairs of symbols, whether T accepts the convolution of u and v [17].

MTDDs restrict 2SLPs by forbidding unbalanced derivation trees. The derivation tree of an MTDD results from unfolding the rules in (1); it is a tree, where every non-leaf node has exactly four children and every root-leaf path has the same length.

Tensor Circuits. In [2,7], the authors investigated the problems of evaluating tensor formulas and tensor circuits. Let us restrict to the latter. A tensor circuit is a circuit where the gates evaluate to matrices over a semiring and the following operations are used: matrix addition, matrix multiplication, and tensor product. Recall that the tensor product of two matrices $A = (a_{i,j})_{1 \leq i \leq m, 1 \leq i \leq m}$ and B is the matrix

$$A \otimes B = \begin{pmatrix} a_{1,1}B & \cdots & a_{1,m}B \\ \vdots & & \vdots \\ a_{n,1}B & \cdots & a_{n,m}B \end{pmatrix}$$

It is a $(mk \times nl)$-matrix if B is a $(k \times l)$-matrix. In [2] it is shown among other results that computing the output value of a scalar tensor circuit (i.e., a tensor circuit that yields a (1×1)-matrix) over the natural numbers is complete for the counting class #EXP. An MTDD$_+$ over $\mathbb{Z}$ can be seen as a tensor circuit that (i) does not use matrix multiplication and (ii) where for every tensor product the left factor is a (2×2)-matrix. To see the correspondence, note that

$$\begin{pmatrix} A_{1,1} & A_{1,2} \\ A_{2,1} & A_{2,2} \end{pmatrix} = \begin{pmatrix} 1 & 0 \\ 0 & 0 \end{pmatrix} \otimes A_{1,1} + \begin{pmatrix} 0 & 1 \\ 0 & 0 \end{pmatrix} \otimes A_{1,2} + \begin{pmatrix} 0 & 0 \\ 1 & 0 \end{pmatrix} \otimes A_{2,1} + \begin{pmatrix} 0 & 0 \\ 0 & 1 \end{pmatrix} \otimes A_{2,2}$$

$$\begin{pmatrix} a_{1,1} & a_{1,2} \\ a_{2,1} & a_{2,2} \end{pmatrix} \otimes B = \begin{pmatrix} a_{1,1}B & a_{1,2}B \\ a_{2,1}B & a_{2,2}B \end{pmatrix}$$

Each of the matrices $a_{i,j}B$ can be generated from B and $-B$ using $\log|a_{i,j}|$ many additions (here we use the fact that the underlying semiring is $\mathbb{Z}$).

3   Preliminaries

We consider matrices over a semiring $(S, +, \cdot)$ with $(S, +)$ a finitely generated commutative monoid with unit 0. The unit of the monoid $(S, \cdot)$ is 1. We assume that $0 \cdot a =$

$a \cdot 0 = 0$ for all $a \in S$. Hence, if $|S| > 1$, then $1 \neq 0$ ($0 = 1$ implies $a = 1 \cdot a = 0 \cdot a = 0$ for all $a \in S$). With $S^{n \times n}$ we denote the set of all $(n \times n)$-matrices over S.

All time bounds in this paper implicitly refer to the RAM model of computation with a logarithmic cost measure for arithmetical operations on integers, where arithmetic operations on n-bit numbers need time $O(n)$. For a number $n \in \mathbb{Z}$ let us denote with $\mathrm{bin}(n)$ its binary encoding.

We assume that the reader has some basic background in complexity theory, in particular we assume that the reader is familiar with the classes NP, coNP, and PSPACE. A function $f : \{0,1\}^* \to \{0,1\}^*$ belongs to the class $\mathsf{FSPACE}(s(n))$ (resp. $\mathsf{FTIME}(s(n))$) if f can be computed on a deterministic Turing machine in space (resp., time) $s(n)$.[2] As usual, only the space on the working tapes is counted. Moreover, the output is written from left to right on the output tape, i.e., in each step the machine either outputs a new symbol on the output tape, in which case the output head moves one cell to the right, or the machine does not output a new symbol in which case the output head does not move. Let $\mathsf{FP} = \bigcup_{k \geq 1} \mathsf{FTIME}(n^k)$ and $\mathsf{FPSPACE} = \bigcup_{k \geq 1} \mathsf{FSPACE}(n^k)$. Note that for a function $f \in \mathsf{FPSPACE}$ we have $|f(w)| \leq 2^{|w|^{O(1)}}$ for every input.

The counting class $\#\mathsf{P}$ consists of all functions $f : \{0,1\}^* \to \mathbb{N}$ for which there exists a nondeterministic polynomial time Turing machine M with input alphabet Σ such that for all $x \in \Sigma^*$, $f(x)$ is the number of accepting computation paths of M for input x. If we replace nondeterministic polynomial time Turing machines by nondeterministic polynomial space Turing machines (resp. nondeterministic logspace Turing machines), we obtain the class $\#\mathsf{PSPACE}$ [15] (resp. $\#\mathsf{L}$ [1]). Note that for a mapping $f \in \#\mathsf{PSPACE}$, the number $f(x)$ may grow doubly exponential in $|x|$, whereas for $f \in \#\mathsf{P}$, the number $f(x)$ is bounded singly exponential in $|x|$. Ladner [15] has shown that a mapping $f : \Sigma^* \to \mathbb{N}$ belongs to $\#\mathsf{PSPACE}$ if and only if the mapping $x \mapsto \mathrm{bin}(f(x))$ belongs to $\mathsf{FPSPACE}$. One cannot expect a corresponding result for the class $\#\mathsf{P}$: If for every function $f \in \#\mathsf{P}$ the mapping $x \mapsto \mathrm{bin}(f(x))$ belongs to FP, then by Toda's theorem [28] the polynomial time hierarchy collapses down to P. For $f \in \#\mathsf{L}$, the mapping $x \mapsto \mathrm{bin}(f(x))$ belongs to NC^2 and hence to $\mathsf{FP} \cap \mathsf{FSPACE}(\log^2(n))$ [1, Thm. 4.1]. The class GapL (resp., GapP, GapPSPACE) consists of all differences of two functions in $\#\mathsf{L}$ (resp. $\#\mathsf{P}$, $\#\mathsf{PSPACE}$). From Ladner's result [15] it follows easily that a function $f : \{0,1\}^* \to \mathbb{Z}$ belongs to GapPSPACE if and only if the mapping $x \mapsto \mathrm{bin}(f(x))$ belongs to $\mathsf{FPSPACE}$, see also [12, Thm. 6].

Logspace reductions between functions can be defined analogously to the language case: If $f, g : \{0,1\}^* \to X$ with $X \in \{\mathbb{N}, \mathbb{Z}\}$, then f is logspace reducible to g if there exists a function $h \in \mathsf{FSPACE}(\log n)$ such that $f(x) = g(h(x))$ for all x. Toda [27] has shown that computing the determinant of a given integer matrix is GapL-complete.

4   Succinct Matrix Representations

In this section, we introduce several succinct matrix representations. We formally define multi-terminal decision diagrams and their extension by the addition operation. Moreover, we briefly discuss the representation of matrices by boolean circuits.

---

[2] The assumption that the input and output alphabet of f is binary is made here to make the definitions more readable; the extension to arbitrary finite alphabets is straightforward.

4.1 Multi-Terminal Decision Diagrams

Fix a semiring $(S, +, \cdot)$ with $(S, +)$ a finitely generated commutative monoid, and let $\Gamma \subseteq S$ be a finite generating set for $(S, +)$. Thus, every element of S can be written as a finite sum $\sum_{a \in \Gamma} n_a a$ with $n_a \in \mathbb{N}$. A *multi-terminal decision diagram G with addition (MTDD$_+$) of height h* is a triple (N, P, A_0), where N is a finite set of variables which is partitioned into non-empty sets N_i $(0 \leq i \leq h)$, $N_h = \{A_0\}$ (A_0 is called the *start variable*), and P is a set of rules of the following three forms:

- $A \rightarrow \begin{pmatrix} A_{1,1} & A_{1,2} \\ A_{2,1} & A_{2,2} \end{pmatrix}$ with $A \in N_i$ and $A_{1,1}, A_{1,2}, A_{2,1}, A_{2,2} \in N_{i-1}$ for some $1 \leq i \leq h$
- $A \rightarrow A_1 + A_2$ with $A, A_1, A_2 \in N_i$ for some $0 \leq i \leq h$
- $A \rightarrow a$ with $A \in N_0$ and $a \in \Gamma \cup \{0\}$

Moreover, for every variable $A \in N$ there is exactly one rule with left-hand side A, and the relation $\{(A, B) \in N \times N \mid B$ occurs in the right-hand side for $A\}$ is acyclic. If $A \in N_i$ then we say that A has height i. The MTDD$_+$ G is called an *MTDD* if for every addition rule $(A \rightarrow A_1 + A_2) \in P$ we have $A, A_1, A_2 \in N_0$. In other words, only scalars are allowed to be added. Since we assume that $(S, +)$ is generated by Γ, this allows to produce arbitrary elements of S as matrix entries. For every $A \in N_i$ we define a square matrix $\mathsf{val}(A)$ of dimension 2^i in the obvious way by unfolding the rules. Moreover, let $\mathsf{val}(G) = \mathsf{val}(A_0)$ for the start variable A_0 of G. This is a $(2^h \times 2^h)$-matrix. The size of a rule $A \rightarrow a$ with $a \in \Gamma \cup \{0\}$ is 1, all other rules have size $\log |N|$. The size $|G|$ of the MTDD$_+$ G is the sum of the sizes of its rules; this is up to constant factors the length of the binary coding of G. An MTDD$_+$ G of size $n \log n$ can represent a $(2^n \times 2^n)$-matrix. Note that only square matrices whose dimension is a power of 2 can be represented. Matrices not fitting this format can be filled up appropriately, depending on the purpose.

An MTDD, where all rules have the form $A \rightarrow a \in \Gamma \cup \{0\}$ or $A \rightarrow B + C$ generates an element of the semiring S. Such an MTDD is an arithmetic circuit in which only input gates and addition gates are used, and is called a $+$-*circuit* in the following. In case the underlying semiring is $\mathbb{Z}$, a $+$-circuit with n variables can produce a number of size 2^n, and the binary encoding of this number can be computed in time $\mathcal{O}(n^2)$ from the $+$-circuit (since, we need n additions of numbers with at most n bits). In general, for a $+$-circuit over the semiring S, we can compute in quadratic time numbers n_a $(a \in \Gamma)$ such that $\sum_{a \in \Gamma} n_a \cdot a$ is the semiring element to which the $+$-circuit evaluates to.

Note that the notion of an MTDD$_+$ makes sense for commutative monoids, since we only used the addition of the underlying semiring. But soon, we want to multiply matrices, for which we need a semiring. Moreover, the notion of an MTDD$_+$ makes sense in any dimension, here we only defined the 2-dimensional case.

Example 1. It is straightforward to produce the unit matrix I_{2^n} of dimension 2^n by an MTDD of size $O(n \log n)$:

$$A_0 \rightarrow 1, \quad 0_0 \rightarrow 0, \quad A_j \rightarrow \begin{pmatrix} A_{j-1} & 0_{j-1} \\ 0_{j-1} & A_{j-1} \end{pmatrix}, \quad 0_j \rightarrow \begin{pmatrix} 0_{j-1} & 0_{j-1} \\ 0_{j-1} & 0_{j-1} \end{pmatrix} \quad (1 \leq j \leq n).$$

(the start variable is A_n here). In a similar way, one can produce the lower triangular $(2^n \times 2^n)$-matrix, where entries on the diagonal and below are 1. To produce the $(2^n \times 2^n)$-matrix over $\mathbb{Z}$, where all entries in the k-th row are k, we need the following rules:

$$E_0 \to 1, \quad E_j \to \begin{pmatrix} E_{j-1} + E_{j-1} & E_{j-1} + E_{j-1} \\ E_{j-1} + E_{j-1} & E_{j-1} + E_{j-1} \end{pmatrix} \quad (1 \le j \le n)$$

$$C_0 \to 1, \quad C_j \to \begin{pmatrix} C_{j-1} & C_{j-1} \\ C_{j-1} + E_{j-1} & C_{j-1} + E_{j-1} \end{pmatrix} \quad (1 \le j \le n).$$

Here, we are bit more liberal with respect to the format of rules, but the above rules can be easily brought into the form from the general definition of an $MTDD_+$. Note that E_j generates the $(2^j \times 2^j)$-matrix with all entries equal to 2^j, and that C_n generates the desired matrix.

Note that the matrix from the last example cannot be produced by an MTDD of polynomial size, since it contains an exponential number of different matrix entries (for the same reason it cannot be produced by an 2SLP [3]). This holds for any non-trivial semiring.

Theorem 1. *For any semiring with at least two elements, $MTDD_+$ are exponentially more succinct than MTDDs.*

Proof. For simplicity we argue with MTDDs in dimension 1 (which generate vectors). We must have $1 \ne 0$ in S. Let $m, d > 0$ be such that $m = 2^d$. For $0 \le i \le m - 1$ let A_i such that $\mathrm{val}(A_i)$ has length m, the i-th entry is 1 (the first entry is the 0-th entry) and all other entries are 0. Moreover, let B_i such that $\mathrm{val}(B_i)$ is the concatenation of 2^i copies of $\mathrm{val}(A_i)$. Let C_0 produce the 0-vector of length $m = 2^d$, and for $0 \le i \le m - 1$ let $C_{i+1} \to (C_i, C_i + B_i)$. Then $\mathrm{val}(C_m)$ is of length 2^{d+m} and consists of the concatenation of all binary strings of length m. This MTDD$_+$ for this vector is of size $O(m^2 \log m)$, whereas an equivalent MTDD must have size at least 2^m, since for every binary string of length m there must exist a nonterminal. $\square$

The following result shows that the matrix product of two MTDD-represented matrices may be incompressible with MTDDs.

Theorem 2. *For any semiring with at least two elements there exist MTDDs G_n and H_n of the same height n and size $O(n^2 \log n)$ such that $\mathrm{val}(G_n) \cdot \mathrm{val}(H_n)$ can only be represented by an MTDD of size at least 2^n.*

On the other hand, the product of two $MTDD_+$-represented matrices can be represented by a polynomially sized $MTDD_+$:

Theorem 3. *For $MTDD_+$ G_1 and G_2 of the same height one can compute in time $O(|G_1| \cdot |G_2|)$ an $MTDD_+$ G of size $O(|G_1| \cdot |G_2|)$ with $\mathrm{val}(G) = \mathrm{val}(G_1) \cdot \mathrm{val}(G_2)$.*

For the proof, we compute from G_1 and G_2 a new $MTDD_+$ G that contains for all variables A of G_1 and B of G_2 of the same height a variable (A, B) such that $\mathrm{val}_G(A, B) = \mathrm{val}_{G_1}(A) \cdot \mathrm{val}_{G_2}(B)$.

The following proposition presents several further matrix operations that can be easily implemented in polynomial time for an $MTDD_+$-represented input matrix.

Proposition 1. *Let G, H be a* MTDD$_+$ *with* $|G| = n$, $|H| = m$, *and* $1 \le i, j \le 2^{\mathsf{height}(G)}$

(1) An MTDD$_+$ *for the transposition of* $\mathsf{val}(G)$ *can be computed in time* $O(n)$.

(2) $+$-circuits for the sum of all entries of $\mathsf{val}(G)$ *and the trace of* $\mathsf{val}(G)$ *can be computed in time* $O(n)$.

(3) A $+$-circuit for the matrix entry $\mathsf{val}(G)_{i,j}$ *can be computed in time* $O(n)$.

(4) MTDD$_+$ *of size* $O(n \cdot m)$ *for the tensor product* $\mathsf{val}(G) \otimes \mathsf{val}(H)$ *(which includes the scalar product) and the element-wise (Hadamard) product* $\mathsf{val}(G) \circ \mathsf{val}(H)$ *(assuming* $\mathsf{height}(G) = \mathsf{height}(H)$*) can be computed in time* $O(n \cdot m)$.

4.2 Boolean Circuits

Another well-studied succinct representation are boolean circuits [13]. A boolean circuit with n inputs represents a binary string of length 2^n, namely the string of output values for the 2^n many input assignments (concatenated in lexicographic order). In a similar way, we can use circuits to encode large matrices. We propose two alternatives:

A boolean circuit $C(\overline{x}, \overline{y}, \overline{z})$ with $|\overline{x}| = m$ and $|\overline{y}| = |\overline{z}| = n$ encodes a $(2^n \times 2^n)$-matrix $M_{C,2}$ with integer entries bounded by 2^{2^m} that is defined as follows: For all $\overline{a} \in \{0,1\}^m$ and $\overline{b}, \overline{c} \in \{0,1\}^n$, the $\overline{a}$-th bit (in lexicographic order) of the matrix entry at position $(\overline{b}, \overline{c})$ in M_C is 1 if and only if $C(\overline{a}, \overline{b}, \overline{c}) = 1$.

Note that in contrast to MTDD$_+$, the size of an entry in $M_{C,2}$ can be doubly exponential in the size of the representation C (this is the reason for the index 2 in $M_{C,2}$). The following alternative is closer to MTDD$_+$: A boolean circuit $C(\overline{x}, \overline{y})$ with $|\overline{x}| = |\overline{y}| = n$ and m output gates encodes a $(2^n \times 2^n)$-matrix $M_{C,1}$ with integer entries bounded by 2^m that is defined as follows: For all $\overline{a}, \overline{b} \in \{0,1\}^n$, $C(\overline{a}, \overline{b})$ is the binary encoding of the entry at position $(\overline{a}, \overline{b})$ in M_C.

Circuit representations for matrices are at least as succinct as MTDD$_+$. More precisely, from a given MTDD$_+$ G one can compute in logspace a Boolean circuit C such that $M_{C,1} = \mathsf{val}(G)$. This is a direct corollary of Proposition 1(3) (stating that a given entry of an MTDD$_+$-represented matrix can be computed in polynomial time) and the fact that polynomial time computations can be simulated by boolean circuits. Recently, it was shown that checking whether for a given circuit C the determinant of the matrix $M_{C,1}$ vanishes is PSPACE-complete [14]. An algebraic version of this result for the algebraic complexity class VPSPACE is shown in [20]. Thm. 6 from Sec. 6 will strengthen the result from [14] to MTDD-represented matrices.

5 Testing Equality

In this section, we consider the problem of testing equality of MTDD$_+$-represented matrices. For this, we do not need the full semiring structure, but we only need the finitely generated additive monoid $(S, +)$. We will show that equality can be checked in polynomial time if $(S, +)$ is cancellative and coNP-complete otherwise.

First we consider the case of a finitely generated abelian group. The proof of the following lemma involves only basic linear algebra.

Lemma 1. *Let $a_{i,1}x_1 + \cdots + a_{i,n}x_n = 0$ for $1 \leq i \leq m \leq n + 1$ be equations over a torsion-free abelian group A, where $a_{i,1}, \ldots, a_{i,n} \in \mathbb{Z}$, and the variables $x_1, \ldots, x_n$ range over A. One can determine in time polynomial in n and $\max\{\log |a_{i,j}| \mid 1 \leq i \leq m, 1 \leq j \leq n\}$ an equivalent set of at most n linear equations.*

Recall that the *exponent* of an abelian group A is the smallest integer k (if it exists) such that $kg = 0$ for all $g \in A$. The following result is shown in [25]:

Lemma 2. *Let $k \geq 2$ and let A be an abelian group of exponent k. Let $a_{i,1}x_1 + \cdots + a_{i,n}x_n = 0$ for $1 \leq i \leq m \leq n + 1$ be equations, where $a_{i,1}, \ldots, a_{i,n} \in \mathbb{Z}$, and the variables $x_1, \ldots, x_n$ range over A. Then one can determine in time polynomial in n, $\log(k)$, and $\max\{\log |a_{i,j}| \mid 1 \leq i \leq m, 1 \leq j \leq n\}$ an equivalent set of at most n linear equations.*

Proof. We can consider the coefficients $a_{i,j}$ as elements from $\mathbb{Z}_k$. By [25] we can compute the Howell normal form of the matrix $(a_{i,j})_{1 \leq i \leq n+1, 1 \leq j \leq n} \in \mathbb{Z}_k^{(n+1) \times n}$ in polynomial time. The Howell normal form is an $(n \times n)$-matrix with the same row span (a subset of the module $\mathbb{Z}_k^n$) as the original matrix, and hence defines an equivalent set of linear equations. □

Theorem 4. *Let G be an MTDD_+ over a finitely generated abelian group S. Given two different variables A_1, A_2 of the same height, it is possible to check $\mathsf{val}(A_1) = \mathsf{val}(A_2)$ in time polynomial in $|G|$.*

Proof. Since every finitely generated group is a finite direct product of copies of $\mathbb{Z}$ and $\mathbb{Z}_k$ ($k \geq 2$), it suffices to prove the theorem only for these groups.

Consider the case $S = \mathbb{Z}$. The algorithm stores a system of m equations (m will be bounded later) of the form $a_{i,1}B_1 + \cdots + a_{i,k}B_k = 0$, where all $B_1, \ldots, B_k$ are pairwise different variables of the same height h. We treat the variables $B_1, \ldots, B_k$ as variables that range over the torsion-free abelian group $\mathbb{Z}^{2^h \times 2^h}$. We start with the single equation $A_1 - A_2 = 0$. We use the rules of G to transform the system of equations into another system of equations whose variables have strictly smaller height. Assume the current height is $h > 1$. We iterate the following steps until only variables of height $h - 1$ occur in the equations:

Step 1. Standardize equations: Transform all equations into the form $a_1B_1 + \cdots + a_mB_m = 0$, where the B_i are different variables and the a_i are integers.

Step 2. Reduce the number of equations, using Lemma 1 applied to the torsion-free abelian group $\mathbb{Z}^{2^h \times 2^h}$.

Step 3. If a variable A of height h occurs in the equations, and the rule for A has the form $A \to A_1 + A_2$, then replace every occurrence of A in the equations by $A_1 + A_2$.

Step 4. If none of steps 1–3 applies to the equations, then only rules of the form

$$A \to \begin{pmatrix} A_{1,1} & A_{1,2} \\ A_{2,1} & A_{2,2} \end{pmatrix} \tag{2}$$

are applicable to a variable A (of height h) occurring in the equations. Applying all possible rules of this form for the current height results in a set of equations where all variables are (2×2)-matrices over variables of height $h - 1$ (like the right-hand side of (2)). Hence, every equation can be decomposed into 4 equations, where all variables are variables of height $h - 1$.

If the height of all variables is finally 0, then only rules of the form $A \to a$ are applicable. In this case, replace all variables by the corresponding integers, and check whether all resulting equations are valid or not. If all equations hold, then the input equation holds, i.e., $\mathsf{val}(A_1) = \mathsf{val}(A_2)$. Otherwise, if at least one equation is not valid, then $\mathsf{val}(A_1) \neq \mathsf{val}(A_2)$.

The number of variables in the equations is bounded by the number of variables of G. An upper bound on the absolute value of the coefficients in the equations is $2^{|G|}$, since only iterated addition can be performed to increase the coefficients. Lemma 1 shows that the number of equations after step 2 above is at most $|G|$, (the bound for the number of different variables).

For the case $S = \mathbb{Z}_k$ the same procedure works, we only have to use Lemma 2 instead of Lemma 1. $\qquad\qquad\square$

Corollary 1. *Let M be a finitely generated cancellative commutative monoid. Given an MTDD_+ G over M and two variables A_1 and A_2 of G, one can check $\mathsf{val}(A_1) = \mathsf{val}(A_2)$ in time polynomial in $|G|$.*

Proof. A cancellative commutative monoid M embeds into its Grothendieck group A, which is the quotient of $M \times M$ by the congruence defined by $(a, b) \equiv (c, d)$ if and only if $a + d = c + b$ in M. This is an abelian group, which is moreover finitely generated if M is finitely generated. Hence, the result follows from Thm. 1. $\qquad\square$

Let us now consider non-cancellative commutative monoids:

Theorem 5. *Let M be a non-cancellative finitely generated commutative monoid. It is coNP-complete to check $\mathsf{val}(A_1) = \mathsf{val}(A_2)$ for a given MTDD_+ G over M and two variables A_1 and A_2 of G.*

Proof. We start with the upper bound. Let $\{a_1, \ldots, a_k\}$ be a finite generating set of M. Let G be an MTDD_+ over M and let A_1 and A_2 two variables of G. Assume that A_1 and A_2 have the same height h. It suffices to check in polynomial time for two given indices $1 \leq i, j \leq 2^h$ whether $\mathsf{val}(A_1)_{i,j} \neq \mathsf{val}(A_2)_{i,j}$. From $1 \leq i, j \leq 2^h$ we can compute +-circuits for the matrix entries $\mathsf{val}(A_1)_{i,j}$ and $\mathsf{val}(A_2)_{i,j}$. From these circuits we can compute numbers $n_1, \ldots, n_k, m_1, \ldots, m_k \in \mathbb{N}$ in binary representation such that $\mathsf{val}(A_1)_{i,j} = n_1 a_1 + \cdots + n_k a_k$ and $\mathsf{val}(A_2)_{i,j} = m_1 a_1 + \cdots + m_k a_k$. Now we can use the following result from [26]: There is a semilinear subset $S \subseteq \mathbb{N}^{2k}$ (depending only on our fixed monoid M) such that for all $x_1, \ldots, x_k, y_1, \ldots, y_k \in \mathbb{N}$ we have: $x_1 a_1 + \cdots + x_k a_k = y_1 a_1 + \cdots + y_k a_k$ if and only if $(x_1, \ldots, x_k, y_1, \ldots, y_k) \in S$. Hence, we have to check, whether $v =: (n_1, \ldots, n_k, m_1, \ldots, m_k) \in S$. The semilinear set S is a finite union of linear sets. Hence, we can assume that S is linear itself. Let

$$S = \{v_0 + \lambda_1 v_1 + \cdots + \lambda_l v_l \mid \lambda_1, \ldots, \lambda_l \in \mathbb{N}\},$$

where $v_0, \ldots, v_l \in \mathbb{N}^{2k}$. Hence, we have to check, whether there exist $\lambda_1, \ldots, \lambda_l \in \mathbb{N}$ such that $v = v_0 + \lambda_1 v_1 + \cdots \lambda_l v_l$. This is an instance of integer programming in the fixed dimension $2k$, which can be solved in polynomial time [16].

For the lower bound we take elements $x, y, z \in M$ such that $x \neq y$ but $x+z = y+z$. These elements exist since M is not cancellative. We use an encoding of 3SAT from [3]. Take a 3CNF formula $C = \bigwedge_{i=1}^{m} C_i$ over n propositional variables $x_1, \ldots, x_n$, and let $C_i = (\alpha_{j_1} \vee \alpha_{j_2} \vee \alpha_{j_3})$, where $1 \leq j_1 < j_2 < j_3 \leq n$ and every α_{j_k} is either x_{j_k} or $\neg x_{j_k}$. For every $1 \leq i \leq m$ we define an MTDD G_i as follows: The variables are $A_0, \ldots, A_n$, and $B_0, \ldots, B_{n-1}$, where B_i produces the vector of length 2^i with all entries equal to 0 (which corresponds to the truth value true, whereas $z \in M$ corresponds to the truth value false). For the variables $A_0, \ldots, A_n$ we add the following rules: For every $1 \leq j \leq n$ with $j \notin \{j_1, j_2, j_3\}$ we take the rule $A_j \to (A_{j-1}, A_{j-1})$. For every $j \in \{j_1, j_2, j_3\}$ such that $\alpha_j = x_j$ (resp. $\alpha_j = \neg x_j$) we take the rule

$$A_j \to (A_{j-1}, B_{j-1}) \ (\text{resp. } A_j \to (B_{j-1}, A_{j-1})).$$

Finally add the rule $A_0 \to z$ and let A_n be the start variable of G_i. Moreover, let G (resp. H) be the 1-dimensional MTDD that produces the vector consisting of 2^n many x-entries (resp. y-entries). Then, $\text{val}(G) + \text{val}(G_1) + \cdots + \text{val}(G_m) = \text{val}(H) + \text{val}(G_1) + \cdots + \text{val}(G_m)$ if and only if C is unsatisfiable.   □

It is worth noting that in the above proof for coNP-hardness, we use addition only at the top level in a non-nested way.

6 Computing Determinants and Matrix Powers

In this section we present several completeness results for MTDDs over the rings $\mathbb{Z}$ and $\mathbb{Z}_n$ ($n \geq 2$). It turns out that over these rings, computing determinants, iterated matrix products, or matrix powers are infeasible for MTDD-represented input matrices, assuming standard assumptions from complexity theory. All completeness results in this section are formulated for MTDDs, but they remain valid if we add addition. In fact, all upper complexity bounds in this section even hold for matrices that are represented by circuits as explained in Sec. 4.2.

The value $\det(\text{val}(G))$ for an MTDD G may be of doubly exponential size (and hence needs exponentially many bits): The diagonal $(2^n \times 2^n)$-matrix with 2's on the diagonal has determinant 2^{2^n}. We first show that checking whether the determinant of an MTDD-represented matrix over any of the rings $\mathbb{Z}$ or $\mathbb{Z}_n$ ($n \geq 2$) vanishes is PSPACE-complete, and that computing the determinant over $\mathbb{Z}$ is GapPSPACE-complete:

Theorem 6. *The following holds for every ring $S \in \{\mathbb{Z}\} \cup \{\mathbb{Z}_n \mid n \geq 2\}$:*

(1) The set $\{G \mid G$ is an MTDD over $S, \det(\text{val}(G)) = 0\}$ is PSPACE-complete.
(2) The function $G \mapsto \det(\text{val}(G))$ with G an MTDD over $\mathbb{Z}$ is GapPSPACE-complete.

To prove this result we use a reduction of Toda showing that computing the determinant of an explicitly given integer matrix is GapL-complete [27]. We apply this reduction to configuration graphs of polynomial space bounded Turing machines, which are of exponential size. It turns out that the adjacency matrix of the configuration graph of a polynomial space bounded machine can be produced by a small MTDD (with terminal entries 0 and 1). This was also shown in [9, proof of Thm. 7] in the context of OBDDs.

Note that the determinant of a diagonal matrix is zero if and only if there is a zero-entry on the diagonal. This can be easily checked in polynomial time for a diagonal matrix produced by an MTDD. For MTDD$_+$ (actually, for a sum of several MTDD-represented matrices) we can show NP-completeness of this problem:

Theorem 7. *It is* NP-*complete to check* $\det(\text{val}(G_1) + \cdots + \text{val}(G_k)) = 0$ *for given MTDDs* $G_1, \ldots, G_k$ *that produce diagonal matrices of the same dimension.*

Our NP-hardness proof uses again the 3SAT encoding from [3] that we applied in the proof of Thm. 5.

Let us now discuss the complexity of iterated multiplication and powering. Computing a specific entry, say at position $(1,1)$, of the product of n explicitly given matrices over $\mathbb{Z}$ (resp., $\mathbb{N}$) is known to be complete for GapL (resp., #L) [27]. Corresponding results hold for the computation of the $(1,1)$-entry of a matrix power A^n, where n is given in unary notation. As usual, these problems become exponentially harder for matrices that are encoded by boolean circuits (see Sec. 4.2). Let us briefly discuss two scenarios (recall the matrices $M_{C,1}$ and $M_{C,2}$ defined from a circuit in Sec. 4.2).

Definition 1. *For a tuple* $\overline{C} = (C_1, \ldots, C_n)$ *of boolean circuits we can define the matrix product* $M_{\overline{C}} = \prod_{i=1}^{n} M_{C_i,1}$.

Lemma 3. *The function* $\overline{C} \mapsto (M_{\overline{C}})_{1,1}$, *where every matrix* $M_{C_i,1}$ *is over* $\mathbb{N}$ *(resp.,* $\mathbb{Z}$*), belongs to* #P *(resp.,* GapP*).*

Definition 2. *A boolean circuit* $C(\overline{w}, \overline{x}, \overline{y}, \overline{z})$ *with* $k = |\overline{w}|$, $m = |\overline{x}|$, *and* $n = |\overline{y}| = |\overline{z}|$ *encodes a sequence of* 2^k *many* $(2^n \times 2^n)$-*matrices: For every bit vector* $\overline{a} \in \{0,1\}^k$, *define the circuit* $C_{\overline{a}} = C(\overline{a}, \overline{x}, \overline{y}, \overline{z})$ *and the matrix* $M_{\overline{a}} = M_{C_{\overline{a}},2}$. *Finally, let* $M_C = \prod_{\overline{a} \in \{0,1\}^k} M_{\overline{a}}$ *be the product of all these matrices.*

Lemma 4. *The function* $C(\overline{w}, \overline{x}, \overline{y}, \overline{z}) \mapsto M_C$ *belongs to* FPSPACE.

Lemmas 3 and 4 yield the upper complexity bounds in the following theorem. For the lower bounds we use again succinct versions of Toda's techniques from [27], similar to the proof of Thm. 6.

Theorem 8. *The following holds:*

(1) The function $(G, n) \mapsto (\text{val}(G)^n)_{1,1}$ *with* G *an MTDD over* $\mathbb{N}$ *(resp.* $\mathbb{Z}$*) and* n *a unary encoded number is complete for* #P *(resp.,* GapP*).*

(2) The function $(G, n) \mapsto (\text{val}(G)^n)_{1,1}$ *with* G *an MTDD over* $\mathbb{N}$ *(resp.* $\mathbb{Z}$*) and* n *a binary encoded number is* #PSPACE-*complete (resp.,* GapPSPACE-*complete).*

By Thm. 8, there is no polynomial time algorithm that computes for a given MTDD G and a unary number n a boolean circuit (or even an MTDD$_+$) for the power $\text{val}(G)^n$, unless #P = FP.

By [27] and Thm. 8, the complexity of computing a specific entry of a matrix power A^n covers three different counting classes, depending on the representation of the matrix A and the exponent n (let us assume that A is a matrix over $\mathbb{N}$):

- #L-complete, if A is given explicitly and n is given unary.
- #P-complete, if A is given by an MTDD and n is given unary.
- #PSPACE-complete, if A is given by an MTDD and n is given binary.

Let us also mention that in [6,12,22] the complexity of evaluating iterated matrix products and matrix powers in a fixed dimension is studied. It turns out that multiplying a sequence of $(d \times d)$-matrices over $\mathbb{Z}$ in the fixed dimension $d \geq 3$ is complete for the class GapNC^1 (the counting version of the circuit complexity class NC^1) [6]. It is open whether the same problem for matrices over $\mathbb{N}$ is complete for $\#\mathsf{NC}^1$. Moreover, the case $d = 2$ is open too. Matrix powers for matrices in a fixed dimension can be computed in TC^0 (if the exponent is represented in unary notation) using the Cayley-Hamilton theorem [22]. Finally, multiplying a sequence of $(d \times d)$-matrices that is given succinctly by a boolean circuit captures the class $\mathsf{FPSPACE}$ for any $d \geq 3$ [12].

For the problem, whether a power of an MTDD-encoded matrix is zero (a variant of the classical mortality problem) we can finally show the following:

Theorem 9. *It is* coNP-*complete (resp.,*PSPACE-*complete) to check whether* $\mathsf{val}(G)^m$ *is the zero matrix for a given MTDD G and a unary (resp., binary) encoded number m.*

7 Conclusion and Future Work

We studied algorithmic problems on matrices that are given by multi-terminal decision diagrams enriched by the operation of matrix addition. Several important matrix problems can be solved in polynomial time for this representation, e.g., equality checking, computing matrix entries, matrix multiplication, computing the trace, etc. On the other hand, computing determinants, matrix powers, and iterated matrix products are computationally hard. For further research, it should be investigated whether the polynomial time problems, like equality test, belong to NC.

References

1. Àlvarez, C., Jenner, B.: A very hard log-space counting class. Theor. Comput. Sci. 107, 3–30 (1993)
2. Beaudry, M., Holzer, M.: The complexity of tensor circuit evaluation. Computational Complexity 16(1), 60–111 (2007)
3. Berman, P., Karpinski, M., Larmore, L.L., Plandowski, W., Rytter, W.: On the complexity of pattern matching for highly compressed two-dimensional texts. J. Comput. Syst. Sci. 65, 332–350 (2002)
4. Bertoni, A., Choffrut, C., Radicioni, R.: Literal shuffle of compressed words. In: Ausiello, G., Karhumäki, J., Mauri, G., Ong, L. (eds.) TCS 2008. IFIP, vol. 273, pp. 87–100. Springer, Heidelberg (2008)
5. Bryant, R.E.: Graph-based algorithms for boolean function manipulation. IEEE Trans. Computers 35(8), 677–691 (1986)
6. Caussinus, H., McKenzie, P., Thérien, D., Vollmer, H.: Nondeterministic NC^1 computation. Journal of Computer and System Sciences 57(2), 200–212 (1998)
7. Damm, C., Holzer, M., McKenzie, P.: The complexity of tensor calculus. Computational Complexity 11(1-2), 54–89 (2002)

8. Eppstein, D., Goodrich, M.T., Sun, J.Z.: Skip quadtrees: Dynamic data structures for multi-dimensional point sets. Int. J. Comput. Geometry Appl. 18, 131–160 (2008)

9. Feigenbaum, J., Kannan, S., Vardi, M.Y., Viswanathan, M.: The complexity of problems on graphs represented as obdds. Chicago J. Theor. Comput. Sci. (1999)

10. Fujii, H., Ootomo, G., Hori, C.: Interleaving based variable ordering methods for ordered binary decision diagrams. In: Proc. ICCAD 1993, pp. 38–41. IEEE Computer Society (1993)

11. Fujita, M., McGeer, P.C., Yang, J.C.-Y.: Multi-terminal binary decision diagrams: An efficient data structure for matrix representation. Formal Methods in System Design 10(2/3), 149–169 (1997)

12. Galota, M., Vollmer, H.: Functions computable in polynomial space. Inf. Comput. 198(1), 56–70 (2005)

13. Galperin, H., Wigderson, A.: Succinct representations of graphs. Inform. and Control 56, 183–198 (1983)

14. Grenet, B., Koiran, P., Portier, N.: On the complexity of the multivariate resultant. J. Complexity 29(2), 142–157 (2013)

15. Ladner, R.E.: Polynomial space counting problems. SIAM J. Comput. 18, 1087–1097 (1989)

16. Lenstra, H.: Integer programming with a fixed number of variables. Mathematics of Operations Research 8, 538–548 (1983)

17. Lohrey, M.: Leaf languages and string compression. Inf. Comput. 209, 951–965 (2011)

18. Lohrey, M.: Algorithmics on SLP-compressed strings: a survey. Groups Complex. Cryptol. 4, 241–299 (2012)

19. Lohrey, M., Schmidt-Schauß, M.: Processing Succinct Matrices and Vectors. arXiv (2014), http://arxiv.org/abs/1402.3452

20. Malod, G.: Succinct algebraic branching programs characterizing non-uniform complexity classes. In: Owe, O., Steffen, M., Telle, J.A. (eds.) FCT 2011. LNCS, vol. 6914, pp. 205–216. Springer, Heidelberg (2011)

21. Meinel, C., Theobald, T.: Algorithms and Data Structures in VLSI Design: OBDD - Foundations and Applications. Springer (1998)

22. Mereghetti, C., Palano, B.: Threshold circuits for iterated matrix product and powering. ITA 34(1), 39–46 (2000)

23. Plandowski, W.: Testing equivalence of morphisms in context-free languages. In: van Leeuwen, J. (ed.) ESA 1994. LNCS, vol. 855, pp. 460–470. Springer, Heidelberg (1994)

24. Samet, H.: The Design and Analysis of Spatial Data Structures. Addison-Wesley (1990)

25. Storjohann, A., Mulders, T.: Fast algorithms for linear algebra modulo N. In: Bilardi, G., Pietracaprina, A., Italiano, G.F., Pucci, G. (eds.) ESA 1998. LNCS, vol. 1461, pp. 139–150. Springer, Heidelberg (1998)

26. Taĭclin, M.A.: Algorithmic problems for commutative semigroups. Dokl. Akda. Nauk SSSR 9(1), 201–204 (1968)

27. Toda, S.: Counting problems computationally equivalent to computing the determinant. Technical Report CSIM 91-07, Tokyo University of Electro-Communications (1991)

28. Toda, S.: PP is as hard as the polynomial-time hierarchy. SIAM J. Comput. 20, 865–877 (1991)

29. Valiant, L.G.: Completeness classes in algebra. In: Proc. STOC 1979, pp. 249–261. ACM (1979)

30. Veith, H.: How to encode a logical structure by an OBDD. In: Proc. 13th Annual IEEE Conference on Computational Complexity, pp. 122–131. IEEE Computer Society (1998)

31. Wegener, I.: The size of reduced OBDD's and optimal read-once branching programs for almost all boolean functions. IEEE Trans. Computers 43(11), 1262–1269 (1994)

Constraint Satisfaction
with Counting Quantifiers 2

Barnaby Martin[1,*] and Juraj Stacho[2]

[1] School of Science and Technology, Middlesex University
The Burroughs, Hendon, London NW4 4BT, U.K.
[2] IEOR Department, Columbia University,
500 West 120th Street, New York, NY 10027, United States

Abstract. We study *constraint satisfaction problems* (CSPs) in the presence of counting quantifiers $\exists^{\geq j}$, asserting the existence of j distinct witnesses for the variable in question. As a continuation of our previous (CSR 2012) paper [11], we focus on the complexity of undirected graph templates. As our main contribution, we settle the two principal open questions proposed in [11]. Firstly, we complete the classification of clique templates by proving a full trichotomy for all possible combinations of counting quantifiers and clique sizes, placing each case either in **P**, **NP**-complete or **PSPACE**-complete. This involves resolution of the cases in which we have the single quantifier $\exists^{\geq j}$ on the clique $\mathbb{K}_{2j}$. Secondly, we confirm a conjecture from [11], which proposes a full dichotomy for $\exists$ and $\exists^{\geq 2}$ on all finite undirected graphs. The main thrust of this second result is the solution of the complexity for the infinite path which we prove is a polynomial-time solvable problem. By adapting the algorithm for the infinite path we are then able to solve the problem for finite paths, and then trees and forests. Thus as a corollary to this work, combining with the other cases from [11], we obtain a full dichotomy for $\exists$ and $\exists^{\geq 2}$ quantifiers on finite graphs, each such problem being either in **P** or **NP**-hard. Finally, we persevere with the work of [11] in exploring cases in which there is dichotomy between **P** and **PSPACE**-complete, and contrast this with situations in which the intermediate **NP**-completeness may appear.

1 Introduction

The *constraint satisfaction problem* CSP($\mathcal{B}$), much studied in artificial intelligence, is known to admit several equivalent formulations, two of the best known of which are the query evaluation of primitive positive (pp) sentences – those involving only existential quantification and conjunction – on $\mathcal{B}$, and the homomorphism problem to $\mathcal{B}$ (see, e.g., [9]). The problem CSP($\mathcal{B}$) is **NP**-complete in general, and a great deal of effort has been expended in classifying its complexity for certain restricted cases. Notably it is conjectured [7,4] that for all fixed $\mathcal{B}$, the problem CSP($\mathcal{B}$) is in **P** or **NP**-complete. While this has not been settled in general, a number of partial results are known – e.g. over structures of size

---

* The author was supported by EPSRC grant EP/L005654/1.

E.A. Hirsch et al. (Eds.): CSR 2014, LNCS 8476, pp. 259–272, 2014.

at most three [13,3] and over smooth digraphs [8,1]. A popular generalization of the CSP involves considering the query evaluation problem for *positive Horn logic* – involving only the two quantifiers, $\exists$ and $\forall$, together with conjunction. The resulting *quantified constraint satisfaction problems* $\mathrm{QCSP}(\mathcal{B})$ allow for a broader class, used in artificial intelligence to capture non-monotonic reasoning, whose complexities rise to **PSPACE**-completeness.

In this paper, we continue the project begun in [11] to study counting quantifiers of the form $\exists^{\geq j}$, which allow one to assert the existence of at least j elements such that the ensuing property holds. Thus on a structure $\mathcal{B}$ with domain of size n, the quantifiers $\exists^{\geq 1}$ and $\exists^{\geq n}$ are precisely $\exists$ and $\forall$, respectively.

We study variants of $\mathrm{CSP}(\mathcal{B})$ in which the input sentence to be evaluated on $\mathcal{B}$ (of size $|B|$) remains positive conjunctive in its quantifier-free part, but is quantified by various counting quantifiers.

For $X \subseteq \{1, \ldots, |B|\}$, $X \neq \emptyset$, the X-$\mathrm{CSP}(\mathcal{B})$ takes as input a sentence given by a conjunction of atoms quantified by quantifiers of the form $\exists^{\geq j}$ for $j \in X$. It then asks whether this sentence is true on $\mathcal{B}$.

In [11], it was shown that X-$\mathrm{CSP}(\mathcal{B})$ exhibits trichotomy as $\mathcal{B}$ ranges over undirected, irreflexive cycles, with each problem being in either **L**, **NP**-complete or **PSPACE**-complete. The following classification was given for cliques.

Theorem 1. [11] *For $n \in \mathbb{N}$ and $X \subseteq \{1, \ldots, n\}$:*

  (i) *X-$CSP(\mathbb{K}_n)$ is in **L** if $n \leq 2$ or $X \cap \{1, \ldots, \lfloor n/2 \rfloor\} = \emptyset$.*
 (ii) *X-$CSP(\mathbb{K}_n)$ is **NP**-complete if $n > 2$ and $X = \{1\}$.*
(iii) *X-$CSP(\mathbb{K}_n)$ is **PSPACE**-complete if $n > 2$ and either $j \in X$ for $1 < j < n/2$ or $\{1, j\} \subseteq X$ for $j \in \{\lceil n/2 \rceil, \ldots, n\}$.*

Precisely the cases $\{j\}$-$\mathrm{CSP}(\mathbb{K}_{2j})$ are left open here. Of course, $\{1\}$-$\mathrm{CSP}(\mathbb{K}_2)$ is *graph 2-colorability* and is in **L**, but for $j > 1$ the situation was very unclear, and the referees noted specifically this lacuna.

In this paper we **settle this question**, and find the surprising situation that $\{2\}$-$\mathrm{CSP}(\mathbb{K}_4)$ is in **P** while $\{j\}$-$\mathrm{CSP}(\mathbb{K}_{2j})$ is **PSPACE**-complete for $j \geq 3$. The algorithm for the case $\{2\}$-$\mathrm{CSP}(\mathbb{K}_4)$ is specialized and non-trivial, and consists in iteratively constructing a collection of forcing triples where we proceed to look for a contradiction.

As a **second focus** of the paper, we continue the study of $\{1, 2\}$-$\mathrm{CSP}(H)$. In particular, we focus on finite undirected graphs for which a dichotomy was proposed in [11]. As a fundamental step towards this, we first investigate the complexity of $\{1, 2\}$-$\mathrm{CSP}(\mathbb{P}_\infty)$, where $\mathbb{P}_\infty$ denotes the infinite undirected path. We find tractability here in describing a particular unique obstruction, which takes the form of a special walk, whose presence or absence yields the answer to the problem. Again the algorithm is specialized and non-trivial, and in carefully augmenting it, we construct another polynomial-time algorithm, this time for all finite paths. This then proves the following theorem.

Theorem 2. $\{1, 2\}$-$CSP(\mathbb{P}_n)$ *is in **P**, for each $n \in \mathbb{N}$.*

A corollary of this is the following **key result**.

Corollary 1. $\{1,2\}$-*CSP(H)* *is in* **P**, *for each forest H.*

Combined with the results from [8,11], this allows us to observe a dichotomy for $\{1,2\}$-CSP(H) as H ranges over undirected graphs, each problem being either in **P** or **NP**-hard, in turn **settling a conjecture** proposed in [11].

Corollary 2. *Let H be a graph.*
(i) $\{1,2\}$-*CSP(H) is in* **P** *if H is a forest or is bipartite with a 4-cycle,*
(ii) $\{1,2\}$-*CSP(H) is* **NP**-*hard in all other cases.*

In [11], the main preoccupation was in the distinction between **P** and **NP**-hard. Here we concentrate our observations to show situations in which we have sharp dichotomies between **P** and **PSPACE**-complete. In particular, for bipartite graphs, we are able to strengthen the above results in the following manner.

Theorem 3. *Let H be a bipartite graph.*
(i) $\{1,2\}$-*CSP(H) is in* **P** *if H is a forest or is bipartite with a 4-cycle,*
(ii) $\{1,2\}$-*CSP(H) is* **PSPACE**-*complete in all other cases.*

Note that this cannot be strengthened further for non-bipartite graphs, since there are **NP**-complete cases (for instance when H is the octahedron $\mathbb{K}_{2,2,2}$) and the situation regarding the **NP**-complete cases is less clear.

Taken together, our work seems to indicate a rich and largely uncharted complexity landscape that these types of problems constitute. The associated combinatorics to this landscape appears quite complex and the absence of a simple algebraic approach is telling. We will return to the question of algebra in the final remarks of the paper.

The paper is structured as follows. In §2, we describe a characterization and a polynomial time algorithm for $\{2\}$-CSP($\mathbb{K}_4$). In §3, we show **PSPACE**-hardness for $\{n\}$-CSP($\mathbb{K}_{2n}$) for $n \geq 3$. In §4, we characterize $\{1,2\}$-CSP for the infinite path $\mathbb{P}_\infty$ and describe the resulting polynomial algorithm. Then, in §5, we generalize this to finite paths and prove Theorem 2 and associated corollaries. Subsequently, in §6, we discuss the **P/PSPACE**-complete dichotomy of bipartite graphs, under $\{1,2\}$-CSP. Finally in §7, we illustrate some situations in which the intermediate **NP**-completeness arises by discussing cases with loops on vertices. We conclude the paper in §8 by giving some final thoughts.

1.1   Preliminaries

Our proofs use the game characterization and structural interpretation from [11]. For completeness, we summarize it here. This is as follows.

Given an input Ψ for X-CSP($\mathcal{B}$), we define the following game $\mathscr{G}(\Psi,\mathcal{B})$:

Definition 1. *Let* $\Psi := Q_1 x_1 Q_2 x_2 \ldots Q_m x_m \ \psi(x_1, x_2, \ldots, x_m)$. *Working from the outside in, coming to a quantified variable* $\exists^{\geq j} x$, *the* <u>Prover</u> *(female) picks a subset* B_x *of j elements of B as witnesses for x, and an* <u>Adversary</u> *(male) chooses one of these, say* b_x, *to be the value of x, denoted by* $f(x)$.

Prover wins if f is a homomorphism to $\mathcal{B}$, i.e., if $\mathcal{B} \models \psi(f(x_1), f(x_2), \ldots, f(x_m))$.

Lemma 1. *Prover has a winning strategy in the game $\mathscr{G}(\Psi, \mathcal{B})$ iff $\mathcal{B} \models \Psi$.*

Definition 2. *Let H be a graph. For an instance Ψ of X-CSP(H):*

- *define $\mathcal{D}_\psi$ to be the graph whose vertices are the variables of Ψ and edges are between variables v_i, v_j for which $E(v_i, v_j)$ appears in Ψ.*
- *denote $\prec$ the total order of variables of Ψ as they are quantified in the formula (from left to right).*

We follow the customary graph-theoretical notation with $V(G)$, $E(G)$ denoting the vertex set and edge set of a graph G, and $\mathbb{K}_n$, $\mathbb{C}_n$, and $\mathbb{P}_n$ denoting respectively the complete graph (clique), the cycle, and the path on n vertices.

2   Algorithm for $\{2\}$-CSP($\mathbb{K}_4$)

Theorem 4. *$\{2\}$-CSP($\mathbb{K}_4$) is decidable in polynomial time.*

The template $\mathbb{K}_4$ has vertices $\{1, 2, 3, 4\}$ and all possible edges between distinct vertices. Consider the instance Ψ of $\{2\}$-CSP($\mathbb{K}_4$) as a graph $G = \mathcal{D}_\psi$ together with a linear ordering $\prec$ on $V(G)$ (see Definition 2).

We iteratively construct the following three sets: R^+, R^-, and F. The set F will be a collection of unordered pairs of vertices of G, while R^+ and R^- will consist of unordered triples of vertices. (For simplicity we write $xy \in F$ in place of $\{x, y\} \in F$, and write $xyz \in R^+$ or R^- in place of $\{x, y, z\} \in R^+$ or R^-.)

The meaning of these sets is as follows. A pair $xy \in F$ where $x \prec y$ indicates that Prover in order to win must offer values so that the value $f(x)$ chosen by Adversary for x is different from the value $f(y)$ chosen for y. A triple $xyz \in R^+$ where $x \prec y \prec z$ indicates that if Adversary chose $f(x) \neq f(y)$, then Prover must offer one (or both) of $f(x), f(y)$ for z. A triple $xyz \in R^-$ where $x \prec y \prec z$ tells us that Prover must offer values different from $f(x), f(y)$ if $f(x) \neq f(y)$.

With this, we describe how to iteratively compute the three sets F, R^+, R^-. We start by initializing the sets as follows: $F = E(G)$ and $R^+ = R^- = \emptyset$. Then we perform the following rules as long as possible:

(X1) If there are $x, y, z \in V(G)$ such that $\{x, y\} \prec z$ where $xz, yz \in F$, then add xyz into R^-.

(X2) If there are vertices $x, y, w, z \in V(G)$ such that $\{x, y, w\} \prec z$ with $wz \in F$ and $xyz \in R^-$, then add xyw into R^+.

(X3) If there are $x, y, w, z \in V(G)$ such that $\{x, y, w\} \prec z$ with $wz \in F$ and $xyz \in R^+$, then if $\{x, y\} \prec w$, then add xyw into R^-
$\qquad\qquad\qquad\qquad\qquad\qquad\qquad\qquad$ else add xw and yw into F.

(X4) If there are vertices $x, y, w, z \in V(G)$ such that $\{x, w\} \prec y \prec z$ with $xyz \in R^+$ and $wyz \in R^-$, then add xw into F, and add xwy into R^+.

(X5) If there are vertices $x, y, w, z \in V(G)$ such that $\{x, y, w\} \prec z$ where either $xyz, wyz \in R^+$, or $xyz, wyz \in R^-$, then add xyw into R^+.

(X6) If there are vertices $x, y, q, w, z \in V(G)$ such that $\{x, y, w\} \prec q \prec z$ where either $xyz, wqz \in R^+$, or $xyz, wqz \in R^-$, then add xyw and xyq into R^+.

(X7) If there are vertices $x, y, q, w, z \in V(G)$ such that $\{x, y, w\} \prec q \prec z$ where either $xyz \in R^+$ and $wqz \in R^-$, or $xyz \in R^-$ and $wqz \in R^+$, then add xyq into R^-, and if $\{x, y\} \prec w$, also add xyw into R^-, else add xw and yw into F.

Theorem 5. *The following are equivalent:*

(i) $\mathbb{K}_4 \models \Psi$

(ii) Prover has a winning strategy in $\mathscr{G}(\Psi, \mathbb{K}_4)$.

(iii) Prover can play so that in every instance of the game, the resulting mapping $f : V(G) \rightarrow \{1, 2, 3, 4\}$ satisfies the following properties:

   (S1) For every $xy \in F$, we have: $f(x) \neq f(y)$.

   (S2) For every $xyz \in R^+$ such that $x \prec y \prec z$:
   $$\text{if } f(x) \neq f(y), \text{ then } f(z) \in \{f(x), f(y)\}.$$

   (S3) For every $xyz \in R^-$ such that $x \prec y \prec z$:
   $$\text{if } f(x) \neq f(y), \text{ then } f(z) \notin \{f(x), f(y)\}.$$

(iv) there is no triple xyz in R^+ such that $x \prec y \prec z$ and (see Fig. 1)

   – $xz \in F$ or $yz \in F$,

   – or $xwz \in R^-$ for some $w \prec z$ (possibly $w = y$),

   – or $ywz \in R^-$ for some $y \prec w \prec z$.

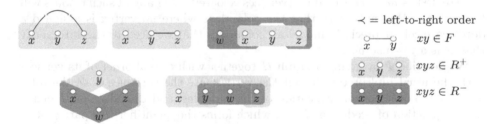

Fig. 1. Forbidden configurations from item (iv) of Theorem 5

Proof. (Sketch) (i) $\iff$ **(ii)** is by definition. **(iii)**$\Rightarrow$**(ii)** is implied by the fact that $F \supseteq E(G)$, and that by (iii) Prover can play to satisfy (S1). Thus in every instance of the game the mapping f is a homomorphism of G to $\mathbb{K}_4 \Rightarrow$ **(ii)**.

Then to complete the proof, we show the implications **(ii)**$\Rightarrow$**(iii)**, **(iii)**$\Rightarrow$**(iv)**, and **(iv)**$\Rightarrow$**(iii)**. This is done by analysis of possible cases.

For **(iii)**$\Rightarrow$**(iv)**, we show that in the presence of the obstruction from (iv), Adversary can play to violate (iii). For **(iv)**$\Rightarrow$**(iii)**, we let Prover make choices to satisfy (iii), first for triples in R^+, then triples in R^-, and finally edges in F. Assuming (iv), this will be a winning strategy. For **(ii)**$\Rightarrow$**(iii)**, we consider the vertex v where (iii) fails and choose v to be largest with respect to the order $\prec$. Assuming (ii) will imply an earlier such a vertex and lead to a contradiction. $\square$

With this characterization, we can now prove Theorem 4 as follows.

Proof. (Theorem 4) By Theorem 5, it suffices to construct the sets F, R^+, and R^-, and check the conditions of item (iv) of the said theorem. This can clearly be accomplished in polynomial time, since each of the three sets contains at most n^3 elements, where n is the number of variables in the input formula, and elements are only added (never removed) from the sets. Thus either a new pair (triple) needs to be added as follows from one of the rules (X1)-(X7), or we can stop and the output the resulting sets.   □

3   Hardness of $\{n\}$-CSP($\mathbb{K}_{2n}$) for $n \geq 3$

Theorem 6. $\{n\}$-*CSP($\mathbb{K}_{2n}$) is* **PSPACE**-*complete for all* $n \geq 3$.

The template $\mathbb{K}_{2n}$ consists of vertices $\{1, 2, \ldots, 2n\}$ and all possible edges between distinct vertices. We shall call these vertices *colours*. We describe a reduction from the **PSPACE**-complete [2] problem QCSP($\mathbb{K}_n$)=$\{1, n\}$-CSP($\mathbb{K}_n$) to $\{n\}$-CSP($\mathbb{K}_{2n}$). Consider an instance of QCSP($\mathbb{K}_n$), namely a formula Ψ where

$$\Psi = \exists^{\geq b_1} v_1 \, \exists^{\geq b_2} v_2 \, \ldots \, \exists^{\geq b_N} v_N \, \psi$$

where each $b_i \in \{1, n\}$. As usual (see Definition 2), let G denote the graph $\mathcal{D}_\psi$ with vertex set $\{v_1, \ldots, v_N\}$ and edge set $\{v_i v_j \mid E(v_i, v_j)$ appears in $\psi\}$.

We construct an instance Φ of $\{n\}$-CSP($\mathbb{K}_{2n}$) with the property that Ψ is a yes-instance of QCSP($\mathbb{K}_n$) if and only if Φ is a yes-instance of $\{n\}$-CSP($\mathbb{K}_{2n}$).

In short, we shall model the n-colouring using $2n - 1$ colours, $n - 1$ of which will treated as *don't care* colours (vertices coloured using any of such colours will be ignored). We make sure that the colourings where no vertex is assigned a don't-care colour precisely model all colourings that we need to check to verify that Ψ is a yes-instance.

We describe Φ by giving a graph H together with a total order of its vertices with the usual interpretation that the vertices are the variables of Φ, the total order is the order of quantification of the variables, and the edges of H define the conjunction of predicates $E(\cdot, \cdot)$ which forms the quantifier-free part ϕ of Φ.

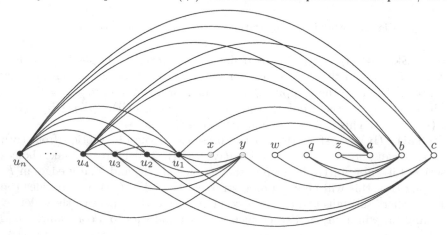

Fig. 2. The edge gadget (here, as an example, x is an $\exists$ vertex while y is a $\forall$ vertex)

We start constructing H by adding the vertices $v_1, v_2, \ldots, v_N$ and no edges. Then we add new vertices $u_1, u_2, \ldots, u_n$ and make them pairwise adjacent.

We make each v_i adjacent to u_1, and if $b_i = n$ (i.e. if v_i was quantified $\forall$), then we also make v_i adjacent to $u_2, u_3, \ldots, u_n$.

We complete H by introducing for each edge $xy \in E(G)$, a gadget consisting of new vertices w, q, z, a, b, c with edges wa, wb, qb, qc, za, zb, and we connect this gadget to the rest of the graph as follows: we make x adjacent to a, make y adjacent to b, make a adjacent to u_1, make c adjacent to u_1, u_2, u_3, and make each of a, b, c adjacent to $u_4, \ldots, u_n$. We refer to Figure 2 for an illustration.

The total order of $V(H)$ first lists $u_1, u_2, \ldots, u_n$, then $v_1, v_2, \ldots, v_N$ (exactly in the same order as quantified in Ψ), and then lists the remaining vertices of each gadget, in turn, as depicted in Figure 2 (listing w, q, z, a, b, c in this order).

We consider the game $\mathscr{G}(\Phi, \mathbb{K}_{2n})$ of Prover and Adversary played on Φ where Prover and Adversary take turns, for each variable in Φ in the order of quantification, respectively providing a set of n colours and choosing a colour from the set. Prover wins if this process leads to a proper $2n$-colouring of H (no adjacent vertices receive the same colour), otherwise Prover loses and Adversary wins. The formula Φ is a yes-instance if and only if Prover has a winning strategy.

Without loss of generality (up to renaming colours), we may assume that the vertices $u_1, u_2, \ldots, u_n$ get assigned colours $n+1, n+2, \ldots, 2n$, respectively, i.e. each u_i gets colour $n+i$. (The edges between these vertices make sure that Prover must offer distinct colours while Adversary has no way of forcing a conflict, since there are $2n$ colours available.)

The claim of Theorem 6 will then follow from the following two lemmas.

Lemma 2. *If Adversary is allowed to choose for the vertices x, y in the edge gadget (Figure 2) the same colour from $\{1, 2, \ldots, n\}$, then Adversary wins. If Adversary is allowed to choose $n + 1$ for x or y, then Adversary also wins.*

In all other cases, Prover wins.

Lemma 3. *Φ is a yes-instance of $\{n\}$-$CSP(\mathbb{K}_{2n})$ if and only if Ψ is a yes-instance of $QCSP(\mathbb{K}_n)$.*

We finish the proof by remarking that the construction of Φ is polynomial in the size of Ψ (in fact the reduction is in **L**). Thus, since $QCSP(\mathbb{K}_n)$ is **PSPACE**-hard, so is $\{n\}$-$CSP(\mathbb{K}_{2n})$. This completes the proof of Theorem 6.

4   Algorithm for $\{1, 2\}$-$CSP(\mathbb{P}_\infty)$

We consider the infinite path $\mathbb{P}_\infty$ to be the graph whose vertex set is $\mathbb{Z}$ and whose edges are $\{ij \ : \ |i - j| = 1\}$. An instance to $\{1, 2\}$-$CSP(\mathbb{P}_\infty)$ is a graph $G = \mathcal{D}_\psi$, a total order $\prec$ on $V(G)$, and a function $\beta : V(G) \to \{1, 2\}$ where

$$\Psi := \exists^{\geq \beta(v_1)} v_1 \ \exists^{\geq \beta(v_2)} v_2 \ \cdots \ \exists^{\geq \beta(v_n)} v_n \bigwedge_{v_i v_j \in E(G)} E(v_i, v_j)$$

We write $X \prec Y$ if $x \prec y$ for each $x \in X$ and each $y \in Y$. Also, we write $x \prec Y$ in place of $\{x\} \prec Y$. A *walk* of G is a sequence $x_1, x_2, \ldots, x_r$ of vertices of G

where $x_i x_{i+1} \in E(G)$ for all $i \in \{1, \ldots, r-1\}$. A walk $x_1, \ldots, x_r$ is a *closed walk* if $x_1 = x_r$. Write $|Q|$ to denote the *length* of the walk Q (number of edges on Q).

Definition 3. *If $Q = x_1, \ldots, x_r$ is a walk of G, we define $\lambda(Q)$ as follows:*

$$\lambda(Q) = |Q| - 2 \sum_{i=2}^{r-1} \left(\beta(x_i) - 1 \right)$$

Put differently, we assign weights to the vertices of G, with weight $+1$ assigned to each $\exists^{\geq 2}$ node, and weight -1 to each $\exists^{\geq 1}$ node; the value $\lambda(Q)$ is then simply the total weight of all inner nodes in the walk Q.

Definition 4. *A walk $x_1, \ldots, x_r$ of G is a* <u>looping walk</u> *if $x_1 \neq x_r$ and if $r \geq 3$*
(i) $\{x_1, x_r\} \prec \{x_2, \ldots, x_{r-1}\}$, and
(ii) there is $\ell \notin \{1, r\}$ such that both $x_1, \ldots, x_\ell$ and $x_\ell, \ldots, x_r$ are looping walks.

The above is a recursive definition. Note that endpoints of a looping walk are distinct and never appear in the interior of the walk. Other vertices, however, may appear on the walk multiple times as long as the walk obeys (ii). Notably, it is possible that the same vertex is one of $x_2, \ldots, x_{\ell-1}$ as well as one of $x_{\ell-1}, \ldots, x_{r-1}$ where ℓ is as defined in (ii). See Figure 3 for examples.

Using looping walks, we define a notion of "distance" in G that will guide Prover in the game.

Definition 5. *For vertices $u, v \in V(G)$, define $\delta(u, v)$ to be the following:*
$$\min \left\{ \lambda(Q) \ \middle| \ Q = x_1, \ldots, x_r \text{ is a looping walk of } G \text{ where } x_1 = u \text{ and } x_r = v \right\}.$$
If no looping walk between u and v exists, define $\delta(u, v) = \infty$.

In other words, $\delta(u, v)$ denotes the smallest λ-value of a looping walk between u and v. Note that $\delta(u, v) = \delta(v, u)$, since the definition of a looping walk does not prescribe the order of the endpoints of the walk.

The main structural obstruction in our characterization of is the following.

Definition 6. *A* <u>bad walk</u> *of G is a looping walk $Q = x_1, \ldots, x_r$ of G such that $x_1 \prec x_r$ and $\lambda(Q) \leq \beta(x_r) - 2$.*

4.1   Characterization

Theorem 7. *Suppose that G is a bipartite graph. Then the following statements are equivalent.*
(I) $\mathbb{P}_\infty \models \Psi$
(II) Prover has a winning strategy in $\mathscr{G}(\Psi, \mathbb{P}_\infty)$.
(III) Prover can play $\mathscr{G}(\Psi, \mathbb{P}_\infty)$ so that in every instance of the game, the resulting mapping f satisfies the following for all $u, v \in V(G)$ with $\delta(u, v) < \infty$:
$$|f(u) - f(v)| \leq \delta(u, v), \qquad\qquad (\star)$$
$$f(u) + f(v) + \delta(u, v) \text{ is an even number.} \qquad\qquad (\triangle)$$
(IV) There are no $u, v \in V(G)$ where $u \prec v$ such that $\delta(u, v) \leq \beta(v) - 2$.
(V) There is no bad walk in G.

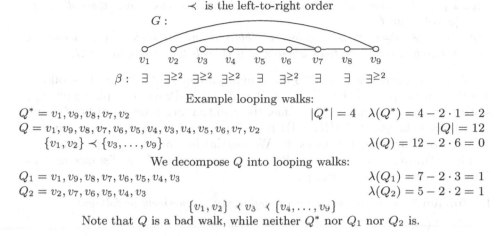

$\prec$ is the left-to-right order

Example looping walks:

$Q^* = v_1, v_9, v_8, v_7, v_2$ $\qquad |Q^*| = 4 \quad \lambda(Q^*) = 4 - 2 \cdot 1 = 2$

$Q = v_1, v_9, v_8, v_7, v_6, v_5, v_4, v_3, v_4, v_5, v_6, v_7, v_2$ $\qquad |Q| = 12$

$\{v_1, v_2\} \prec \{v_3, \ldots, v_9\}$ $\qquad \lambda(Q) = 12 - 2 \cdot 6 = 0$

We decompose Q into looping walks:

$Q_1 = v_1, v_9, v_8, v_7, v_6, v_5, v_4, v_3$ $\qquad \lambda(Q_1) = 7 - 2 \cdot 3 = 1$

$Q_2 = v_2, v_7, v_6, v_5, v_4, v_3$ $\qquad \lambda(Q_2) = 5 - 2 \cdot 2 = 1$

$[v_1, v_2] \prec v_3 \prec [v_4, \ldots, v_9]$

Note that Q is a bad walk, while neither Q^* nor Q_1 nor Q_2 is.

Fig. 3. Examples of looping walks

Proof. (Sketch) We prove the claim by considering individual implications. The equivalence (I)⇔(II) is proved as Lemma 1. The equivalence (IV)⇔(V) follows immediately from the definitions of $\delta(\cdot, \cdot)$ and bad walk. The other implications are proved as follows. For (III)⇒(II), we show that Prover's strategy described in (III) is a winning strategy. For (II)⇒(III), we show that every winning strategy must satisfy the conditions of (III). For (III)⇒(IV), we show that having vertices $u \prec v$ with $\delta(u, v) \leq \beta(v) - 2$ allows Adversary to win, by playing along the bad walk defined by vertices u, v. Finally, for (IV)⇒(III), assuming no bad pair u, v, we describe a Prover's strategy satisfying (III). □

We conclude this section by remarking that the values $\delta(u, v)$ can be easily computed in polynomial time by dynamic programming. This allows us to test conditions of the above theorem and thus decide $\{1, 2\}$-CSP($\mathbb{P}_\infty$) in polytime.

5   Algorithm for $\{1, 2\}$-CSP($\mathbb{P}_n$)

The path $\mathbb{P}_n$ has vertices $\{1, 2, \ldots, n\}$ and edges $\{ij \; : \; |i - j| = 1\}$.

Let Ψ be an instance of $\{1, 2\}$-CSP($\mathbb{P}_n$). As usual, let G be the graph $\mathcal{D}_\psi$ corresponding to Ψ, and let $\prec$ be the corresponding total ordering of $V(G)$.

For simplicity, let us assume that G is connected and bipartite with white and black vertices forming the bipartition. (If it is not bipartite, there is no solution; if disconnected, we solve the problem independently on each component.)

We start with a warmup lemma.

Lemma 4. *Assume* $\mathbb{P}_\infty \models \Psi$. *Let* f *be the first vertex in the ordering* $\prec$. *Then*

(i) $\mathbb{P}_1 \models \Psi \iff G$ *is the single* $\exists^{\geq 1}$ *vertex* f.

(ii) $\mathbb{P}_2 \models \Psi \iff G$ *does not contain* $\exists^{\geq 2}$ *vertex except possibly for* f.

(iii) $\mathbb{P}_3 \models \Psi \iff$ *all* $\exists^{\geq 2}$ *vertices in* G *have the same colour.*

(iv) $\mathbb{P}_4 \models \Psi \iff$ all $\exists^{\geq 2}$ vertices in G are pairwise non-adjacent except possibly for f.

(v) $\mathbb{P}_5 \models \Psi \iff$ there is colour C (black or white) such that each edge xy between two $\exists^{\geq 2}$ vertices where $x \prec y$ is such that x has colour C.

We now expand this lemma to the general case of $\{1,2\}$-$\mathrm{CSP}(\mathbb{P}_n)$ as follows. Recall that we proved that $\mathbb{P}_\infty \models \Psi$ if and only if Prover can play $\mathscr{G}(\Psi, \mathbb{P}_\infty)$ so that in every instance of the game, the resulting mapping f satisfies $(\star)$ and $(\triangle)$. In fact the proof of (III)$\Rightarrow$(II) from Theorem 7 shows that every winning strategy of Prover has this property. We use this fact in the subsequent text.

The following value $\gamma(v)$ will allow us to keep track of the distance of $f(v)$ from the center of the (finite) path.

Definition 7. For each vertex v we define $\gamma(v)$ recursively as follows:

$$\gamma(v) = 0 \qquad \text{if } v \text{ is first in the ordering } \prec$$

$$\text{else} \quad \gamma(v) = \beta(v) - 1 + \max\left\{0, \max_{u \prec v}\left(\gamma(u) - \delta(u,v) + \beta(v) - 1\right)\right\}$$

Lemma 5. Let M be a real number. Suppose that $\mathbb{P}_\infty \models \Psi$ and that Prover plays a winning strategy in the game $\mathscr{G}(\Psi, \mathbb{P}_\infty)$. Then Adversary can play so that the resulting mapping f satisfies $|f(v) - M| \geq \gamma(v)$ for every vertex $v \in V(D_\psi)$.

Lemma 6. Let M be a real number. Suppose that $\mathbb{P}_\infty \models \Psi$. Then there exists a winning strategy for Prover such that in every instance of the game the resulting mapping f satisfies $|f(v) - M| \leq \gamma(v) + 1$ for every $v \in V(\mathcal{D}_\psi)$.

With these tools, we can now prove a characterization of the case of even n.

Theorem 8. Let $n \geq 4$ be even. Assume that $\mathbb{P}_\infty \models \Psi$. Then TFAE.

(I) $\mathbb{P}_n \models \Psi$.

(II) Prover has a winning strategy in the game $\mathscr{G}(\Psi, \mathbb{P}_n)$.

(III) There is no vertex v with $\gamma(v) \geq \frac{n}{2}$.

Proof. Note first that since n is even, we may assume, without loss of generality, the first vertex in the ordering is quantified $\exists^{\geq 1}$. If not, we can freely change its quantifier to $\exists^{\geq 1}$ without affecting the satisfiability of the intance.

(I)$\Leftrightarrow$(II) is by Lemma 1. For (II)$\Rightarrow$(III), assume there is v with $\gamma(v) \geq \frac{n}{2}$ and Prover has a winning strategy in $\mathscr{G}(\Psi, \mathbb{P}_n)$. This is also a winning strategy in $\mathscr{G}(\Psi, \mathbb{P}_\infty)$. This allows us to apply Lemma 5 for $M = \frac{n+1}{2}$ to conclude that Adversary can play against Prover so that $|f(v) - \frac{n+1}{2}| = |f(v) - M| \geq \gamma(v) \geq \frac{n}{2}$. Thus either $f(v) \geq \frac{2n+1}{2} > n$ or $f(v) \leq \frac{1}{2} < 1$. But then $f(v) \notin \{1, \ldots, n\}$ contradicting our assumption that Prover plays a winning strategy.

For (III)$\Rightarrow$(II), assume that $\gamma(v) \leq \frac{n}{2} - 1$ for all vertices v. We apply Lemma 6 for $M = \frac{n+1}{2}$. This tells us that Prover has a winning strategy on $\mathscr{G}(\Psi, \mathbb{P}_\infty)$ such that in every instance of the game, if f is the resulting mapping, the mapping satisfies $|f(v) - \frac{n+1}{2}| \leq \gamma(v) + 1$ for every vertex v. From this we conclude that $f(v) \geq \frac{n+1}{2} - \gamma(v) - 1 \geq \frac{n+1}{2} - \frac{n}{2} = \frac{1}{2}$ and that $f(v) \leq \frac{2n+1}{2} = n + \frac{1}{2}$. Therefore $f(v) \in \{1, 2, \ldots, n\}$ confirming that f is a valid homomorphism to $\mathbb{P}_n$. $\qquad\square$

This generalizes to odd n with a subtle twist. Define $\gamma'(v)$ using same recursion as $\gamma(v)$ except set $\gamma'(v) = \beta(v) - 1$ if v is first in $\prec$. Note that $\gamma'(v) \geq \gamma(v)$.

Theorem 9. *Let $n \geq 5$ be odd. Assume that $\mathbb{P}_\infty \models \Psi$ and that the vertices of $\mathcal{D}_\psi$ are properly coloured with colours black and white. Then TFAE.*

(I) $\mathbb{P}_n \models \Psi$.
(II) Prover has a winning strategy in the game $\mathscr{G}(\Psi, \mathbb{P}_n)$.
(III) There are no vertices u, v with $\gamma'(u) \geq \frac{n-1}{2}$ and $\gamma'(v) \geq \frac{n-1}{2}$ such that u is black and v is white.

Now to derive Theorem 2, it remains to observe that the values $\gamma(v)$ and $\gamma'(v)$ can be calculated using dynamic programming in polynomial time.

5.1 Proofs of Corollaries 1 and 2

In this section, we sketch proofs of the two corollaries.

For Corollary 1, we want to decide $\{1, 2\}$-CSP(Π) when Π is a forest. Let Ψ be a given instance to this problem, and let $G = \mathcal{D}_\psi$ be the corresponding graph.

First, we note that we may assume that H is a tree. This follows easily (with a small caveat mentioned below) as the connected components of G have to be mapped to connected components of H. Therefore with H being a tree, we first claim that if Ψ is a yes-instance, then Ψ is also a yes-instance to $\{1, 2\}$-CSP($\mathbb{P}_\infty$). To conclude this, it can be shown that the condition (III) of Theorem 7 can be generalized to trees by replacing the absolute value in the condition $(\star)$ by the distance in H, and by using a proper colouring of H instead of parity in $(\triangle)$. This implies that no two vertices u, v are mapped in H farther away than $\delta(u, v)$. So a bad walk cannot exist and Ψ is a yes-instance of $\{1, 2\}$-CSP($\mathbb{P}_\infty$).

A similar argument allows us to generalize Theorems 8 and 9 to trees. Namely, in an optimal strategy Adversary will play **away** from some vertex, while Prover will play **towards** some vertex. The absolute values will again be replaced by distances in H. From this we conclude that Adversary can force each v to be assigned to a vertex in H which is at least $\gamma'(v)$ or $\gamma(v)$ away from the center vertex, resp. center edge of H. In summary, this then proves the following.

Corollary 3. *Let H be a tree. Let P be a longest path in H. Then Ψ is a yes-instance of $\{1,2\}$-CSP(H) if and only if Ψ is a yes-instance of $\{1,2\}$-CSP(P).*

This can be phrased more generally for forests in a straightforward manner. The only caveat is that if two components contain a longest path with odd number of vertices, then we can make the first vertex in the instance an $\exists^{\geq 1}$ vertex without affecting the satisfiability, because if it is $\exists^{\geq 2}$, we let Adversary choose which midpoint of the two longest paths to use (and either choice is fine).

Finally, to prove Corollary 2, we note that $\{1, 2\}$-CSP(H) is **NP**-hard for non-bipartite H, since $\{1\}$-CSP(H) is as famously proved in [8]. For bipartite H, the problem is in **P** if H is a forest (Corollary 1) or if H contains a 4-cycle (Proposition 10 in [11]). For bipartite graphs of larger girth, the problem is actually **PSPACE**-complete as we prove in the next section (Proposition 1).

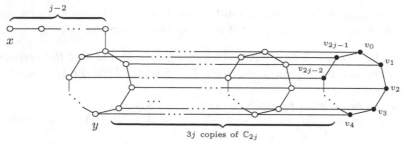

Fig. 4. The gadget for the case when H contains a cycle $\mathbb{C}_{2j}$

6   Proof of Theorem 3

In this section, we prove the **P** / **PSPACE** dichotomy for $\{1,2\}$-CSP(H) for bipartite graphs H as stated in Theorem 3. We have already discussed the polynomial cases in the previous section. It remains to discuss the hardness.

Proposition 1. *If H is a bipartite graph whose smallest cycle is $\mathbb{C}_{2j}$ for $j \geq 3$, then $\{1,2\}$-CSP(H) is* **PSPACE**-*complete.*

Proof. We reuse the reduction from [11] used to prove Theorem 1. We briefly discuss the key steps. The reduction is from QCSP($\mathbb{K}_j$). Let Ψ be an input formula for QCSP($\mathbb{K}_j$). We begin by considering the graph $\mathcal{D}_\psi$ to which we add a disjoint copy $W = \{w_1, \ldots, w_{2j}\}$ of $\mathbb{C}_{2j}$. Then we replace every edge $(x,y) \in \mathcal{D}_\psi$ with a gadget shown in Figure 4, where the black vertices are identified with W. Finally, for $\forall$ variables v of Ψ, we add a new path $z_1, z_2, \ldots, z_j$ where $z_j = v$.

The resulting graph defines the quantifier-free part of θ of our desired formula Θ. The quantification in Θ is as follows. The outermost quantifiers are $\exists^{\geq 2}$ for variables $w_1, \ldots, w_{2j}$. Then we move inwards through the quantifier order of Ψ; when we encounter an existential variable v, we apply $\exists^{\geq 1}$ to it in Θ. When we encounter a $\forall$ variable v, we apply $\exists^{\geq 2}$ to the path $z_1, z_2, \ldots, z_j$ constructed for v, in that order. All the remaining variables are then quantified $\exists^{\geq 1}$.

As proved in [11], the cycle $\mathbb{C}_{2j}$ models Θ if and only if $\mathbb{K}_j$ models Ψ. We now adjust this to the bipartite graph H. There are three difficulties arising from simply using the above construction as it is.

Firstly, assume the variables $w_1, \ldots, w_{2j}$ are mapped to a fixed copy C of $\mathbb{C}_{2j}$ in H. We need to ensure that variables x, y derived from the original instance Ψ are also mapped to C. For y variables in our gadget one can check this must be true – the successive cycles in the edge gadget may never deviate from C, since H contains no 4-cycle. For x variables off on the pendant this might not be true. To fix this, we insist that Ψ contains an atom $E(x,y)$ iff it also contains $E(y,x)$; QCSP($\mathbb{K}_j$) remains **PSPACE**-complete on such instances [2].

Secondly, we need to check that Adversary has freedom to assign any value from C to each $\forall$ variable v. Consider $z_1, \ldots, z_j$, the path associated with v. As long as Prover offers values for $z_1, \ldots, z_j$ from C, Adversary has freedom to chose any value for $v = z_j$. If on the other hand Prover offers for one of $z_1, \ldots, z_j$, say for z_i, a value not on C, then Adversary can choose all subsequent $z_{i+1}, \ldots, z_j$ to also be mapped outside C, since H has no cycle shorter than $\mathbb{C}_{2j}$. Thus $v = z_j$ is mapped outside C, but we already ensured that this does not happen.

Finally, we discuss how to ensure that W is mapped to a copy of $\mathbb{C}_{2j}$. Since each vertex in W is quantified $\exists^{\geq 2}$, Adversary can force this by always choosing a value not seen already when going through each of $w_1, \ldots, w_{2j}$ in turn. If this is not possible (both offered values have been seen), this gives rise to a cycle in H shorter than $\mathbb{C}_{2j}$. In conclusion, if Adversary maps W to a cycle, then Prover must play exclusively on this cycle, thus solving QCSP($\mathbb{K}_j$). If Adversary maps W to a subpath of $\mathbb{C}_{2j}$, then Prover can play to win (regardless whether Φ is a yes- or no- instance). So the situation is just like with $\{1,2\}$-CSP($\mathbb{C}_{2j}$).   □

7   Partially Reflexive Graphs

In this section, we briefly list some results for graphs allowing self-loops on some vertices (so-called *partially reflexive* graphs). Our understanding of these cases is rather limited and some recent results [10,12] suggest that a simple dichotomy is very unlikely. Nonetheless, some cases might still be of further interest.

First, we consider the class of undirected graphs with a single dominating vertex w which is also a self-loop.

Proposition 2. *If H has a reflexive dominating vertex w and $H \setminus \{w\}$ contains a loop or is irreflexive bipartite, then $\{1,2\}$-CSP(H) is in* **P**.

Proposition 3. *If H has a reflexive dominating vertex w and $H \setminus \{w\}$ is irreflexive non-bipartite, then $\{1,2\}$-CSP(H) is* **NP**-*complete.*

Corollary 4. *If H has a reflexive dominating vertex, then $\{1,2\}$-CSP(H) is either in* **P** *or is* **NP**-*complete.*

It follows from Proposition 3 that there is a partially reflexive graph on four vertices, $\mathbb{K}_4$ with a single reflexive vertex, so that the corresponding $\{1,2\}$-CSP is **NP**-complete. We can argue this phenomenem is not visible on smaller graphs.

Proposition 4. *Let H be a (partially reflexive) graph on at most three vertices, then either $\{1,2\}$-CSP(H) is in* **P** *or it is* **PSPACE**-*complete.*

8   Final Remarks

In this paper we have settled the major questions left open in [11] and it might reasonably be said we have now concluded our preliminary investigations into constraint satisfaction with counting quantifiers. Of course there is still a wide vista of work remaining, not the least of which is to improve our **P**/ **NP**-hard dichotomy for $\{1,2\}$-CSP on undirected graphs to a **P**/ **NP**-complete / **PSPACE**-complete trichotomy (if indeed the latter exists). The absence of a similar trichotomy for QCSP, together with our reliance on [8], suggests this could be a challenging task. Some more approachable questions include lower bounds for $\{2\}$-CSP($\mathbb{K}_4$) and $\{1,2\}$-CSP($\mathbb{P}_\infty$). For example, intutition suggests these might be **NL**-hard (even **P**-hard for the former). Another question would be to study X-CSP$(\mathbb{P}_\infty)$, for $\{1,2\} \not\subseteq X \subset \mathbb{N}$.

Since we initiated our work on constraint satisfaction with counting quantifiers, a possible algebraic approach has been published in [5,6]. It is clear reading our expositions that the combinatorics associated with our counting quantifiers is complex, and unfortunately the same seems to be the case on the algebraic side (where the relevant "expanding" polymorphisms have not previously been studied in their own right). At present, no simple algebraic method, generalizing results from [2], is known for counting quantifiers with majority operations. This would be significant as it might help simplify our tractability result of Theorem 2. So far, only the Mal'tsev case shows promise in this direction.

References

1. Barto, L., Kozik, M., Niven, T.: The CSP dichotomy holds for digraphs with no sources and no sinks (a positive answer to a conjecture of Bang-Jensen and Hell). SIAM Journal on Computing 38(5), 1782–1802 (2009)
2. Börner, F., Bulatov, A.A., Chen, H., Jeavons, P., Krokhin, A.A.: The complexity of constraint satisfaction games and QCSP. Inf. Comput. 207(9), 923–944 (2009)
3. Bulatov, A.: A dichotomy theorem for constraint satisfaction problems on a 3-element set. J. ACM 53(1), 66–120 (2006)
4. Bulatov, A., Krokhin, A., Jeavons, P.G.: Classifying the complexity of constraints using finite algebras. SIAM Journal on Computing 34, 720–742 (2005)
5. Bulatov, A.A., Hedayaty, A.: Counting predicates, subset surjective functions, and counting csps. In: 42nd IEEE International Symposium on Multiple-Valued Logic, ISMVL 2012, pp. 331–336 (2012)
6. Bulatov, A.A., Hedayaty, A.: Galois correspondence for counting quantifiers. CoRR abs/1210.3344 (2012)
7. Feder, T., Vardi, M.: The computational structure of monotone monadic SNP and constraint satisfaction: A study through Datalog and group theory. SIAM Journal on Computing 28, 57–104 (1999)
8. Hell, P., Nešetřil, J.: On the complexity of H-coloring. Journal of Combinatorial Theory, Series B 48, 92–110 (1990)
9. Kolaitis, P.G., Vardi, M.Y.: A logical Approach to Constraint Satisfaction. In: Finite Model Theory and Its Applications. Texts in Theoretical Computer Science. An EATCS Series. Springer-Verlag New York, Inc. (2005)
10. Madelaine, F., Martin, B.: QCSP on Partially Reflexive Cycles – The Wavy Line of Tractability. In: Bulatov, A.A., Shur, A.M. (eds.) CSR 2013. LNCS, vol. 7913, pp. 322–333. Springer, Heidelberg (2013)
11. Madelaine, F., Martin, B., Stacho, J.: Constraint Satisfaction with Counting Quantifiers. In: Hirsch, E.A., Karhumäki, J., Lepistö, A., Prilutskii, M. (eds.) CSR 2012. LNCS, vol. 7353, pp. 253–265. Springer, Heidelberg (2012)
12. Martin, B.: QCSP on partially reflexive forests. In: Lee, J. (ed.) CP 2011. LNCS, vol. 6876, pp. 546–560. Springer, Heidelberg (2011)
13. Schaefer, T.J.: The complexity of satisfiability problems. In: Proceedings of STOC 1978, pp. 216–226 (1978)

Dynamic Complexity of Planar 3-Connected Graph Isomorphism

Jenish C. Mehta*

jenishc@gmail.com

Abstract. Dynamic Complexity (as introduced by Patnaik and Immerman [14]) tries to express how hard it is to *update* the solution to a problem when the input is changed *slightly*. It considers the *changes* required to some stored data structure (possibly a massive database) as small quantities of data (or a tuple) are inserted or deleted from the database (or a structure over some vocabulary). The main difference from previous notions of dynamic complexity is that instead of treating the update quantitatively by finding the the time/space trade-offs, it tries to consider the update *qualitatively*, by finding the *complexity class* in which the update can be expressed (or made). In this setting, DynFO, or Dynamic First-Order, is one of the smallest and the most natural complexity class (since SQL queries can be expressed in First-Order Logic), and contains those problems whose solutions (or the stored data structure from which the solution can be found) can be updated in First-Order Logic when the data structure undergoes small changes.

Etessami [7] considered the problem of isomorphism in the dynamic setting, and showed that Tree Isomorphism can be decided in DynFO. In this work, we show that isomorphism of Planar 3-connected graphs can be decided in DynFO$^+$ (which is DynFO with some polynomial precomputation). We maintain a canonical description of 3-connected Planar graphs by maintaining a database which is accessed and modified by First-Order queries when edges are added to or deleted from the graph. We specifically exploit the ideas of Breadth-First Search and Canonical Breadth-First Search to prove the results. We also introduce a novel method for canonizing a 3-connected planar graph in First-Order Logic from Canonical Breadth-First Search Trees.

1 Introduction

Consider the problem LIS(A) of finding the longest increasing subsequence of a sequence (or array) of n numbers A. The "template" dynamic programming polynomial time solution proceeds by subsequently finding and storing LIS(A$[1{:}i]$) - the longest increasing subsequence of numbers from 1 to i that necessarily ends with the i'th number. LIS(A$[1{:}i+1]$) is found, given LIS(A$[1{:}1]$) to LIS(A$[1{:}i]$), by simply finding the maximum sequence formed by possibly appending A$[i+1]$ to the largest subsequence from LIS(A$[1{:}1]$) to LIS(A$[1{:}i]$).

---

* This work was done while the author was interning at Chennai Mathematical Institute in May-June, 2010

E.A. Hirsch et al. (Eds.): CSR 2014, LNCS 8476, pp. 273–286, 2014.
© Springer International Publishing Switzerland 2014

This *paradigm* of dynamic programming (or incremental thinking), of *storing* information using polynomial space, and *updating* it to get the required results, is neatly captured in the Dynamic Complexity framework introduced by Patnaik and Immerman [14]. Broadly, Dynamic Complexity tries to measure or express how hard it is to *update* some stored information, so that some required query can be answered. For instance, for some graph problem, like reachability, it tries to measure (or express) how hard it is to update some stored information when an edge is inserted or deleted from the graph, so that the required query, like reachability between two vertices s and t, can be answered easily from the stored information. Essentially, it asks how hard is one step of induction, or how hard it is to update one step of some recurrence.

This Dynamic Complexity framework (as in [14]) differs from other notions in two ways. For some problem (say a graph theoretic problem like colorability or reachability), the traditional notions of the dynamic complexity try to measure the amount of *time* and *space* required to make some update to the problem (like inserting/deleting edges from a graph or inserting/deleting tuples from a database), and the trade-offs between the two. Dynamic Complexity, instead tries to measure (or express) the resources required for an update *qualitatively*. Hence, it tries to measure an update by the *complexity class* in which it lies, rather than the explicit time/space requirements. For any static complexity class C, informally, the dynamic complexity class DynC consists of the set of problems, such that any *update* (to be defined formally later) to the problem can be expressed in the compexity class C. A bit more formally, a language L is in the dynamic complexity class DynC if we can maintain a tuple of relations (say T) for deciding the language in C, such that after any insertion or deletion of a tuple to the relations, they can be effectively updated in the complexity class C (updation is required so that even after the insertion/deletion of the tuple, they decide the same language L).

Another difference is that it treats the complexity classes in a Descriptive manner (using the language of Finite Model Theory) rather than the standard Turing manner (defined by tapes and movement of pointers). Since Descriptive Complexity tries to measure the hardness of *expressing* a problem rather than the hardness of *finding* a solution to the problem, Dynamic Complexity tries to measure how hard it is to *express* an update to some problem. Though, since either definition - Descriptive or Turing - leads to complexity classes with the same expressive power, any of the definitions remain valid.

Consider the dynamic complexity class DynP (or DynFO(LFP)). Intuitively, it permits storage of a polynomial amount of information (generated in polynomial time), so that (for some problem) the information during any update can be modified in P. Observe that the above problem of LIS(A) lies in DynP, since at every stage we stored a polynomial amount of information, and the *update* step took polynomial time to modify the information.

Although we do not consider relations between static and dynamic complexity classes here, it is worth mentioning that DynP=P (under a suitable notion of a reduction). Hence, unless P=NP, it is not possible to store some polynomial amount of information (generated in polynomial time), so that insertion of a single edge in a graph or a single clause in a 3-SAT expression (over a fixed set of variables), leads to finding whether the graph is 3-colorable or whether the 3-SAT expression is satisfiable. As another illustration, for the NP-complete problem of finding the longest path between any two vertices in an n-vertex undirected graph, even if we are given any kind of (polynomial) information[1], including the longest path between all possible pairs of vertices in the given graph, it is not possible to find the *new* longest path between any pair of vertices when a single edge is inserted to the graph (unless P=NP). This means that NP-complete problems are even hard to simply update, i.e, even a small update to an NP-complete problem cannot be done in polynomial time. The reader is referred to [9] for complete problems for DynFO and for reductions among problems in the dynamic setting.

Although a dynamic programming solution to any problem is in effect a DynP solution, the class DynP is less interesting since it is essentially same as P. More interesting classes are primarily the dynamic versions of smaller circuit complexity classes inside P, like DynNC^1, DynTC^0, etc. The most interesting, and perhaps the smallest dynamic complexity class, is DynFO. Intuitively, DynFO or Dynamic First-Order is the set of problems for which a polynomial sized database of information can be stored to answer the problem query (like reachability), such that after any insertion/deletion of a tuple, the database can be updated using merely a FO query (i.e. in First-Order Logic). A problem being in DynFO means that any updation to the problem is extremely easy in some sense.

Another reason why DynFO is important is because it is closely related to practice. A limitation of static complexity classes is that they are not appropriate for systems where large amounts of data are to be queried. Most real-life problems are dynamic, extending over extremely long periods of time, manipulating stored data. In such systems, it is necessary that small perturbations to massive quantities of data can be computed very fast, instead of processing the data from scratch. Consider for instance, a massive code that is dynamically compiled. We would expect that the compilation, as letters are typed, should be done very fast, since only a small part of the program is modified with every letter. Hence, for huge continually changing databases (or Big-Data), it is not feasible to re-compute a query all over again when a new tuple is inserted or deleted to/from the database. For the problems in DynFO, since an SQL query is essentially a FO Query, an SQL query can update the database without computing everything again. This is very useful in dynamic settings. A nice exposition on DynFO in this respect can be found in [15].

One basic problem considered in this setting is that of Reachability. In [14], it was shown that Undirected Reachability (which is in the static class L), lies in the complexity class DynFO. Note how a simple class like FOL, which does not even contain parity, becomes powerfully expressive in the dynamic setting.

---

[1] By polynomial information, we mean information that has been generated in polynomial time, and after the insertion of an edge, it can be regenerated (in polynomial time) so as to allow insertion of another edge, and so on ad infinitum.

Hesse [8] showed that Directed Reachablity lies in DynTC^0. Also, Dong and Su [6] further showed that Directed Rechability for acyclic graphs lies in DynFO.

The Graph Isomorphism problem (of finding a bijection between the vertex sets of two graphs such that the adjacencies are preserved) has so far been elusive to algorithmic efforts and has not yet yielded a better than subexponential ($2^{o(n)}$) time static algorithm. The general problem is in NP, and also in SPP (Arvind and Kurur [1]). Thus, various special cases have been considered, one important case being restriction to planar graphs. Hopcroft and Wong [10] showed that Planar Graph Isomorphism can be decided in linear time. In a series of works, it was further shown that Tree Isomorphism is in L (Lindell [12]), 3-connected Planar Graph Isomorphism is in L (Datta et. al. [3]) and finally, Planar Graph Isomorphism is in L (Datta et. al. [4]).

Etessami considered the problem of isomorphism in the dynamic setting. It was shown in [7] that Tree Isomorphism can be decided in DynFO.

In this work, we consider a natural extension and show that isomorphism for Planar 3-connected graphs can be decided in DynFO (with some polynomial precomputation). Our method of showing this is different from that in [7]. The main technical tool we employ is that of Canonical Breadth-First Search trees (abbreviated CBFS tree), which were used by Thierauf and Wagner [16] to show that 3-connected Planar Graph Isomorphism lies in UL. We also introduce a novel method for finding the canon of a 3-connected Planar graph from Canonical Breadth-First Search trees in First-Order Logic (FOL). We finally compare the canons of the two graphs to decide on isomorphism.

Our main results are:

1. Breadth-First Search for undirected graphs is in DynFO
2. Isomorphism for Planar 3-connected graphs is in DynFO^+

DynFO^+ is exactly same as DynFO, except that it allows some polynomial precomputation, which is necessary until enough edges are inserted so that the graph becomes 3-connected. Note that this is the best one can hope for, due to the requirement of 3-connectivity.

In Section 3, we prove Result 1. In Section 4, we prove Result 2. In Section 5, we introduce a novel method of canonizing a planar 3-connected graph in FOL from Canonical Breadth-First Search trees. Finally, we conclude with open problems and scope for future work. All the proofs, diagrams, detailed preliminaries, and First-Order queries can be found in the extended version of the paper [13].

2   Preliminaries

The reader is referred to [5] for the graph-theoretic definitions, to [11] for the definitions on Finite-Model Theory, and to [11] or [14] for definitions on Dynamic Complexity.

Let E_v be the set of edges incident to v. A permutation π_v on E_v that has only one cycle is called a rotation. A rotation scheme for a graph G is a set

π of rotations, $\pi = \{\pi_v \mid v \in V$ and π_v is a rotation on $E_v\}$. Let π^c be the set of inverse rotations, $\pi^c = \{\pi_v^c \mid v \in V\}$. A rotation scheme π describes an embedding of graph G in the plane. For 3-connected planar graphs, we shall asssume that π is the set of anti-clockwise rotations around each vertex, and π^c is the set of clockwise rotations around every vertex. Whitney [17] showed that π and π^c are the only two rotations for 3-connected planar graphs.

We shall refer to the following theorem at certain places, and we make it explicit here, which can be proven using Ehrenfeucht-Fraisse Games [11]:

Theorem 1. *Transitive Closure is not in* FOL $=$ *uniform* AC^0.

Definition 1. *For any static complexity class* C, *we define its dynamic version,* DynC *as follows: Let* $\rho = \langle R_1^{a_1}, ..., R_s^{a_s}, c_1, ..., c_t \rangle$, *be any vocabulary and* $S \subseteq STRUC(\rho)$ *be any problem.*

Let $R_{n,\rho} = \{ins(i, a'),\ del(i, a'),\ set(j, a) \mid 1 \le i \le s,\ a' \in \{0, ..., n-1\}^{a_i},$ $1 \le j \le t\}$ *be the request to insert/delete tuple* a' *into/from the relation* R_i, *or set constant* c_j *to* a.

Let $eval_{n,\rho} : R_{n,\rho}^* \to STRUC(\rho)$ *be the evaluation of a sequence or stream of requests. Define* $S \in$ DynC *iff there exists another problem* $T \subset STRUC(\tau)$ *(over some vocabulary* τ*) such that* $T \in$ C *and there exist maps* f *and* g:

$$f : R_{n,\rho}^* \to STRUC(\tau),\ g : STRUC(\tau) \times R_{n,\rho} \to STRUC(\tau)$$

satisfying the following properties:

1. **(Correctness)** *For all* $r' \in R_{n,\rho}^*$, $(eval_{n,\rho}(r') \in S) \Leftrightarrow (f(r') \in T)$
2. **(Update)** *For all* $s \in R_{n,\rho}$, *and* $r' \in R_{n,\rho}^*$, $f(r's) = g(f(r'), s)$
3. **(Bounded Universe)** $||f(r')|| = ||eval_{n,\rho}(r')||^{O(1)}$
4. **(Initialisation)** *The functions* g *and the initial structure* $f(\emptyset)$ *are computable in* C *as functions of* n.

Our main aim is to define the update function g (over some vocabulary τ). If condition (4) is relaxed, to the extent that the initializing function f may be polynomially computable (before any insertion or deletion of tuples begin), the resulting class is DynC$^+$, that is DynC with polynomial precomputation.

3   Breadth-First-Search in DynFO

In this section, we shall show that Breadth-First-Search (abbreviated BFS) for any arbitrary undirected graph lies in DynFO. More specifically, we shall show that there exists a set of relations, such that using those relations, finding the minimum distance between any two points in a graph can be done through FOL, and the set of all the points at a particular distance from a given point can be retrieved through a FO query, in any arbitrary undirected graph. Also, the modification of the relations can be carried out using FOL, during insertion or deletion of edges.

The definitions and terminologies regarding BFS can be found in any standard textbook on algorithms, like [2].

The main idea is to maintain the BFS tree from *each* vertex in the graph. This idea is important, because it will be extended in the next section. To achieve this, we shall maintain the following relations:

- $Level(v, x, l)$, implying that the vertex x is at level l in the BFS tree of vertex v (A vertex x is said to be at level l in the BFS tree of v if the distance between x and v is l);
- $BFSEdge(v, x, y)$, meaning that the edge (x, y) of the graph is in the BFS tree rooted at v;
- $Path(v, x, y, z)$, meaning that vertex z is on the path from x to y, in BFS tree of v. Also
- $Edge(x, y)$ will denote all the edges present in the entire graph.

Note that it is sufficient to maintain the *Level* relation to query the length of the shortest path between any two vertices. We maintain the *BFSEdge* and *Path* relations only if we want the actual shortest path between any two vertices.

These relations form the vocabulary τ as in Definition 1.

3.1   Maintaining $Level(v, x, l)$, $BFSEdge(v, x, y)$, $Path(v, x, y, z)$

We shall first focus on the $Level(v, x, l)$ relation, since it will give us the tools required for the other two relations.

In maintaining this relation, we are effectively maintaining the shortest distances between every pair of vertices. We will need to understand how the various BFS trees behave during insertion and deletion of edges before we write down the queries.

We will use the following notations from this section onwards. Let $path_v(\alpha, \beta)$ denote the set of edges in the path from vertex α to β, in the BFS tree of v. Let $|path_v(\alpha, \beta)|$ denote its size. Hence $Level(v, x, l)$ means $|path_v(v, x)| = l$. Let $level_v(x)$ denote the level of vertex x in BFS tree of v. Hence, $Level(v, x, l) \Leftrightarrow level_v(x) = l$. Also, we shall succinctly denote the edge from a to b by $\{a, b\}$. The vertices which are not connected to v will not appear in any tuple in the BFS-tree of v.

Note that any path can be split into two disjoint paths. For instance, $path_v(a, b) = path_v(a, d) \cup path_v(d, b)$ for any vertex d on $path_v(a, b)$, simply because there is only one path in a tree between any two vertices.

insert(a, b). Due to the insertion of edge $\{a, b\}$, various paths in many BFS trees will change. We will show that many of the paths do not change, and these can be used to update the shortest paths that do change.

We shall see how to modify level of some vertex x in the BFS tree of some vertex v. But before we proceed, we'll need the following important lemma:

Lemma 1. *After the insertion of an edge $\{a, b\}$, the level of a vertex x cannot change both in the BFS trees of a and b.*

Since the level of vertex x remains invariant in atleast one BFS tree, this fact can be used to modify the level of (and subsequently even the paths to) x using this invariant. This fact will be crucial in the queries that we write next.

To update the *BFSEdge* and *Path* relations, since we will create the new shortest path by joining together two different paths, we need to ensure that these paths are disjoint.

Without loss of generality, let $|path_b(b, x)| \leq |path_a(a, x)|$.

Lemma 2. *If any vertex t is on $path_b(b, x)$ and on $path_v(v, a)$, then the shortest path from v to x does not change after insertion of the edge $\{a, b\}$*

The proofs of the above lemmas and the corresponding queries can be found in the extended version of the paper.

delete(a, b). Consider now the deletion of some edge $\{a, b\}$ from the graph. If it is present in the BFS tree of some vertex v, the removal of the edge splits the tree into two different trees. Let $R_1 = \{u \mid v, u \text{ are connected in } V_G \backslash \{a, b\}\}$, and $R_2 = \{u \mid u \notin R_1\}$. We find the set $PR = \{(p, r) \mid p \in R_1 \land r \in R_2 \land Edge(p, r)\}$, where PR is the set of edges in the graph that connect the trees R_1 and R_2. The new path to x will be a path from v to p in the BFS-tree of v, edge $\{p, r\}$, and path from r to x in the BFS-tree of r; and $\{p, r\}$ will be chosen to yield the shortest such path, and we will choose $\{p, r\}$ to be the lexicographically smallest amongst all such edges that yield the shortest path.

The only thing we need to address is the fact that the path from r to x in the BFS tree of r does not pass through the edge $\{a, b\}$.

Lemma 3. *When an edge $\{a, b\}$ separates a set of vertices R_2 from the BFS tree of v, and r and x are vertices belonging to R_2, then $path_r(r, x)$ cannot pass through edge $\{a, b\}$*

Remark 1. An important observation is that the above lemma holds only for the "undirected" case. It fails for the directed case, implying that the same relations cannot be used for BFS in directed graphs. To see a simple counter-example, note that there can be a directed edge from r to a in the directed case, and in that case, the shortest path from r to x can pass through (a, b).

Also note that for every vertex x in R_1, the shortest path from v to x remains the same, since removal of an edge cannot *decrease* the shortest distance.

Remark 2. Note that although we pick the new paths for every vertex in the set R_2 in parallel, we need to ensure that the paths picked are consistent, i.e. the paths form a tree and no cycle is formed. This is straightforward to see, since if a cycle is formed, it is possible to pick another path for some vertex that came earlier in the lexicographic ordering. Hence, our queries are consistent.

This leads us to the following theorem:

Theorem 2. *Breadth-First-Search for an undirected graph is in DynFO.*

4   3-Connected Planar Graph Isomorphism

The ideas and the techniques hitherto developed were for general undirected graphs. Now onwards, our relations would no longer hold for general graphs, and we restrict ourselves to 3-connected and planar graphs.

We shall now show how to maintain a canonical description of a 3-connected planar graph in DynFO. To achieve this end, we shall maintain Canonical Breadth-First Search (abbreviated CBFS) trees similar to the ones used by Thierauf and Wagner [16].

4.1   Canonical Breadth-First Search Trees

We have defined CBFS trees for the sake of completeness in the extended version of the paper, and the reader is strongly urged to read it.

We maintain a CBFS tree, denoted by $[v, v_e]$, from *each* vertex v in the graph, for *each* edge (v, v_e) used as the starting embedding edge. This set of CBFS trees will help us in maintaining the necessary relations, during insertions and deletions, for isomorphism.

Modifying our previous conventions, let $path_{v,v_e}(\alpha, \beta)$ denote the path from vertex α to β, in the CBFS tree $[v, v_e]$. Let Least Common Ancestor (LCA) of x and y, $lca_{v,v_e}(x, y)$, denote that vertex d which is on $path_{v,v_e}(v, a)$ and $path_{v,v_e}(v, x)$ and whose level is *maximum* amongst all such vertices. Denote the embedded number of vertex x around vertex u by $emnum_u(x)$, i.e. $emnum_u(x) = \pi_u(x)$. Let $parent_{v,v_e}(u, u_p)$ be true if u_p is the parent of u in $[v, v_e]$. Denote by $emnum_{v,v_e}(u, x) = (\pi_u(x) - \pi_u(u_p)) \bmod d_u$, the embedded number of some vertex x around u in $[v, v_e]$ if u_p (the parent of u in $[v, v_e]$) has been assigned the number 0.

Since the embedding π is unique, given a vertex and an edge incident on it, the entire CBFS tree is fixed. As such, given v, v_e, the *length* of the shortest path to a vertex x is fixed. The actual path is decided by the following definition.

Definition 2. *Let $<_c$ denote a canonical ordering on paths. Let $P_1 = path_{v,v_e}(v, x_1)$ and $P_2 = path_{v,v_e}(v, x_2)$. Let $d = lca_{v,v_e}(x_1, x_2)$, d_1 a vertex on P_1 and d_2 a vertex on P_2, and (d, d_1) and (d, d_2) edges in the graph.*
Then $P_1 <_c P_2$ if:

- $|P_1| < |P_2|$ *or*
- $|P_1| = |P_2|$ *and* $emnum_{v,v_e}(d, d_1) < emnum_{v,v_e}(d, d_2)$

We shall now see how to maintain the CBFS trees $[v, v_e]$. We maintain the following relations:

- $Emb(v, x, n_x)$, meaning that the vertex x is in the neighbourhood of v, and the edge (v, x) around v has the embedded number n_x;

– $Face(f, x, y, z)$, meaning that the vertex z is in the anti-clockwise path from vertex x to vertex y, around the face labelled f.
  Note that since the number of faces in a planar graph with n vertices can be more than n, we should label the face with a 2-tuple instead of a single symbol; but we do not do this since it adds unnecessary technicality without adding any new insight. If required, all the queries can be maintained for the faces labelled as two tuples $f = (f_1, f_2)$.
– $Level(v, x, l)$, meaning that the vertex x is at level l in the BFS tree of v. This is exactly as in the general case.
– $CBFSEdges(v, v_e, s, t)$, where (s, t) is an edge in the CBFS tree $[v, v_e]$.
– $CPath(v, v_e, x, y, z)$ denoting that z is on the path from x to y in $[v, v_e]$.

4.2 Maintaining $Emb(v, x, n_x)$ and $Face(f, x, y, z)$

These two relations define the embedding of the graph in the plane. We assume throughout in this section that the embedded numbering is the anti-clockwise one, and note that the same relations that we maintain for the anti-clockwise embedding can be maintained for the clockwise embedding.

Lemma 4. *In a 3-connected planar graph G, two distinct vertices not connected by an edge cannot both lie on two distinct faces unless G is a cycle.*

As a corrollary to the above lemma, we get:

Corollary 1. *In a 3-connected planar graph G, two distinct vertices when connected by an edge splits one face into two new faces, creating exactly 1 new face.*

Now, due to the above theorem, we can update the Emb and the $Face$ relations.

insert(a, b). Any edge $\{a, b\}$ that is inserted lies on a particular face, say f. Consider the edges from vertex a. Since a lies on the face f, exactly two edges from a will lie on the boundary of f. Let these two edges, considered anti-clockwise be e_1 and e_2, having the embedded numbering n_1 and n_2, respectively. Note that $n_2 = (n_1 + 1) \bmod d_a$, where d_a is the degree of a. This is because if this was not so, there would be some other edge in the anti-clockwise direction between e_1 and e_2, which would mean either we have selected a wrong face or the wrong edges e_1 and e_2.

Hence, when we insert the new edge $\{a, b\}$, we can give $\{a, b\}$ the embedded number n_2, and all the other edges around a which have an embedded number more than n_2, can be incremented by 1. Similarly we do this for b.

delete(a, b). Note that since we expect the graph to be 3-connected and planar once the edge (a, b) is removed, by the converse of Lemma 4 above, exactly two faces will get merged. As such, our queries now will be the exact opposite to those for insertion.

Rotating and Flipping the Embedding. We will show how to rotate or flip the embedding of the graph in FOL if required, as it will be necessary for further sections.

The type of rotation that we will accomplish in this section is as follows: In any given CBFS tree $[v, v_e]$, for every vertex x, we rotate the embedding around x until its parent gets the least embedding number, number 0 (that is the 0'th number in the ordering). For the root vertex v which has no parent, we give v_e the least embedding number.

This scheme is like a *'normal'* form for ordering the edges around any vertex, or *'normalizing'* the embedding. We show in this section that this can be done in FOL. Also, flipping the ordering from anti-clockwise to clockwise is (very easily) in FOL.

We shall create the following relation: $Emb_p(v, v_e, t, x, n_x)$, which will mean that in the CBFS tree $[v, v_e]$, for some vertex t, if the parent of t is t_p, and if the edge (t, t_p) (or the vertex t_p) is given the embedded number 0, then the edge (t, x) (or the vertex x) gets the embedded number n_x.

Note that our relation Emb was independent of any particular CBFS tree, since it depended only on the structure of the 3-connected planar graph and not on any CBFS tree we chose. But Emb_p depends on the chosen CBFS tree. Another thing to note is that we do not maintain the relation Emb_p in our vocabulary τ, since it can be easily created in FOL from the rest of the relations whenever required.

We create the relation Emb_p in the following manner. In every CBFS tree $[v, v_e]$, for every vertex t, we find the degree (d_t) and the parent (t_p) of t, and the embedded number n_p of t_p. Then for every vertex x in the neighbourhood of t with embedded numbering n_x, we do $n_x = (n_x - n_p) \bmod d_t$.

We also create the relation Emb_f which will contain the flipped or the clockwise embedding π^c.

A note said throughout in the manuscript is necessary to repeat here. Though the parent of v is null in $[v, v_e]$, we allow the parent of v to be v_e, so as to keep the queries neater. If this convention is not required, then the special case of the parent of v can be handled easily by modifying the queries.

This shows that the embedding can be *flipped* and *normalized* in FOL. We conclude the following:

Theorem 3. *The embedding of a 3-connected planar graph can be maintained, normalized and flipped in DynFO.*

4.3   Maintaining $CBFSEdges(v, v_e, s, t)$ and $CPath(v, v_e, x, y, z)$

In this section, we show how to maintain the final two relations via insertions and deletions of tuples that will help us to decide the isomorphism of two graphs. The relations are almost completely similar to the ones used for Breadth-First Search in the previous section. The only difference which arises is due to the uniqueness of the paths in Canonical Breadth-First Search Trees. We do not rewrite the $Level(v, x, l)$ since it will be exactly similar to the general BFS case.

insert(a, b). The CBFS tree is unique if the path to every vertex x from the root vertex v is uniquely defined. How shall we choose the unique path? First, we consider the paths with the shortest length. This is exactly same as in Breadth-First Search seen in the previous section. But unlike BFS, where we chose the shortest path arbitrarily (that is by lexicographic ordering during insertion/deletion), we will very precisely choose one of the paths from the set of shortest paths, by Definition 2. Intuitively, Definition 2 chooses the path based on its orientation according to the embedding π.

An important observation from Definition 2 is the following: *Distance has preference over Orientation.* This means that if there are two paths P_1 and P_2 from v to x in $[v, v_e]$ (due to the insertion of an edge which created a cycle in the tree $[v, v_e]$), though $P_2 <_c P_1$, the path P_1 will be chosen if $|P_1| < |P_2|$ irrespective of the canonical ordering $<_c$.

Consider some $[v, v_e]$. During insertion of $\{a, b\}$, let the old path (from v to some x) be P_1 and assume that the new path P_2 passes through (a, b). If $|P_1| < |P_2|$ or $|P_1| = |P_2| \wedge P_1 <_c P_2$, the path to x does not change, and all the edges and tuples to x in the old relations will belong to the new relations. If $|P_2| < |P_1|$ or $|P_1| = |P_2| \wedge P_2 <_c P_1$, the path to x changes. In this case, the new path will be from v to a in $[v, v_e]$, the edge $\{a, b\}$ (from a to b), and the path from b to x in $[b, b_e]$. The way we choose b_e is as follows: We find the set of vertices C that are adjacent to b and are at $level_v(b) + 1$. Since a will be the parent of b in $[v, v_e]'$, we rotate the embedding around b until a gets the value 0, and the choose b_e to be the vertex in C that gets the least embedding number.

To check for the condition $P_1 <_c P_2$, we do the following: We create the set Emb_p so that the parent of each vertex has the least embedding number. Let the path P_a denote the path from v to a, which will be a subset of P_2. We choose the vertex which is the least common ancestor of a and x, say $d = lca_{a,x}$, and normalize the embedding so that d_p, the parent of d, gets the embedding number 0. Existence of $lca_{a,x}$ is guaranteed since v lies on both P_1 and P_a. Now consider the edge $e_1 = (lca_{a,x}, d_1)$ on P_1 and $e_2 = (lca_{a,x}, d_2)$ on P_2. Since the embedding is normalized, we see which edge gets the smaller embedding number around the vertex $lca_{a,x}$. The path on which that edge lies will be the lesser ordered path according to $<_c$. It is nice to pause here for a moment and observe that this was possible since the embedding was 'normalized', otherwise it would not have been possible.

One more thing needs to be shown. In Lemma 1, we proved that for any vertex x, its level cannot change both in the BFS trees of a and b. In the previous case of BFS, as per our algorithm, the level not being changed implied the path not being changed. But that is not the case in CBFS trees. In CBFS trees, the level not changing may still imply that the path changes (due to the $<_c$ ordering on paths). Hence, it may be possible that though the level of the vertex x changes in only one of the CBFS trees, its actual path changes in *both* the CBFS trees. We need to show that this is not possible. And the reason this is necessary is because (just like the previous case) the updation of the path will depend on one specific path to vertex x in the CBFS tree of a or b which has not changed.

Lemma 5. *After the insertion of edge $\{a, b\}$, the path to any vertex x, cannot change in both the CBFS trees $[a, a_e]$ and $[b, b_e]$ for all a_e, b_e.*

***delete*(a, b).** For the deletion operation, we choose the edge from PR_{min} based on the $<_c$ relation. Note that when some edge $\{a, b\}$ is deleted, the path to some vertex x in $[v, v_e]$ cannot change if $\{a, b\}$ does not lie on the path. Other things remain exactly similar to the general case.

5   Canonization and Isomorphism Testing

In [16], the 3-connected planar graph is canonized from Canonical Breadth-First Search trees by using Depth-First Search (DFS). Performing DFS or any method that employs computing the transitive closure in any manner cannot be used here since it would not be possible in FOL, and most of the known methods of canonization seem to require computing the transitive closure. Note that a canon is required for condition 1 in Definition 1 to hold. What we seek is a method to canonize the graph, which depends only on the properties of vertices that can be inferred globally.

To achieve this, we shall label each vertex with a vector. Though the label will not be succinct now, it will be possible to create it in FOL.

Essentially, the canon for a vertex x in some CBFS tree $[v, v_e]$ will be *a set of tuples (l, h) of the levels and (normalized) embedding numbers of ancestors of* x.

Definition 3. *Let canon for each vertex x in $[v, v_e]$ be represented by $Canon_{v,v_e}(x)$. Then,*

$$Canon_{v,v_e}(x) = \{(l, h) : \exists q, q_p, \ C \wedge L \wedge P \wedge H\}$$

where, $C :\ CPath(v, v_e, v, x, q)$, $L :\ l = level_v(q)$, $P :\ parent_{v,v_e}(q, q_p)$, $H :\ h = emnum_{v,v_e}(q_p, q)$

Lemma 6. *For any CBFS tree $[v, v_e]$, for any two vertices x and y, $x = y \Leftrightarrow Canon(x) = Canon(y)$*

It is now easy to canonize each of the CBFS trees in FOL. Once each vertex has a canon, each edge is also uniquely numbered. The main idea is this: A canon will in itself encode all the necessary properties of the vertex, and the set of canons of all vertices become the signature of the graph, preserving edges. The main advantage of Definition 3 is that the canon of the graph can be generated in FOL. It's worthwhile to observe how this neatly beats the otherwise inevitable computation of transitive closure (Theorem 1) to canonize the graph.

Hence, two 3-connected planar graphs G and H are isomorphic if and only if for some CBFS tree $[g, g_e]$ of G, there is a CBFS tree $[h, h_e]$, such that:

- $\forall x \exists y, \ (x \in G \land y \in H \land (Canon(x) = Canon(y)))$ and
- $\forall x_1, x_2, \ ((Edge(x_1, x_2) \in G) \Leftrightarrow (Edge(Canon(x_1), Canon(x_2)) \in H))$

H implies either H with the embedding ρ or H with flipped embedding ρ^{-1}. It is evident that if the graphs are isomorphic, there will be some CBFS tree in G and H whose canons will be equivalent in the above sense. If the graphs are not isomorphic, no canon of any CBFS tree could be equivalent, since it would then directly give a bijection between the vertices of the graph that preserves the edges, which would be a contradiction.

Since we still need to precompute all the relations before the condition of 3-connectivity is reached, isomorphism of G and H is in DynFO$^+$. This brings us to the main conclusion of this section:

Theorem 4. *3-connected planar graph isomorphism is in DynFO$^+$*

Conclusion. We have shown that Breadth-First Search for undirected graphs is in DynFO and Planar 3-connected Graph Isomorphism is in DynFO$^+$. A natural extension is to show that Planar Graph Isomorphism is in DynFO. Though even parallel algorithms for this problem are known [4], the ideas cannot be directly employed because of myriad problems arising due to automorphisms of the bi/tri-connected component trees (which are used in [4]), and various subroutines that require computing the transitive closure. In spite of these shortcomings, we strongly believe that Planar Graph Isomorphism is in DynFO, though the exact nature of the queries still remains open.

Acknowledgments. The author sincerely thanks Samir Datta for fruitful discussions and critical comments on all topics ranging from the problem statement to the preparation of the final manuscript.

References

1. Arvind, V., Kurur, P.P.: Graph isomorphism is in SPP. In: Proceedings of the Forty-Third Annual IEEE Symposium on Foundations of Computer Science, pp. 743–750. IEEE (2002)
2. Cormen, T.H., Stein, C., Rivest, R.L., Leiserson, C.E.: Introduction to Algorithms, 2nd edn. McGraw-Hill Higher Education (2001)
3. Datta, S., Limaye, N., Nimbhorkar, P.: 3-connected Planar Graph Isomorphism is in Logspace. arXiv:0806.1041 (2008)
4. Datta, S., Limaye, N., Nimbhorkar, P., Thierauf, T., Wagner, F.: Planar Graph Isomorphism is in Logspace. In: Proceedings of the Twenty-Fourth Annual IEEE Conference on Computational Complexity, pp. 203–214. IEEE (2009)
5. Diestel, R.: Graph Theory. Springer (2005)
6. Dong, G.Z., Su, J.W.: Incremental and Decremental evaluation of Transitive Closure by First-Order queries. Information and Computation 120(1), 101–106 (1995)
7. Etessami, K.: Dynamic Tree Isomorphism via First-Order Updates to a Relational Database. In: Proceedings of the Seventeenth ACM SIGACT-SIGMOD-SIGART Symposium on Principles of Database Systems, pp. 235–243. ACM (1998)

8. Hesse, W.: The Dynamic Complexity of Transitive Closure is in DynTC^0. In: Proceedings of the Eighth International Conference on Database Theory. Citeseer (2002)
9. Hesse, W.M.: Dynamic Computational Complexity. Computer Science (2003)
10. Hopcroft, J.E., Wong, J.-K.: Linear time algorithm for Isomorphism of Planar graphs (preliminary report). In: Proceedings of the Sixth Annual ACM Symposium on Theory of Computing, pp. 172–184. ACM (1974)
11. Immerman, N.: Descriptive Complexity. Springer (1999)
12. Lindell, S.: A Logspace algorithm for Tree Canonization. In: Proceedings of the Twenty-Fourth Annual ACM Symposium on Theory of Computing, pp. 400–404. ACM (1992)
13. Mehta, J.C.: Dynamic Complexity of Planar 3-connected Graph Isomorphism. arXiv (2013), http://arxiv.org/abs/1312.2141
14. Patnaik, S., Immerman, N.: Dyn-FO (preliminary version): A Parallel, Dynamic Complexity Class. In: Proceedings of the Thirteenth ACM SIGACT-SIGMOD-SIGART Symposium on Principles of Database Systems, pp. 210–221. ACM (1994)
15. Schwentick, T.: Perspectives of Dynamic Complexity. In: Libkin, L., Kohlenbach, U., de Queiroz, R. (eds.) WoLLIC 2013. LNCS, vol. 8071, pp. 33–33. Springer, Heidelberg (2013)
16. Thierauf, T., Wagner, F.: The isomorphism problem for Planar 3-connected graphs is in Unambiguous Logspace. Theory of Computing Systems 47(3), 655–673 (2010)
17. Whitney, H.: A set of topological invariants for graphs. American Journal of Mathematics 55(1), 231–235 (1933)

Fast Approximate Computations with Cauchy Matrices, Polynomials and Rational Functions*

Victor Y. Pan

Department of Mathematics and Computer Science
Lehman College and the Graduate Center of the City University of New York
Bronx, NY 10468 USA
victor.pan@lehman.cuny.edu
http://comet.lehman.cuny.edu/vpan/

Abstract. The papers [18], [9], [29], and [28] combine the techniques of the Fast Multipole Method of [15], [8] with the transformations of matrix structures, traced back to [19]. The resulting numerically stable algorithms approximate the solutions of Toeplitz, Hankel, Toeplitz-like, and Hankel-like linear systems of equations in nearly linear arithmetic time, versus the classical cubic time and the quadratic time of the previous advanced algorithms. We extend this progress to decrease the arithmetic time of the known numerical algorithms from quadratic to nearly linear for computations with matrices that have structure of Cauchy or Vandermonde type and for the evaluation and interpolation of polynomials and rational functions. We detail and analyze the new algorithms, and in [21] we extend them further.

Keywords: Cauchy matrices, Fast Multipole Method, HSS matrices, Vandermonde matrices, Polynomial evaluation; Rational evaluation, Interpolation.

1 Introduction

The numerically stable algorithms of [18], [9], [29], and [28] approximate the solution of Toeplitz, Hankel, Toeplitz-like, and Hankel-like linear systems of equations in nearly linear arithmetic time versus the classical cubic time and the previous record quadratic time of [14]. All five cited papers first transform the matrix structures of Toeplitz and Hankel types into the structure of Cauchy type, which is a special case of the general technique proposed in [19]. Then [14] exploits the invariance of the Cauchy matrix structure in row and column interchange, whereas the other four papers apply numerically stable FMM to operate efficiently with HSS approximation of the basic Cauchy matrix. "HSS" and "FMM" are the acronyms for "Hierarchically Semiseparable" and "Fast Multipole Method", respectively. "Historically HSS representation is just a special case of the representations commonly exploited in the FMM literature" [7].

---

* Some preliminary results of this paper have been presented at CASC 2013. Our research has been supported by the NSF Grant CC 1116736 and the PSC CUNY Awards 64512–0042 and 65792–0043.

E.A. Hirsch et al. (Eds.): CSR 2014, LNCS 8476, pp. 287–299, 2014.

Our present paper extends the algorithms of [18], [9], [29], and [28] to computations with Cauchy and Vandermonde matrices, namely to approximation of their products by a vector and of the solution of linear systems of equations with these matrices, which also covers approximate multipoint polynomial and rational evaluation and interpolation. The arithmetic time of the known numerical approximation algorithms for all these tasks is quadratic [4], [3], and we decrease it to nearly linear.

As in the papers [18], [9], [29], and [28], we approximate Cauchy matrices by HSS matrices and exploit the HSS matrix structure. As in these papers our basic computational blocks are the numerically stable FFT and FMM algorithms, which have been efficiently implemented on both serial and parallel computers [16], [1], [6]. Unlike the cited papers, however, we treat a large subclass of Cauchy matrices $C = (\frac{1}{s_i - t_j})_{i,j=0}^{n-1}$ (we call them CV matrices because they are linked to Vandermonde matrices via FFT-based unitary transformations) rather than just the single CV matrix involved in the fast Toeplitz solvers. For that matrix, $\{s_0, \ldots, s_{n-1}\}$ is the set of the nth roots of unity, and $\{t_0, \ldots, t_{n-1}\}$ is the set of the other $(2n)$-th roots of unity, but for a CV matrix C only the knots $\{t_0, \ldots, t_{n-1}\}$ are assumed to be equally spaced on the unit circle, whereas $\{s_0, \ldots, s_{n-1}\}$ is an unrestricted set of n knots. We still yield the desired HSS approximation of CV matrices by exploiting a proper partition of the complex plane into congruent sectors sharing the origin 0. To decrease the cost of computing this approximation and of subsequent computations with HSS matrices, we handle the harder and so far untreated case where the diagonal blocks are rectangular and have row indices that pairwise overlap. We detail and analyze our algorithms. In [21] we extend them to other classes of structured matrices.

We refer the reader to the papers and books [11], [13], [18], [7], [9], [25], [26], [29], [27], [28], [2], [6], [15], [10], [8], [17], [23], and the bibliography therein on FMM, HSS, and Matrix Compression (e.g., Nested Dissection) algorithms.

We organize our paper as follows. In the next section we recall some basic results on computations with general matrices. In Section 3 we study polynomial and rational evaluation and interpolation as computations with Vandermonde and Cauchy matrices. In Sections 4 and 5 we extend the known results on HSS matrix computations. in Section 6 we apply these results to treat CV matrices. In Section 7 we discuss extensions and implementation. In Section 8 we summarize our study. Because of the space limitation we leave to [22] COLORED FIGURES, demonstrations by examples, proofs, details, and comments.

2   Definitions and Auxiliary Results

We measure the computational complexity by the number of arithmetic operations performed in the field $\mathbb{C}$ of complex numbers with no error and hereafter referred to as *ops*. $|\mathcal{S}|$ denotes the cardinality of a set $\mathcal{S}$. $M = (m_{i,j})_{i,j=0}^{m-1,n-1}$ is an $m \times n$ matrix. M^T is its transpose, M^H is its Hermitian transpose. $\mathcal{C}(B)$ and $\mathcal{R}(B)$ are the index sets of the rows and columns of its submatrix B, respectively. For two sets $\mathcal{I} \subseteq \{1, \ldots, m\}$ and $\mathcal{J} \subseteq \{1, \ldots, n\}$ define the submatrix

$M(\mathcal{I}, \mathcal{J}) = (m_{i,j})_{i \in \mathcal{I}, j \in \mathcal{J}}$. $\mathcal{R}(B) = \mathcal{I}$ and $\mathcal{C}(B) = \mathcal{J}$ if and only if $B = M(\mathcal{I}, \mathcal{J})$. Write $M(\mathcal{I}, .) = M(\mathcal{I}, \mathcal{J})$ when $\mathcal{J} = \{1, \ldots, n\}$. Write $M(., \mathcal{J}) = M(\mathcal{I}, \mathcal{J})$ when $\mathcal{I} = \{1, \ldots, m\}$. $(B_0 \ldots B_{k-1})$ and $(B_0 \mid \ldots \mid B_{k-1})$ denote a $1 \times k$ block matrix with k blocks $B_0, \ldots, B_{k-1}$, whereas $\mathrm{diag}(B_0, \ldots, B_{k-1}) = \mathrm{diag}(B_j)_{j=0}^{k-1}$ is a $k \times k$ block diagonal matrix with k diagonal blocks $B_0, \ldots, B_{k-1}$, possibly rectangular. $O = O_{m,n}$ is the $m \times n$ matrix filled with zeros. $I = I_n$ is the $n \times n$ identity matrix. M is a $k \times l$ *unitary* matrix if $M^H M = I_l$ or $M M^H = I_k$. An $m \times n$ matrix M has a nonunique *generating pair* (F, G^T) of a *length* ρ if $M = F G^T$ for two matrices $F \in \mathbb{C}^{m \times \rho}$ and $G \in \mathbb{C}^{n \times \rho}$. The rank of a matrix is the minimum length of its generating pairs. An $m \times n$ matrix is *nonsingular* or *regular* if it has full rank $\min\{m, n\}$. A matrix M has a rank at least ρ if and only if it has a nonsingular $\rho \times \rho$ submatrix $M(\mathcal{I}, \mathcal{J})$, and if so, then $M = M(., \mathcal{J}) M(\mathcal{I}, \mathcal{J})^{-1} M(\mathcal{I}, .)$. This expression defines a *generating triple* $(M(., \mathcal{J}), M(\mathcal{I}, \mathcal{J})^{-1}, M(\mathcal{I}, .))$ and two generating pairs $(M(., \mathcal{J}), M(\mathcal{I}, \mathcal{J})^{-1} M(\mathcal{I}, .)$ and $(M(., \mathcal{J}) M(\mathcal{I}, \mathcal{J})^{-1}, M(\mathcal{I}, .)$ for a matrix M of a length ρ. We call such pairs and triples *generators*. One can obtain some generators of the minimum length for a given matrix by computing its SVD or its less costly rank revealing factorizations such as ULV and URV factorizations in [9], [29], and [28], where the factors are unitary, diagonal or triangular. $\alpha(M)$ and $\beta(M)$ denote the arithmetic complexity of computing the vectors $M\mathbf{u}$ and $M^{-1}\mathbf{u}$, respectively, maximized over all vectors $\mathbf{u}$ and minimized over all algorithms, and we write $\beta(M) = \infty$ when the matrix M is singular. The straightforward algorithm supports the following bound.

Theorem 1. $\alpha(M) \leq 2(m+n)\rho - \rho - m$ for an $m \times n$ matrix M given with its generating pair of a length ρ.

$||M|| = ||M||_2$ denotes the spectral norm of an $m \times n$ matrix $M = (m_{i,j})_{i,j=0}^{m-1,n-1}$. We also write $|M| = \max_{i,j} |m_{i,j}|$, $||M|| \leq \sqrt{mn}|M|$. If a matrix U is unitary, then $||U|| = 1$ and $||MU|| = ||UM|| = ||M||$. A vector $\mathbf{u}$ is unitary if and only if $||\mathbf{u}|| = 1$, and if this holds we call it a *unit vector*. A matrix $\tilde{M}$ is an ϵ-*approximation* of a matrix M if $|\tilde{M} - M| \leq \epsilon$. The ϵ-*rank* of a matrix M denotes the integer $\min_{|\tilde{M} - M| \leq \epsilon} \mathrm{rank}(\tilde{M})$. An ϵ-*basis* for a linear space $\mathbb{S}$ of dimension k is a set of vectors that ϵ-approximate the k vectors of a basis for this space. An ϵ-generator of a matrix is a generator of its ϵ-approximation. $\alpha_\epsilon(M)$ and $\beta_\epsilon(M)$ replace the bounds $\alpha(M)$ and $\beta(M)$ when we ϵ-approximate the vectors $M\mathbf{u}$ and $M^{-1}\mathbf{u}$ instead of evaluating them. The *numerical rank* of a matrix M, which we denote $\mathrm{nrank}(M)$, is its ϵ-rank for a small ϵ. A matrix M is *ill conditioned* if its rank exceeds its numerical rank.

3   Polynomial and Rational Evaluation and Interpolation As Operations with Structured Matrices

$V = V_{\mathbf{s}} = (s_i^j)_{i,j=0}^{m-1,n-1}$ and $C = C_{\mathbf{s},\mathbf{t}} = \left(\frac{1}{s_i - t_j}\right)_{i,j=0}^{m-1,n-1}$ denote $m \times n$ *Vandermonde* and *Cauchy* matrices, respectively. Some authors define Vandermonde matrices as the transposes V^T (rather than the above matrices V).

Problem 1. Vandermonde-by-vector multiplication.
INPUT: $m + n$ complex scalars $p_0, \ldots, p_{n-1}; s_0, \ldots, s_{m-1}$.
OUTPUT: n complex scalars $v_0, \ldots, v_{m-1}$ satisfying

$$V\mathbf{p} = \mathbf{v} \text{ for } V = V_{\mathbf{s}} = (s_i^j)_{i,j=0}^{m-1,n-1}, \ \mathbf{p} = (p_j)_{j=0}^{n-1}, \text{ and } \mathbf{v} = (v_i)_{i=0}^{m-1}. \quad (1)$$

Problem 2. The solution of a Vandermonde linear system.
INPUT: $2n$ complex scalars $v_0, \ldots, v_{n-1}; s_0, \ldots, s_{n-1}$, the last n of them distinct.
OUTPUT: n complex scalars $p_0, \ldots, p_{n-1}$ satisfying equation (1) for $m = n$.

Problem 3. Cauchy-by-vector multiplication.
INPUT: $2m + n$ complex scalars $s_0, \ldots, s_{m-1}; t_0, \ldots, t_{n-1}; v_0, \ldots, v_{m-1}$.
OUTPUT: m complex scalars $v_0, \ldots, v_{m-1}$ satisfying

$$C\mathbf{u} = \mathbf{v} \text{ for } C = C_{\mathbf{s},\mathbf{t}} = \left(\frac{1}{s_i - t_j}\right)_{i,j=0}^{m-1,n-1}, \ \mathbf{u} = (u_j)_{j=0}^{n-1}, \text{ and } \mathbf{v} = (v_i)_{i=0}^{m-1}. \quad (2)$$

Problem 4. The solution of a Cauchy linear system of equations.
INPUT: $3n$ complex scalars $s_0, \ldots, s_{n-1}; t_0, \ldots, t_{n-1}; v_0, \ldots, v_{n-1}$, the first $2n$ of them distinct.
OUTPUT: n complex scalars $u_0, \ldots, u_{n-1}$ satisfying equation (2) for $m = n$.

The scalars $s_0, \ldots, s_{m-1}, t_0, \ldots, t_{n-1}$ define the Vandermonde and Cauchy matrices $V_{\mathbf{s}}$ and $C_{\mathbf{s},\mathbf{t}}$, are basic for Problems 1–4, and are said to be the *knots*. We can define a Cauchy matrix up to shifting its knots and scaling them by constants because $aC_{a\mathbf{s},a\mathbf{t}} = C_{\mathbf{s},\mathbf{t}}$ and $C_{\mathbf{s}+a\mathbf{e},\mathbf{t}+a\mathbf{e}} = C_{\mathbf{s},\mathbf{t}}$ for $a \neq 0$ and $\mathbf{e} = (1, \ldots, 1)^T$.

Theorem 2. *(i) An $m \times n$ Vandermonde matrix $V_{\mathbf{s}} = (s_i^j)_{i,j=0}^{m-1,n-1}$ has full rank if and only if all m knots $s_0, \ldots, s_{m-1}$ are distinct. (ii) An $m \times n$ Cauchy matrix $C_{\mathbf{s},\mathbf{t}} = \left(\frac{1}{s_i - t_j}\right)_{i,j=0}^{m-1,n-1}$ is well defined if and only if its two knot sets $s_0, \ldots, s_{m-1}$ and $t_0, \ldots, t_{n-1}$ share no elements. (iii) If this matrix is well defined, then it has full rank if and only if all its $m + n$ knots $s_0, \ldots, s_{m-1}, t_0, \ldots, t_{n-1}$ are distinct and also (iv) if and only if all its submatrices have full rank.*

Problems 1–4 are equivalent to polynomial and rational multipoint evaluation and interpolation, e.g., in Problem 1 we evaluate the polynomial $\sum_{i=0}^{n-1} p_i x^i$ at the $n - 1$ knots $s_0, \ldots, s_{n-1}$, whereas in Problem 1 we interpolate to this polynomial from its values at the $n - 1$ knots. The known solution algorithms use either $O((m+n)\log^2(m+n))$ ops, allowing extended precision [20, Sections 3.1–3.6], or order of $(m + n)^2$ ops performed numerically, with rounding to a fixed (e.g., standard IEEE double) precision [4], [3].

We decrease this quadratic bound to nearly linear. At first we recall a special case where the solution is well known. Suppose $s_i = \omega^i$ are the nth roots of 1, $\omega = \omega_n = \exp(2\pi\sqrt{-1}/n)$, $i = 0, \ldots, n - 1$, and $V_{\mathbf{s}} = (\omega^{ij})_{i,j=0}^{n-1}$. Write $\Omega = \frac{1}{\sqrt{n}}(\omega^{ij})_{i,j=0}^{n-1})$ and note that $\Omega^H \Omega = I_n$, that is $\Omega = \Omega^T$ and $\Omega^H = \Omega^{-1} = \frac{1}{\sqrt{n}}(\omega^{-ij})_{i,j=0}^{n-1}$ are unitary matrices. Then the Generalized *FFT* (Fast Fourier transform) and the Generalized Inverse FFT yield numerically stable solutions of Problems 1 and 2 by using $O(n \log(n))$ ops, and this solution can be extended to Problems 3 and 4 (cf. [5, Sections 1.2 and 3.4], [20, Sections 2.2, 2.3, and Problem 2.4.2]). Now write $\mathbf{t} = (f\omega^j)_{j=0}^{n-1}$, $V_t = \sqrt{n}\Omega \operatorname{diag}(f^j)_{j=0}^{n-1}$, $V_t^{-1} = \frac{1}{\sqrt{n}} \operatorname{diag}(f^{-j})_{j=0}^{n-1}\Omega^H$, and $C_{\mathbf{s},f} = C_{\mathbf{s},t} = (\frac{1}{s_i - f\omega^j})_{i,j=0}^{n-1}$ and obtain from [20, equations (3.6.5)–(3.6.7)] that

$$C_{\mathbf{s},f} = \sqrt{n} \operatorname{diag}\left(\frac{f^{n-1}}{s_i^n - f^n}\right)_{i=0}^{m-1} V_{\mathbf{s}} \operatorname{diag}(f^{-j})_{j=0}^{n-1}\Omega^H \operatorname{diag}(\omega^{-j})_{j=0}^{n-1}, \qquad (3)$$

$$V_{\mathbf{s}} = \frac{f^{1-n}}{\sqrt{n}} \operatorname{diag}\left(s_i^n - f^n\right)_{i=0}^{m-1} C_{\mathbf{s},f} \operatorname{diag}(\omega^j)_{j=0}^{n-1}\Omega \operatorname{diag}(f^j)_{j=0}^{n-1}, \qquad (4)$$

$$V_{\mathbf{s}}^{-1} = \sqrt{n} \operatorname{diag}(f^{-j})_{j=0}^{n-1}\Omega^H \operatorname{diag}(\omega^{-j})_{j=0}^{n-1} C_{\mathbf{s},f}^{-1} \operatorname{diag}\left(\frac{f^{n-1}}{s_i^n - f^n}\right)_{i=0}^{n-1} \text{ for } m = n. \qquad (5)$$

These equations link Vandermonde matrices $V_{\mathbf{s}}$ and their inverses to the Cauchy matrices with the knot set $\mathcal{S} = \{s_i = f\omega^i, \ i = 0, \dots, n-1\}$ (for $f \neq 0$), which we call *CV matrices* and denote $C_{\mathbf{s},f}$. The equations also link Problems 1 and 2 to Problems 3 and 4.

By means of the transposition of these equations and the substitution of the vectors $\mathbf{t} \to \mathbf{s}$, we link the transposed Vandermonde matrices V_t^T to the matrices $C_{f,\mathbf{t}} = (\frac{1}{f\omega^i - t_j})_{i,j=0}^{n-1}$, for $f \neq 0$, which we call the *CV^T matrices*.

4   HSS Matrices

Definition 1. *(Cf. Figure 1.) Let $M = (M_0 \mid \dots \mid M_{k-1})$ be a $1 \times k$ block matrix with k block columns M_q, each partitioned into a diagonal block Σ_q and a basic neutered block column N_q, $q = 0, \dots, k-1$ (cf. [18, Section 1]). A matrix given with its diagonal blocks is basically ρ-neutered (resp. basically (ϵ, ρ)-neutered) if all its basic neutered block columns have ranks (resp. ϵ-ranks) at most ρ.*

Definition 2. *(Cf. Figure 1.) Fix two positive integers l and q such that $l + q \leq k$ and merge the l basic block columns $M_q, \dots, M_{q+l-1}$, the l diagonal blocks $\Sigma_q, \dots, \Sigma_{q+l-1}$, and the l basic neutered block columns $N_q, \dots, N_{q+l-1}$ into their union $M_{q,l} = M(., \cup_{j=0}^{l-1}\mathcal{C}(\Sigma_{q+j}))$, their diagonal union $\Sigma_{q,l}$, and their neutered union $N_{q,l}$, respectively, such that $\mathcal{R}(\Sigma_{q,l}) = \cup_{j=0}^{l-1}\mathcal{R}(\Sigma_{q+j})$ and the block column $M_{q,l}$ is partitioned into the diagonal union $\Sigma_{q,l}$ and the neutered union $N_{q,l}$.*

Define *recursive merging* of all diagonal blocks $\Sigma_0, \dots, \Sigma_{k-1}$ by a binary tree whose leaves are associated to these blocks and whose every internal vertex is the union of its two children. For every vertex v define the sets $L(v)$ and $R(v)$ of its left and right descendants, respectively. A binary tree is *balanced* if $0 \leq |L(v)| - |R(v)| \leq 1$ for all its vertices v. Such a tree identifies *balanced merging* of its leaves, in our case the diagonal blocks. We can uniquely define a balanced tree with n leaves by removing the $2^{l(n)} - n$ rightmost leaves of the complete binary tree that has $2^{l(n)}$ leaves for $l(n) = \lceil \log_2(n) \rceil$. All leaves of the resulting *heap structure* with n leaves lie in its two lowest levels.

Definition 3. *A block matrix is a* balanced ρ-HSS *matrix if it is basically ρ-neutered at every stage of balanced merging of its diagonal blocks, that is if all neutered unions of its basic neutered block columns defined at every stage of the process of balance merging have ranks at most ρ. By replacing ranks with ϵ-ranks we define* balanced (ϵ, ρ)-HSS *matrices.*

Next we bound $\alpha(M)$ and $\beta(M)$ by adjusting the algorithms of [9, Sections 3 and 4], [29], and [28], devised for a distinct matrix class.

Theorem 3. *Assume a balanced ρ-HSS matrix M with $m_q \times n_q$ diagonal blocks Σ_q, $q = 0, \ldots, k - 1$, having $s = \sum_{q=0}^{k-1} m_q n_q$ entries overall and write $l = \lceil \log_2(k) \rceil$, $m = \sum_{q=0}^{k-1} m_q$, $n = \sum_{q=0}^{k-1} n_q$, $m_+ = \max_{q=0}^{k-1} m_q$, $n_+ = \max_{q=0}^{k-1} n_q$, and $s = \sum_{q=0}^{k-1} m_q n_q$, and so $s \leq \min\{m_+ n, m n_+\}$. Then*

$$\alpha(M) < 2s + (m + 4(m + n)\rho)l. \tag{6}$$

If also $m = n$ and the matrix M is nonsingular, then

$$\beta(M) = O(n_+ s + (n_+^2 + \rho n_+ + l\rho^2)n + (k\rho + n)\rho^2). \tag{7}$$

The bounds (6) and (7) also hold if the matrix M is the transpose of a balanced ρ-HSS matrix and has $n_q \times m_q$ diagonal blocks Σ_q for $q = 0, \ldots, k - 1$.

Corollary 1. *Let $k\rho = O(n)$ and $n_+ + \rho = O(\log^2(n))$ under the assumptions of Theorem 3. Then $\alpha(M) = O((m + n) \log(n))$ and $\beta(M) = O(n \log^3(n))$.*

5   Extension to Tridiagonal Blocks

We wish to approximate CV matrices by balanced ρ-HSS -matrices, but this only works when we extend this class. We are going to do this and to extend Theorem 3 and Corollary 1 accordingly.

Given a block matrix M with diagonal blocks $\Sigma_0, \ldots, \Sigma_{k-1}$, we first glue together its lower and upper block boundaries. Then each diagonal block, including the two extremal blocks Σ_0 and Σ_{k-1}, has exactly two *neighboring blocks* in its basic block column, given by the pair of its subdiagonal and superdiagonal blocks. Define the *tridiagonal blocks* $\Sigma_0^{(c)}, \ldots, \Sigma_{k-1}^{(c)}$ of sizes $m_q^{(c)} \times n_q$ by combining such triples of blocks where $m_q^{(c)} = m_{q-1 \bmod k} + m_q + m_{q+1 \bmod k}$, $q = 0, \ldots, k - 1$. Write $m^{(c)} = \sum_{q=0}^{k-1} m_q^{(c)}$ and note that $m^{(c)} = 3m$ because the number of rows in each of the three block diagonals sums to m. Therefore $s^{(c)} = \sum_{q=0}^{k-1} m_q^{(c)} n_q \leq m^{(c)} n_+ \leq 3m n_+$. The complements of the tridiagonal blocks in their basic block columns are also blocks, which we call *admissible* (cf. [2]). These blocks play the role of basic neutered block columns of Definition 1, which become blocks after gluing the two block boundaries (see Figure 2).

Working with tridiagonal rather than diagonal blocks, we extend our definitions of recursive and balanced merging, unions of blocks, the basically ρ-neutered, and balanced ρ-HSS matrices M (cf. [22] and our Definitions 1, 2, and 3), as well as basically (ϵ, ρ)-neutered and balanced (ϵ, ρ)-HSS matrices M, and we call such matrices *extended*. Let us also extend Theorem 3.

Theorem 4. *Suppose the matrix M in Theorem 3 is replaced by an $m \times n$ extended balanced ρ-HSS matrix $M^{(c)}$, whereas the integer parameters $m^{(c)} = \sum_{q=0^{k-1}} m_q^{(c)} = 3m$ and $s^{(c)} = \sum_{q=0^{k-1}} m_q^{(c)} n_q \leq 3mn_+$ replace m and s, respectively, in bounds (6) on $\alpha(M)$ and (7) on $\beta(M)$. (i) Then bound (6) is extended and (ii) for $m = n$ and under some nondegeneracy assumption for the input matrix bound (7) is extended as well.*

Corollary 2. *Under the assumptions of Theorem 4 suppose that $k\rho = O(n)$ and $n_+ + \rho = O(\log(n))$. Then $\alpha(M) = O((m+n)\log^2(n))$ and $\beta(M) = O(n\log^3(n))$.*

Remark 1. An extended balanced HSS process supporting Theorem 4 can fail numerically without some additional assumptions on the input matrix (see [22]).

6 Approximation of the CV Matrices by Extended ρ-HSS Matrices and Algorithmic Implications

6.1 Small-Rank Approximation of Certain Cauchy Matrices

Definition 4. *(See [9, page 1254].) For a separation bound $\theta < 1$ and a complex separation center c, two complex points s and t are (θ, c)-separated from one another if $|\frac{t-c}{s-c}| \leq \theta$. Two sets of complex numbers S and T are (θ, c)-separated from one another if every two points $s \in S$ and $t \in T$ are (θ, c)-separated from one another. $\delta_{c,S} = \min_{s \in S} |s - c|$ denotes the distance from the center c to the set S.*

Lemma 1. *(See [24] and [9, equation (2.8)].) Suppose two complex values s and t are (θ, c)-separated from one another for $0 \leq \theta < 1$ and a complex center c and write $q = \frac{t-c}{s-c}$, $|q| \leq \theta$. Then for every positive integer ρ we have*

$$\frac{1}{s-t} = \frac{1}{s-c} \sum_{h=0}^{\rho-1} \frac{(t-c)^h}{(s-c)^h} + \frac{q_\rho}{s-c} \text{ where } |q_\rho| = \frac{|q|^\rho}{1-|q|} \leq \frac{\theta^\rho}{1-\theta}. \tag{8}$$

Proof. $\frac{1}{s-t} = \frac{1}{s-c}\frac{1}{1-q}$, $\frac{1}{1-q} = (\sum_{h=0}^{\rho-1} q^h + \sum_{h=\rho}^{\infty} q^h) = (\sum_{h=0}^{\rho-1} q^h + \frac{q^\rho}{1-q})$.

Theorem 5. *(Cf. [9, Section 2.2] and [2].) Suppose two sets of $2n$ distinct complex numbers $S = \{s_0, \ldots, s_{m-1}\}$ and $T = \{t_0, \ldots, t_{n-1}\}$ are (θ, c)-separated from one another for $0 < \theta < 1$ and a global complex center c. Define the Cauchy matrix $C = (\frac{1}{s_i-t_j})_{i,j=0}^{m-1,n-1}$ and write $\delta = \delta_{c,S} = \min_{i=0}^{m-1} |s_i - c|$ (cf. Definition 4). Fix a positive integer ρ and define the $m \times \rho$ matrix $F = (1/(s_i - c)^{\nu+1})_{i,\nu=0}^{m-1,\rho-1}$ and the $n \times \rho$ matrix $G = ((t_j - c)^\nu)_{j,\nu=0}^{n-1,\rho-1}$. (We can compute these matrices by using $(m + n)\rho + m$ arithmetic operations.) Then*

$$C = FG^T + E, \quad |E| \leq \frac{\theta^\rho}{(1-\theta)\delta}. \tag{9}$$

Proof. Apply (8) for $s = s_i$, $t = t_j$, and all pairs (i, j) to deduce (9).

Remark 2. Assume an $m \times n$ Cauchy matrix $C = (\frac{1}{s_i - t_j})_{i,j=0}^{m-1,n-1}$ with $m + n$ distinct knots $s_0, \ldots, s_{m-1}, t_0, \ldots, t_{n-1}$. Then $\mathrm{rank}(C) = \min\{m, n\}$ (cf. Theorem 2). Further assume that the sets $\mathcal{S} = \{s_0, \ldots, s_{m-1}\}$ and $\mathcal{T} = \{t_0, \ldots, t_{n-1}\}$ are (θ, c)-separated from one another for a global complex center c and $0 < \theta < 1$ such that the value $(1 - \theta)\delta/\sqrt{mn}$ is not small. Then by virtue of the theorem the matrix C, having full rank, can be closely approximated by a matrix FG^T of a smaller rank $\rho < \min\{m, n\}$, and therefore is ill conditioned. Furthermore if we have such (θ, c)-separation just for a $k \times l$ submatrix $C_{k,l}$ of the matrix C (this implies that $\mathrm{nrank}(C_{k,l}) \leq \rho$), then it follows that $\mathrm{nrank}(C) \leq m - k + n - l + \rho$. Consequently if $m - k + n - l + \rho < \min\{m, n\}$, then again the matrix C is ill conditioned. This class of ill conditioned Cauchy matrices contains a large subclass of CV and CV^T matrices. In particular a CV matrix is ill conditioned if all its knots s_i or all knots s_i of its submatrix of a large size lie far enough from the unit circle $\{z : |z| = 1\}$, because in this case the origin serves as a global center for the matrix or the submatrix.

6.2   Block Partition of a Cauchy Matrix

Generally neither CV matrix nor its submatrices of a large size have global separation centers. So we approximate a CV matrix by an extended balanced ρ-HSS matrix for a bounded integer ρ rather than by a low-rank matrix. We first fix a reasonably large integer k and then partition the complex plane into k congruent sectors sharing the origin 0 to induce a *uniform k-partition* of the knot sets $\mathcal{S}$ and $\mathcal{T}$ and thus a block partition of the associated Cauchy matrix. In the next subsection we specialize such partitions to the case of a CV matrix.

Definition 5. $A(\phi, \phi') = \{z = \exp(\psi\sqrt{-1}) : 0 \leq \phi \leq \psi < \phi' \leq 2\pi\}$ *is the semi-open arc of the unit circle* $\{z : |z| = 1|\}$ *having length* $\phi' - \phi$ *and the endpoints* $\tau = \exp(\phi\sqrt{-1})$ *and* $\tau' = \exp(\phi'\sqrt{-1})$. $\Gamma(\phi, \phi') = \{z = r\exp(\psi\sqrt{-1}) : r \geq 0, \ 0 \leq \phi \leq \psi < \phi' \leq 2\pi\}$ *is the semi-open sector bounded by the two rays from the origin to the two endpoints of the arc.* $\bar{\Gamma}(\phi, \phi')$ *denotes the exterior (that is the complement) of this sector.*

Fix a positive integer l_+, write $k = 2^{l_+}$, $\phi_q = 2q\pi/k$, and $\phi'_q = \phi_{q+1 \bmod k}$, partition the unit circle $\{z : |z| = 1|\}$ by k equally spaced points $\phi_0, \ldots, \phi_{k-1}$ into k semi-open arcs $A_q = A(\phi_q, \phi'_q)$, each of the length $2\pi/k$, and define the semi-open sectors $\Gamma_q = \Gamma(\phi_q, \phi'_q)$ for $q = 0, \ldots, k - 1$. Now assume the polar representation $s_i = |s_i|\exp(\mu_i\sqrt{-1})$ and $t_j = |t_j|\exp(\nu_j\sqrt{-1})$, and reenumerate the knots in the counter-clockwise order of the angles μ_i and ν_j beginning with the sector $\Gamma(\phi_0, \phi'_0)$ and breaking ties arbitrarily. Induce the block partition of a Cauchy matrix $C = (C_{p,q})_{p,q=0}^{k-1}$ and its partition into basic block columns $C = (C_0 \mid \ldots \mid C_{k-1})$ such that $C_{p,q} = \left(\frac{1}{s_i - t_j}\right)_{s_i \in \Gamma_p, t_j \in \Gamma_q}$ and $C_q = \left(\frac{1}{s_i - t_j}\right)_{s_i \in \{0, \ldots, n-1\}, t_j \in \Gamma_q}$ for $p, q = 0, \ldots, k - 1$. Now for every q define the diagonal block $\Sigma_q = C_{q,q}$, its two neighboring blocks $C_{q-1 \bmod k, q}$ and $C_{q+1 \bmod k, q}$, the tridiagonal block $\Sigma_q^{(c)}$ (made up of the block C_q and its two neighbors), and the admissible block $N_q^{(c)}$, which complements the tridiagonal block $\Sigma_q^{(c)}$ in its basic block column C_q.

6.3 $(0.5, c_q)$-Separation of the Extended Diagonal and Admissible Blocks of a CV Matrix

The following lemma can be readily verified (cf. Figure 3).

Lemma 2. *Suppose* $0 \leq \chi \leq \phi \leq \eta < \phi' < \chi' \leq \pi/2$ *and write* $\tau = \exp(\phi\sqrt{-1})$, $c = \exp(\eta\sqrt{-1})$, *and* $\tau' = \exp(\phi'\sqrt{-1})$. *Then* $|c - \tau| = 2\sin((\eta - \phi)/2)$ *and the distance from the point* c *to the sector* $\bar{\Gamma}(\chi, \chi')$ *is equal to* $\sin(\psi)$, *for* $\psi = \min\{\eta - \chi, \chi' - \eta\}$.

Now let C be actually a CV matrix $C_{s,f}$ for a fixed complex f such that $|f| = 1$, and so $t_j = f\omega_k^j$ for $\omega_k = \exp(2\pi\sqrt{-1}/k)$, $j = 0, \ldots, n-1$. In this case all knots t_j are lying on the arcs $\mathcal{A}_q$ and each arc contains $\lceil n/k \rceil$ or $\lfloor n/k \rfloor$ knots. Apply Lemma 2 for $\phi = \phi_q$, $c = c_q$, $\phi' = \phi_q'$, $\chi = 2\phi_q - \phi_q'$, and $\chi' = 2\phi_q' - \phi_q$, and obtain the following results.

Theorem 6. *Assume a uniform k-partition of the knot sets of a CV matrix above for $k \geq 12$. Let Γ_q' denote the union of the sector Γ_q and its two neighbors on both sides, that is $\Gamma_q' = \Gamma_{q-1 \bmod k} \cup \Gamma_q \cup \Gamma_{q+1 \bmod k}$, let $\bar{\Gamma}_q'$ denote its exterior, and let c_q denote the midpoints of the arcs $\mathcal{A}_q = \mathcal{A}(\phi_q, \phi_q')$ for $q = 0, \ldots, k-1$. Then for every q the arc $\mathcal{A}_q$ and the sector $\bar{\Gamma}_q'$ are $(0, c_q)$-separated for $\theta = 2\sin((\phi_q' - \phi_q)/4)/\sin(\phi_q' - \phi_q)$.*

Recall that $x/\sin x \approx 1$ as $x \approx 0$, and therefore $\theta = 2\sin\frac{\phi'-\phi}{4}/\sin(\phi' - \phi) \approx 0.5$ as $\phi' - \phi \approx 0$, and hereafter assume that the integer k is large enough such that $\theta \approx 0.5$. Furthermore observe that for every q the admissible block $N_q^{(c)}$ is defined by the knots t_j lying on the arc $\mathcal{A}_q$ and the knots s_i lying in the sector $\bar{\Gamma}_q'$, apply Theorem 5, and obtain the following result.

Corollary 3. *Assume that for a sufficiently large integer k, $2k < n$, a uniform k-partition of the knot sets of an $m \times n$ CV matrix C defines admissible blocks $N_0^{(c)}, \ldots, N_{k-1}^{(c)}$. Then for every q the block $N_q^{(c)}$ has the ϵ_q-rank at most ρ_q such that bound (9) holds for $\theta \approx 0.5$, $|E| = \epsilon_q$, $\rho = \rho_q$, and $\delta = \min_{i \in \bar{\Gamma}_q'} |s_i - c_q|$.*

6.4 Approximation of a CV matrix by a Balanced ρ-HSS Matrix. The Complexity of Approximate Computations with CV and CV^T Matrices

The angles $2\pi/k$ of the k congruent sectors $\Gamma_0, \ldots, \Gamma_{k-1}$ are recursively doubled in every merging. So Lemma 2 implies that $\delta \leq \delta_h = \sin(3\pi 2^h/k)$ after the hth merging, $h = 1, \ldots, l$. Choose the integers $k = 2^{l_+}$ and $l < l_+$ such that the integer $k/2^l = 2^{l_+ - l}$ is reasonably large, to support separation with parameters θ about 0.5 or less at all stages of recursive merging. Then $\delta_h \approx 3\pi 2^h/k$, and $\delta_{h+1}/\delta_h \approx 2$ for all h. Now Corollary 3 implies the following result.

Theorem 7. *The CV matrix C of Corollary 3 is an extended balanced (ϵ, ρ)-HSS matrix when the values ϵ and ρ are linked by bound (9) for $\theta \approx 0.5$, $|E| = \epsilon$, $\delta = \delta_h \approx 3\pi 2^h/k$, and $h = 0, \ldots, l$.*

Combine Corollary 2 with this theorem applied for $k = 2^{l_+}$ of order $n/\log(n)$, for ρ and $\log(1/\epsilon)$ of order $\log(n)$, and for $l < l_+$ such that the integer $l_+ - l$ is reasonably large (verify that the assumptions of the corollary are satisfied), and obtain the following result.

Theorem 8. *Assume an $m \times n$ CV matrix C and let $\epsilon > 0$ and $\log(1/\epsilon) = O(\log(n))$. Then $\alpha_\epsilon(C) = O((m + n)\log^2(n))$. If in addition $m = n$ and the matrix C is ϵ-approximated by an extended balanced ρ-HSS matrix satisfying certain nondegeneration assumptions (see [22] on details), then $\beta_\epsilon(C) = O(n\log^3(n))$.*

Because of the dual role of the rows and columns in our constructions we can readily extend all our results from CV matrices C to CV^T matrices C^T.

Corollary 4. *The estimates of Theorem 8 also hold for a CV^T matrix C.*

Remark 3. Suppose we extend diagonal blocks to v-diagonal blocks for an odd integer $v > 3$. How would this change our complexity bounds? The separation parameter θ would increase by a factor of v, but the implied decrease of the cost bound would be offset by the increase of the overall numbers of the entries in the diagonal blocks.

7   Extensions and Implementation

Next we employ equations (3)–(5) to extend Theorem 8 to computations with Vandermonde matrices, their transposes, and polynomials.

Theorem 9. *Suppose that we are given two positive integers m and n, a positive ϵ, and a vector $\mathbf{s} = (s_i)_{i=0}^{m-1}$ defining an $m \times n$ Vandermonde matrix $V = V_{\mathbf{s}}$. Write $s_+ = \max_{i=0}^{m-1} |s_i|$ and let $\log(1/\epsilon) = O(\log(m + n) + n\log(s_+))$.*
   (i) Then $\alpha_\epsilon(V) + \alpha_\epsilon(V^T) = O((m + n)(\log^2(m + n) + n\log(s_+))$.
   (ii) Suppose that in addition $m = n$ and for some complex f, $|f| = 1$, the matrix $C_{\mathbf{s},f}$ of equation (3) is approximated by an extended balanced (ϵ, ρ)-HSS matrix satisfying certain nondegeneration assumptions. Then $\beta_\epsilon(V) + \beta_\epsilon(V^T) = O(n\log^3(n))$.
   (iii) The latter bounds on $\alpha_\epsilon(V)$ and $\beta_\epsilon(V)$ can be applied also to the solution of Problems 1 and 2 of Section 3, respectively.

The term $n\log(s_+)$ is dominated and can be removed from the bounds on $\log(1/\epsilon)$ and $\alpha_\epsilon(V) + \alpha_\epsilon(V^T)$ when $s_+ = 1 + O(\frac{\log^2(m+n)}{n})$.
   Various extensions to computations with the more general class of Cauchy-like matrices and with rational functions are covered in [21] and [22], whereas the recipes in [9], [26], [29], and [28] simplify the implementation of the proposed

algorithms dramatically. In particular to implement our algorithms one can compute the centers c_q and the admissible blocks $\widehat{N}_q$ of bounded ranks throughout the merging process, but one can avoid a large part of these computations by following the papers [9], [26], [29], and [28]. They bypass the computation of the centers c_q and immediately compute the HSS generators for the admissible blocks $\widehat{N}_q$, defined by HSS trees. The length (size) of the generators at every merging stage (represented by a fixed level of the tree) can be chosen equal to the available upper bound on the numerical ranks of these blocks or can be adapted empirically.

8   Conclusions

The papers [18], [9], [29], and [28] combine the advanced FMM/HSS techniques with a transformation of matrix structures (traced back to [19]) to devise numerically stable algorithms that compute approximate solution of Toeplitz, Hankel, Toeplitz-like, and Hankel-like linear systems of equations in nearly linear arithmetic time (versus cubic time of the classical numerical algorithms). We yield similar results for multiplication of Vandermonde and Cauchy matrices by a vector and the solution of linear systems of equations with these matrices (with the extensions to polynomial and rational evaluation and interpolation). The resulting decrease of the running time of the known approximation algorithms is by order of magnitude, from quadratic to nearly linear. Our study provides new insight into the subject and the background for further advances in [21], which include the extension of our results to Cauchy-like matrices and further acceleration of the known approximation algorithms in the case of Toeplitz inputs. The FMM can help decrease similarly our cost bound (6) (cf. [2]).

APPENDIX: Three Figures

In Figures 1 and 2 we mark by black color the diagonal blocks and by dark grey color the basic neutered block columns.

In Figure 1 the pairs of smaller diagonal blocks (marked by grey color) are merged into their diagonal unions, each made up of four smaller blocks, marked by grey and black colors.

In Figure 2 admissible blocks are shown by grey color, each grey diagonal block has two black neighboring blocks, and the triples of grey and black blocks form tridiagonal blocks.

In Figure 3 we show an arc of the unit circle $\{z : |z| = 1|\}$ and the five line intervals $[0, \tau]$, $[0, c]$, $[0, \tau']$, $[\tau, c]$, and $[c, \tau]$. We also show the two line intervals bounding the intersection of the sector $\Gamma(\psi, \psi')$ and the unit disc $D(0, 1)$ as well as the two perpendiculars from the center c onto these two bounding line intervals.

References

1. Bracewell, R.: The Fourier Transform and Its Applications, 3rd edn. McGraw-Hill, New York (1999)
2. Börm, S.: Efficient Numerical Methods for Non-local Operators: $\mathcal{H}^2$-Matrix Compression, Algorithms and Analysis. European Math. Society (2010)
3. Bella, T., Eidelman, Y., Gohberg, I., Olshevsky, V.: Computations with Quasiseparable Polynomials and Matrices. Theoretical Computer Science 409(2), 158–179 (2008)
4. Bini, D.A., Fiorentino, G.: Design, Analysis, and Implementation of a Multiprecision Polynomial Rootfinder. Numer. Algs. 23, 127–173 (2000)
5. Bini, D., Pan, V.Y.: Polynomial and Matrix Computations, Volume 1: Fundamental Algorithms. Birkhäuser, Boston (1994)
6. Barba, L.A., Yokota, R.: How Will the Fast Multipole Method Fare in Exascale Era? SIAM News 46(6), 1–3 (2013)
7. Chandrasekaran, S., Dewilde, P., Gu, M., Lyons, W., Pals, T.: A Fast Solver for HSS Representations via Sparse Matrices. SIAM J. Matrix Anal. Appl. 29(1), 67–81 (2006)
8. Carrier, J., Greengard, L., Rokhlin, V.: A Fast Adaptive Algorithm for Particle Simulation. SIAM J. Scientific Computing 9, 669–686 (1998)
9. Chandrasekaran, S., Gu, M., Sun, X., Xia, J., Zhu, J.: A Superfast Algorithm for Toeplitz Systems of Linear Equations. SIAM J. Matrix Anal. Appl. 29, 1247–1266 (2007)
10. Dutt, A., Gu, M., Rokhlin, V.: Fast Algorithms for Polynomial Interpolation, Integration, and Differentiation. SIAM Journal on Numerical Analysis 33(5), 1689–1711 (1996)
11. Dewilde, P., van der Veen, A.: Time-Varying Systems and Computations. Kluwer Academic Publishers, Dordrecht (1998)
12. Eidelman, Y., Gohberg, I.: A Modification of the Dewilde–van der Veen Method for Inversion of Finite Structured Matrices. Linear Algebra and Its Applications 343, 419–450 (2002)
13. Eidelman, Y., Gohberg, I., Haimovici, I.: Separable Type Representations of Matrices and Fast Algorithms. Birkhäuser (2013)
14. Gohberg, I., Kailath, T., Olshevsky, V.: Fast Gaussian Elimination with Partial Pivoting for Matrices with Displacement Structure. Mathematics of Computation 64, 1557–1576 (1995)
15. Greengard, L., Rokhlin, V.: A Fast Algorithm for Particle Simulation. Journal of Computational Physics 73, 325–348 (1987)
16. Gentelman, W., Sande, G.: Fast Fourier Transform for Fun and Profit. Full Joint Comput. Conference 29, 563–578 (1966)
17. Lipton, R.J., Rose, D., Tarjan, R.E.: Generalized Nested Dissection. SIAM J. on Numerical Analysis 16(2), 346–358 (1979)
18. Martinsson, P.G., Rokhlin, V., Tygert, M.: A Fast Algorithm for the Inversion of Toeplitz Matrices. Comput. Math. Appl. 50, 741–752 (2005)
19. Pan, V.Y.: On Computations with Dense Structured Matrices, Math. of Computation, 55(191), 179–190 (1990); Also in Proc. Intern. Symposium on Symbolic and Algebraic Computation (ISSAC 1989), 34–42. ACM Press, New York (1989)
20. Pan, V.Y.: Structured Matrices and Polynomials: Unified Superfast Algorithms. Birkhäuser/Springer, Boston/New York (2001)

21. Pan, V.Y.: Transformations of Matrix Structures Work Again, accepted by Linear Algebra and Its Applications and available in arxiv:1311.3729[math.NA]
22. Pan, V.Y.: Fast Approximation Algorithms for Computations with Cauchy Matrices and Extensions, in Tech. Report TR 2014005, PhD Program in Comp. Sci., Graduate Center, CUNY (2014),
http://tr.cs.gc.cuny.edu/tr/techreport.php?id=469
23. Pan, V.Y., Reif, J.: Fast and Efficient Parallel Solution of Sparse Linear Systems. SIAM J. on Computing 22(6), 1227–1250 (1993)
24. Rokhlin, V.: Rapid Solution of Integral Equations of Classical Potential Theory. Journal of Computational Physics 60, 187–207 (1985)
25. Vandebril, R., Van Barel, M., Mastronardi, N.: Matrix Computations and Semiseparable Matrices: Linear Systems, vol. 1. The Johns Hopkins University Press, Baltimore (2007)
26. Xia, J.: On the Complexity of Some Hierarchical Structured Matrix Algorithms. SIAM J. Matrix Anal. Appl. 33, 388–410 (2012)
27. Xia, J.: Randomized Sparse Direct Solvers. SIAM J. Matrix Anal. Appl. 34, 197–227 (2013)
28. Xia, J., Xi, Y., Cauley, S., Balakrishnan, V.: Superfast and Stable Structured Solvers for Toeplitz Least Squares via Randomized Sampling. SIAM J. Matrix Anal. and Applications 35, 44–72 (2014)
29. Xia, J., Xi, Y., Gu, M.: A Superfast Structured Solver for Toeplitz Linear Systems via Randomized Sampling. SIAM J. Matrix Anal. Appl. 33, 837–858 (2012)

First-Order Logic on CPDA Graphs

Paweł Parys*

University of Warsaw, Warsaw, Poland
parys@mimuw.edu.pl

Abstract. We contribute to the question about decidability of first-order logic on configuration graphs of collapsible pushdown automata. Our first result is decidability of existential FO sentences on configuration graphs (and their ε-closures) of collapsible pushdown automata of order 3, restricted to reachable configurations. Our second result is undecidability of the whole first-order logic on configuration graphs which are not restricted to reachable configurations, but are restricted to constructible stacks. Our third result is decidability of first-order logic on configuration graphs (for arbitrary order of automata) which are not restricted to reachable configurations nor to constructible stacks, under an alternative definition of stacks, called annotated stacks.

1 Introduction

Already in the 70's, Maslov [1, 2] generalized the concept of pushdown automata to higher-order pushdown automata (n-PDA) by allowing the stack to contain other stacks rather than just atomic elements. In the last decade, renewed interest in these automata has arisen. They are now studied not only as acceptors of string languages, but also as generators of graphs and trees. Knapik et al. [3] showed that trees generated by deterministic n-PDA coincide with trees generated by *safe* order-n recursion schemes (safety is a syntactic restriction on the recursion scheme). Driven by the question of whether safety implies a semantical restriction to recursion schemes (which was recently proven [4, 5]), Hague et al. [6] extended the model of n-PDA to order-n collapsible pushdown automata (n-CPDA) by introducing a new stack operation called collapse (earlier, panic automata [7] were introduced for order 2), and proved that trees generated by n-CPDA coincide with trees generated by all order-n recursion schemes.

In this paper we concentrate on configuration graphs of these automata. In particular we consider their ε-closures, whose edges consist of an unbounded number of transitions rather than just single steps. The ε-closures of n-PDA graphs form precisely the Caucal hierarchy [8–10], which is defined independently in terms of MSO-interpretations and graph unfoldings. These results imply that the graphs have decidable MSO theory, and invite the question about decidability of logics in ε-closures of n-CPDA graphs.

---

* The author holds a post-doctoral position supported by Warsaw Center of Mathematics and Computer Science. Work supported by the National Science Center (decision DEC-2012/07/D/ST6/02443).

E.A. Hirsch et al. (Eds.): CSR 2014, LNCS 8476, pp. 300–313, 2014.

Unfortunately there is even a 2-CPDA graph that has undecidable MSO theory [6]. Kartzow showed that the ε-closures of 2-CPDA graphs are tree automatic [11], thus they have decidable first-order theory. This topic was further investigated by Broadbent [12–15]. He proved that for order 3 (and higher) the FO theory starts to be undecidable. This can be made more precise. Let n_m-CPDA denote an n-CPDA in which we allow collapse links only of one order m. First-order theory is undecidable already on:

- n_m-CPDA graphs restricted to reachable configurations,[1] when $n \geq 3$, and $3 \leq m \leq n$, and the formula is Σ_2, and
- n_m-CPDA graphs restricted to reachable configurations,[1] when $n \geq 4$, and $2 \leq m \leq n - 2$, and the formula is Σ_1, and
- ε-closures[2] of 3_2-CPDA graphs, when the formula is Σ_2, and
- 3 CPDA graphs not restricted to reachable configurations (nor to stacks which are constructible from the empty one by a sequence of stack operation).

On the other side, Broadbent gives some small decidability results:

- for $n = 2$, FO is decidable even when extended by transitive closures of quantifier free formulae;
- FO is decidable on 3_2-CPDA graphs restricted to reachable configurations;
- Σ_1 formulae are decidable on ε-closures of n_n-CPDA graphs (for each n), and of 3_2-CPDA graphs.

In the current paper we complement this picture by three new results (answering questions stated by Broadbent). First, we prove that the existential (Σ_1) FO sentences are decidable on ε-closures of 3-CPDA graphs. This is almost proved in [15]: it holds under the assumption that the 3-CPDA is *luminous*, which means that after removing all order-3 collapse links from two different reachable configurations, they are still different (that is, the targets of such links are uniquely determined by the structure of the stack). We prove that each 3-CPDA can be turned into an equivalent luminous one. The question whether Σ_1 formulae are decidable for n_{n-1}-CPDA and $n_{n,n-1}$-CPDA (allowing links of orders n and $n - 1$) where $n \geq 4$, both with and without ε-closure, remains open.

Second, we prove (contrarily to the Broadbent's conjecture) that first-order logic is undecidable on 4-CPDA graphs not restricted to reachable configurations, but restricted to stacks constructible from the empty one by a sequence of stack operations (although not necessarily ever constructed by the particular CPDA in question). Our reduction is similar to the one showing undecidability of 3-CPDA graphs not restricted to reachable configurations nor to constructible stacks.

Third, we prove that first-order logic is decidable (for each n) on n-CPDA graphs not restricted to reachable configurations nor to constructible stacks, when stacks are represented as *annotated stacks*. This is an alternative representation of stacks of n-CPDA (defined independently in [16] and [17]), where

---

[1] Thus for their ε-closures as well.
[2] For ε-closures, it does not change anything whether we restrict to reachable configurations or not.

in an atomic element, instead of an order-k link, we keep an order-k stack; the collapse operation simply recalls this stack stored in the topmost element. In the constructible case, annotated and CPDA stacks amount to the same thing (although the annotated variant offers some conveniences in expressing certain proofs), but in the unconstructible case there is an important difference. Whilst with an unconstructible CPDA stack each link is constrained to point to some stack below its source, in an annotated stack it can point to an arbitrary stack, completely unrelated to the original one. This shows up when we go back through a pop edge: in the classical case links in the appended stack point (potentially anywhere) inside our original stack, so we can use them to inspect any place in the stack. On the other hand, in the annotated case we can append an arbitrary stack, which does not give us any new information: in first-order logic we can refer only locally to some symbols near the top of the stack.

2 Preliminaries

We give a standard definition of an n-CPDA, using the "annotated stack" representation of stacks. We choose this representation because of Section 5, in which we talk about all configurations with such stacks. For Sections 3 and 4 we could choose the standard representation (with links as numbers) as well.

Given a number n (the order of the CPDA) and a stack alphabet Γ, we define the set of stacks as the smallest set satisfying the following. If $1 \leq k \leq n$ and $s_1, s_2, \ldots, s_m$ for $m \geq 1$ are $(k-1, n)$-stacks, then the sequence $[s_1, s_2, \ldots, s_m]$ is a (k, n)-stack. If $a \in \Gamma$, and $1 \leq k \leq n$, and s is a (k, n)-stack or $s = []$ (the "empty stack", which, according to our definition, is not a stack), then (a, k, s) is a $(0, n)$-stack. We sometimes use "k-stack" instead of "(k, n)-stack" when n is clear from the context or meaningless.

A 0-stack (a, l, t) is also called an *atom*; it has label $\mathsf{lb}((a, l, t)) := a$ and link t of order l. In a k-stack $s = [s_1, s_2, \ldots, s_m]$, the top of the stack is on the right. We define $|s| := m$, called the *height* of s, and $\mathsf{pop}(s) := [s_1, \ldots, s_{m-1}]$ (which is equal to $[]$ if $m = 1$). For $0 \leq i \leq k$, $\mathsf{top}^i(s)$ denotes the topmost i-stack of s.

An n-CPDA has the following operations on an (n, n)-stack s:

- pop^k, where $1 \leq k \leq n$, removes the topmost $(k-1)$-stack (undefined when $|\mathsf{top}^k(s)| = 1$);
- $\mathsf{push}^1_{a,l}$, where $1 \leq l \leq n$ and $a \in \Gamma$, pushes on the top of the topmost 1-stack the atom $(a, l, \mathsf{pop}(\mathsf{top}^l(s)))$;
- push^k, where $2 \leq k \leq n$, duplicates the topmost $(k-1)$-stack inside the topmost k-stack;
- $\mathsf{collapse}$, when $\mathsf{top}^0(s) = (a, l, t)$, replaces the topmost l-stack by t (undefined when $t = []$);
- rew_a, where $a \in \Gamma$, replaces the topmost atom (b, l, t) by (a, l, t).

Denote the set of all these operations as $\Theta^n(\Gamma)$. Operation rew_a is not always present in definitions of CPDA, but we add it following [15].

A *position* is an n-tuple $x = (p_n, \ldots, p_1)$ of natural numbers. The atom at position x in an n-stack s is the p_1-th 0-stack in the p_2-th 1-stack in ... in the p_n-th $(n-1)$-stack of s. We say that x is a position of s, if such atom exists. For an n-stack s and a position x in s, we define $s_{\leq x}$ as the stack obtained from s by a sequence of pop operations, in which the topmost atom is at position x.

An (n,n)-stack s is called *constructible* if it can be obtained by a sequence of operations in $\Theta^n(\Gamma)$ from a stack with only one atom $(a, 1, [])$ for some $a \in \Gamma$. It is not difficult to see that when restricted to constructible stacks, our definition of stacks coincides with the classical one.

Proposition 1. *Let s be a constructible n-stack, and x a position of an atom (a, l, t) in s. Then t is a proper prefix of $\mathsf{top}^l(s_{\leq x})$, that is, $t = [t_1, \ldots, t_m]$ and $\mathsf{top}^l(s_{\leq x}) = [t_1, \ldots, t_{m'}]$ with $m < m'$.*

An n-CPDA $\mathcal{A}$ is a tuple $(\Sigma, \Pi, Q, q_0, \Gamma, \bot_0, \Delta, \Lambda)$, where Σ is a finite set of transition labels; Π is a finite set of configuration labels; Q is a finite set of control states containing the initial state q_0; Γ is a finite stack alphabet containing the initial stack symbol $\bot_0$; $\Delta \subseteq Q \times \Gamma \times \Sigma \times \Theta^n(\Gamma) \times Q$ is a transition relation; $\Lambda \subseteq Q \times \Gamma \times \Pi$ is a predicate relation.

A configuration of $\mathcal{A}$ is a pair (q, s) where q is a control state and s is an (n,n)-stack. Such a configuration satisfies a predicate $b \in \Pi$ just in case $(q, \mathsf{lb}(\mathsf{top}^0(s)), b) \in \Lambda$. For $c \in \Sigma$, we say that $\mathcal{A}$ can c-transition from (q, s) to $(q', \theta(s))$, written $(q, s) \xrightarrow{c} (q', \theta(s))$, if and only if $(q, \mathsf{lb}(\mathsf{top}^0(s)), c, \theta, q') \in \Delta$. For a language L over Σ we write $(q, s) \xrightarrow{L} (q', s')$ when (q', s') can be reached from (q, s) by a sequence of transitions such that the word of their labels is in L. The initial configuration of $\mathcal{A}$ is $(q_0, \bot)$, where $\bot$ is the stack containing only one atom which is $(\bot_0, 1, [])$.

We define three graphs with Π-labelled nodes and Σ-labelled directed edges. The graph $\mathcal{G}^{ano}(\mathcal{A})$ has as nodes all configurations, $\mathcal{G}^{con}(\mathcal{A})$ only configurations (q, s) in which s is constructible, and $\mathcal{G}(\mathcal{A})$ only configurations (q, s) such that $(q_0, \bot) \xrightarrow{\Sigma^*} (q, s)$. In all cases we have a c-labelled edge from (q, s) to (q', s') when $(q, s) \xrightarrow{c} (q', s')$. Assuming that $\varepsilon \in \Sigma$, we can define the ε-closure of a graph $\mathcal{G}$: it contains only those nodes of $\mathcal{G}$ which have some incoming edge not labeled by ε, and two nodes are connected by a c-labelled edge (where $c \neq \varepsilon$) when in $\mathcal{G}$ they are related by $\xrightarrow{\varepsilon^* c}$. We denote the ε-closure of $\mathcal{G}(\mathcal{A})$ as $\mathcal{G}_{/\varepsilon}(\mathcal{A})$.

We consider first-order logic (FO) on graphs as it is standardly defined, with a unary predicate for each symbol in Π and a binary relation for each symbol in Σ, together with a binary equality symbol. A formula is Σ_1, if it is of the form $\exists x_1 \ldots \exists x_k . \varphi$, where φ is without quantifiers.

3   Luminosity for 3-CPDA

The goal of this section is to prove the following theorem.

Theorem 2. *Given a Σ_1 first-order sentence φ and a 3-CPDA $\mathcal{A}$, it is decidable whether φ holds in $\mathcal{G}_{/\varepsilon}(\mathcal{A})$.*

In [15] (Theorem 5, and the comment below) this is proven under the restriction to 3-CPDA $\mathcal{A}$ which are luminous. It remains to show that each 3-CPDA $\mathcal{A}$ can be turned into a luminous 3-CPDA $\mathcal{A}'$ for which $\mathcal{G}_{/\varepsilon}(\mathcal{A}) = \mathcal{G}_{/\varepsilon}(\mathcal{A}')$.

Let us recall the definition of luminosity. For an (n, n)-stack s, we write $stripln(s)$ to denote the (n, n)-stack that results from deleting all order-n links from s (that is, changing atoms (a, n, p) into $(a, n, [])$; of course we perform this stripping also inside all links). An n-CPDA $\mathcal{A}$ is *luminous* whenever for every two configurations (q, s), (q', s') in the ε-closure with $stripln(s) = stripln(s')$ it holds $s = s'$.

For example, the two 2-stacks

$$[[(a, 1, []), (b, 1, [])], [(a, 1, []), (b, 2, s_1)], [(a, 1, []), (b, 2, s_1)]]] \qquad \text{and}$$
$$[[(a, 1, []), (b, 1, [])], [(a, 1, []), (b, 2, s_1)], [(a, 1, []), (b, 2, s_2)]]]$$

with $s_1 = [[(a, 1, []), (b, 1, [])]]$ and $s_2 = [[(a, 1, []), (b, 1, [])], [(a, 1, []), (b, 2, s_1)]]$ become identical if the links are removed. One has to add extra annotations to the stack to tell them apart without links.

We explain briefly why luminosity is needed in the decidability proof in [15]. The proof reduces the order of the CPDA by one (a configuration of an n-CPDA is represented as a sequence of configurations in an $(n-1)$-CPDA), at the cost of creating a more complicated formula. This reduction allows to deal with the operational aspect of links (that is, with the collapse operation). However, there is also the problem of preserving identities, to which first-order logic is sensitive. For this reason, the reduction would be incorrect, if by removing links from two different configurations, suddenly they would become equal.

Let us emphasize that we are not trying to simulate the operational behavior of links in a 3-CPDA after removing them. We only want to construct another 3-CPDA with the same $\mathcal{G}_{/\varepsilon}$, which still uses links of order-3, but such that $stripln(s) = stripln(s')$ implies $s = s'$.

Our construction is quite similar to that from [15] (which works for such n-CPDA which only have links of order n). The key idea which allows to extend it to 3-CPDA which also have links of order 2, is to properly assign the value of "generation" (see below) to atoms with links of order 2.

Fix a 3-CPDA $\mathcal{A}$ with a stack alphabet Γ. W.l.o.g. we assume that $\mathcal{A}$ "knows" what is the link order in each atom, and that it does not perform collapse on links of order 1. We will construct a luminous 3-CPDA $\mathcal{A}'$ with stack alphabet

$$\Gamma' = \Gamma \times \{1{>}, 1{=}, 1{<}\} \times \{2{>}, 2{=}, 2{<}, \neg 2\} \times \{3{\geq}, 3{<}, \neg 3\}.$$

To obtain luminosity, it would be enough to mark for each atom (in particular for atoms with links of order 3), whether it was created at its position, or copied from the 1-stack below, or copied from the 2-stack below. Of course we cannot do this for each atom independently, since when a whole stack is copied, we cannot change markers in all its atoms; thus some markers are needed also on top of 1-stacks and 2-stacks.

There is an additional difficulty that all markers should be placed as a function of a stack, not depending on how the stack was constructed (otherwise one node in $\mathcal{G}_{/\varepsilon}(\mathcal{A})$ would be transformed into several nodes in $\mathcal{G}_{/\varepsilon}(\mathcal{A}')$). Thus when an atom is created by $\mathsf{push}^1_{a,l}$ we cannot just mark it as created here, since equally well an identical atom could be copied from a stack below. However, an atom with a link pointing to the 3-stack containing all the 2-stacks below cannot be a copy from the previous 2-stack. We can also be sure about this for some atoms with links of order 2, namely those whose link target already contains an atom with such "fresh" link of order 3. For these reasons, for each k-stack s (for $0 \leq k \leq 2$), including $s = []$, we define $gn(s)$, the *generation* of s:

$$gn([]) := 0,$$
$$gn([s_1, \ldots, s_m]) := \max(0, \max_{1 \leq i \leq m} gn(s_i)),$$
$$gn((a, k, t)) := \begin{cases} |t| + 1 & \text{if } k = 3, \\ gn(t) & \text{if } k = 2, \\ -1 & \text{if } k = 1. \end{cases}$$

Intuitively, $gn(s)$ is a lower bound for the height of the 3-stack of the CPDA at the moment when s was last modified (or created). For convenience, the generation of an atom with a link of order 1 is smaller than the generation of any k-stack for $k > 0$, and the generation of any atom with a link of order 3 is greater than the generation of the empty stack.

For each constructible 3-stack s over Γ we define its marked variant $mar(s)$, which is obtained by adding markers at each position x of s as follows.

- Let $i \in \{1, 2\}$ and $r \in \{>, =, <\}$, or $i = 3$ and $r \in \{\geq, <\}$. If x is the topmost position in its $(i-1)$-stack (always true for $i = 1$), we put marker ir at x if

$$gn(\mathsf{pop}(\mathsf{top}^i(s_{<x}))) \ r \ gn(\mathsf{top}^{i-1}(s_{<x})).$$

- Assume that x is not topmost in its 1-stack, and the position directly above it has assigned marker $1<$. Let t be the atom just above x, and let y be the highest position in $s_{\leq x}$ (in the lexicographic order) such that $gn(\mathsf{top}^2(s_{\leq y})) < gn(t)$. We put marker $2r$ at x if

$$gn(\mathsf{pop}(\mathsf{top}^2(s_{\leq y}))) \ r \ gn(\mathsf{top}^1(s_{\leq y}));$$

- If no marker of the form $2r$ (or $3r$) is placed at x, we put there $\neg 2$ (respectively, $\neg 3$).
- Recall that when the atom at x is (a, l, t), then t is a proper prefix of $\mathsf{top}^l(s_{\leq x})$. We attach the markers in t so that this property is preserved, than is in the same way as in $\mathsf{top}^l(s_{\leq x})$.

For example, the marker $2<$ is placed at the top of some 1-stack to say that the generation of this 1-stack is greater than of all the 1-stacks below it, in the same 2-stack.

In the second item, notice that y always will be found, even inside the topmost 2-stack of $s_{\leq x}$. Intuitively, when an atom from a new generation is placed above y, in y we keep his $2r$ marker. This is needed to reproduce the $2r$ marker when

y again becomes the topmost position. Necessarily, the marker from y will be also present at positions x which are copies of y. Notice however that when we remove an atom at position x using pop^1, and then we reproduce an identical atom using $push^1_{a,k}$, the $2r$ marker has to be written there again (*mar* should be a function of the stack). For this reason the x containing the $2r$ marker from y is not necessarily a copy of y: we store the marker in the highest atom below an atom from the higher generation. See Figure 1 for an example.

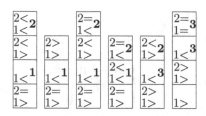

Fig. 1. An example 2-stack (one out of many in a 3-stack). It grows from left to right. We indicate all $1r$ and $2r$ markers, as well as the generation of atoms (bold; no number for generation -1). To calculate the $2<$ marker at positions $(1,3)$, $(3,3)$, and $(4,2)$ we have used position $(1,3)$ as y. Observe the atom of generation 2 above an atom of generation 3; this is possible for an atom with a link of order 2.

The key property is that the markers can be updated by a CPDA. We will say that a CPDA *defines a path* if from each configuration there is at most one transition available.

Lemma 3. *Let $\theta \in \Theta^n(\Gamma)$ be a stack operation. Then there exists a 3-CPDA $\mathcal{A}_\theta$ which defines a path, with stack alphabet Γ' and two distinguished states q_0, q_1, such that for each constructible 3-stack s:*

- *if $\theta(s)$ exists, then there is a unique configuration with state q_1 reachable by $\mathcal{A}_\theta$ from $(q_0, mar(s))$; the stack in this configuration is $mar(\theta(s))$;*
- *if θ cannot be applied to s, no configuration with state q_1 is reachable by $\mathcal{A}_\theta$ from $(q_0, mar(s))$.*

Additionally, $\mathcal{A}_\theta$ does not use the collapse operation for $\theta \neq$ collapse.

Proof (sketch). This is a tedious case analysis. In most cases we just have to apply a local change of markers. For a push, we update markers in the previously topmost atom (depending on markers which were previously there), then we perform the push, and then we update markers in the new topmost atom. For pop^k or collapse, we perform this operation, and then we update markers in the atom which became topmost, depending on markers in this atom, and in the atom which was topmost previously.

There is one exception from this schema, during such $push^1_{a,k}$ operation which increases the generation of the topmost 1-stack, but not of the topmost 2-stack. In this situation in the previously topmost atom we should place a $2r$ marker, the same as in the atom just below the bottommost atom having the highest generation in the 2-stack. This information is not available locally; to find this atom (and the marker in it), we copy the topmost 2-stack (push^3), we destructively search for this atom (which is easy using the markers), and then we remove the garbage using pop^3. □

Lemma 4. *Let s and s' be constructible 3-stacks such that $stripln(mar(s)) = stripln(mar(s'))$. Then $s = s'$.*

Proof (sketch). We prove this by induction, so we can assume that s is equal to s' everywhere except its topmost atom. Only the situation when $\mathsf{top}^0(s)$ has a link of order 3 is nontrivial; then we have to prove that the generation of the topmost atoms of s and s' is the same. We notice that $gn(\mathsf{top}^0(s)) = gn(\mathsf{top}^1(s))$. We have several cases. When the topmost atom is marked by $2{=}$, its generation is $gn(\mathsf{pop}(\mathsf{top}^2(s)))$, which is determined by the part below $\mathsf{top}^0(s)$. When it is marked by $2{<}$ and $3{<}$, its generation is $|s|$. When it is marked by $2{<}$ and by $3{\geq}$, this atom was necessarily copied from the 2-stack below (and has the same generation as the corresponding atom there). Finally, when it is marked by $2{>}$, this atom was necessarily copied from the 1-stack below (here, or in some 2-stack below). □

Having these two lemmas it is easy to conclude. We construct $\mathcal{A}'$ from $\mathcal{A}$ as follows. The initial stack of $\mathcal{A}'$ should be $mar(\bot)$. Whenever $\mathcal{A}$ wants to apply a c-labelled transition with operation θ and final state q, $\mathcal{A}'$ simulates the automaton $\mathcal{A}_\theta$ using ε-transitions, and then changes state to q using a c-labelled transition. Then $\mathcal{G}_{/\varepsilon}(\mathcal{A})$ is isomorphic to $\mathcal{G}_{/\varepsilon}(\mathcal{A}')$: a configuration (q, s) corresponds to $(q, mar(s))$. Moreover, by Lemma 4 the CPDA is luminous (notice that the ε-closure contains only configurations with stack of the form $mar(s)$).

4   Unreachable Configurations with Constructible Stack

In this section we prove that the FO theory is undecidable for configuration graphs without the restriction to reachable configurations, but when we allow only constructible stacks (contrarily to the conjecture stated in [15]). On the other hand, in the next section we show decidability when one also allows stacks which are unconstructible. Let us recall that the FO theory is known [15] to be undecidable, if one also allows stacks which are unconstructible, but for the classical definition of stacks (links represented as numbers, pointing to substacks). Our proof goes along a similar line, but additional care is needed to ensure that the stacks used in the reduction are indeed constructible. For this reason we need to use stacks of order 4 (while [15] uses stacks of order 3).

To be precise, we prove our undecidability result for the graph $\mathcal{G}^{con}(\mathcal{A})$, where $\mathcal{A}$ is the 4-CPDA which has a single-letter stack alphabet $\{\star\}$, one state, and for each stack operation θ a θ-labelled transition performing operation θ. Since there is only one state we identify a configuration with the $(4, 4)$-stack it contains.

Theorem 5. *FO is undecidable on $\mathcal{G}^{con}(\mathcal{A})$.*

We reduce from the first-order theory of finite graphs, which is well-known to be undecidable [18]. A finite graph $G = (V, E)$ consists of a finite domain V of nodes over which there is a binary irreflexive and symmetric relation E of edges. We will use the domain of $\mathcal{G}^{con}(\mathcal{A})$ to represent all possible finite graphs.

First we observe that in first-order logic we can determine the order of the link in the topmost atom. That is, for $1 \leq k \leq 4$ we have a formula $link^k(s)$ which is true in configurations s such that $\mathsf{top}^0(s) = (\star, k, t)$ with $t \neq []$. The formulae are defined by

$$link^k(s) := \bigwedge_{1 \leq i < k} \neg link^i(s) \wedge \exists t.(s \xrightarrow{\text{collapse}} t \wedge eq^k(s, t)),$$

where $eq^k(s, t)$ states that s and t differ at most in their topmost k-stacks, that is $eq^4(s, t) := true$, and for $1 \leq k \leq 3$,

$$eq^k(s, t) := \exists u.(s \xrightarrow{\mathsf{pop}^{k+1}} u \wedge t \xrightarrow{\mathsf{pop}^{k+1}} u) \vee (eq^{k+1}(s, t) \wedge \neg \exists u.(s \xrightarrow{\mathsf{pop}^{k+1}} u)).$$

Next, we define two sets of substacks of a 4-stack s which can be easily accessed in FO. The set $vis^4(s)$ contains s and the stacks t for which in $\mathsf{top}^3(s)$ there is the atom $(\star, 4, t)$. The set $vis^3(s)$ contains s and the stacks t for which $\mathsf{pop}(s) = \mathsf{pop}(t)$ and in $\mathsf{top}^2(s)$ there is the atom $(\star, 3, \mathsf{top}^3(t))$. When s is constructible, the property that $t \in vis^k(s)$ (for $k \in \{3, 4\}$) can be expressed by the FO formula

$$vis^k(s, t) := \exists u.(u \xrightarrow{\mathsf{pop}^k} s \wedge link^k(u) \wedge u \xrightarrow{\text{collapse}} t).$$

To every constructible 4-stack s we assign a finite graph $G(s)$ as follows. Its nodes are $V := vis^4(s)$. Two nodes $t, u \in V$ are connected by an edge when $\mathsf{top}^0(v) = (\star, 4, u)$ for some $v \in vis^3(t)$, or $\mathsf{top}^0(v) = (\star, 4, t)$ for some $v \in vis^3(u)$.

Lemma 6. *For each non-empty finite graph G there exists a constructible $(4, 4)$-stack s_G (in the domain of $\mathcal{G}^{con}(\mathcal{A})$) such that G is isomorphic to $G(s_G)$.*

Proof. Suppose that $G = (V, E)$ where $V = \{1, 2, \ldots, k\}$. The proof is by induction on k. If $k = 1$, as s_G we just take the (constructible) 4-stack consisting of one atom $(\star, 1, [])$. Assume that $k \geq 2$. For $1 \leq i < k$, let G_i be the subgraph of G induced by the subset of nodes $\{1, 2, \ldots, i\}$, and let $s_i := s_{G_i}$ be the stack corresponding to G_i obtained by the induction assumption. We will have $\mathsf{pop}^4(s_G) = s_{k-1}$, and $\mathsf{top}^3(s_G) = t_k$, where 3-stacks t_i for $0 \leq i \leq k$ are defined by induction as follows. We take $t_0 = []$. For $i > 0$ we take take $\mathsf{pop}(t_i) = t_{i-1}$, and the topmost 2-stack of t_i consists of one or two 1-stacks. Its first 1-stack is

$$[(\star, 1, []), (\star, 4, s_1), (\star, 4, s_2), \ldots, (\star, 4, s_{k-1}), (\star, 3, t_0), (\star, 3, t_1), \ldots, (\star, 3, t_{i-1})].$$

If $(i, k) \notin E$ we only have this 1-stack; if $(i, k) \in E$, in $\mathsf{top}^2(t_i)$ we also have the 1-stack

$$[(\star, 1, []), (\star, 4, s_1), (\star, 4, s_2), \ldots, (\star, 4, s_i)].$$

We notice that $vis^4(s_G)$ contains stacks $s_1, s_2, \ldots, s_{k-1}, s_G$, and $vis^3(s_G)$ contains all stacks obtained from s_G by replacing its topmost 3-stack by t_i for some $i \geq 1$. It follows that $G(s_G)$ is isomorphic to G.

It is also easy to see that s_G is constructible. We create it out of s_{k-1} by performing push^4 and appropriately changing the topmost 3-stack. Notice that the bottommost 1-stack of $top^3(s_{k-1})$ starts with $(\star, 1, [])$, $(\star, 4, s_1)$, $(\star, 4, s_2)$, $\ldots$, $(\star, 4, s_{k-2})$. We uncover this prefix using a sequence of pop^i operations. We append $(\star, 4, s_{k-1})$ and $(\star, 3, t_0)$ by push$^1_{\star,4}$ and push$^1_{\star,3}$. If $(1, k) \in E$, we create the second 1-stack using push^2 and a sequence of pop^1. This already gives the first 2-stack. To append each next (i-th) 2-stack, we perform push^3; we remove the second 1-stack if it exists using pop^2; we append $(\star, 3, t_{i-1})$ using push$^1_{\star,3}$; if necessary we create the second 1-stack using push^2 and a sequence of pop^1. $\square$

We have a formula stating that two nodes x, y of $G(s)$ are connected by an edge:

$$E(x, y) := \exists z.(vis^3(x, z) \wedge link^4(z) \wedge z \xrightarrow{\text{collapse}} y) \vee$$
$$\vee \exists z.(vis^3(y, z) \wedge link^4(z) \wedge z \xrightarrow{\text{collapse}} x).$$

Given any sentence φ over finite graphs, we construct a formula $\varphi'(s)$ by replacing all occurrences of the atomic binary predicate xEy with the formula $E(x, y)$ from above, and relativising all quantifiers binding a variable x to $vis^4(s, x)$. Then for each constructible $(4, 4)$-stack s, φ holds in $G(s)$ if and only if $\varphi'(s)$ holds in $\mathcal{G}^{con}(\mathcal{A})$. Thus φ holds in some finite graph if and only if it holds in the empty graph or $\exists s.\varphi'(s)$ holds in $\mathcal{G}^{con}(\mathcal{A})$. This completes the reduction and hence the proof of Theorem 5, since it is trivial to check whether φ holds in the empty graph.

5   Unreachable Configurations with Annotated Stack

In this section we prove decidability of first order logic in the graph of all configurations, not restricted to constructible stacks.

Theorem 7. *Given a first-order sentence φ and a CPDA $\mathcal{A}$, it is decidable whether φ holds in $\mathcal{G}^{ano}(\mathcal{A})$.*

For the rest of the section fix a CPDA $\mathcal{A}$ of order n, with stack alphabet Γ. The key idea of the proof is that an FO formula can inspect only a small topmost part of the stack, and check equality of the parts below. Thus instead of valuating variables into stacks, it is enough to describe how the top of the stack looks like, and which stacks below are equal. When the size of the described top part of the stack is fixed, there are only finitely many such descriptions. For each quantifier in the FO sentence we will be checking all possible descriptions of fixed size (of course the size of the described part has to decrease with each next variable). To formalize this we define generalized stacks.

Consider the following operations on stacks:

- for each $k \in \{1, \ldots, n\}$ operation $\mathsf{first}^k(\cdot)$ which takes a $(k-1)$-stack s and returns the k-stack $[s]$,
- for each $k \in \{1, \ldots, n\}$ operation $\mathsf{app}^k(\cdot, \cdot)$ which takes a k-stack $[s_1, \ldots, s_m]$ and a $(k-1)$-stack s, and returns the k-stack $[s_1, \ldots, s_m, s]$,
- for each $a \in \Gamma$ and $k \in \{1, \ldots, n\}$ operation $\mathsf{cons}(a, k, [])$ (without arguments) which returns the 0-stack $(a, k, [])$,
- for each $a \in \Gamma$ and $k \in \{1, \ldots, n\}$ operation $\mathsf{cons}(a, k, \cdot)$ which takes a k-stack s and returns the 0-stack (a, k, s).

We notice that stacks can be seen as elements of the free multisorted algebra with these operations and no generators (we have $n+1$ sorts, one for each order of stacks). In the proof we need elements of the free multisorted algebra with these operations and some generators: for each sort k we have an infinite set of constants, denoted $x_1^k, x_2^k, \ldots$. Elements of this algebra will be called *generalized stacks*. Thus a generalized stack is a stack in which we have replaced some prefixes of some stacks by constants. Generalized stacks will be denoted by uppercase letters.

For each generalized stack S and each $d \in \mathbb{N}$ we define the set $\mathsf{ts}_{=d}(S)$ of stacks. These are substacks of S which are at "distance" exactly d from the top. The definition is inductive: we take $\mathsf{ts}_{=0}(S) := \{S\}$,

$$\mathsf{ts}_{=1}(S) := \begin{cases} \{T\} & \text{if } S = \mathsf{first}^k(T), \\ \{T, U\} & \text{if } S = \mathsf{app}^k(T, U), \\ \emptyset & \text{if } S = \mathsf{cons}(a, k, []), \\ \{T\} & \text{if } S = \mathsf{cons}(a, k, T), \\ \emptyset & \text{if } S \text{ is a constant,} \end{cases}$$

and $\mathsf{ts}_{=d+1}(S) := \bigcup_{T \in \mathsf{ts}_{=d}(S)} \mathsf{ts}_{=1}(T)$ for $d \geq 1$. Moreover we define $\mathsf{ts}_{\leq d}(S) := \bigcup_{e \leq d} \mathsf{ts}_{=e}(S)$ and for $d \in \mathbb{N} \cup \{\infty\}$, $\mathsf{ts}_{<d}(S) := \bigcup_{e < d} \mathsf{ts}_{=e}(S)$.

A *valuation* is a (partial) function v mapping constants to stacks, preserving the order. Such v can be generalized to a homomorphism $\bar{v}$ mapping generalized stacks to stacks. Obviously, to compute $\bar{v}(S)$ it is enough to define v only on constants appearing in S.

In FO we can also talk about equality of stacks, so we are interested in valuations which applied to different generalized stacks give different stacks. This is described by a relation $\hookrightarrow_d$ which is defined as follows. Let $S_1, \ldots, S_m$ (for $m \geq 0$) be generalized stacks, and $s_1, \ldots, s_m$ stacks, and $d \in \mathbb{N}$. Then we say that $(S_1, \ldots, S_m) \hookrightarrow_d (s_1, \ldots, s_m)$ if there exists a valuation v such that

- $s_i = \bar{v}(S_i)$ for each i, and
- no element of $\bigcup_i \mathsf{ts}_{<d}(S_i)$ is a constant (that is, all constants are at depth at least d), and
- for each $T, U \in \bigcup_i \mathsf{ts}_{\leq d}(S_i)$ such that $\bar{v}(T) = \bar{v}(U)$, it holds $T = U$.

Example 8. Consider the following 2-stack:

$$s := [[(a, 1, []), (b, 1, []), (c, 1, [])], [(a, 1, []), (b, 1, []), (c, 1, [])]].$$

It can be written as:

$$\mathsf{app}^2\Big(\mathsf{first}^2\Big(\mathsf{app}^1(\mathsf{app}^1(\mathsf{first}^1(\mathsf{cons}(a,1,[])),\mathsf{cons}(b,1,[])),\mathsf{cons}(c,1,[]))\Big),$$
$$\mathsf{app}^1(\mathsf{app}^1(\mathsf{first}^1(\mathsf{cons}(a,1,[])),\mathsf{cons}(b,1,[])),\mathsf{cons}(c,1,[]))\Big).$$

It holds

$$(\mathsf{app}^2(\mathsf{first}^2(\mathsf{app}^1(x^1,\mathsf{cons}(c,1,[]))),\mathsf{app}^1(x^1,\mathsf{cons}(c,1,[])))) \hookrightarrow_2 (s),$$

where the valuation maps x^1 into $\mathsf{app}^1(\mathsf{first}^1(\mathsf{cons}(a,1,[])),\mathsf{cons}(b,1,[]))$. On the other hand it does not hold that $(\mathsf{app}^2(\mathsf{first}^2(y^1),\mathsf{app}^1(x^1,\mathsf{cons}(c,1,[])))) \hookrightarrow_2 (s)$; the problem is that the two 1-stacks of s were equal, while they are different in this generalized 2-stack. This shows that we cannot just cut our stack at one, fixed depth, and place constants in all places at this depth. In fact, for some stacks, we need to place some constants exponentially deeper than other constants. As a consequence, our algorithm will be nonelementary (we have to increase d exponentially with each quantifier).

When a formula (having already some generalized stacks assigned to its free variables) starts with a quantifier, as a value of the quantified variable we want to try all possible generalized stacks which are of a special form, as described by the following definition. Let $S_1, \ldots, S_m, S_{m+1}$ (for $m \geq 0$) be generalized stacks, let $d \in \mathbb{N}$, and $d' := d + 2^{d+1}$. We say that S_{m+1} is d-normalized with respect to $(S_1, \ldots, S_m)$ if

- no element of $\mathsf{ts}_{<d}(S_{m+1})$ is a constant, and
- each element of $\mathsf{ts}_{=d}(S_{m+1})$ is
  - a "fresh" constant, i.e. not belonging to $\bigcup_{i\leq m}\mathsf{ts}_{<\infty}(S_i)$, or
  - an element of $\bigcup_{i\leq m}\mathsf{ts}_{\leq d'}(S_i)$, or
  - an element of $\mathsf{ts}_{<d}(S_{m+1})$.

The key point is that for fixed $S_1, \ldots, S_m$ there are only finitely many d-normalized generalized stacks S_{m+1} (up to renaming of fresh constants), so we can try all of them. The next two lemmas say that to consider d-normalized generalized stacks is exactly what we need.

Lemma 9. *Let $S_1, \ldots, S_m$ (for $m \geq 0$) be generalized stacks, let $s_1, \ldots, s_m$ and s_{m+1} be stacks, let $d \in \mathbb{N}$ and $d'' := d + 2^{d+2}$. Assume that $(S_1, \ldots, S_m) \hookrightarrow_{d''} (s_1, \ldots, s_m)$. Then there exists a generalized stack S_{m+1} d-normalized with respect to $(S_1, \ldots, S_m)$ and such that $(S_1, \ldots, S_m, S_{m+1}) \hookrightarrow_d (s_1, \ldots, s_m, s_{m+1})$.*

Proof (sketch). Let $d' := d + 2^{d+1}$ (this is the d' used in the definition in d-normalization, and is smaller than d''). Let v be a valuation witnessing that $(S_1, \ldots, S_m) \hookrightarrow_{d''} (s_1, \ldots, s_m)$, i.e. such that $s_i = \overline{v}(S_i)$ for each $i \leq m$. For each stack $s \in \mathsf{ts}_{\leq d}(s_{m+1})$ we define by induction a generalized stack $\mathsf{repl}(s)$:

– if $s \in \text{ts}_{<d}(s_{m+1})$, we take

$$\text{repl}(s) := \begin{cases} \text{first}^k(\text{repl}(t)) & \text{if } s = \text{first}^k(t), \\ \text{app}^k(\text{repl}(t), \text{repl}(u)) & \text{if } s = \text{app}^k(t, u), \\ \text{cons}(a, k, []) & \text{if } s = \text{cons}(a, k, []), \\ \text{cons}(a, k, \text{repl}(t)) & \text{if } s = \text{cons}(a, k, t); \end{cases}$$

– otherwise, if $s = \overline{v}(S)$ for some $S \in \bigcup_{i \leq m} \text{ts}_{\leq d'}(S_i)$, we take $\text{repl}(s) := S$;
– otherwise, we take $\text{repl}(s) := x_s$, where x_s is a fresh constant.

At the end we take $S_{m+1} := \text{repl}(s_{m+1})$. It remains to check in detail that such S_{m+1} satisfies all parts of the definitions. Notice that there can exist stacks s which are simultaneously in $\text{ts}_{<d}(s_{m+1})$ and $\text{ts}_{=d}(s_{m+1})$ (so it is not true that we apply one of the last two cases to each stack at depth d). $\qquad\square$

Lemma 10. *Let $S_1, \ldots, S_m$ and S_{m+1} (for $m \geq 0$) be generalized stacks, let $s_1, \ldots, s_m$ be stacks, let $d \in \mathbb{N}$ and $d'' := d + 2^{d+2}$. Assume that $(S_1, \ldots, S_m) \hookrightarrow_{d''} (s_1, \ldots, s_m)$, and that S_{m+1} is d-normalized with respect to $(S_1, \ldots, S_m)$. Then there exists a stack s_{m+1} such that $(S_1, \ldots, S_m, S_{m+1}) \hookrightarrow_d (s_1, \ldots, s_m, s_{m+1})$.*

Proof (sketch). It is enough to map the constants appearing in S_{m+1} but not in S_i for $i \leq m$ into "fresh" stacks, such that none of them is a substack of any other nor of any s_i for $i \leq m$ (the latter is easy to obtain by taking these stacks to be bigger than all s_i). $\qquad\square$

We also easily see that the atomic FO formulae can evaluated on the level of generalized stacks related by the $\hookrightarrow_{n+1}$ to the actual stacks.

Lemma 11. *Let S, T be generalized n-stacks, and let s, t be n-stacks. Assume that $(S, T) \hookrightarrow_{n+1} (s, t)$. Then taking as input S and T (even not knowing s and t) one can compute:*

– $\text{lb}(\text{top}^0(s))$,
– *whether $s = t$, and*
– *for any stack operation $\theta \in \Theta^n(\Gamma)$, whether it holds $\theta(s) = t$.*

Using the last three lemmas we can check whether an FO sentence holds in $\mathcal{G}^{ano}(\mathcal{A})$. Indeed, for each quantifier we check all possible generalized stacks which are d-normalized with respect to the previously fixed variables, for big enough d (depending on the quantifier rank of the formula, so that the induction works fine), and we deal with atomic formulae using Lemma 11.

References

1. Maslov, A.N.: The hierarchy of indexed languages of an arbitrary level. Soviet Math. Dokl. 15, 1170–1174 (1974)
2. Maslov, A.N.: Multilevel stack automata. Problems of Information Transmission 12, 38–43 (1976)

3. Knapik, T., Niwiński, D., Urzyczyn, P.: Higher-order pushdown trees are easy. In: Nielsen, M., Engberg, U. (eds.) FOSSACS 2002. LNCS, vol. 2303, pp. 205–222. Springer, Heidelberg (2002)

4. Parys, P.: Collapse operation increases expressive power of deterministic higher order pushdown automata. In: Schwentick, T., Dürr, C. (eds.) STACS. LIPIcs, vol. 9, pp. 603–614. Schloss Dagstuhl, Leibniz-Zentrum fuer Informatik (2011)

5. Parys, P.: On the significance of the collapse operation. In: LICS, pp. 521–530. IEEE (2012)

6. Hague, M., Murawski, A.S., Ong, C.-H.L., Serre, O.: Collapsible pushdown automata and recursion schemes. In: LICS, pp. 452–461. IEEE Computer Society (2008)

7. Knapik, T., Niwiński, D., Urzyczyn, P., Walukiewicz, I.: Unsafe grammars and panic automata. In: Caires, L., Italiano, G.F., Monteiro, L., Palamidessi, C., Yung, M. (eds.) ICALP 2005. LNCS, vol. 3580, pp. 1450–1461. Springer, Heidelberg (2005)

8. Caucal, D.: On infinite terms having a decidable monadic theory. In: Diks, K., Rytter, W. (eds.) MFCS 2002. LNCS, vol. 2420, pp. 165–176. Springer, Heidelberg (2002)

9. Cachat, T.: Higher order pushdown automata, the Caucal hierarchy of graphs and parity games. In: Baeten, J.C.M., Lenstra, J.K., Parrow, J., Woeginger, G.J. (eds.) ICALP 2003. LNCS, vol. 2719, pp. 556–569. Springer, Heidelberg (2003)

10. Carayol, A., Wöhrle, S.: The Caucal hierarchy of infinite graphs in terms of logic and higher-order pushdown automata. In: Pandya, P.K., Radhakrishnan, J. (eds.) FSTTCS 2003. LNCS, vol. 2914, pp. 112–123. Springer, Heidelberg (2003)

11. Kartzow, A.: Collapsible pushdown graphs of level 2 are tree-automatic. Logical Methods in Computer Science 9(1) (2013)

12. Broadbent, C.H.: On collapsible pushdown automata, their graphs and the power of links. PhD thesis, University of Oxford (2011)

13. Broadbent, C.H.: Prefix rewriting for nested-words and collapsible pushdown automata. In: Czumaj, A., Mehlhorn, K., Pitts, A., Wattenhofer, R. (eds.) ICALP 2012, Part II. LNCS, vol. 7392, pp. 153–164. Springer, Heidelberg (2012)

14. Broadbent, C.H.: The limits of decidability for first order logic on CPDA graphs. In: Dürr, C., Wilke, T. (eds.) STACS. LIPIcs, vol. 14, pp. 589–600. Schloss Dagstuhl, Leibniz-Zentrum fuer Informatik (2012)

15. Broadbent, C.H.: On first-order logic and CPDA graphs. Accepted to Theory of Computing Systems

16. Broadbent, C.H., Carayol, A., Hague, M., Serre, O.: A saturation method for collapsible pushdown systems. In: Czumaj, A., Mehlhorn, K., Pitts, A., Wattenhofer, R. (eds.) ICALP 2012, Part II. LNCS, vol. 7392, pp. 165–176. Springer, Heidelberg (2012)

17. Kartzow, A., Parys, P.: Strictness of the collapsible pushdown hierarchy. In: Rovan, B., Sassone, V., Widmayer, P. (eds.) MFCS 2012. LNCS, vol. 7464, pp. 566–577. Springer, Heidelberg (2012)

18. Trachtenbrot, B.: Impossibility of an algorithm for the decision problem in finite classes. Doklady Akad. Nauk. 70, 569–572 (1950)

Recognizing Two-Sided Contexts in Cubic Time

Max Rabkin

Saarland University
max.rabkin@gmail.com

Abstract. Barash and Okhotin ("Grammars with two-sided contexts",
Tech. Rep. 1090, Turku Centre for Computer Science, 2013) recently
introduced conjunctive grammars with two-sided contexts, and gave a
variant of Valiant's algorithm which recognizes the languages they gen-
erate in $O(|G|^2 \cdot n^{3.3727})$ time and $O(|G| \cdot n^2)$ space. We use a new nor-
mal form and techniques from logic programming to improve this to
$O(|G| \cdot n^3)$, without increasing the space usage.

1 Introduction

Barash and Okhotin (2012) introduced grammars with one-sided contexts, extending context-free and conjunctive grammars with quantifiers allowing productions to depend on their context on the left-hand side. They later extended
this to allow context on both sides (Barash and Okhotin, 2013).

This notion of context is purely syntactic, and therefore quite different to formulations of context in terms of rewriting systems, such as the context-sensitive
grammars. To determine whether a substring is matched by a non-terminal in a
grammar with two-sided contexts, one need only examine the string. In a classical context-sensitive, this question does not make sense: one must consider not
only the string itself, but also the appropriate step of a particular derivation.
Derivations in grammars with two-sided contexts, like in context-free grammars,
do not require any notion of time-steps. This means one can draw meaningful
parse trees (actually directed acyclic graphs) for derivations.

In the first paper, Barash and Okhotin gave a cubic-time algorithm for recognizing the languages generated by these grammars which is similar to the
Cocke-Kasami-Younger (CKY) algorithm for context-free languages. In the second, they gave a recognition algorithm in the style of Valiant (1975) which takes
$O(|G| \cdot n^2)$ space and $O(|G|^2 \cdot n^{\omega+1})$ time, where $O(n^\omega)$ is the complexity of
multiplying $n \times n$ boolean matrices. The best known bound is $2 \leq \omega < 2.3727$
(Williams, 2012). Their algorithm works on grammars in *binary normal form*; it
is not known whether this form can be achieved with sub-exponential blow-up.

We give an algorithm which takes only $O(|G| \cdot n^3)$ time. This algorithm is
derived from the definition of two-sided contexts using deduction systems: we
give a normal form (*separated normal form*) for the grammars such that the
corresponding deduction can be efficiently computed using standard techniques.
However, some specialization is required to keep the space usage quadratic. Our

E.A. Hirsch et al. (Eds.): CSR 2014, LNCS 8476, pp. 314–324, 2014.

normal form allows ε-rules and unit rules, and so can be achieved with only linear increase in the size of the grammar.

The speed-up can be seen as arising from the improved resolution of dependencies between sub-problems. In the context-free case, the syntactic properties of a string (i.e., the non-terminals which match it) depend only on its substrings. With left contexts, the properties of a string-in-context $u\langle v \rangle w$ depend on the substrings of uv. In either case, we can resolve the dependencies by computing the properties of substrings in a fixed order. As Barash and Okhotin (2013) noted, the dependencies between substrings are more complicated in the case of grammars with two-sided contexts; and therefore their algorithm requires $O(|N| \cdot n)$ passes to ensure all properties have been recognized, where N is the set of non-terminals in the grammar. In our algorithm the dependencies are resolved implicitly, so only a single pass is required.

2   Definitions

Informally, a string in context is simply a string considered as a part of a larger string. Formally, a string-in-context is a triple of strings $u\langle v \rangle w$, where u and w are the left and right context of v. That is, $u\langle v \rangle w$ is v seen as a substring of uvw. Concatenation of strings-in-context must respect contexts: the concatenation of $u\langle v \rangle v'w$ and $uv\langle v' \rangle w$ is $u\langle vv' \rangle w$, but is not defined for pairs that are not of this form. In particular, our concept of contexts cover the whole string: the left context includes everything before the substring, and the right context includes everything after.

For example, if $x = he\langle llo \rangle world$ and $y = hello\,\langle wor \rangle ld$, then we have $xy = he\langle llo\;wor \rangle ld$. On the other hand, $he\langle llo \rangle world$ and $hello\,\langle ear \rangle th$ cannot be concatenated. A string with empty contexts, such as $\varepsilon\langle hello \rangle\varepsilon$, cannot be a strict substring of any string.

We give here an informal description of grammars with two-sided contexts; we will give a formal definition below. These grammars are similar to context-free grammars but allow rules with conjunction and context quantifiers. Non-terminals and sentential forms of the grammar should be considered as properties of terminal strings-in-context; we will not use rewriting systems. For example, the rule $A \to BC$ should be read as meaning that if x and y have properties B and C respectively, then xy has property A.

A string-in-context has the property $\alpha\ \&\ \beta$ if it has properties α and β. There are four context quantifiers: $\lhd\alpha$ denotes the property of strings which are preceded by a string with property α, i.e., $u\langle v \rangle w$ has property $\lhd\alpha$ if $\varepsilon\langle u \rangle vw$ has property α; $\trianglelefteq$ denotes a property of a string including its left context, i.e., $u\langle v \rangle w$ has property $\trianglelefteq\alpha$ if $\varepsilon\langle uv \rangle w$ has property α; the right context quantifiers, $\rhd$ and $\trianglerighteq$, have symmetrical interpretations. For example, a rule $A \to BC\ \&\ \rhd D$ means that a string has property A if it is of the form BC and is followed by a string with property D.

Definition 1. *A grammar with two-sided contexts is a tuple* $G = (\Sigma, N, R, S)$ *where* Σ *is the terminal alphabet,* N *is the non-terminal alphabet,* $S \in N$ *is the* *start symbol, and R is a set of rules of the form*

$$A \to \alpha_1 \ \& \ \cdots \ \& \ \alpha_m,$$

where $\alpha_1, \ldots, \alpha_m$ are strings in $(\Sigma \cup N)^*$, each possibly preceded by a quantifier from $\{\triangleleft, \trianglelefteq, \triangleright, \trianglerighteq\}$. Each α_i is called a conjunct.

We retain the term *non-terminal* due to its familiarity, but in the absence of rewriting systems perhaps *atomic property* would be more descriptive.

We define the semantics grammars with two-sided contexts by means of deduction systems. We will use atoms of the form $[\alpha, u\langle v \rangle w]$ where $u, v, w \in \Sigma^*$ and α has the form of a right-hand side of rule.

Definition 2. *Let $G = (\Sigma, N, R, S)$ be a grammar with two-sided contexts. Then we create a deduction system $\vdash_G$ with the axiom schemes:*

$$\vdash_G [\varepsilon, u\langle \varepsilon \rangle w]$$
$$\vdash_G [a, u\langle a \rangle w] \qquad\qquad a \in \Sigma$$

and the following schemes for deduction rules:

$$[\alpha, u\langle v \rangle w] \vdash_G [A, u\langle v \rangle w] \qquad A \to \alpha \in R$$
$$[\alpha, u\langle v_1 \rangle v_2 w], [\beta, uv_1 \langle v_2 \rangle w] \vdash_G [\alpha\beta, u\langle v_1 v_2 \rangle w]$$
$$[\alpha, u\langle v \rangle w], [\beta, u\langle v \rangle w] \vdash_G [\alpha \ \& \ \beta, u\langle v \rangle w]$$
$$[\alpha, \varepsilon\langle u \rangle vw] \vdash_G [\triangleleft\alpha, u\langle v \rangle w]$$
$$[\alpha, \varepsilon\langle uv \rangle w] \vdash_G [\trianglelefteq\alpha, u\langle v \rangle w]$$
$$[\alpha, uv\langle w \rangle \varepsilon] \vdash_G [\triangleright\alpha, u\langle v \rangle w]$$
$$[\alpha, u\langle vw \rangle \varepsilon] \vdash_G [\trianglerighteq\alpha, u\langle v \rangle w].$$

We define the language of α as

$$L_G(\alpha) = \{u\langle v \rangle w : u, v, w \in \Sigma^* \text{ and } \vdash_G [\alpha, u\langle v \rangle w]\}$$

and the language of G as

$$L(G) = \{v \in \Sigma^* : \varepsilon\langle v \rangle \varepsilon \in L_G(S)\}.$$

If we forbid rules with context quantifiers, the deduction systems $\vdash_G$ defines the conjunctive languages (Okhotin, 2013); if rules with conjunctions are also forbidden, then $\vdash_G$ defines the ordinary context-free languages.

3 Examples

The following grammar recognises the language $\{x\$x : x \in \{a, b\}^*\}$.

Example 3. Let $G = (\{a, b, \$\}, \{S, J, A, L, C_a, C_b\}, R, S)$ where R contains the following rules

$$S \to \$$$
$$S \to LSA$$
$$J \to \varepsilon$$
$$J \to AJ$$
$$A \to \sigma$$
$$\left. \begin{array}{l} L \to \sigma \ \& \ \rhd C_\sigma \\ C_\sigma \to \$J\sigma \\ C_\sigma \to AC_\sigma A \end{array} \right\} \sigma \in \{a, b\}.$$

If we ignore the context condition on the rule for L, this grammar simply matches $\{x\$y : |x| = |y|$ and $x, y \in \{a, b\}^*\}$. However, C_σ matches strings of the form $x\$y\sigma z$ where $|x| = |z|$, so the context condition on the rule for L ensures that, for every σ in the left-hand part, a σ also appears in the corresponding position (counting from the end) of the right-hand part. Thus $L(G) = \{x\$x : x \in \{a, b\}^*\}$.

The following example derivation shows that $a\$a \in L(G)$:

$$\vdash_G [a, \varepsilon\langle a\rangle\$a]$$
$$\vdash_G [\$, a\langle\$\rangle a]$$
$$\vdash_G [a, a\$\langle a\rangle\varepsilon]$$
$$\vdash_G [\varepsilon, a\$\langle\varepsilon\rangle a]$$
$$[\varepsilon, a\$\langle\varepsilon\rangle a] \vdash_G [J, a\$\langle\varepsilon\rangle a]$$
$$[\$, a\langle\$\rangle a], [J, a\$\langle\varepsilon\rangle a] \vdash_G [\$J, a\langle\$\rangle a]$$
$$[\$J, a\langle\$\rangle a], [a, a\$\langle a\rangle\varepsilon] \vdash_G [\$Ja, a\langle\$a\rangle\varepsilon]$$
$$[\$Ja, a\langle\$a\rangle\varepsilon] \vdash_G [C_a, a\langle\$a\rangle\varepsilon]$$
$$[C_a, a\langle\$a\rangle\varepsilon] \vdash_G [\rhd C_a, \varepsilon\langle a\rangle\$a]$$
$$[a, \varepsilon\langle a\rangle\$a], [\rhd C_a, \varepsilon\langle a\rangle\$a] \vdash_G [a \ \& \ \rhd C_a, \varepsilon\langle u\rangle\$a]$$
$$[a \ \& \ \rhd C_a, \varepsilon\langle a\rangle\$a] \vdash_G [L, \varepsilon\langle a\rangle\$a]$$
$$[\$, a\langle\$\rangle a] \vdash_G [S, a\langle\$\rangle a]$$
$$[L, \varepsilon\langle a\rangle\$a], [S, a\langle\$\rangle a] \vdash_G [LS, \varepsilon\langle a\$\rangle a]$$
$$[a, a\$\langle a\rangle\varepsilon] \vdash_G [A, a\$\langle a\rangle\varepsilon]$$
$$[LS, \varepsilon\langle a\$\rangle a], [A, a\$\langle a\rangle\varepsilon] \vdash_G [LSA, \varepsilon\langle a\$a\rangle\varepsilon]$$
$$[LSA, \varepsilon\langle a\$a\rangle\varepsilon] \vdash_G [S, \varepsilon\langle a\$a\rangle\varepsilon].$$

This derivation corresponds to the parse tree in Fig. 1.

Barash and Okhotin (2013, Sect. 5) give an upper bound for the running time of their Algorithm 1, but do not show that this is tight. To show that our algorithm is strictly faster, we give a grammar for which their algorithm requires $\Theta(n)$ passes.

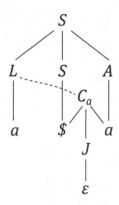

Fig. 1. A parse tree for $a\$a$

Example 4. Let $G = (\{a\}, \{A, S\}, R, S)$ where R contains the following rules:

$$S \to a$$
$$S \to AS$$
$$A \to a \,\&\, \rhd S.$$

This grammar is in both binary and separated normal forms, and matches the language aa^*.

In a run of Barash and Okhotin's Algorithm 1, each pass examines the substrings of w from left to right (by ending position). However, to deduce anything in G, we must work from right to left. Therefore, the algorithm can only make one deduction in each pass, but $\Theta(n)$ deductions are required.

4 Normal Form

The recognition algorithm of Barash and Okhotin (2013) requires grammars to be in a normal form similar to Chomsky normal form, called *binary normal form*.

Our algorithm will not be adversely affected by ε-productions, but cannot efficiently handle productions with more than one concatenation (e.g. $A \to BC \,\&\, \lhd DE$ or $A \to BCD$), so we will use a normal form which excludes these cases.

Definition 5. *A grammar with two-sided contexts* $G = (\Sigma, N, R, S)$ *is said to be in* separated normal form (SNF) *if each rule in R is in one of the following forms:*

$$A \to \varepsilon$$
$$A \to a$$
$$A \to QB$$
$$A \to B_1 B_2$$
$$A \to B_1 \,\&\, \cdots \,\&\, B_m$$

where a is a terminal symbol, $A, B, B_1, \ldots, B_m$ ($m \geq 1$) are non-terminals, and $Q \in \{\lhd, \unlhd, \rhd, \unrhd\}$, and $m \geq 1$.

The size $|G|$ of a grammar G (which we shall use as a complexity parameter) is the total number of symbols appearing in its rules.

Our algorithm could be made to work with a much more general normal form but this form will simplify its presentation. A grammar with two-sided contexts can easily be transformed into SNF using standard techniques.

Theorem 6. *If $G = (\Sigma, N, R, S)$ is a grammar with two-sided contexts then it can be effectively transformed into an equivalent grammar G' in separated normal form with $|G'| \in O(|G|)$.*

For a grammar G, we will call the equivalent SNF grammar generated by the above procedure SNF(G).

Ultimately, we are only interested in proofs of atoms where the property α is a single non-terminal (in particular, $[S, \varepsilon\langle x\rangle\varepsilon]$). However, the deduction system $\vdash_G$ can prove many unnecessary atoms not of this form. Therefore we give an alternative system which only uses atoms of this form, by combining the derivation steps used to obtain them. For example, when there is a rule $A \to BC$, instead of the proof $[B, x], [C, y] \vdash_G [BC, xy] \vdash_G [A, xy]$, we can prove $[B, x], [C, y] \vdash'_G [A, xy]$ directly.

Definition 7. *Let G be a grammar in SNF. Then $\vdash'_G$ is the deduction system with the following schemes:*

$$\vdash'_G [A, u\langle\varepsilon\rangle w] \qquad A \to \varepsilon$$

$$\vdash'_G [A, u\langle a\rangle w] \qquad A \to a,\ a \in \Sigma$$

$$[B_1, u\langle v_1\rangle v_2 w], [B_2, uv_1\langle v_2\rangle w] \vdash'_G [A, u\langle v_1 v_2\rangle w] \qquad A \to B_1 B_2$$

$$[B_1, u\langle v\rangle w], \ldots, [B_m, u\langle v\rangle w] \vdash'_G [A, u\langle v\rangle w] \qquad A \to B_1 \ \& \ \cdots \ \& \ B_m$$

$$[B, \varepsilon\langle u\rangle vw] \vdash'_G [A, u\langle v\rangle w] \qquad A \to \lhd B$$

$$[B, \varepsilon\langle uv\rangle w] \vdash'_G [A, u\langle v\rangle w] \qquad A \to \unlhd B$$

$$[B, uv\langle w\rangle \varepsilon] \vdash'_G [A, u\langle v\rangle w] \qquad A \to \rhd B$$

$$[B, u\langle vw\rangle \varepsilon] \vdash'_G [A, u\langle v\rangle w] \qquad A \to \unrhd B$$

In fact, the system $\vdash'_G$ is exactly the system of Barash and Okhotin (2013, Definition 2) restricted to SNF; they give a single scheme for deduction rules, with a complex side-condition, which entails all of ours.

Lemma 8. *If $G = (\Sigma, N, R, S)$ is a grammar in SNF, $A \in N$, and x is a string-in-context over Σ, then $\vdash_G [A, x]$ if and only if $\vdash'_G [A, x]$.*

Proof. Completeness of $\vdash'_G$ with respect to $\vdash_G$ can be shown by a straightforward induction on the length of proofs in $\vdash_G$.

Soundness can be seen from the fact that every axiom in $\vdash'_G$ is obtained from an axiom of $\vdash_G$ followed by a single deduction step, and every deduction rule is obtained by combining rules of $\vdash_G$.

5 Algorithm

To recognize a string x of length n, we will construct a set of Horn clauses from a grammar G in SNF of size $O(|G| \cdot n^3)$. We can then apply a linear-time algorithm for Horn satisfiability (Dowling and Gallier, 1984). The Horn clauses are derived from the deduction rules used to define the semantics of a grammar with two-sided contexts, interpreted as logical implications, except that we only include those which relate to substrings(-in-context) of x. The restriction to SNF ensures the size of this set is at most cubic in n.

The idea of parsing by interpreting grammar rules as logical implications is due to Colmerauer and Kowalski (Kowalski, 1979, Chapter 3). Shieber et al. (1995) noted that when this interpretation is applied to context-free grammars in Chomsky normal form, one can obtain an efficient algorithm similar to the CKY algorithm.

We define a set of axioms from which we can prove that a string belongs to the language of a given grammar with two-sided contexts.

We use the symbols $\Rightarrow$ and $\wedge$ for logical implication and conjunction, respectively. In a logical formula of the form $\phi_1 \wedge \cdots \wedge \phi_m \Rightarrow \psi$, we call $\phi_1, \ldots, \phi_m$ the *antecedents* and ψ the *consequent*. We will sometimes consider a formula consisting of a single propositional atom ψ to be an implication with consequent ψ and no antecedents.

Definition 9. *Let* $G = (\Sigma, N, R, S)$ *be a grammar with two-sided contexts in SNF, and let* $x \in \Sigma^*$ *have length* n. *We will use only atoms of the form* $[A, u\langle v \rangle w]$ *where* $A \in N$ *and* $uvw = x$.

Construct $\mathrm{AXIOMS}(G, x) = \bigcup_{r \in R} f(r)$, *where*

$$f(A \to \varepsilon) = \{[A, u\langle \varepsilon \rangle w] : uw = x\}$$

$$f(A \to a) = \{[A, u\langle a \rangle w] : uaw = x\}$$

$$f(A \to B_1 B_2) = \{[B_1, u\langle v_1 \rangle v_2 w] \wedge [B_2, uv_1 \langle v_2 \rangle w] \Rightarrow [A, u\langle v_1 v_2 \rangle w]$$
$$: uv_1 v_2 w = x\}$$

$$f(A \to B_1 \, \& \, \cdots \, \& \, B_m) = \{[B_1, u\langle v \rangle w] \wedge \cdots \wedge [B_m, u\langle v \rangle w] \Rightarrow [A, u\langle v \rangle w]$$
$$: uvw = x\}$$

$$f(A \to \lhd B) = \{[B, \varepsilon\langle u \rangle vw] \Rightarrow [A, u\langle v \rangle w] : uvw = x\}$$

$$f(A \to \unlhd B) = \{[B, \varepsilon\langle uv \rangle w] \Rightarrow [A, u\langle v \rangle w] : uvw = x\}$$

$$f(A \to \rhd B) = \{[B, uv\langle w \rangle \varepsilon] \Rightarrow [A, u\langle v \rangle w] : uvw = x\}$$

$$f(A \to \unrhd B) = \{[B, u\langle vw \rangle \varepsilon] \Rightarrow [A, u\langle v \rangle w] : uvw = x\}$$

and $\mathrm{Ax}(G, x) = \bigcup_{r \in R'} f(r)$, *where* $R' \subset R$ *is the set of rules not of the form* $A \to B_1 B_2$.

$\mathrm{AXIOMS}(G, x)$ is essentially a restatement of the definition of $\vdash'_G$. Thus, by Lemma 8, $u\langle v \rangle w \in L_G(A)$ if and only if $\mathrm{AXIOMS}(G, uvw) \vdash [A, u\langle v \rangle w]$.

From each rule of G we constructed a set of $O(n^3)$ axioms, so the total size of $\mathrm{AXIOMS}(G, x)$ is $O(|G| \cdot n^3)$ atoms. The only rules which lead to a cubic number of axioms are those of the form $A \to B_1 B_2$, so $\mathrm{Ax}(G, x)$ has size $O(|G| \cdot n^2)$.

Each axiom is a Horn clause, and the set of atoms which can be deduced from a set of Horn clauses can be computed in linear time (Dowling and Gallier, 1984, Algorithm 2). Combining such a procedure with the SNF transformation and $\text{AXIOMS}(G, x)$ yields Algorithm 1 for recognizing the language of a grammar with two-sided contexts.

Algorithm 1.

---

    function $\text{HORNCONSEQUENCES}(\Gamma)$
        Requires: Γ is a set of Horn clauses
        Returns: $\{\psi : \Gamma \vdash \psi, \psi \text{ is an atom}\}$

        $Q \leftarrow \{\psi : \psi \in \Gamma, \psi \text{ is an atom}\}$ — queue of atoms to resolve
        $P \leftarrow Q$ — set of atoms that have been deduced
        while $Q \neq \emptyset$ **do**
            remove any atom ϕ from Q
            mark every occurrence of ϕ in Γ
            for $\phi_1 \wedge \cdots \wedge \phi_m \Rightarrow \psi \in \Gamma$ with $\phi_1, \ldots, \phi_m$ all marked **do**
                remove $\phi_1 \wedge \cdots \wedge \phi_m \Rightarrow \psi$ from Γ
                if $\psi \notin P$ **then**
                    add ψ to Q
                    add ψ to P
        return P

    function $\text{RECOGNIZE}(G, x)$
        $(\Sigma, N, R, S) \leftarrow \text{SNF}(G)$
        $\Gamma \leftarrow \text{AXIOMS}((\Sigma, N, R, S), x)$
        return $[S, \varepsilon\langle x \rangle \varepsilon] \in \text{HORNCONSEQUENCES}(\Gamma)$

Intuitively, this algorithm deduces each atom as soon as it is possible to do so: if we have an implication, we deduce its consequent as soon as all its antecedents have been deduced. In contrast, the CKY-style algorithm essentially considers each axiom in a fixed order.

If implemented using the appropriate data structures, this algorithm requires $O(|G| \cdot n^3)$ time, but it is not space efficient: the set Γ requires $O(|G| \cdot n^3)$ space.

We will therefore specialise HORNCONSEQUENCES to represent concatenation rules separately: Algorithm 2 simulates Algorithm 1 on the set $\text{Ax}(G, x)$ but handles concatenations separately, ensuring only quadratic space usage. We will give a more concrete description of this algorithm, to demonstrate that it has the desired complexity.

Each set $\text{starts}[A][i]$ is intended to contain the atoms corresponding to a match of non-terminal A starting from position i; $\text{ends}[A][i]$ does the same for ending positions. These are needed to handle concatenations. For efficiency, $u\langle v \rangle w$ should be represented as the pair $(|u|, |uv|)$ rather than as a triple of strings; this reduces size of atoms from $O(n)$ to $O(\log n)$, and allows string manipulation to be replaced by arithmetic.

Lemma 10. *Let $G = (\Sigma, N, R, S)$ be a grammar in SNF, $x \in \Sigma^*$ and ψ an atom. Then* $\text{AXIOMS}(G, x) \vdash \psi$ *if and only if* $\text{PARSESNF}(G, x)$ *adds ψ to P.*

Proof. We prove the forward direction (completeness) by induction on the length of derivations.

It is clear that an atom is added to P if and only if $\text{PROVED}([A, u\langle v \rangle w])$ is called, and if it is, then $[A, u\langle v \rangle w]$ is added exactly once to P, Q, $\text{starts}[A][|u|]$ and $\text{ends}[A][|uv|]$; and it is eventually removed from Q in line 23.

If ψ can be deduced from $\text{AXIOMS}(G, x)$ in zero steps, then it is an element of $\text{AXIOMS}(G, x)$ and is added to the sets by line 21.

Now let $\phi_1 \wedge \cdots \wedge \phi_m \Rightarrow \psi \in \text{AXIOMS}(G, x)$, and suppose for induction that $\text{AXIOMS}(G, x) \vdash \phi_j$ and $\text{PROVED}(\phi_j)$ has been called for each $j \in 1, \ldots, m$. If $\phi_1 \wedge \cdots \wedge \phi_m \Rightarrow \psi = \text{axiom}[i] \in \text{AX}(G, x)$, then each ϕ_j is eventually removed from Q, and at those points $\text{antecedents}[i]$ is decremented; since $\text{antecedents}[i]$ was initialized to m, it reaches 0, and PROVED is called on $\text{consequent}[i] = \psi$.

Otherwise, $\phi_1 \wedge \cdots \wedge \phi_m \Rightarrow \psi$ arose from a concatenation rule $A \rightarrow B_1 B_2$, so it has the form $[B_1, u\langle v_1 \rangle v_2 w] \wedge [B_2, uv_1 \langle v_2 \rangle w] \Rightarrow [A, u\langle v_1 v_2 \rangle w]$. By the induction hypothesis and the definition of PROVED, eventually $[B_1, u\langle v_1 \rangle v_2 w] \in \text{ends}[B_1][|uv_1|]$ and $[B_2, uv_1 \langle v_2 \rangle w] \in \text{starts}[B_2][|uv_1|]$. When both of these have been removed from Q, $\text{PROVED}([A, u\langle v_1 v_2 \rangle w])$ will be called on line 30 or 33.

By induction, every atom which can be deduced from $\text{AXIOMS}(G, x)$ is eventually added to P.

In the reverse direction, it is not hard to see that the algorithm makes no unsound inferences, and therefore the algorithm is sound.

6   Evaluation

Lemma 11. $\text{PARSESNF}(G, x)$ *has running time* $O(|G| \cdot |x|^3)$

Proof. All the initialisation clearly takes $O(|G| \cdot |x|^2)$ time.

Lines 19, 25, 30 and 33 are in each of the innermost loops.

Line 19 and line 25 are each run at most once for each appearance of an atom as an antecedent in $\text{AX}(G, x)$, i.e. $O(|G| \cdot |x|^2)$ times.

Lines 30 and 33 are run at most once for each tuple (r, u, v, v', w') where r is a rule of the form $A \rightarrow BC$ and $uvv'w' = x$. There are $O(|G| \cdot |x|^3)$ such tuples.

Lemma 12. $\text{PARSESNF}(G, x)$ *requires* $O(|G| \cdot |x|^2)$ *space.*

Proof. The algorithm treats $O(|G| \cdot |x|^2)$ atoms; each of these can appear at most once in each of P, Q, Starts and Ends.

There is one entry in $\text{appearances}[\phi]$ for each appearance of ϕ as an antecedent in $\text{AX}(G, x)$, so the total size of the appearances array is at most the size of $\text{AX}(G, x)$. There is one entry in each of antecedents and consequent for each axiom in $\text{AX}(G, x)$.

Since $|\text{AX}(G, x)| \in O(|G| \cdot |x|^2)$, and the variables not mentioned above use only constant space, the total space usage is $O(|G| \cdot |x|^2)$.

Algorithm 2.

1: **function** PARSESNF(G, x)
2:     **Requires:** $G = (\Sigma, N, R, S)$ in SNF, $x \in \Sigma^*$
3:     **Returns:** $\{\psi : \text{AXIOMS}(G, x) \vdash \psi, \psi \text{ is an atom}\}$

4:     $Q \leftarrow \emptyset$ — queue of atoms to resolve
5:     $P \leftarrow \emptyset$ — atoms that have been deduced
6:     $\text{starts}[A][i] \leftarrow \emptyset$ **for** $A \in N$ and $i \in \{0, \ldots, |x|\}$
7:     $\text{ends}[A][i] \leftarrow \emptyset$ **for** $A \in N$ and $i \in \{0, \ldots, |x|\}$
8:     $\text{appearances}[\phi] \leftarrow \emptyset$ **for** each atom ϕ — axioms where ϕ is an antecedent

9:     **procedure** PROVED$(\psi = [A, u\langle v \rangle w])$
10:         **if** $\psi \notin P$ **then**
11:             add ψ to Q and P
12:             add ψ to $\text{starts}[A][|u|]$ and $\text{ends}[A][|uv|]$

13:     $\text{axiom}[1, \ldots, k] \leftarrow \text{AX}(G, x)$
14:     **for** $i \in \{1, \ldots, k\}$ **do**
15:         $(\phi_1 \wedge \cdots \wedge \phi_m \Rightarrow \psi) \leftarrow \text{axiom}[i]$
16:         $\text{antecedents}[i] \leftarrow m$ — number of unresolved antecedents
17:         $\text{consequent}[i] \leftarrow \psi$
18:         **for** $j \in \{1, \ldots, m\}$ **do**
19:             add i to $\text{appearances}[\phi_j]$
20:         **if** $m = 0$ **then**
21:             PROVED(ψ)

22:     **while** $Q \neq \emptyset$ **do**
23:         remove any atom $\phi = [B, u\langle v \rangle w]$ from Q
24:         **for** $i \in \text{appearances}[\phi]$ **do**
25:             $\text{antecedents}[i] \leftarrow \text{antecedents}[i] - 1$
26:             **if** $\text{antecedents}[i] = 0$ **then**
27:                 PROVED$(\text{consequent}[i])$
28:         **for** $A \rightarrow BC \in R$ **do**
29:             **for** $[C, uv\langle v' \rangle w'] \in \text{starts}[C][|uv|]$ **do**
30:                 PROVED$([A, u\langle vv' \rangle w'])$
31:         **for** $A \rightarrow CB \in R$ **do**
32:             **for** $[C, u'\langle v' \rangle vw] \in \text{ends}[C][|u|]$ **do**
33:                 PROVED$([A, u'\langle v'v \rangle w])$
34:     **return** P

35: **function** RECOGNIZE(G, x)
36:     $G' = (\Sigma, N, R, S) \leftarrow \text{SNF}(G)$
37:     **return** $[S, \varepsilon\langle x \rangle \varepsilon] \in \text{PARSESNF}(G', x)$

Combining PARSESNF with the SNF transformation, we obtain our main result.

Theorem 13. *The language of a grammar with two-sided contexts G can be recognized in $O(|G| \cdot n^3)$ time and $O(|G| \cdot n^2)$ space.*

Proof. By Lemma 10, 11 and 12, the function RECOGNISE from Algorithm 2 recognizes such languages with the desired complexity.

Parse trees correspond to proofs of $[S, \varepsilon \langle x \rangle \varepsilon]$, so a parse forest can be obtained by modifying PARSESNF to record all the ways in which each atom is proved.

Acknowledgements. The author is grateful to Alexander Okhotin and the anonymous reviewers for their suggestions on improving the presentation of this paper.

References

Barash, M., Okhotin, A.: Defining contexts in context-free grammars. In: Dediu, A.-H., Martín-Vide, C. (eds.) LATA 2012. LNCS, vol. 7183, pp. 106–118. Springer, Heidelberg (2012)

Barash, M., Okhotin, A.: Grammars with two-sided contexts. Tech. Rep. 1090, Turku Centre for Computer Science (2013),
http://tucs.fi/publications/view/?pub_id=tBaOk13b

Dowling, W.F., Gallier, J.H.: Linear-time algorithms for testing the satisfiability of propositional Horn formulae. The Journal of Logic Programming 1(3), 267–284 (1984)

Kowalski, R.: Logic for problem-solving. North-Holland Publishing Co. (1979),
http://www.doc.ic.ac.uk/~rak/

Okhotin, A.: Conjunctive and boolean grammars: the true general case of the context-free grammars. Computer Science Review 9, 27–59 (2013)

Shieber, S.M., Schabes, Y., Pereira, F.C.N.: Principles and implementation of deductive parsing. The Journal of Logic Programming 24(1-2), 3–36 (1995)

Valiant, L.G.: General context-free recognition in less than cubic time. Journal of Computer and System Sciences 10(2), 308–315 (1975)

Williams, V.V.: Multiplying matrices faster than Coppersmith-Winograd. In: Proceedings of the 44th Symposium on Theory of Computing, STOC 2012, pp. 887–898. ACM (2012)

A Parameterized Algorithm for Packing Overlapping Subgraphs

Jazmín Romero and Alejandro López-Ortiz

David R. Cheriton School of Computer Science, University of Waterloo, Canada

Abstract. Finding subgraphs with arbitrary overlap was introduced as the k-H-Packing with t-Overlap problem in [10]. Specifically, does a given graph G have at least k induced subgraphs each isomorphic to a graph H such that any pair of subgraphs share at most t vertices? This problem has applications in the discovering of overlapping communities in real networks. In this work, we introduce the first parameterized algorithm for the k-H-Packing with t-Overlap problem when H is an arbitrary graph of size r. Our algorithm combines a bounded search tree with a greedy localization technique and runs in time $O(r^{rk}k^{(r-t-1)k+2}n^r)$, where $n = |V(G)|$, $r = |V(H)|$, and $t < r$. Applying similar ideas we also obtain an algorithm for packing sets with possible overlap which is a version of the k-Set Packing problem.

1 Introduction

Discovering communities in large and complex networks such as social, citation, or biological networks has been of interest on the last decades. A *community* is a part of the network in which the nodes are more highly interconnected to each other than to the rest. For example, a community can represent a group of friends in social networks or a protein complex in biological networks. Naturally, one person can have different groups of friends, and one protein can belong to more than one protein complex. Therefore, in realistic scenarios, communities can share members. The problem of finding communities with possible overlap was formalized as the k-H-Packing with t-Overlap problem in [10].

In the k-H-Packing with t-Overlap problem, the goal is to find at least k induced subgraphs (the communities) in a graph G (the network) such that each subgraph is isomorphic to a graph H (a community model) and each pair of subgraphs overlap in at most t vertices (the shared members)[10][1].

The k-H-Packing with t-Overlap problem is NP-Complete [10]. Therefore, we are interested in the design of algorithms that provide a solution in $f(k)n^{O(1)}$ running time, i.e., *fixed-parameter algorithms* or *FPT-algorithms*. In other words, the running time of a fixed-parameter algorithm is polynomial in the input size n but possibly exponential or worse in a specified parameter k, usually the size of the solution. Thus, our fundamental goal is to explore how the overlap

---

[1] To follow standard notation with packing and isomorphism problems, the meaning of the graphs G and H have been exchanged with respect to their meaning in [10].

E.A. Hirsch et al. (Eds.): CSR 2014, LNCS 8476, pp. 325–336, 2014.
© Springer International Publishing Switzerland 2014

influences the complexity of this problem. Interestingly enough, according to sociological studies, well-defined communities should usually have low overlap [1] which naturally leads to a fixed-parameter treatment in a practical setting.

The formal definition of our studied problem is as follows.

The k-H-Packing with t-Overlap problem

Input: A graph G and non-negative integers k and t, where $t < |V(H)|$.

Parameter: k

Question: Does G contain at least k induced subgraphs $\mathcal{K} = \{Q_1^*, \ldots, Q_k^*\}$ where each Q_i^* is isomorphic to a graph H and $|V(Q_i^*) \cap V(Q_j^*)| \leq t$, for any pair Q_i^*, Q_j^*?

The number of vertices in G and H will be denoted as n and r, respectively. Note that r and t are constants.

Related Results. The study of the parameterized complexity of the k-H-Packing with t-Overlap problem was initiated in [10]. The authors introduced a clique-crown decomposition for this problem when H is a clique of size r. With that decomposition, an algorithm that reduces any instance of the problem to a size of $2(rk - r)$ is achieved (i.e., a *kernel* of size $2(rk - r)$). As far as we know, there are no other parameterized results for this problem.

The k-H-Packing with t-Overlap problem is a generalization of the well-studied *k-H-Packing problem* which seeks for vertex-disjoint subgraphs instead of overlapping subgraphs. The parameterized complexity of packing triangles and stars of n leaves was studied in [4] and [9], respectively. Fellows et al. [5] provided an $O(n + 2^{|V(H)|k})$ running time algorithm for packing an arbitrary graph H. The latest result is a kernel of size $O(k^{|V(H)|-1})$ for packing an arbitrary graph H by Moser [8].

Another problem related to the k-H-packing with t-Overlap, when H is a clique, is the *cluster editing problem*. This problem consists of modifying a graph G by adding or deleting edges such that the modified graph is composed of a vertex-disjoint union of cliques. Some works have considered overlap in the cluster editing problem [2,3]. Fellows et al. [3] allow that each vertex of the modified graph can be contained in at most s maximal cliques.

The problem of finding one community of size at least r in a given network is also related to the k-H-Packing with t-Overlap problem. The most studied community models are cliques and some relaxations of cliques. Parameterized complexity results for this problem can be found in [6,7,11]. Overlap has not yet been considered under this setting.

Our Results. In this work, we provide an $O(r^{rk} k^{(r-t-1)k+2} n^r)$ running time algorithm for the k-H-packing with t-Overlap for any arbitrary graph H of size r and any overlap value $t < r$. Our algorithm is a non-trivial generalization of the search tree algorithm to find disjoint triangles presented by Fellows et al. [4]. In addition, we introduce a novel analysis to handle the need of overlap, and as we shall see an overlap of as little as one vertex significantly increases the complexity of the problem. Even though the k-H-packing problem (the vertex-disjoint version) is well studied, our search tree algorithm is the first one to consider variable overlap between subgraphs.

The classical k-Set Packing problem asks for at least k pairwise disjoint sets from a given collection of sets. We introduce a variant of this problem that would allow overlap between the sets. We show how our search tree algorithm would work for this variant as well. To the best of our knowledge, this variant of the k-set packing problem has not been studied before.

This paper is organized as follows. In Section 2, we introduce the terminology and notation used in the paper. Section 3 describes the details of our search tree algorithm. In Section 4, we explain how the search tree algorithm can be applied for a variant of the k-Set Packing problem. Finally, Section 5 states the conclusion of this work.

2 Terminology and Notation

Graphs in this paper are undirected and simple. For a graph G, we denote as $V(G)$ and $E(G)$ its sets of vertices and edges, respectively. The notation $G[L]$ represents the subgraph induced in G by a set of vertices L. Two subgraphs S and P are vertex-disjoint if they do not share vertices, i.e., $V(S) \cap V(P) = \emptyset$. Otherwise, we say that S and P *overlap* in $|V(S) \cap V(P)|$ vertices. We extend this terminology when S and P are sets of vertices instead.

An H-*Packing with* t-*Overlap* $\mathcal{L}$ is a set of induced subgraphs $\mathcal{L} = \{Q_1, Q_2, ...\}$ of G where each Q_i is isomorphic to a graph H, and every pair Q_i, Q_j overlap in at most t vertices. The size of the H-Packing with t-Overlap is the number of subgraphs in $\mathcal{L}$. If $|\mathcal{L}| \geq k$ then $\mathcal{L}$ is a k-H-Packing with t-Overlap, also called a k-*solution*. We use the letter $\mathcal{K}$ to represent a k-solution. An H-Packing with t-Overlap $\mathcal{M}$ is *maximal*, if there is no isomorphic subgraph to H in G that can be added to $\mathcal{M}$. A maximal packing with t-Overlap will be called a *maximal solution*. For simplicity, a subgraph isomorphic to H is called an H-*subgraph*.

For any pair of disjoint sets of vertices S and P such that $|S| + |P| = r$, we say that P is a *sponsor* of S (or vice versa) if $G[S \cup P]$ is an H-subgraph. Sponsors(S) is the set of sponsors of S in $V(G)$. We use the term *complete* S to represent the selection of a sponsor P in Sponsors(S) to update S as $S \cup P$. The resulting H-subgraph $G[S \cup P]$ is called an H-*completed subgraph*, and it is denoted as $S \cdot P$. Figure 1 shows an instance of the 4-K_4-Packing with 1-Overlap ($H = K_4$). In this instance, the sets of vertices $\{1, 3\}$ and $\{9, 10\}$ form a K_4 $G[\{1, 3, 9, 10\}]$; thus, the set $\{9, 10\}$ is a sponsor of the set $\{1, 3\}$. Other sponsors of $\{1, 3\}$ are $\{10, 11\}$ and $\{2, 4\}$.

Let $\mathbf{Q}$ be a collection of k sets of vertices each set of size at most r. We say that $\mathbf{Q}$ *is completed into a* k-*solution*, if each set in $\mathbf{Q}$ is completed into an H-subgraph such that any pair of H-completed subgraphs overlap in at most t vertices. The sponsors that complete the sets in $\mathbf{Q}$ are called *feasible sponsors*.

3   An FPT-Algorithm for the H-Packing with t-Overlap Problem

The key point of the FPT-algorithm is based on the following lemma which is a generalization of Observation 2 in [4]. This lemma proves how a maximal solution intersects with a k-solution of the graph G, assuming that G has one.

Lemma 1. *Let $\mathcal{M}$ and $\mathcal{K}$ be a maximal H-Packing with t-Overlap and a k-H-Packing with t-Overlap, respectively. We claim that any $Q^* \in \mathcal{K}$ overlaps with some $Q \in \mathcal{M}$ in at least $t+1$ vertices, i.e., $|V(Q^*) \cap V(Q)| \geq t+1$. Furthermore, there is no pair $Q_i^*, Q_j^* \in \mathcal{K}$ for $i \neq j$ that overlap in the same set of vertices with Q i.e., $V(Q_i^*) \cap V(Q) \neq V(Q_j^*) \cap V(Q)$.*

Proof. Assume by contradiction that there is an H-subgraph $Q^* \in \mathcal{K}$ such that for any H-subgraph $Q \in \mathcal{M}$, the overlap between them is at most t, i.e., $|V(Q^*) \cap V(Q)| \leq t$. However, in this case, we could add Q^* to $\mathcal{M}$, and $\mathcal{M} \cup Q^*$ is an H-Packing with t-Overlap contradicting the assumption of the maximality of $\mathcal{M}$.

To prove the second part of the lemma, assume by contradiction that there is a pair $Q_i^*, Q_j^* \in \mathcal{K}$ that overlap in the same set of vertices for all $Q \in \mathcal{M}$. However, by the first part of the lemma, we know that there is at least one $Q \in \mathcal{M}$ such that $|V(Q_i^*) \cap V(Q)| \geq t + 1$. This would imply that $|V(Q_i^*) \cap V(Q_j^*)| \geq t + 1$, a contradiction since the k-H-Packing with t-Overlap problem does not allow overlap greater than t. $\square$

Lemma 1 states that every H-subgraph of a k-solution $\mathcal{K}$ overlaps in at least $t + 1$ vertices with some H-subgraph of a maximal solution $\mathcal{M}$. Let us call this intersection of $t + 1$ vertices a *feasible seed*. A feasible seed is shared only by a unique pair composed of an H-subgraph of $\mathcal{M}$ and an H-subgraph of $\mathcal{K}$. Thus, a feasible seed is contained in only one H-subgraph of a k-solution $\mathcal{K}$. Note that a subset of at most t vertices of a feasible seed could belong to more than one H-subgraph of a k-solution. Based on that we have the following observation.

Observation 1. If the graph G has a k-H-packing with t-Overlap $\mathcal{K}$, each H-subgraph in $\mathcal{K}$ has at least one feasible seed. Furthermore, each feasible seed has at least one feasible sponsor.

The left-side of Figure 1 shows an example of this intersection for the 4-K_4-Packing with 1-Overlap problem ($k = 4$, $t = 1$). The two K_4's of the maximal solution are indicated by solid lines while the four K_4's of the k-solution are indicated with solid and dashed lines. Edges of the graph that do not belong to any of these solutions are indicated in light gray. The seeds in the example have size $t + 1 = 2$. A collection of four feasible seeds is $\{\{1,3\}, \{3,4\}, \{5,6\}, \{6,8\}\}$; the vertices of these seeds are filled in the figure.

We now proceed to describe the algorithm for the k-H-Packing with t-Overlap problem. First, we obtain a maximal solution $\mathcal{M}$ of G. If the number of H-subgraphs in $\mathcal{M}$ is at least k then $\mathcal{M}$ is a k-H-Packing with t-Overlap and the

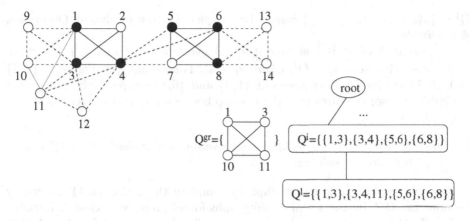

Fig. 1. On the left side, an intersection of a k-solution with a maximal solution of the $4\text{-}K_4$-Packing with 1-Overlap problem ($k = 4$ and $t = 1$). The seeds in this example are of size $t + 1 = 2$. To the right, part of the search tree corresponding to the instance to the left.

algorithm stops. Otherwise, we want to find a k-solution using k feasible seeds (Observation 1).

Since we do not know if a set of $t + 1$ vertices of an H-subgraph $Q \in \mathcal{M}$ is a feasible seed, we would need to consider all the distinct sets of $t + 1$ vertices from $V(Q)$. To avoid confusion, we call these sets simply *seeds*. Two seeds are distinct if they differ by at least one vertex. Therefore, seeds can overlap in at most t vertices. The set of all possible seeds from all H-subgraphs of $\mathcal{M}$ is called *the universe of seeds*. Observe that there are not duplicate seeds in this universe. Otherwise, there would be at least one pair of H subgraphs of $\mathcal{M}$ with the same seed implying that they overlap in at least $t + 1$ vertices. Now, we need to select k feasible seeds from that universe and complete them into a k-solution.

To do that, we create a search tree where at each node i there is a collection $\mathbf{Q^i}$ of k sets of vertices. Each set represents an H-subgraph that would be part of the k-solution. Initially, the root has a child i for each possible selection of k seeds from the universe of seeds. The collection $\mathbf{Q^i}$ is initialized with these k seeds, i.e., $\mathbf{Q^i} = \{s_1^i, \ldots, s_k^i\}$. Since we are trying all possible selections of k seeds, at least one child i should have k feasible seeds each one contained in a different H-subgraph of a k-solution, assuming G has one. In that case, $\mathbf{Q^i}$ can be completed into a k-solution.

The right side of Figure 1 shows one child of the root of the search tree created with the maximal solution in the left. This child has the collection $\mathbf{Q^i} = \{\{1, 3\}, \{3, 4\}, \{5, 6\}, \{6, 8\}\}$.

Next, for each child i of the root, the goal is to try to complete the collection $\mathbf{Q^i}$ into a k-H-Packing with t-Overlap. To do that, we need to find a feasible sponsor for each seed in $\mathbf{Q^i}$.

Before explaining how a seed can be completed, we next introduce a simple way to discard some of the sponsors that a seed $s_j^i \in \mathbf{Q^i}$ could have. We say that a sponsor A of the seed s_j^i *is ineligible* to complete s_j^i if the collection

$\mathbf{Q^i} = \{s_1^i, \ldots, s_j^i \cup A, \ldots, s_k^i\}$ cannot be completed into a k-solution. Otherwise, A is *eligible*.

A sponsor A of s_j^i is ineligible if $s_j^i \cdot A$ overlaps in at least $t+1$ vertices with some other seed $s_l^i \in \mathbf{Q^i}$, where $s_l^i \neq s_j^i$. For example, in Figure 1, $\{2,4\}$ and $\{5,7\}$ are ineligible sponsors of $\{1,3\}$ and $\{6,8\}$, respectively. Discarding ineligible sponsors ensures that the overlap between any pair of seeds is at any stage at most t.

Lemma 2. *If for a seed $s_j^i \in \mathbf{Q^i}$ all its sponsors are ineligible then $\mathbf{Q^i}$ cannot be completed into a k-solution.*

Now, we explain how we attempt to complete the collection $\mathbf{Q^i}$ in greedy fashion. Let $\mathbf{Q^{gr}}$ be the set of H-subgraphs found greedily at child i. Initially, $\mathbf{Q^{gr}} = \emptyset$. At iteration j, greedy searches an eligible sponsor A for s_j^i such that $s_j^i \cdot A$ overlaps in at most t vertices with every H-subgraph in $\mathbf{Q^{gr}}$. If such eligible sponsor exists, greedy adds $s_j^i \cdot A$ to $\mathbf{Q^{gr}}$, i.e., $\mathbf{Q^{gr}} = \mathbf{Q^{gr}} \cup s_j^i \cdot A$; if not, greedy stops. If all the seeds of $\mathbf{Q^i}$ were completed then we have a k-H-Packing with t-Overlap.

If the greedy algorithm cannot find a k-H-Packing with t-Overlap, then the next step will be to increase the size of one of the seeds of $\mathbf{Q^i}$ by one vertex. Let s_j^i be the seed in $\mathbf{Q^i}$ that could not be completed by the greedy algorithm. Greedy could not complete s_j^i because for each eligible sponsor A of s_j^i, the H-subgraph $s_j^i \cdot A$ overlaps in more than t vertices with at least one H-subgraph in $\mathbf{Q^{gr}}$. For example, in Figure 1, greedy completed $\{1,3\}$ with the sponsor $\{10,11\}$, and $\{1,3\} \cdot \{10,11\}$ is added to $\mathbf{Q^{gr}}$. After that greedy cannot complete $\{3,4\}$. The seed has only one eligible sponsor $\{11,12\}$ but $\{3,4\} \cdot \{11,12\}$ overlaps in two vertices with $\{1,3\} \cdot \{10,11\}$ in $\mathbf{Q^{gr}}$. Note that the sponsor $\{1,2\}$ of the seed $\{3,4\}$ is ineligible.

If $\mathbf{Q^i}$ can be completed into a k-solution, then at least one of the eligible sponsors of s_j^i is feasible. We do not know which one it is, but we are certain that this feasible sponsor shares some vertices with at least one H-subgraph in $\mathbf{Q^{gr}}$. We will use this intersection of vertices to find such a feasible sponsor.

Let us denote as $I(\mathbf{Q^{gr}}, \mathsf{EligibleSponsors}(s_j^i))$ the set of vertices that are shared between each eligible sponsor of s_j^i and each H-subgraph in $\mathbf{Q^{gr}}$. We will increase the size of the seed s_j^i by one vertex by creating a child l of the node i for each vertex $v_l \in I(\mathbf{Q^{gr}}, \mathsf{EligibleSponsors}(s_j^i))$.

The collection of seeds at child l, $\mathbf{Q^l}$ is the same as the collection of its parent i with the update of the seed s_j^i as $s_j^i \cup v_l$, i.e., $\mathbf{Q^l} = \{s_1^i, \ldots, s_{j-1}^i, s_j^i \cup v_l, s_{j+1}^i, \ldots, s_k^i\}$. After that, the greedy algorithm is repeated at the collection $\mathbf{Q^l}$ of child l ($\mathbf{Q^{gr}}$ starts empty again). In the example of Figure 1, there is one child of the node i where the seed $\{3,4\}$ is updated with the vertex 11. Observe that after this update, the sponsor $\{10,11\}$ is ineligible to complete $\{1,3\}$.

The algorithm stops attempting to complete $\mathbf{Q^i}$ or some collection of a descendant of the node i when there are no eligible sponsors for one of the seeds (Lemma 2). Otherwise, one of the leaves of the tree would have a k-H-Packing with t-Overlap. In the example of Figure 1, one leaf of the search tree would complete the collection into the solution $\mathcal{K} = \{\{1,3\} \cdot \{9,10\}, \{3,4\} \cdot \{11,12\}, \{5,6\} \cdot \{4,7\}, \{6,8\} \cdot \{13,14\}\}$.

3.1   Correctness

The next basic lemma will help us to prove that the algorithm is correct.

Lemma 3. *If A is a sponsor of the seed s_j^i then $A \backslash X$ is a sponsor of $s_j^i \cup X$, for any $X \subset A$.*

A node i of the tree is *feasible*, if its collection $\mathbf{Q^i}$ is composed of k feasible seeds, and each feasible seed is contained in a different H-subgraph of a k-solution $\mathcal{K}$. That is, $\mathbf{Q^i} = \{s_1^i, \ldots, s_k^i\}$, $\mathcal{K} = \{Q_1^*, \ldots, Q_k^*\}$, and $s_j^i \subseteq V(Q_j^*)$ for $1 \leq j \leq k$. A child of the root that is feasible is called a *feasible child*. By Observation 1, the collection $\mathbf{Q^i}$ of a feasible node can be completed into a k-solution.

Lemma 4. *The Feasible Path. If the graph G has a k-solution then there is at least one path P on the subtree rooted at a feasible child where each node in P is feasible.*

Proof. The lemma states that there is a path $P = < i_1, i_2, \ldots, i_m >$ such that each node i_l in P has the collection $\mathbf{Q^{i_l}} = \{s_1^{i_l}, \ldots, s_k^{i_l}\}$ where $s_j^{i_l} \subseteq V(Q_j^*)$ for $1 \leq j \leq k$ and $1 \leq l \leq m$. Note that $i_1 = i$.

We prove this claim by induction on the number of levels. At level 1, the first node of the path is a feasible child of the root and the claim follows.

By Observation 1, each feasible seed in $\mathbf{Q^i}$ has at least one feasible sponsor. Let $\{A_1^*, \ldots, A_k^*\}$ be the set of feasible sponsors where A_j^* is the feasible sponsor of s_j^i, i.e., $\mathcal{K} = \{s_1^i \cdot A_1^*, \ldots, s_k^i \cdot A_k^*\}$.

Next we show that for the remaining nodes of P, the seeds are updated only with vertices from the feasible sponsors.

Let us suppose that the greedy algorithm failed to complete a seed $s_j^{i_1}$ at level 1. The seed $s_j^{i_1}$ has at least one feasible sponsor A_j^*. Since A_j^* is feasible then it is eligible. Greedy failed to complete $s_j^{i_1}$, if the H-subgraph formed with $s_j^{i_1}$ and each eligible sponsor of $s_j^{i_1}$ (including the feasible one A_j^*) overlaps in more than t vertices with an H-subgraph completed by greedy (i.e., in the set $\mathbf{Q^{gr}}$). Therefore at level 2, there is at least one child of the node i_1 where the seed $s_j^{i_1}$ is updated with one vertex from the feasible sponsor A_j^*. That is, $\mathbf{Q^{i_2}} = \{s_1^{i_1}, \ldots, s_j^{i_1} \cup v^*, \ldots, s_k^{i_1}\}$ where $v^* \in A_j^*$, and the lemma follows.

Now, let us assume that the lemma is true up to the level $h-1$. We next show that lemma holds for level h. Let $< i_2, \ldots, i_{h-1} >$ be the subpath of P where one seed in each node is updated with one vertex from the feasible sponsors.

By contradiction, suppose that at level $h-1$ greedy could not complete the seed $s_j^{i_{h-1}}$ but there is no child of the node i_{h-1} such that $s_j^{i_{h-1}}$ is updated with a vertex from the feasible sponsor.

Let us suppose that U^* is the set of vertices that has been added to $s_j^{i_1}$ during the levels $1, \ldots, h-1$. By our assumption, the seed $s_j^{i_1}$ is feasible, and it has been updated only with vertices from the feasible sponsor. Therefore, $U^* \subset A_j^*$.

By Lemma 3, $A_j^* \backslash U^*$ is a sponsor of $s_j^{i_{h-1}}$. Therefore, the only way that none of the children of i_{h-1} would update $s_j^{i_{h-1}}$ with a vertex from $A_j^* \backslash U^*$ is if $A_j^* \backslash U^*$ is not an eligible sponsor of $s_j^{i_{h-1}}$.

However, this would imply that the H-subgraph $s_j^{i_{h-1}} \cdot (A_j^* \backslash U^*) = s_j^{i_1} \cdot A_j^*$ overlaps in more than t vertices with some feasible seed $s_l^{i_1} \in \mathbf{Q^{i_1}}$. This contradicts our assumption that the collection $\mathbf{Q^{i_1}}$ can be completed into a solution (feasible child). $\qquad\square$

Lemma 5. *The collection of seeds of the last node l (a leaf) of a feasible path is a k-solution.*

Proof. By Lemma 4, we know that the collection of seeds of every node in a feasible path is updated only with vertices from the feasible sponsors. Therefore, eventually a seed is completed into an H-subgraph of the k-solution. Let us suppose that $k-m$ seeds, where $m \geq 1$, were completed in this way.

We claim now that greedy finds an eligible sponsor for the remaining m seeds. Assume by contradiction that greedy failed to complete one of these m seeds and suppose it is s_j^l. Since the collection of a feasible child can be completed into a solution, and the node l is in a feasible path, then s_j^l should have at least one eligible sponsor. In that case, the next step would be to create children of the node l updating the seed s_j^l, contradicting the assumption that the node l is a leaf. $\qquad\square$

Theorem 1. *The search tree algorithm finds one k-H-Packing with t-Overlap, if the graph G has at least one.*

Proof. Since we are creating a child of the root for each possible selection of k seeds from the universe of seeds, at least one child is feasible. By Lemma 4, there is at least one feasible path that starts at this child. By Lemma 5, a k-solution is given in the collection of the last node of this path. $\qquad\square$

3.2 Analysis

Lemma 6. *The root has at most $\left(\frac{e^2 r}{t+1}\right)^{k(t+1)}$ children.*

Proof. There are $\binom{r}{t+1}$ distinct sets of $t+1$ vertices (seeds) from the set of vertices of an H-subgraph. Since $|\mathcal{M}| \leq k-1$ then there are at most $(k-1)\left(\frac{er}{t+1}\right)^{t+1}$ seeds in the universe of seeds.

From the root of the search tree, we create a node i for each possible selection of k seeds of the universe of seeds. Therefore, the root has $\binom{|universe\,of\,seeds|}{k}$ children. Hence, $\binom{(k-1)\left(\frac{er}{t+1}\right)^{t+1}}{k} \leq \left(\frac{e(k-1)\left(\frac{er}{t+1}\right)^{t+1}}{k}\right)^k \leq \left(\frac{e^2 r}{t+1}\right)^{k(t+1)}$. $\qquad\square$

Lemma 7. *The height of the search tree is at most $(r - t - 1)k - 1$.*

Proof. If greedy cannot complete the collection $\mathbf{Q^i}$ of a node i, then we create at least one child of i. In this new child, the first seed not completed by greedy, let's say s^i_j, is updated with one vertex. Since at the first level $|s^i_j| = t + 1$, then s^i_j could be completed into an H-subgraph in at most $r - (t + 1)$ levels (which are not necessarily consecutive).

By Lemma 5, at most $k - 1$ H-subgraphs are completed in this way since the last k H-subgraph is completed in greedy fashion. Therefore, we need at most $(r - t - 1)k - 1$ levels to complete the k seeds in $\mathbf{Q^i}$.   □

Lemma 8. *A node i at level h can have at most $r(k - 1)$ children if $h \leq (r - t - 2)k$ and at most $r(k - m - 1)$ where $m = h - (r - t - 2)k$ if $h > (r - t - 2)k$.*

Proof. There is a child of i for each vertex in $I(\mathbf{Q^{gr}}, \text{EligibleSponsors}(s^i_j))$, where s^i_j is the first seed not completed by greedy. This is the set of vertices shared by each eligible sponsor of s^i_j with the H-subgraphs completed by greedy, i.e., $\mathbf{Q^{gr}}$. Therefore, $I(\mathbf{Q^{gr}}, \text{EligibleSponsors}(s^i_j)) \leq r|\mathbf{Q^{gr}}|$.

The greedy algorithm needs to complete only the seeds in $\mathbf{Q^i}$ that are not already completed H-subgraphs. Therefore, $|\mathbf{Q^{gr}}| \leq |\mathbf{Q^i}| - m$ where m is the number of seeds of $\mathbf{Q^i}$ that are completed H-subgraphs.

Assuming all the seeds of $\mathbf{Q^i} - m$ were completed by greedy but the last one, i.e., $|\mathbf{Q^{gr}}| \leq |\mathbf{Q^i}| - m - 1$, then $I(\mathbf{Q^{gr}}, \text{EligibleSponsors}(s^i_j)) < r(|\mathbf{Q^i}| - m - 1) \leq r(k - m - 1)$.

Now, we need to determine how many seeds of $\mathbf{Q^i}$ are already H-subgraphs at level h, i.e., the value of m. Since a seed s^i_j can be completed into an H-subgraph in at most $r - (t + 1)$ levels but these are not necessarily consecutive, then we cannot guarantee that $\frac{h}{r - (t+1)}$ is the number of H-subgraphs at level h. Therefore, in the worst-case one vertex is added to each seed of $\mathbf{Q^i}$ level by level. In this way, in at most $(r - t - 2)k$ levels every seed of $\mathbf{Q^i}$ could have $r - 1$ vertices, and at level $(r - t - 2)k + 1$ we could obtain the first seed completed into an H-subgraph. After that level, the remaining seeds of $\mathbf{Q^i}$ are completed into H-subgraphs by adding one vertex.   □

Theorem 2. *The k-H-Packing with t-Overlap can be solved in $O(r^{rk}k^{(r-t-1)k+2}n^r)$ time.*

Proof. By Lemmas 7 and 8, the size of the tree is

$$\left(\frac{e^2r}{t+1}\right)^{k(t+1)} \prod_{i=1}^{(r-t-2)k} r(k-1) + \prod_{i=1}^{k-r+t} r(k-i)$$

$$< \left(\frac{e^2r}{t+1}\right)^{k(t+1)} (r(k-1))^{(r-t-1)k-1}.$$

A maximal solution $\mathcal{M}$ can be computed in time $O(krn^r)$, which is also the required time to compute the list of sponsors of the seeds. The greedy algorithm runs in $O(k^2rn^r)$.   □

3.3   Overlapping vs. Disjoint Subgraphs

The number of children that every node can have (Lemma 8) is substantially reduced when overlap is not allowed between the H-subgraphs. At level h, h vertices have been added to the seeds of the collection $\mathbf{Q^i}$. When overlap is not allowed, these vertices cannot update any other seed in $\mathbf{Q^i}$, and there are at most $(r-1-h)(k-1)$ children of a node i.

If we increase the overlap, even as little as $t = 1$, the argument above cannot be applied. At level 1, one vertex has been added to some seed $s_j^i \in \mathbf{Q^i}$, let us suppose it is the vertex v. Since $t = 1$, this vertex could be added to a different seed at level 2. Assume for example that in the $1, \ldots, k$ levels, the vertex v has been added to the seeds $s_1^i, \ldots, s_k^i$, respectively. Therefore at the $k + 1$ level, the vertex v cannot be added to any other seed in the collection $\mathbf{Q^i}$. However at that level, we cannot distinguish between the same vertex added to $k + 1$ different sets, or $k + 1$ different vertices added to the same set, or a combination of both.

4   An FPT-Algorithm for the k-Set Packing with t-Overlap Problem

In this section, we propose an algorithm for the k-Set Packing with t-Overlap problem, which is defined as follows.

The k-Set Packing with t-Overlap problem
Instance: A collection $\mathcal{S}$ of distinct sets, each of size at least $t + 1$ and at most r, drawn from a universe $\mathcal{U}$. In addition, there is no pair Q_i, Q_j in $\mathcal{S}$ such that $Q_i \subset Q_j$. Non-negative integers k and t, where $t < r$.
Parameter: A non-negative integer k.
Question: Does $\mathcal{S}$ contain at least k sets $\mathcal{K} = \{Q_1^*, \ldots, Q_k^*\}$ where $|Q_i^* \cap Q_j^*| \leq t$, for any pair Q_i^*, Q_j^*?

The classical k-Set Packing problem asks for at least k pairwise disjoint sets. Therefore, the k-Set Packing with t-Overlap problem generalizes that condition of the problem.

We now apply the search tree algorithm of Section 3 to this problem. First, Lemma 1 can be restated as follows.

Lemma 9. *Let $\mathcal{M}$ and $\mathcal{K}$ be a maximal Set Packing with t-Overlap and a k-Set Packing with t-Overlap, respectively. We claim that any $Q^* \in \mathcal{K}$ overlaps with some $Q \in \mathcal{M}$ in at least $t+1$ elements, i.e., $|Q^* \cap Q| \geq t+1$. Furthermore, there is no pair $Q_i^*, Q_j^* \in \mathcal{K}$ for $i \neq j$ that overlaps in the same set of elements with $Q \in \mathcal{M}$, i.e., $Q_i^* \cap Q \neq Q_j^* \cap Q$.*

Since the size of each set in $\mathcal{S}$ is at least $t + 1$ and at most r, and there is no pair Q_i, Q_j in $\mathcal{S}$ such that $Q_i \subset Q_j$, then this lemma follows by similar arguments as Lemma 1.

Once again, we have that sets from a k-Set Packing with t-Overlap $\mathcal{K}$ share some elements with sets from a maximal k-Set Packing with t-Overlap $\mathcal{M}$. Thus, we can use the notion of feasible seeds to find a k-Set Packing with t-Overlap $\mathcal{K}$.

For this problem, a *seed* is a subset of size $t + 1$ from a set in $\mathcal{M}$. In this way, the universe of seeds is the set of all possible seeds for each set in $\mathcal{M}$.

Now, given a collection $\mathbf{Q}$ of k seeds we want *to complete it* into a k-Set Packing with t-Overlap. That is, we want to add elements from $\mathcal{U}$ to each seed in $\mathbf{Q}$ such that each updated seed is a set of $\mathcal{S}$, and the overlap between any pair of updated seeds is at most t. In this sense, the term *sponsor* of a seed s is redefined as a set of elements from $\mathcal{U}$ that updates s as a set of $\mathcal{S}$. Specifically, we say that A is a sponsor of s, if $|s \cap A| = 0$ and $s \cup A \in \mathcal{S}$.

The set $\mathsf{Sponsors}(s)$ can be computed as follows. For each set $Q \in \mathcal{S}$, if $s \subset Q$ then a sponsor of s is $Q \backslash s$. Once the set of sponsors for each seed is computed, the bounded search tree algorithm follows with minor differences. For example, now $\mathbf{Q^{gr}}$ would be a collection of sets each of size at most r instead of H-completed subgraphs. In the same way, the rule to discard ineligible sponsors can be applied.

Theorem 3. *The search tree algorithm finds one k-Set Packing with t-Overlap assuming there is at least one.*

Since the sets in $\mathcal{S}$ have size at most r and $|\mathcal{U}| = n$, then $O(n^r)$ is still an upper bound for the number of sponsors of the seeds. Thus, the running time of the algorithm follows as well by Theorem 2.

Theorem 4. *The k-Set Packing with t-Overlap can be solved in* $O(r^{rk} k^{(r-t-1)k+2} n^r)$ *time.*

5   Conclusion

We have introduced the first fixed-parameter algorithm for packing subgraphs with arbitrary overlap (the k-H-Packing with t-Overlap problem). We have also provided an insight of the difficulty of packing overlapping subgraphs rather than vertex-disjoint subgraphs. As discussed in Section 3.3, even overlap of at most one vertex substantially complicates the analysis of the algorithm. On the other hand, we show that the algorithm is applicable for a generalized version of the k-Set Packing problem.

Many leads arise from the results presented here. Naturally, the first one is to improve of the running time of the algorithm. The second is the design of data reduction rules to decrease the number of children of the root or to bound the number of sponsors of the seeds. Finally, alternative parameters besides the number of the subgraphs could be considered as well.

Acknowledgments. We would like to thank our referees for their invaluable comments to improve the presentation of this paper.

References

1. Adamcsek, B., Palla, G., Farkas, I., Derenyi, I., Vicsek, T.: Cfinder: locating cliques and overlapping modules in biological networks. Bioinformatics 22(8), 1021–1023 (2006)
2. Damaschke, P.: Fixed-parameter tractable generalizations of cluster editing. In: The 6th International Conference on Algorithms and Complexity (CIAC), pp. 344–355 (January 2006)
3. Fellows, M., Guo, J., Komusiewicz, C., Niedermeier, R., Uhlmann, J.: Graph-based data clustering with overlaps. Discrete Optimization 8(1), 2–17 (2011)
4. Fellows, M., Heggernes, P., Rosamond, F., Sloper, C., Telle, J.: Finding k disjoint triangles in an arbitrary graph. In: The 30th Workshop on Graph-Theoretic Concepts in Computer Science (WG), pp. 235–244 (2004)
5. Fellows, M., Knauer, C., Nishimura, N., Ragde, P., Rosamond, F., Stege, U., Thilikos, D., Whitesides, S.: Faster fixed-parameter tractable algorithms for matching and packing problems. Algorithmica 52(2), 167–176 (2008)
6. Hartung, S., Komusiewicz, C., Nichterlein, A.: On structural parameterizations for the 2-club problem. In: van Emde Boas, P., Groen, F.C.A., Italiano, G.F., Nawrocki, J., Sack, H. (eds.) SOFSEM 2013. LNCS, vol. 7741, pp. 233–243. Springer, Heidelberg (2013)
7. Komusiewicz, C., Sorge, M.: Finding dense subgraphs of sparse graphs. In: 7th International Symposium on Parameterized and Exact Computation (IPEC), pp. 242–251 (2012)
8. Moser, H.: A problem kernelization for graph packing. In: Nielsen, M., Kučera, A., Miltersen, P.B., Palamidessi, C., Tůma, P., Valencia, F. (eds.) SOFSEM 2009. LNCS, vol. 5404, pp. 401–412. Springer, Heidelberg (2009)
9. Prieto, E., Sloper, C.: Looking at the stars. Theoretical Computer Science 351(3), 437–445 (2006)
10. Romero, J., López-Ortiz, A.: The $\mathcal{G}$-packing with t-overlap problem. In: Pal, S.P., Sadakane, K. (eds.) WALCOM 2014. LNCS, vol. 8344, pp. 114–124. Springer, Heidelberg (2014)
11. Schäfer, A., Komusiewicz, C., Moser, H., Niedermeier, R.: Parameterized computational complexity of finding small-diameter subgraphs. Optimization Letters 6(5), 883–891 (2012)

Crossing-Free Spanning Trees in Visibility Graphs of Points between Monotone Polygonal Obstacles

Julia Schüler and Andreas Spillner

Department of Mathematics and Computer Science, University of Greifswald
{julia.schueler,andreas.spillner}@uni-greifswald.de

Abstract. We consider the problem of deciding whether or not a geometric graph has a crossing-free spanning tree. This problem is known to be NP-hard even for very restricted types of geometric graphs. In this paper, we present an $O(n^5)$ time algorithm to solve this problem for the special case of geometric graphs that arise as visibility graphs of a finite set of n points between two monotone polygonal obstacles. In addition, we give a combinatorial characterization of those visibility graphs induced by such obstacles that have a crossing-free spanning tree. As a byproduct, we obtain a family of counterexamples to the following conjecture by Rivera-Campo: A geometric graph has a crossing-free spanning tree if every subgraph obtained by removing a single vertex has a crossing-free spanning tree.

Keywords: geometric graph, crossing-free spanning tree, polygonal obstacle.

1 Introduction

A *geometric graph* is a graph whose vertices are points in the plane. Two distinct edges $\{u_1, v_1\}$ and $\{u_2, v_2\}$ in such a graph *cross* if the straight line segments $\overline{u_1 v_1}$ and $\overline{u_2 v_2}$ have a point in common that is not an endpoint of both edges. A subgraph of a geometric graph is *crossing-free* if it does not contain any crossing edges (cf. Fig. 1(a)). Rivera-Campo [21] gave the following sufficient condition for the existence of a crossing-free spanning tree in a geometric graph $G = (V, E)$ with $n \geq 5$ vertices:

(I_5) For every 5-element subset $U \subseteq V$, the induced subgraph $G[U]$ has a crossing-free spanning tree.

He conjectured that the constant 5 in condition (I_5) can be replaced by $n - 1$ which, in turn, would imply that it can be replaced by any $k \in \{2, 3, \ldots, n\}$, yielding a family of conditions (I_k). Moreover, he showed that condition (I_k) is indeed sufficient for the existence of a crossing-free spanning tree for all $k \in \{2, 3, \ldots, n\}$ if the vertex set of the geometric graph is in convex position.

In this paper, we present, for every $n \geq 16$, a geometric graph that satisfies condition (I_{n-1}) but does not have a crossing-free spanning tree. We obtained these counterexamples as a byproduct when exploring computationally tractable

E.A. Hirsch et al. (Eds.): CSR 2014, LNCS 8476, pp. 337–350, 2014.
© Springer International Publishing Switzerland 2014

(a)   (b)   (c)

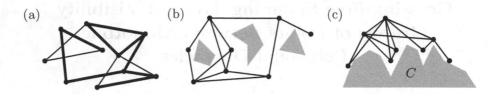

Fig. 1. (a) A geometric graph G. The bold edges yield a crossing-free subgraph of G, but G has no crossing-free spanning tree. (b) The geometric graph $(V, E(\mathcal{O}))$ induced on a set V of points by a collection $\mathcal{O}$ of three polygonal obstacles (drawn shaded). (c) A geometric graph $G = (V, E(\{C\}))$ induced by a single monotone obstacle C below V.

variants of the following decision problem CROSSING-FREE SPANNING TREE (CFST): Given a geometric graph G. Does there exist a crossing-free spanning tree in G? Jansen and Woeginger [10] showed that this problem is NP-hard even for geometric graphs with just two different edge lengths or with just two different edge slopes. Here we restrict to geometric graphs $G = (V, E(\mathcal{O}))$ that are induced on a set V of points in the plane by a collection $\mathcal{O}$ of obstacles (cf. Fig. 1(b)). In general, the obstacles $P \in \mathcal{O}$ are pairwise disjoint open polygons with $P \cap V = \emptyset$ and two distinct points $u, v \in V$ form an edge $\{u, v\} \in E(\mathcal{O})$ if the straight line segment $\overline{uv}$ does not intersect any $P \in \mathcal{O}$. To clearly distinguish between the vertices of $G = (V, E(\mathcal{O}))$ and the vertices of the polygons in $\mathcal{O}$, we will refer to the latter as *corners*.

It is easy to see that, without any restrictions on the collection of obstacles, there exists, for every geometric graph $G = (V, E)$, a collection of obstacles $\mathcal{O}$ with $G = (V, E(\mathcal{O}))$. Therefore, to explore potentially tractable special cases of CFST, we restrict to obstacles C that are *monotone* (cf. Fig. 1(c)), that is, the intersection of C with any vertical straight line is connected. A monotone obstacle is *below/above* V if, for all points $v \in V$, the vertical ray emanating downwards/upwards from v intersects C.

In this paper, we present a characterization for when a geometric graph induced above a single monotone obstacle has a crossing-free spanning tree (Section 2). The attempt to generalize this result to all geometric graphs yields the counterexamples to the conjecture by Rivera-Campo mentioned above. In addition, we show that the geometric graphs obtained from collections consisting of precisely two monotone obstacles, one below and one above V, allow to solve CFST in $O(n^5)$ time where n is the number of vertices of the input graph (Section 3). We conclude the paper mentioning some open problems and possible directions for future work. The remainder of this section gives a brief overview over previous and related work.

Crossing-Free Spanning Trees: Knauer et al. [15] showed that the problem CFST is fixed-parameter tractable with respect to the number of pairs of crossing edges in the input graph $G = (V, E)$. Subsequently, Halldórsson et al. [9] improved the run time and, in addition, established that CFST is also fixed-parameter tractable with respect to the number of vertices that lie in the interior of the convex hull of V. Rivera-Campo and Urrutia-Galicia showed that condition (I_3) can be relaxed: It suffices that there are at most $n - 3$ subsets $U \subseteq V$ with $|U| = 3$ for which (i) the interior of the convex hull of U does not contain a vertex $v \in V$ and (ii) the induced subgraph $G[U]$ has no crossing-free spanning tree. A well-known result established by Károlyi et al. [11] states that there is no geometric graph $G = (V, E)$ such that neither G nor its complement $G^c = (V, \binom{V}{2} - E)$ has a crossing-free spanning tree. A complete characterization of the minimal geometric graphs G such that G^c has no crossing-free spanning tree is given by Keller et al. [12].

Obstacle Representations: Given an abstract graph $G' = (V', E')$, an *obstacle representation* of G' is a geometric graph $G = (V, E(\mathcal{O}))$ induced by a collection $\mathcal{O}$ of obstacles on a set V of points such that G' is isomorphic to G. Alpert et al. [1] showed that every outerplanar graph has an obstacle representation $(V, E(\mathcal{O}))$ with $|\mathcal{O}| = 1$ and they present a family of graphs that require an arbitrarily large number of obstacles to represent them. Subsequently, Pach and Sariöz [20] showed that there are even bipartite graphs that require an arbitrarily large number of obstacles, Mukkamala et al. [19] presented a lower bound of $\Omega(n/\log n)$ on the number of obstacles needed in the worst case for representing a graph with n vertices and Koch et al. [16] gave a characterization of the biconnected graphs that have a representation where all obstacles lie in the unbounded region of the plane outside of the induced geometric graph. Given an obstacle representation $G = (V, E(\mathcal{O}))$ such that $\mathcal{O}$ consists of a single polygon P that has precisely one hole, and all points in V are contained in this hole, Cheng et al. [3] presented an algorithm that, for any two vertices $s, t \in V$, decides whether or not G contains a crossing-free path from s to t. The run time of the algorithm is $O(n^2 m^2)$, where m is the number of corners of P and $n = |V|$. Later, Daescu and Luo [5] improved the run time to $O(n^3 \log m + nm)$.

Monotone Obstacles: A related problem on monotone obstacles that has received considerable attention over the last years is that of *optimal guarding*. In this problem, we want to compute a minimum size set W of points, also called watchmen, on the upper boundary U of a monotone obstacle C such that for every point $p \in U$ there exists a watchman $w \in W$ that *sees* p, that is, the straight line segment $\overline{wp}$ does not intersect C. There was a series of papers [2,4,7,13] presenting constant-factor approximation algorithms for this problem, which was only recently shown to be NP-hard by King and Krohn [14] (see also [17,18] for hardness results on closely related problems). Our inspiration for restricting the problem CFST to geometric graphs induced by monotone obstacles came from the work by Gibson et al. [8] who showed that there is a polynomial time approximation scheme for optimal guarding monotone obstacles while more general versions of this problem are APX-hard [6].

2   A Combinatorial Characterization

Throughout this paper, we assume that the vertices of geometric graphs and the corners of obstacles are in general position, that is, no three of them are collinear and no two of them have the same x- or y-coordinate. Moreover, for any two points p and q in the plane, we will say that p lies to the *left/right* of q if p has a smaller/larger x-coordinate than q. Often, when the intended meaning is clear from the context, we will identify an edge $\{u, v\}$ in a geometric graph and the straight line segment $\overline{uv}$. We denote, for any finite non-empty set V of points in the plane, by $ch_{\text{top}}(V)$ the *upper boundary* of the convex hull of V, that is, the set of all those points that we meet when moving in clockwise direction along the boundary of the convex hull of V from its leftmost to its rightmost point. In addition, for any geometric graph

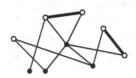

Fig. 2. A geometric graph $G = (V, E)$ with the vertices in V_{top} drawn as empty circles and the edges in E_{top} drawn bold

$G = (V, E)$, we denote by V_{top} the set of those vertices in V that are contained in $ch_{\text{top}}(V)$ and, similarly, by E_{top} the set of those edges in E that are contained in $ch_{\text{top}}(V)$ (cf. Fig. 2). We will use the following observation from [21].

Observation 1. *Let $G = (V, E)$ be a geometric graph with $n \geq 3$ vertices that are in convex position. Then every crossing-free spanning tree of G must contain at least two edges that lie on the boundary of the convex hull of V.*

In addition, we also rely on the following fact that can be viewed as a dual version of Observation 1 and that can easily be verified by induction on the number of vertices of the geometric graph (it clearly holds for graphs with $n = 3$ vertices and, for $n \geq 4$ vertices, any edge naturally partitions the graph into two smaller subgraphs to which the induction hypothesis can be applied).

Observation 2. *Let $G = (V, E)$ be a geometric graph with $n \geq 3$ vertices that are in convex position. If G contains no edges that lie on the boundary of the convex hull of V then every crossing-free spanning subgraph of G has at least three connected components.*

Next, we explore some properties that are very similar in spirit to Rivera-Campo's conditions (I_k) mentioned in the introduction. To formally describe these properties, we first introduce some more notation. Let $G = (V, E)$ be a geometric graph with n vertices and let $k \in \{0, 1, 2, \ldots, n\}$. We say that G is k-*Steiner* if for every k-element subset $K \subseteq V$ there exists a crossing-free subtree $T' = (V', E')$ of G with $K \subseteq V'$. Note that such a tree T' can be viewed as a crossing-free *Steiner tree* in G for the terminal vertices in K.

Our original motivation for looking into crossing-free Steiner trees was to identify interesting families $\mathcal{F}$ of geometric graphs for which there exists a small constant k^* such that a graph $G \in \mathcal{F}$ has a crossing-free spanning tree if and only if G is k^*-Steiner. Then, assuming that there exists a polynomial time algorithm **A** that decides for every $G = (V, E) \in \mathcal{F}$ and every k^*-element subset $K \subseteq V$ whether or not there exists a crossing-free Steiner tree for K in G, we would immediately obtain a polynomial time algorithm for CFST when restricted to the family $\mathcal{F}$. As mentioned in the introduction, at least for $k^* = 2$,

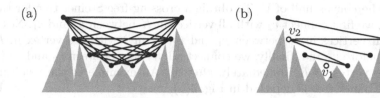

Fig. 3. (a) The geometric graph G_n considered in the proof of Theorem 4 for $n = 8$. (b) A crossing-free Steiner tree in G_8 for a set of $n - 3 = 5$ terminal vertices (non-terminal vertices marked by empty circles).

that is, crossing-free paths between two specified vertices, there exists such an algorithm **A** for geometric graphs induced by certain polygonal obstacles [3,5]. Unfortunately, as we will see below, this overall approach does not even work for geometric graphs induced above a single monotone obstacle because there exist such graphs G that are k-Steiner for arbitrarily large values of k but have no crossing-free spanning tree. First, we present a technical lemma (proof omitted). Based on this lemma, we obtain the above-mentioned characterization.

Lemma 3. *Let $G = (V, E(\{C\}))$ be a connected geometric graph that is induced on a set V of n points above a monotone obstacle C. If G is $(n - |V_{top}|)$-Steiner then G has a crossing-free spanning tree.*

Theorem 4. *A geometric graph $G = (V, E(\{C\}))$ that is induced on a set V of $n \geq 4$ points above a monotone obstacle C has a crossing-free spanning tree if and only if G is $(n - 2)$-Steiner. This equivalence does not hold for any value smaller than $n - 2$.*

Proof: Clearly, if G has a crossing-free spanning tree then it is k-Steiner for any $k \in \{0, 1, 2, \ldots, n\}$. For the converse direction, assume that G is $(n - 2)$-Steiner. Since $n \geq 4$, this implies that G is connected. Moreover, since $|V_{top}| \geq 2$, it also implies that G is $(n - |V_{top}|)$-Steiner. Hence, by Lemma 3, G must have a crossing-free spanning tree.

To see that $n - 2$ is the smallest possible value, fix an arbitrary $n \geq 4$. We consider a graph resulting from a complete geometric graph with n vertices in convex position by removing all except one of the edges that lie on the boundary of the convex hull of its vertex set. Such a geometric graph can easily be obtained as $G_n = (V, E(\{C\}))$ for some suitably chosen set V of n points above some monotone obstacle C (cf. Fig. 3(a)).

First note that, by Observation 1, the graph G_n cannot have a crossing-free spanning tree. Thus, by the equivalence already established, G_n cannot be k-Steiner for any $k \in \{n - 2, n - 1, n\}$. Hence, it remains to show that G_n is $(n - 3)$-Steiner: Consider an arbitrary set $K \subseteq V$ of $n - 3$ terminal vertices. Let v_1, v_2 and v_3 be the non-terminal vertices in $V - K$ numbered in clockwise order around the convex hull of V (the terms between/before/after used in the following always refer to this order). Since $n \geq 4$, we can assume without loss of generality that there is at least one vertex in K between v_3 and v_1 on the boundary of the convex hull of V. To obtain a crossing-free Steiner tree for the vertices in K, we first connect v_2 with all vertices in K between v_3 and v_1. Next, we connect all vertices in K between v_1 and v_2 (if any) with the vertex in K immediately before v_1 and, finally, we connect all vertices between v_2 and v_3 (if any) with the vertex in K immediately after v_3. An example of the resulting crossing-free Steiner tree is depicted in Fig. 3(b). ∎

When exploring the situation for general geometric graphs, we found that there are such graphs with n vertices that have no crossing-free spanning tree even though they are $(n - 1)$-Steiner. Since a crossing-free Steiner tree for an $(n - 1)$-element set K of terminal vertices in such a graph must actually be a crossing-free spanning tree for the subgraph induced by K, this immediately gives counterexamples to Rivera-Campo's conjecture.

Lemma 5. *For all $n \geq 16$, there exists a geometric graph $G = (V, E)$ with n vertices that satisfies condition (I_{n-1}) but has no crossing-free spanning tree.*

Proof: Fix an arbitrary $n \geq 16$. To construct a suitable geometric graph $G = (V, E)$ with n vertices, we first arrange a set V_1 of $n - 8$ points in convex position such that $v_1, v_2, \ldots, v_{n-8}$ is the order of these points along the boundary of the convex hull of V_1 in clockwise direction. We arrange V_1 in such a way that v_1 is the point with largest y-coordinate and v_8 is the point with smallest y-coordinate. Similarly, we arrange a set V_2 of 8 points in convex position. Let $u_1, u_2, \ldots, u_8$ be the order of the points in V_2 in counterclockwise direction along the boundary of the convex hull of V_2. The points are arranged in such a way that u_1 is the point with largest y-coordinate and u_8 is the point with smallest y-coordinate.

Now, for $i \in \{1, 2\}$, let $G_i = (V_i, E_i)$ denote the geometric graph that we obtain from the complete geometric graph on vertex set V_i by removing all edges that lie on the boundary of the convex hull of V_i. Putting

$$E^* = \{\{v_j, u_{j+1}\} : j \in \{1, 3, 5, 7\}\} \cup \{\{v_j, u_{j-1}\} : j \in \{2, 4, 6, 8\}\},$$

we obtain our final geometric graph G with vertex set $V = V_1 \cup V_2$ and edge set $E = E_1 \cup E_2 \cup E^*$. By placing G_2 far enough away to the right of G_1 we ensure that no edge in E^* crosses an edge in $E_1 \cup E_2$ (cf. Fig. 4(a)). Note that the edges in E^* form four disjoint pairs of crossing edges.

We first argue that G has no crossing-free spanning tree: By Observation 2, the restriction of any crossing-free spanning subgraph of G to V_i, $i \in \{1, 2\}$, has at least three connected components. Hence, a crossing-free spanning tree of G would need to use at least five edges that are neither in E_1 nor in E_2. But, by construction of G, a crossing-free subgraph of G can use at most four edges from $E - (E_1 \cup E_2) = E^*$.

It remains to show that for every $(n-1)$-element subset $V' = V - \{w\}$, $w \in V$, there exists a crossing-free spanning tree for the subgraph of G induced by V'. In view of the high degree of symmetry of G, it suffices to consider the case that $w \in V_1$ and, as indicated in Fig. 4(b)-(f), there are only five different types of $(n - 1)$-element subsets V' that result from removing a vertex w from V_1. It is easy to check that, for each of them, a crossing-free spanning tree exists. ∎

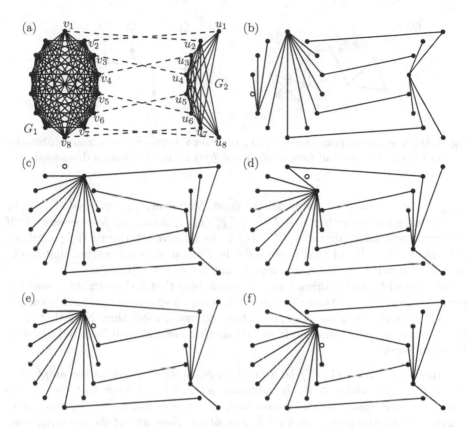

Fig. 4. (a) The geometric graph G constructed in the proof of Lemma 5 for $n = 22$. The eight edges in E^* between the two subgraphs G_1 and G_2 are drawn dashed. (b)-(f) The empty circle marks the vertex $w \in V_1$ that is removed to obtain an $(n-1)$-element subset V'. There is always a crossing-free spanning tree for the subgraph of G induced by V' but its structure depends on the position of w: (b) $w = v_j$ for some $j \in \{9, 10, \ldots, n-8\}$, (c)-(f) $w = v_j$ for some $j \in \{1, 2, 3, 4\}$.

3   A Polynomial Time Algorithm

In this section, we present a polynomial time algorithm for CFST restricted to geometric graphs $G = (V, E(\mathcal{O}))$ where $\mathcal{O}$ consists of precisely two monotone obstacles C_a and C_b that lie above and below V, respectively (cf. Fig. 5(a)).

3.1   Definitions and Key Facts

For any edge $e = \{u, v\} \in E(\mathcal{O})$ we define the set $\overline{R}(e)$ of those vertical rays that emanate downwards from a point on e and contain a point in V. We sort the rays in $\overline{R}(e)$ from left to right and refer to them as $\overline{r}_1, \overline{r}_2, \ldots, \overline{r}_\ell$, $\ell = |\overline{R}(e)|$ (cf. Fig. 5(b)). Similarly, we define the set $\underline{R}(e)$ of those vertical rays that emanate upwards from a point on e and contain a point in V. We refer to them as

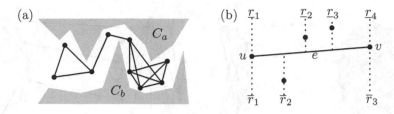

Fig. 5. (a) A geometric graph induced on a point set V between two monotone obstacles C_a and C_b. (b) The vertical rays in $\overline{R}(e)$ and $\underline{R}(e)$ emanating from e downwards and upwards, respectively.

$\underline{r}_1, \underline{r}_2, \ldots, \underline{r}_{\ell'}$, $\ell' = |\underline{R}(e)|$, from left to right. Moreover, for any edge $e = \{u, v\}$ in a crossing-free spanning tree $T' = (V, E')$ of G, a ray $\underline{r} \in \underline{R}(e)$ *crosses* T' if (i) $\underline{r}$ intersects some edge $e' \in E' - \{e\}$ in its relative interior or (ii) $\underline{r}$ contains a vertex $w \in V - \{u, v\}$ that is adjacent in T' to at least one vertex that lies to the left of w and to at least one vertex that lies to the right of w.

We start with two technical lemmas establishing that if there exists a crossing-free spanning tree of G then there also exists such a spanning tree having certain special properties that we will exploit later on in our algorithm. Note that these lemmas also apply to symmetric situations that are obtained by a reflection on the x- or y-axis.

Lemma 6. *Let* $T' = (V, E')$ *be a crossing-free spanning tree of the graph* $G = (V, E(\{C_a, C_b\}))$ *and* $q \in V$. *In addition, let* $g \in E'$ *be such that there exists a ray* $\underline{r} \in \underline{R}(g)$ *that contains* q *and such that* $\underline{r}$ *does not cross any edge of* T' *between its starting point* p *and* q *(cf. Fig. 6(a)). Then one of the following must hold:*

(i) There exists a crossing-free spanning tree of G *that is obtained by removing* g *from* T' *and replacing it by some other edge* g' *in* G *that does not cross the straight line segment* $\overline{pq}$ *(cf. Fig. 6(b)).*

(ii) $\underline{R}(g)$ *contains a ray* $\underline{r}'$ *that does not cross* T' *(cf. Fig. 6(c)).*

Proof: We direct all edges in T' away from q. In the following, we only consider the case that g is directed from its left endpoint u to its right endpoint v (the other case is completely symmetric). Let V_v denote the set of those vertices in V that can be reached from v by following the directed edges of T'.

Now, in addition to C_a and C_b we also view the edges of T' as obstacles and consider the shortest path π from v to q that avoids all these obstacles and is homotopic to the piece-wise linear curve C that we traverse when moving from v along g to p and then upwards along $\underline{r}$ to q. Note that, intuitively, π is obtained by pulling C tight and, therefore, is a polygonal path with vertices $v = v_0, v_1, \ldots, v_\ell = q$. Let i be the smallest index in $\{0, 1, \ldots, \ell\}$ such that $v_i \notin V_v$. Note that this index must exist since $q \notin V_v$. To finish the proof, we distinguish two cases.

Case 1: $v_i \in V - V_v$. Then we replace edge g by the edge $g' = \{v_{i-1}, v_i\}$ and obtain again a crossing-free spanning tree.

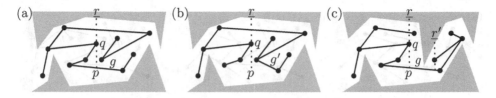

Fig. 6. (a) A crossing-free spanning tree T' as described in the assumptions of Lemma 6. (b) An alternative crossing-free spanning tree obtained from the tree T' in (a) by replacing edge g by g'. (c) A crossing-free spanning tree T' for which there exists a ray $\underline{r}' \in \underline{R}(g)$ that does not cross T'.

Case 2: $v_i \notin V$. Then v_i is a corner of C_a and, by the construction of π, this implies that there exists some $\underline{r}' \in \underline{R}(g)$ that does not cross T'.   ■

The proof of the next lemma uses a similar argument and is omitted.

Lemma 7. *Let $T' = (V, E')$ be a crossing-free spanning tree of the graph $G = (V, E(\{C_a, C_b\}))$ and $g \in E'$. In addition, let $q \in V$ be a vertex that lies on a ray $\underline{r} \in \underline{R}(g)$ that does not cross T' (q may be the left endpoint of g but not the right endpoint). Further, assume that all vertices adjacent to q in T' lie to the right of q. Let h denote the edge incident to q in T' with minimum slope (cf. Fig. 7(a)). Then one of the following must hold:*

(i) There exists a crossing-free spanning tree T'' of G that contains edges g and h, the ray $\underline{r}$ does not cross T'', all vertices adjacent to q in T'' lie to the right of q and there exists a ray $\underline{r}' \in \underline{R}(h) - \{\underline{r}_1\}$ that does not cross T'' (cf. Fig. 7(b)).

(ii) There exists some edge h' incident to q in T' for which there exists a ray $\overline{r}' \in \overline{R}(h')$ that does not cross T' (cf. Fig. 7(c)).

3.2   Types of Subproblems

The facts collected in the previous section suggest that vertical rays and edges of $G = (V, E(\{C_a, C_b\}))$ may be used to partition the given instance of CFST into independent subproblems. In the following we first describe the three types of subproblems that can arise and then how we process each type. Note that, as before, for each type we describe below there are symmetric versions obtained by reflection on the x- or y-axis.

A subproblem of *Type (1)* is defined by an edge e of G and two rays $\underline{r}_i \in \underline{R}(e)$ and $\overline{r}_j \in \overline{R}(e)$ with $\max\{i, j\} > 1$ and, in addition, $i \in \{1, |\underline{R}(e)|\}$ or $j \in \{1, |\overline{R}(e)|\}$ (cf. Fig. 8(a)). Let $\mathcal{R}(e, \underline{r}_i, \overline{r}_j)$ denote the closed region that lies to the right of the piece-wise linear curve that we traverse by moving from $-\infty$ along $\overline{r}_j$ to its starting point, then along e to the starting point of $\underline{r}_i$ and then along $\underline{r}_i$ to $+\infty$. We have to decide whether there exists a crossing-free spanning tree $T' = (V', E')$ for the set V' of those vertices of G that lie in $\mathcal{R}(e, \underline{r}_i, \overline{r}_j)$ such that all edges $e' \in E'$ are completely contained in $\mathcal{R}(e, \underline{r}_i, \overline{r}_j)$ and such that none of these edges crosses e.

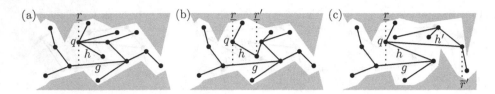

Fig. 7. (a) A crossing-free spanning tree T' as described in the assumptions of Lemma 7. (b) An alternative crossing-free spanning tree T'' obtained from the tree T' in (a). (c) A crossing-free spanning tree T' for which there exists an edge h' incident to q and a ray $\bar{r}' \in \overline{R}(h')$ that does not cross T'.

A subproblem of *Type (2)* is defined by an edge e of G and two rays $\bar{r}_i, \bar{r}_j \in \overline{R}(e)$ with $i \leq j$ (cf. Fig. 8(b)). Let $\mathcal{R}(e, \bar{r}_i, \bar{r}_j)$ denote the closed region that lies to the right of the piece-wise linear curve that we traverse by moving from $-\infty$ along $\bar{r}_i$ to its starting point, then along e to the starting point of $\bar{r}_j$ and then along $\bar{r}_j$ back to $-\infty$. We have to decide whether there exists a crossing-free spanning tree $T' = (V', E')$ for the set V' of those vertices of G that lie in $\mathcal{R}(e, \bar{r}_i, \bar{r}_j)$.

A subproblem of *Type (3)* is defined by two edges e and f of G that have their left endpoint in common and such that the slope of f is less than the slope of e together with a ray $\underline{r}_i \in \underline{R}(f)$ that does not cross e (cf. Fig. 8(c)). Let $\mathcal{R}(e, f, \underline{r}_i)$ denote the closed region that lies to the right of the piece-wise linear curve that we traverse by moving from $+\infty$ along $\underline{r}_i$ to its starting point, then along f to its left endpoint, then along e to its right endpoint and then along the rightmost ray $\underline{r} \in \underline{R}(e)$ to $+\infty$. We have to decide whether there exists a crossing-free spanning tree $T' = (V', E')$ for the set V' of those vertices of G that lie in $\mathcal{R}(e, f, \underline{r}_i)$ such that all edges $e' \in E'$ are completely contained in $\mathcal{R}(e, \underline{r}_i, \bar{r}_j)$.

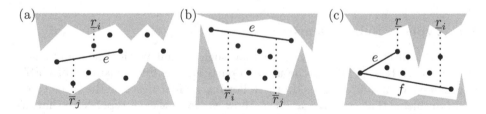

Fig. 8. (a) A subproblem of Type (1). (b) A subproblem of Type (2). (c) A subproblem of Type (3).

It follows immediately from the description of the three types of subproblems above that there are at most $O(n^4)$ subproblems of each type. Thus, to obtain a polynomial time algorithm based on dynamic programming, it suffices to establish that the solution for each of these subproblems can be derived efficiently from the solutions of other, smaller subproblems that have one of the types described above too. This is made more precise in the following three lemmas and the outline of their proofs.

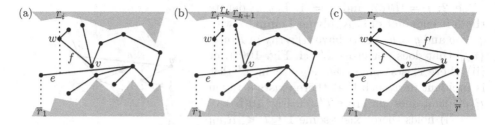

Fig. 9. (a) Processing a subproblem of Type (1), Case 1. (b)-(c) There are two different ways how the problem may be partitioned into two smaller subproblems, but it will always be into a subproblem of Type (1) and a subproblem of Type (2).

Lemma 8. *Suppose there exists a crossing-free spanning tree $T' = (V', E')$ for a subproblem of Type (1) with $|V'| > 1$. Then this subproblem can be partitioned into one or two smaller subproblems of Type (1), (2) or (3) that admit a crossing-free spanning tree. The total number of subproblems that need to be considered is in $O(n)$.*

Proof: We continue to use the notation introduced above to describe a subproblem of Type (1). The focus of the proof will be on describing how the smaller subproblems arise. It is then not hard to check that the total number of relevant smaller subproblems is in $O(n)$. We distinguish four cases.

Case 1: $1 < i < |\underline{R}(e)|$ and $j = 1$. Let w denote the vertex in V' that is contained in $\underline{r}_i$ and let v denote the vertex in V' that lies above or on edge e and for which the slope of the edge $f = \{v, w\}$ is minimum (cf. Fig. 9(a)). Note that f cannot cross any edge of T'. Thus, we assume that $f \in E'$. Then, applying Lemma 7 with $q = w$, $g = e$ and $h = f$, there must (i) exist some ray $\underline{r} \in \underline{R}(f)$ that does not cross T' or (ii) T' contains an edge f' incident to w for which some ray $\overline{r} \in \overline{R}(f')$ does not cross T' (cf. Fig. 9(c)).

If (i) holds, the construction of f implies that there exists some $i \leq k < |\underline{R}(e)|$ such that both $\underline{r}_k \in \underline{R}(e)$ and $\underline{r}_{k+1} \in \underline{R}(e)$ have a nonempty intersection with f but do not cross any other edge of T'. In particular, T' can be partitioned into a subtree T_1' that is a crossing-free spanning tree for the subproblem of Type (2) defined by e, $\underline{r}_i$ and $\underline{r}_k$ and a subtree T_2' that is a crossing-free spanning tree for the subproblem of Type (1) defined by e, $\underline{r}_{k+1}$ and $\overline{r}_1$ (cf. Fig. 9(b)). Note that T_1' and T_2' are linked together by edge f to form T'.

If (ii) holds, we repeatedly apply Lemma 6 to the right endpoint u of e. The existence of edge f' in T' implies that it is possible to replace all edges in T' that are crossed by the ray $\overline{r}_k \in \overline{R}(e)$, $k = |\overline{R}(e)|$. Let $T'' = (V', E'')$ denote the resulting crossing-free spanning tree for V'. This tree can be partitioned into a subtree T_1'' that is a crossing-free spanning tree for the subproblem of Type (2) defined by e, $\overline{r}_1$ and $\overline{r}_k$ and a subtree T_2'' that is a crossing-free spanning tree for the subproblem of Type (1) defined by e, $\underline{r}_i$ and $\overline{r}_k$. Note that T_1'' and T_2'' are glued together at vertex u to form T'.

Case 2: $i = |\underline{R}(e)|$ and $j = 1$. Let w denote the left endpoint of e. Applying Lemma 7 with $q = w$ and $g = h = e$ we have (i) a ray in $\overline{R}(e) - \{\overline{r}_1\}$ that does not cross T' (cf. Fig. 10(a)) or (ii) there exists some edge $e' \neq e$ in T' that is incident to w such that there exists a ray in $\underline{R}(e')$ that does not cross T' (cf. Fig. 10(b)).

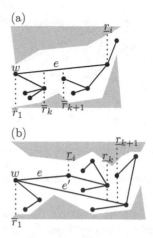

If (i) holds there exists some $1 \leq k < |\overline{R}(e)|$ such that both $\overline{r}_k \in \overline{R}(e)$ and $\overline{r}_{k+1} \in \overline{R}(e)$ do not cross T'. Thus, T' can be partitioned into a subtree T'_1 that corresponds to the subproblem of Type (2) defined by e, $\overline{r}_1$ and $\overline{r}_k$ and a subtree T'_2 that corresponds to the subproblem of Type (1) defined by e, $\overline{r}_{k+1}$ and $\underline{r}_i$. These two subtrees are linked by edge e.

If (ii) holds there exists some $1 < k < |\underline{R}(e')|$ such that both $\underline{r}_k \in \underline{R}(e')$ and $\underline{r}_{k+1} \in \underline{R}(e')$ do not cross T'. Thus, T' can be partitioned into a subtree T'_1 that corresponds to the subproblem of Type (3) defined by e, e' and $\underline{r}_k$ and the subtree T'_2 that corresponds to the subproblem of Type (1) defined by e', $\overline{r}_1$ and $\underline{r}_{k+1}$. These two subtrees are glued together at vertex w.

Fig. 10. Processing a subproblem of Type (1), Case 2. (a) Two resulting smaller subproblems of Type (1) and (2). (b) Two resulting smaller subproblems of Type (1) and (3).

The remaining two cases can be handled in a similar way.   ∎

Subproblems of Type (2) and (3) can be processed in a similar way as described above for subproblems of Type (1) in Case 1:

Lemma 9. *Suppose there exists a crossing-free spanning tree $T' = (V', E')$ for a subproblem of Type (2) with $|V'| > 1$. Then this subproblem can be partitioned into two smaller subproblems of Type (2) that admit a crossing-free spanning tree. The total number of subproblems that need to be considered is in $O(n)$.*

Lemma 10. *Suppose there exists a crossing-free spanning tree $T' = (V', E')$ for a subproblem of Type (3) in which $\underline{r}_i$ does not contain the right endpoint of e. Then this subproblem can be partitioned into two smaller subproblems, one of Type (2) and one of Type (3), that both admit a crossing-free spanning tree. The total number of subproblems that need to be considered is in $O(n)$.*

3.3   Summary of the Algorithm

It is now not difficult to design a dynamic programming algorithm for solving CFST for a geometric graph $G = (V, E(\{C_a, C_b\}))$ induced by two monotone obstacles. Note that in our run time analysis below we assume that G is explicitly given, that is, we ignore the time it would take to compute the edge set $E(\{C_a, C_b\})$ if the input were only the point set V and the monotone obstacles C_a and C_b.

Theorem 11. *There is a dynamic programming algorithm that, for any geometric graph $G = (V, E(\{C_a, C_b\}))$ that is induced on a set V of n points between two monotone obstacles C_a and C_b decides in $O(n^5)$ time whether or not G has a crossing-free spanning tree.*

Proof: The input graph G can be viewed as a family of subproblems of Type (1): Let w be the leftmost vertex in G. We consider each edge e that is incident to w in G and consider the subproblem of Type (1) defined by e, $\underline{r}_2 \in \underline{R}(e)$ and $\bar{r}_1 \in \overline{R}(e)$. Clearly, there exists a crossing-free spanning tree of G if and only if at least one of these subproblems has a crossing-free spanning tree.

As observed in Section 3.2, there are $O(n^4)$ subproblems in total. It follows from Lemmas 8, 9 and 10, that each of these subproblems can be processed in $O(n)$ time assuming that all relevant smaller subproblems have already been solved and the solutions have been recorded in a dynamic programming table. Filling all entries in the table then takes $O(n^5)$ time. ∎

4 Concluding Remarks

In this paper, we have started to explore properties of crossing-free spanning trees in geometric graphs that are induced on a point set in the plane by special types of polygonal obstacles. To illustrate that this may indeed lead to interesting tractable instances of the problem CFST, we showed that for graphs induced between two monotone obstacles it can be solved in polynomial time. In view of the fact that the existence of a crossing-free path between two specified vertices can be decided in polynomial time even for geometric graphs that are induced by a single non-monotone polygonal obstacle, it would be interesting to know whether CFST can also be solved in polynomial time on these more general instances too. Moreover, our counterexamples to Rivera-Campo's conjecture immediately raise the following question: What is the largest number $k^* \in \mathbb{N} - \{0, 1\}$ such that, for all geometric graphs $G = (V, E)$ with $n \geq k^*$ vertices, condition (I_{k^*}) implies the existence of a crossing-free spanning tree in G? Combining the results in this paper with those in [21], we obtain the bounds $5 \leq k^* \leq 14$.

Acknowledgments. We would like to thank Alexander Wolff for initiating the work presented here while hosting the second author at the University of Würzburg. In addition, we would like to thank him and Ivo Vigan as well as the anonymous reviewers for their helpful comments.

References

1. Alpert, H., Koch, C., Laison, J.: Obstacle numbers of graphs. Discrete & Computational Geometry 44, 223–244 (2010)
2. Ben-Moshe, B., Katz, M., Mitchell, J.: A constant-factor approximation algorithm for optimal 1.5D terrain guarding. SIAM Journal on Computing 36, 1631–1647 (2007)

3. Cheng, Q., Chrobak, M., Sundaram, G.: Computing simple paths among obstacles. Computational Geometry 16, 223–233 (2000)
4. Clarkson, K., Varadarajan, K.: Improved approximation algorithms for geometric set cover. Discrete & Computational Geometry 37, 43–58 (2007)
5. Daescu, O., Luo, J.: Computing simple paths on points in simple polygons. In: Ito, H., Kano, M., Katoh, N., Uno, Y. (eds.) KyotoCGGT 2007. LNCS, vol. 4535, pp. 41–55. Springer, Heidelberg (2008)
6. Eidenbenz, S.: Inapproximability results for guarding polygons without holes. In: Chwa, K.-Y., Ibarra, O.H. (eds.) ISAAC 1998. LNCS, vol. 1533, pp. 427–436. Springer, Heidelberg (1998)
7. Elbassioni, K., Krohn, E., Matijević, D., Mestre, J., Ševerdija, D.: Improved approximations for guarding 1.5-dimensional terrains. Algorithmica 60, 451–463 (2011)
8. Gibson, M., Kanade, G., Krohn, E., Varadarajan, K.: An approximation scheme for terrain guarding. In: Dinur, I., Jansen, K., Naor, J., Rolim, J. (eds.) APPROX and RANDOM 2009. LNCS, vol. 5687, pp. 140–148. Springer, Heidelberg (2009)
9. Halldórsson, M., Knauer, C., Spillner, A., Tokuyama, T.: Fixed-parameter tractability for non-crossing spanning trees. In: Dehne, F., Sack, J.-R., Zeh, N. (eds.) WADS 2007. LNCS, vol. 4619, pp. 410–421. Springer, Heidelberg (2007)
10. Jansen, K., Woeginger, G.: The complexity of detecting crossingfree configurations in the plane. BIT 33, 580–595 (1993)
11. Károlyi, G., Pach, J., Tóth, G.: Ramsey-type results for geometric graphs I. Discrete & Computational Geometry 18, 247–255 (1997)
12. Keller, C., Perles, M., Rivera-Campo, E., Urrutia-Galicia, V.: Blockers for non-crossing spanning trees in complete geometric graphs. In: Pach, J. (ed.) Thirty Essays on Geometric Graph Theory, pp. 383–397. Springer (2013)
13. King, J.: A 4-approximation algorithm for guarding 1.5-dimensional terrains. In: Correa, J.R., Hevia, A., Kiwi, M. (eds.) LATIN 2006. LNCS, vol. 3887, pp. 629–640. Springer, Heidelberg (2006)
14. King, J., Krohn, E.: Terrain guarding is NP-hard. SIAM Journal on Computing 40, 1316–1339 (2011)
15. Knauer, C., Schramm, É., Spillner, A., Wolff, A.: Configurations with few crossings in topological graphs. Computational Geometry 37, 104–114 (2007)
16. Koch, A., Krug, M., Rutter, I.: Graphs with plane outside-obstacle representations (2013) available online: arXiv:1306.2978
17. Krohn, E., Nilsson, B.: The complexity of guarding monotone polygons. In: Proc. Canadian Conference on Computational Geoemtry, pp. 167–172 (2012)
18. Krohn, E., Nilsson, B.: Approximate guarding of monotone and rectilinear polygons. Algorithmica 66, 564–594 (2013)
19. Mukkamala, P., Pach, J., Pálvölgyi, D.: Lower bounds on the obstacle number of graphs. The Electronic Journal of Combinatorics 19 (2012)
20. Pach, J., Sarıöz, D.: On the structure of graphs with low obstacle number. Graphs and Combinatorics 27, 465–473 (2011)
21. Rivera-Campo, E.: A note on the existente of plane spanning trees of geometrie graphs. In: Akiyama, J., Kano, M., Urabe, M. (eds.) JCDCG 1998. LNCS, vol. 1763, pp. 274–277. Springer, Heidelberg (2000)

The Connectivity of Boolean Satisfiability: Dichotomies for Formulas and Circuits

Konrad Schwerdtfeger

Institut für Theoretische Informatik, Leibniz Universität Hannover,
Appelstr. 4, 30167 Hannover, Germany
k.w.s@gmx.net

Abstract. For Boolean satisfiability problems, the structure of the solution space is characterized by the solution graph, where the vertices are the solutions, and two solutions are connected iff they differ in exactly one variable. Motivated by research on heuristics and the satisfiability threshold, in 2006, Gopalan et al. studied connectivity properties of the solution graph and related complexity issues for CSPs [3]. They found dichotomies for the diameter of connected components and for the complexity of the *st*-connectivity question, and conjectured a trichotomy for the connectivity question. Their results were refined by Makino et al. [7]. Recently, we were able to establish the trichotomy [15].

Here, we consider connectivity issues of satisfiability problems defined by Boolean circuits and propositional formulas that use gates, resp. connectives, from a fixed set of Boolean functions. We obtain dichotomies for the diameter and the connectivity problems: on one side, the diameter is linear and both problems are in P, while on the other, the diameter can be exponential and the problems are PSPACE complete.

1 Introduction

The Boolean satisfiability problem, as well as many related questions like equivalence, counting, enumeration, and numerous versions of optimization, are of great importance in both theory and applications of computer science.

Common to all these problems is that one asks questions about a Boolean relation given by some short description, e.g. a propositional formula, Boolean circuit, binary decision diagram, or Boolean neural network. For the usual formulas with the connectives $\wedge$, $\vee$ and $\neg$, several generalizations and restrictions have been considered. Most widely studied are Boolean constraint satisfactions problems (*CSPs*), that can be seen as a generalization of formulas in *CNF* (conjunctive normal form), see Definition 2. Another generalization, that we will consider here, are formulas with connectives from an arbitrary fixed set of Boolean functions B, known as *B-formulas*. This concept also applies to circuits, where the allowed gates implement the functions from B, called *B-circuits*. A further extension that allows for shorter representations, and in turn makes many problems harder, are quantifiers, which we will look at in Section 5.

E.A. Hirsch et al. (Eds.): CSR 2014, LNCS 8476, pp. 351–364, 2014.

Here we will investigate the structure of the solution space, which is of obvious relevance to these satisfiability related problems. Indeed, the solution space connectivity is strongly correlated to the performance of standard satisfiability algorithms like WalkSAT and DPLL on random instances: As one approaches the *satisfiability threshold* (the ratio of constraints to variables at which random k-CNF-formulas become unsatisfiable for $k \geq 3$) from below, the solution space fractures, and the performance of the algorithms breaks down [9,8]. These insights mainly came from statistical physics, and lead to the development of the *survey propagation algorithm*, which has much better performance on random instances [8]. This research was ϕ motivation for Gopalan et al. to study connectivity properties of the solution space of Boolean CSPs [3].

While the most efficient satisfiability solvers take CNF-formulas as input, one of the most important applications of satisfiability testing is verification and optimization in Electronic Design Automation (EDA), where the instances derive mostly from digital circuit descriptions [18]. Though many such instances can easily be encoded in CNF, the original structural information, such as signal ordering, gate orientation and logic paths, is lost, or at least obscured. Since exactly this information can be very helpful for solving these instances, considerable effort has been made recently to develop satisfiability solvers that work with the circuit description directly [18], which have far superior performance in EDA applications, or to restore the circuit structure from CNF [2]. This is one major motivation for our study.

A direct application of *st*-connectivity are *reconfiguration problems*, that arise when we wish to find a step-by-step transformation between two feasible solutions of a problem such that all intermediate results are also feasible. Recently, the reconfiguration versions of many problems such as INDEPENDENT-SET, VERTEX-COVER, SET-COVER, GRAPH-k-COLORING, SHORTEST-PATH have been studied [4,5], and many complexity results were obtained, in some cases making use of Gopalan et al.'s results.

Since many of the satisfiability related problems are hard to solve in general (they are NP- or even PSPACE-complete), one has tried to identify easier fragments and to classify restrictions in terms of their complexity. Possibly the best known result is Schaefer's 1978 dichotomy theorem for CSPs, which states that for certain classes of allowed constraints the satisfiability of a CSP is in P, while it is NP-complete for all other classes [13]. Analogously, Gopalan et al. in 2006 classified the complexity of connectivity questions for CSPs in Schaefer's framework. In this paper, we consider the same connectivity issues as Gopalan et al., but for problems defined by Boolean circuits and propositional formulas that use gates, resp. connectives, from a fixed set of Boolean functions.

2   Propositional Formulas and Their Solution Space Connectivity

Definition 1. *An n-ary Boolean relation is a subset of* $\{0,1\}^n$ *(n ≥ 1). The set of solutions of a propositional formula ϕ with n variables defines in a natural way*

*an n-ary Boolean relation R, where the variables are taken in lexicographic order.
The* solution graph $G(\phi)$ *of* ϕ *is the subgraph of the n-dimensional hypercube
graph induced by the vectors in R, i.e., the vertices of $G(\phi)$ are the vectors in R,
and there is an edge between two vectors precisely if they differ in exactly one
position.*

We use $a, b, \ldots$ *to denote vectors of Boolean values and* $x, y, \ldots$ *to denote vectors of variables,* $a = (a_1, a_2, \ldots)$ *and* $x = (x_1, x_2, \ldots)$. *The* Hamming distance
$|a - b|$ *of two Boolean vectors* a *and* b *is the number of positions in which they
differ. If* a *and* b *are solutions of* ϕ *and lie in the same connected component
of $G(\phi)$, we write $d_\phi(a, b)$ to denote the shortest-path distance between a and b.
The diameter of a connected component is the maximal shortest-path distance
between any two vectors in that component. The diameter of $G(\phi)$ is the maximal
diameter of any of its connected components.*

In our proofs for B-formulas and B-circuits, we will use Gopalan et al.'s results
for 3-CNF-formulas, so we also need to introduce some terminology for constraint
satisfaction problems.

Definition 2. *A CNF-formula is a Boolean formula of the form $C_1 \wedge \cdots \wedge C_m$
$(1 \leq m < \infty)$, where each C_i is a clause, that is, a finite disjunction of literals
(variables or negated variables). A k-CNF-formula (k $\geq$ 1) is a CNF-formula
where each C_i has at most k literals.*

*For a finite set of Boolean relations S, a CNF(S)-formula (with constants)
over a set of variables V is a finite conjunction $C_1 \wedge \cdots \wedge C_m$, where each C_i is
a constraint application (constraint for short), i.e., an expression of the form
$R(\xi_1, \ldots, \xi_k)$, with a k-ary relation $R \in S$, and each ξ_j is a variable in V or one
of the constants 0, 1.*

*A k-clause is a disjunction of k variables or negated variables. For $0 \leq i < k$,
let D_i be the set of all satisfying truth assignments of the k-clause whose first i
literals are negated, and let $S_k = \{D_0, \ldots, D_k\}$ Thus, CNF(S_k) is the collection
of k-CNF-formulas.*

Gopalan et al. studied the following two decision problems for CNF(S)-formulas:

- the *connectivity problem* CONN*(S)*: given a CNF(S)-formula ϕ, is $G(\phi)$ connected? (if ϕ is unsatisfiable, then $G(\phi)$ is considered connected)
- the *st-connectivity problem* ST-CONN*(S)*: given a CNF(S)-formula ϕ and
  two solutions s and t, is there a path from s to t in $G(\phi)$?

Lemma 1. *[3, Lemm 3.6]* ST-CONN(S_3) *and* CONN(S_3) *are* PSPACE-*complete.*

Proof. ST-CONN(S_3) and CONN(S_3) are in PSPACE: Given a CNF(S_3)-formula
ϕ and two solutions s and t, we can guess a path of length at most 2^n between
them and verify that each vertex along the path is indeed a solution. Hence
ST-CONN(S_3) is in NPSPACE=PSPACE. For CONN(S_3), by reusing space we
can check for all pairs of vectors whether they are satisfying and, if they both
are, whether they are connected in $G(\phi)$.

We can not state the full proof for the PSPACE-hardness here. It consists of
a direct reduction from the computation of a space-bounded Turing machine M.

The input-string w of M is mapped to a $\mathrm{CNF}(S_3)$-formula and two satisfying assignments s and t, corresponding to the initial and accepting configuration respectively, s.t. s and t are connected in $G(\phi)$ iff M accepts w.   □

Lemma 2. *[3, Lemm 3.7] For $n \geq 2$, there is an n-ary Boolean function f with $f(1,\ldots,1) = 1$ and a diameter of at least $2^{\lfloor \frac{n}{2} \rfloor}$.*

3   Circuits, Formulas, and Post's Lattice

An n-ary *Boolean function* is a function $f : \{0,1\}^n \to \{0,1\}$. Let B be a finite set of Boolean functions.

A B-*circuit* C with input variables $x_1,\ldots,x_n$ is a directed acyclic graph, augmented as follows: Each node (here also called *gate*) with indegree 0 is labeled with an x_i or a 0-ary function from B, each node with indegree $k > 0$ is labeled with a k-ary function from B. The edges (here also called *wires*) pointing into a gate are ordered. One node is designated the output gate.

Given values $a_1,\ldots,a_n \in \{0,1\}$ to $x_1,\ldots,x_n$, C computes an n-ary function f_C as follows: A gate v labeled with a variable x_i returns a_i, a gate v labeled with a function f computes the value $f(b_1,\ldots,b_k)$, where $b_1,\ldots,b_k$ are the values computed by the predecessor gates of v, ordered according to the order of the wires. For a more formal definition see [17].

A B-*formula* is defined inductively: A variable x is a B-formula. If $\phi_1,\ldots,\phi_m$ are B-formulas, and f is an n-ary function from B, then $f(\phi_1,\ldots,\phi_n)$ is a B-formula; here, we identify the function f and the symbol representing it in a formula.

It is easy to see that the functions computable by a B-circuit, as well as the functions definable by a B-formula, are exactly those that can be obtained from B by *superposition*, together with all projections [1]. By superposition, we mean substitution (that is, composition of functions), permutation and identification of variables, and introduction of *fictive variables* (variables on which the value of the function does not depend). This class of functions is denoted by $[B]$. B is *closed* (or said to be a *clone*) if $[B] = B$. A *base* of a clone F is any set B with $[B] = F$.

Already in the early 1920s, Emil Post extensively studied Boolean functions, identified all closed classes, found a finite base for each of them, and detected their inclusion structure [11]. The closed classes form a lattice, called *Post's lattice*, depicted in Figure 1; a table of the bases can be found e.g. in [1], a modern proof e.g. in [19]. The classes are defined as follows:

- BF is the class of all Boolean functions.
- For $a \in \{0,1\}$, an n-ary Boolean function f is called a-*reproducing*, if $f(a,\ldots,a) = a$; the classes R_a contain all a-reproducing functions.
- f is called *monotonic*, if $a_1 \leq b_1,\ldots,a_n \leq b_n$ implies $f(a_1,\ldots,a_n) \leq f(b_1,\ldots,b_n)$; M is the class of all monotonic functions.
- f is called *self-dual*, if $f(x_1,\ldots,x_n) = \overline{f(\overline{x_1},\ldots,\overline{x_n})}$; D is the class of all self-dual functions.

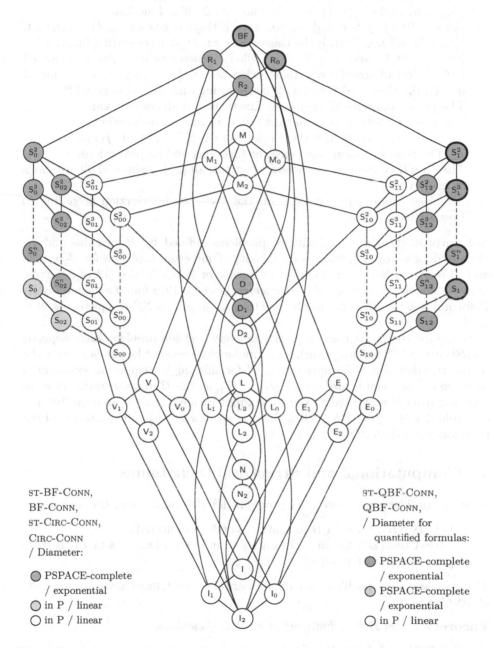

Fig. 1. Post's lattice with our results for the connectivity problems and the diameter. For comparison, the satisfiability problem (without quantifiers) is NP-complete for the bold circled classes, and in P for the other ones.

- f is called *affine*, if $f(x_1, \ldots, x_n) = x_{i_1} \oplus \cdots \oplus x_{i_m} \oplus c$ with $i_1, \ldots, i_m \in \{1, \ldots, n\}$ and $c \in \{0, 1\}$; L is the class of all affine functions.
- For $c \in \{0, 1\}$, f is called *c-separating*, if there exists an $i \in \{1, \ldots, n\}$ s.t. $a_i = c$ for all $\boldsymbol{a} \in f^{-1}(c)$; the classes S_c contain all c-separating functions.
- For $c \in \{0, 1\}$ and $k \geq 2$, f is called *c-separating of degree* k, if for all $U \subseteq f^{-1}(c)$ of size $|U| = k$ there exists an $i \in \{1, \ldots, n\}$ s.t. $a_i = c$ for all $\boldsymbol{a} \in U$; the classes S_c^k contain all c-separating functions of degree k.
- The class E contains the constant functions and all conjunctions.
- The class V contains the constant functions and all disjunctions.
- f is called a *projection*, if there exists an $i \in \{1, \ldots, n\}$ s.t. $f(x_1, \ldots, x_n) = x_i$; The class I contains the constant functions and all projections.
- The class N contains the constant functions, all projections and all negations of projections.
- All other classes are defined from the above by intersection according to Post's lattice.

Not surprisingly, the complexity of problems defined by B-formulas and B-circuits depends on $[B]$, and the complexity of numerous problems for B-circuits and B-formulas has been classified by means of Post's lattice [12,14], starting with satisfiability: Analogously to Schaefer, Lewis in 1978 found a dichotomy for B-formulas [6]; if $[B]$ contains the function $x \wedge \overline{y}$, SAT is NP-complete, else it is in P.

While for B-circuits the complexity of every decision problem solely depends on $[B]$ (up to AC^0 isomorphism), for formulas this need not be the case, since the transformation of a B-formula into a B'-formula might require an exponential increase in the formula size even if $[B] = [B']$, as the B'-representation of some function from B may need to use some input variable more than once [10]. For example, let $h(x, y) = x \wedge \overline{y}$; then there is no shorter $\{h\}$-representation of the function $x \wedge y$ than $h(x, h(x, y))$.

4 Computational and Structural Dichotomies

Now we consider the connectivity problems for B-formulas and B-circuits:

- BF-CONN(B): Given a B-formula ϕ, is $G(\phi)$ connected?
- ST-BF-CONN(B): Given a B-formula ϕ and two solutions $\boldsymbol{s}$ and $\boldsymbol{t}$, is there a path from $\boldsymbol{s}$ to $\boldsymbol{t}$ in $G(\phi)$?

The corresponding problems for circuits are denoted CIRC-CONN(B) resp. ST-CIRC-CONN(B).

Theorem 1. *Let B be a finite set of Boolean functions.*

1. *If $B \subseteq \mathsf{M}$, $B \subseteq \mathsf{L}$, or $B \subseteq \mathsf{S}_0$, then*
   (a) *ST-CIRC-CONN(B) and CIRC-CONN(B) are in P,*
   (b) *ST-BF-CONN(B) and BF-CONN(B) are in P,*
   (c) *the diameter of every function $f \in [B]$ is linear in the number of variables of f.*

2. *Otherwise,*

    (a) ST-CIRC-CONN(B) *and* CIRC-CONN(B) *are* PSPACE-*complete,*

    (b) ST-BF-CONN(B) *and* BF-CONN(B) *are* PSPACE-*complete,*

    (c) there are functions $f \in [B]$ such that their diameter is exponential in the number of variables of f.

The proof follows from the Lemmas in the next subsections. By the following Proposition, we can relate the complexity of B-formulas and B-circuits.

Proposition 1. *Every B-formula can be transformed into an equivalent B-circuit in polynomial time.*

Proof. A B-formula already is a suitable encoding for a special B-circuit with outdegree of at most one.   □

4.1   The Easy Side of the Dichotomy

Lemma 3. *If $B \subseteq$ M, the solution graph of any n-ary function $f \in [B]$ is connected, and $d_f(a, b) = |a - b| \leq n$ for any two solutions a and b.*

Proof. The table of all closed classes of Boolean functions shows that f is monotonic in this case. Thus, either $f = 0$, or $(1, \ldots, 1)$ must be a solution, and every other solution a is connected to $(1, \ldots, 1)$ in $G(\phi)$ since $(1, \ldots, 1)$ can be reached by flipping the variables assigned 0 in a one at a time to 1. Further, if a and b are solutions, b can be reached from a in $|a - b|$ steps by first flipping all variables that are assigned 0 in a and 1 in b, and then flipping all variables that are assigned 1 in a and 0 in b.   □

Lemma 4. *If $B \subseteq$ S$_0$, the solution graph of any function $f \in [B]$ is connected, and $d_f(a, b) \leq |a - b| + 2$ for any two solutions a and b.*

Proof. Since f is 0-separating, there is an i such that $a_i = 0$ for every vector a with $f(a) = 0$, thus every b with $b_i = 1$ is a solution. It follows that every solution t can be reached from any solution s in at most $|s - t| + 2$ steps by first flipping the i-th variable from 0 to 1 if necessary, then flipping all other variables in which s and t differ, and finally flipping back the i-th variable if necessary.   □

Lemma 5. *If $B \subseteq$ L,*

1. ST-CIRC-CONN(B) *and* CIRC-CONN(B) *are in* P,
2. ST-BF-CONN(B) *and* BF-CONN(B) *are in* P,
3. *for any function $f \in [B]$, $d_f(a, b) = |a - b|$ for any two solutions a and b that lie in the same connected component of $G(\phi)$.*

Proof. Since every function $f \in$ L is linear, $f(x_1, \ldots, x_n) = x_{i_1} \oplus \ldots \oplus x_{i_m} \oplus c$, and any two solutions s and t are connected iff they differ only in fictional variables: If s and t differ in at least one non-fictional variable (i.e., an $x_i \in \{x_{i_1}, \ldots, x_{i_m}\}$), to reach t from s, x_i must be flipped eventually, but for every solution a, any vector b that differs from a in exactly one non-fictional variable is no solution. If s and t differ only in fictional variables, t can be reached from s in $|s - t|$ steps by flipping one by one the variables in which they differ.

Since $\{x \oplus y, 1\}$ is a base of L (see Fig. 1 int [1]), every B-circuit $\mathcal{C}$ can be transformed in polynomial time into an equivalent $\{x \oplus y, 1\}$-circuit $\mathcal{C}'$ by replacing each gate of $\mathcal{C}'$ with an equivalent $\{x \oplus y, 1\}$-circuit. Now one can decide in polynomial time whether a variable x_i is fictional by checking for $\mathcal{C}'$ whether the number of "backward paths" from the output gate to gates labeled with x_i is odd, so ST-CIRC-CONN(B) is in P.

$G(\mathcal{C})$ is connected iff at most one variable is non-fictional, thus CIRC-CONN(B) is in P.

By Proposition 1, ST-BF-CONN(B) and BF-CONN(B) are in P also.   □

This completes the proof of the easy side of the dichotomy.

4.2   The Hard Side of the Dichotomy

Proposition 2. ST-CIRC-CONN(B) and CIRC-CONN(B), as well as ST-BF-CONN(B) and BF-CONN(B), are in PSPACE for any finite set B of Boolean functions.

Proof. This follows as in Lemma 1.   □

Proposition 3. For 1-reproducing 3-CNF-formulas, the problems ST-CONN and CONN are PSPACE-complete.

Proof. We chose the variables in the proof of Lemma 1 such that the accepting configuration of the Turing machine corresponds to the $(1, \ldots, 1)$ vector.   □

An inspection of Post's lattice shows that if $B \not\subseteq$ M, $B \not\subseteq$ L, and $B \not\subseteq$ S$_0$, then $[B] \supseteq$ S$_{12}$, $[B] \supseteq$ D$_1$, or $[B] \supseteq$ S$_{02}^k, \forall k \geq 2$, so we have to prove PSPACE-completeness and show the existence of B-formulas with an exponential diameter in these cases.

In the proofs, we will use the following abbreviations: If we have the n variables $x_1, \ldots, x_n$, we write x for $x_1 \wedge \cdots \wedge x_n$ and $\bar{x}$ for $\bar{x}_1 \wedge \cdots \wedge \bar{x}_n$. Also, we write $(x = c_1 \cdots c_n)$ for $x_1 \leftrightarrow c_1 \wedge \cdots \wedge x_n \leftrightarrow c_n$, where $c_1, \ldots, c_n \in \{0, 1\}$ are constants; e.g., we write $(x = 101)$ for $x_1 \wedge \bar{x}_2 \wedge x_3$. Further, we use $x \in \{a, b, \ldots\}$ for $(x = a) \vee (x = b) \vee \ldots$. Finally, if we have two vectors of Boolean values a and b of length n and m resp., we write $a \cdot b$ for their concatenation $(a_1, \ldots, a_n, b_1, \ldots b_m)$.

Lemma 6. If $[B] \supseteq$ S$_{12}$,

1. ST-BF-CONN(B) and BF-CONN(B) are PSPACE-complete,
2. ST-CIRC-CONN(B) and CIRC-CONN(B) are PSPACE-complete,
3. for $n \geq 3$, there is an n-ary function $f \in [B]$ with diameter of at least $2^{\lfloor \frac{n-1}{2} \rfloor}$.

Proof. 1. We reduce the problems for 1-reproducing 3-CNF-formulas to the ones for B-formulas: We map a 1-reproducing 3-CNF-formula ϕ and two solutions s and t of ϕ to a B-formula ϕ' and two solutions s' and t' of ϕ' such that s' and t' are connected in $G(\phi')$ iff s and t are connected in $G(\phi)$, and such that $G(\phi')$ is connected iff $G(\phi)$ is connected.

First for any 1-reproducing formula ψ, we define a connectivity-equivalent formula $T_\psi \in S_{12}$ using the standard connectives, then we show how to transform ϕ into the B-formula ϕ' that will be equivalent to T_ϕ.

Let ψ be a 1-reproducing formula over the variables $x_1, \ldots, x_n$. We define the formula T_ψ over the $n+1$ variables $x_1, \ldots, x_n$ and y as

$$T_\psi = \psi \wedge y,$$

where y is a new variable. All solutions a of $T_\psi(x, y)$ have $a_{n+1} = 1$, so T_ψ is 1-seperating and 0-reproducing. Moreover, T_ψ is still 1-reproducing, and thus in S_{12}. For any two solutions s and t of $\psi(x)$, $s' = s \cdot 1$ and $t' = t \cdot 1$ are solutions of $T_\psi(x, y)$, and it is easy to see that they are connected in $G(T_\psi)$ iff s and t are connected in $G(\psi)$, and that $G(T_\psi)$ is connected iff $G(\psi)$ is connected.

Now we know that for any 1-reproducing 3-CNF-formula ϕ, T_ϕ can be expressed as a B-formula ϕ' since $T_\phi \in S_{12}$. However, the transformation could lead to an exponential increase in the formula size (see Section 3), so we have to show how to construct ϕ' in polynomial time. We do this by parenthesizing the conjunctions of ψ such that we get a tree of $\wedge$'s of depth logarithmic in the size of ϕ, and then replacing each clause C_i with some B-formula ξ_{C_i}, and each expression $\phi_1 \wedge \phi_2$ with a B-formula $\xi_\wedge(\phi_1, \phi_2)$, s.t. the resulting formula is equivalent to T_ϕ. This can increase the formula size by only a polynomial in the original size even if $\xi_\wedge$ uses some input variable more than once. This is a standard-technique for such proofs in Post's framework, see e.g. [1]. Here we easily see that we can simply replace each clause C_i of ϕ with some B-formula equivalent to T_{C_i} and each $\wedge$ with a B-formula equivalent to $T_\wedge$ since $(\psi_1 \wedge y) \wedge (\psi_2 \wedge y) \wedge y \equiv \psi_1 \wedge \psi_2 \wedge y$, but in the next proofs this will not be obvious, so we formalize the procedure.

Let $\phi = C_1 \wedge \cdots \wedge C_n$ be a 1-reproducing 3-CNF-formula. Since ϕ is 1-reproducing, every clause C_i of ϕ is itself 1-reproducing, and we can express T_{C_i} through a B-formula $T_{C_i}^*$. Also, we can express $T_\wedge(x_1, x_2) = x_1 \wedge x_2 \wedge y$ through a B-formula $T_\wedge^*$ since $\wedge$ is 1-reproducing. Now let $\phi' = \text{TR}\left(T_{C_1}^*, \ldots, T_{C_n}^*\right)$, where TR is the following recursive algorithm that takes a list of formulas as input,

Algorithm $\text{TR}(\psi_1, \ldots, \psi_m)$

1. if $m = 1$ return ψ_1
2. else if m is even, return
   $\text{TR}(T_\wedge^*[x_1/\psi_1, x_2/\psi_2], T_\wedge^*[x_1/\psi_3, x_2/\psi_4], \ldots, T_\wedge^*[x_1/\psi_{m-1}, x_2/\psi_m])$
3. else return
   $\text{TR}(T_\wedge^*[x_1/\psi_1, x_2/\psi_2], T_\wedge^*[x_1/\psi_3, x_2/\psi_4], \ldots, T_\wedge^*[x_1/\psi_{m-2}, x_2/\psi_{m-1}], \psi_m).$

Here $\psi[x_i/\xi]$ denotes the formula obtained by substituting the formula ξ for the variable x_i in the formula ψ. Note that in every T_ψ^* we have the *same* variable y.

Since the recursion terminates after a number of steps logarithmic in the number of clauses of ϕ, and every step increases the total formula size by only a constant factor, the algorithm runs in polynomial time. We show $\phi' = T_\phi$ by induction. The basis is clear. Since $T_\psi \equiv T_\psi^*$, it suffices to show that $T_\wedge[x_1/T_{\psi_1}, x_2/T_{\psi_2}] \equiv T_{\psi_1 \wedge \psi_2}$:

$$T_\wedge[x_1/T_{\psi_1}, x_2/T_{\psi_2}] = T_{\psi_1} \wedge T_{\psi_2} \wedge y = (\psi_1 \wedge y) \wedge (\psi_2 \wedge y) \wedge y \equiv \psi_1 \wedge \psi_2 \wedge y = T_{\psi_1 \wedge \psi_2}.$$

2. This follows from 1. by Proposition 1.

3. By Lemma 2 there is an 1-reproducing $(n-1)$-ary function f with diameter of at least $2^{\lfloor \frac{n-1}{2} \rfloor}$. Let f be represented by a formula ϕ; then, T_ϕ represents an n-ary function of the same diameter in S_{12}. □

Lemma 7. *If* $[B] \supseteq D_1$,

1. ST-BF-CONN(B) *and* BF-CONN(B) *are* PSPACE-*complete,*
2. ST-CIRC-CONN(B) *and* CIRC-CONN(B) *are* PSPACE-*complete,*
3. *for* $n \geq 5$, *there is an* n-*ary function* $f \in [B]$ *with diameter of at least* $2^{\lfloor \frac{n-3}{2} \rfloor}$.

Proof. 1. This proof is similar to the previous one, but the construction is more intricate; for every 1-reproducing 3-CNF formula we have to construct a self-dual function s.t. the connectivity is retained. For clarity, we do the construction in two steps.

For a 1-reproducing formula ψ over the n variables $x_1, \ldots, x_n$, we construct a formula $T_\psi^\sim \in D_1$ with three new variables $(y_1, y_2, y_3) = \boldsymbol{y}$,

$$T_\psi^\sim = (\psi(\boldsymbol{x}) \wedge \boldsymbol{y}) \vee \left(\overline{\psi(\overline{\boldsymbol{x}})} \wedge \overline{\boldsymbol{y}} \right) \vee \boldsymbol{y} \in \{100, 010, 001\}.$$

Observe that $T_\psi^\sim(\boldsymbol{x}, \boldsymbol{y})$ is self-dual: for any solution ending with 111, the inverse vector (that ends with 000) is no solution; all vectors ending with 100, 010, or 001 are solutions and their inverses are no solutions. Also, $T_\psi^\sim$ is still 1-reproducing, and it is 0-reproducing since $\overline{\psi(\overline{0 \cdots 0})} \equiv \overline{\psi(1 \cdots 1)} \equiv 0$.

Further, for any two solutions $\boldsymbol{s}$ and $\boldsymbol{t}$ of $\psi(\boldsymbol{x})$, $\boldsymbol{s}' = \boldsymbol{s} \cdot 111$ and $\boldsymbol{t}' = \boldsymbol{t} \cdot 111$ are solutions of $T_\psi^\sim(\boldsymbol{x}, \boldsymbol{y})$ and are connected in $G(T_\psi^\sim)$ iff $\boldsymbol{s}$ and $\boldsymbol{t}$ are connected in $G(\psi)$: Every solution $\boldsymbol{a}$ of ψ corresponds to a solution $\boldsymbol{a} \cdot 111$ of $T_\psi^\sim$, and the connectivity does not change by padding the vectors with 111, and since there are no solutions of $T_\psi^\sim$ ending with 110, 101, or 011, every other solution of $T_\psi^\sim$ differs in at least two variables from the solutions $\boldsymbol{a} \cdot 111$ that correspond to solutions of ψ.

Note that exactly one connected component is added in $G(T_\psi^\sim)$ to the components corresponding to those of $G(\psi)$: It consists of all vectors ending with 000, 100, 010, or 001 (any two vectors ending with 000 are connected e.g. via those ending with 001). It follows that $G(T_\psi^\sim)$ is always unconnected. To fix this, we modify $T_{\tilde{\psi}}$ to a function T_ψ by adding $1 \cdots 1 \cdot 110$ as a solution, thereby connecting $1 \cdots 1 \cdot 111$ (which is always a solution because $T_{\tilde{\psi}}$ is 1-reproducing) with $1 \cdots 1 \cdot 100$, and thereby with the additional component of T_ψ. To keep the function self-dual, we must in turn remove $0 \cdots 0 \cdot 001$, which does not alter the connectivity. Formally,

$$T_\psi = \left(T_{\tilde{\psi}} \vee (\boldsymbol{x} \wedge (\boldsymbol{y} = 110))\right) \wedge \neg(\overline{\boldsymbol{x}} \wedge (\boldsymbol{y} = 001))$$

$$= (\psi(\boldsymbol{x}) \wedge \boldsymbol{y}) \vee \left(\overline{\psi(\overline{\boldsymbol{x}})} \wedge \overline{\boldsymbol{y}}\right) \tag{1}$$

$$\vee \left(\boldsymbol{y} \in \{100, 010, 001\} \wedge \neg(\overline{\boldsymbol{x}} \wedge (\boldsymbol{y} = 001))\right) \vee (\boldsymbol{x} \wedge (\boldsymbol{y} = 110)).$$

Now $G(T_\psi)$ is connected iff $G(\psi)$ is connected.

Next again we use the algorithm TR from the previous proof to transform any 1-reproducing 3-CNF-formula ϕ into a B-formula ϕ' equivalent to T_ϕ, but with the definition (1) of T. Again, we have to show $T_\wedge [x_1/T_{\psi_1}, x_2/T_{\psi_2}] \equiv T_{\psi_1 \wedge \psi_2}$. Here,

$$T_\wedge [x_1/T_{\psi_1}, x_2/T_{\psi_2}] = (T_{\psi_1} \wedge T_{\psi_2} \wedge \boldsymbol{y}) \vee \left(\overline{T_{\psi_1} \wedge T_{\psi_2}} \wedge \overline{\boldsymbol{y}}\right)$$

$$\vee \left(\boldsymbol{y} \in \{100, 010, 001\} \wedge \neg \left(\overline{T_{\psi_1} \wedge T_{\psi_2}} \wedge (\boldsymbol{y} = 001)\right)\right)$$

$$\vee (T_{\psi_1} \wedge T_{\psi_2} \wedge (\boldsymbol{y} = 110)).$$

We consider the parts of the formula in turn: For any formula ξ we have $T_\xi(\boldsymbol{x}_\xi) \wedge \boldsymbol{y} \equiv \xi(\boldsymbol{x}_\xi) \wedge \boldsymbol{y}$ and $T_\xi(\boldsymbol{x}_\xi) \wedge \overline{\boldsymbol{y}} = \overline{\psi(\overline{\boldsymbol{x}_\xi})} \wedge \overline{\boldsymbol{y}}$, where $\boldsymbol{x}_\xi$ denotes the variables of ξ. Using $\overline{T_{\psi_1}(\boldsymbol{x}_{\psi_1}) \wedge T_{\psi_2}(\boldsymbol{x}_{\psi_2})} \wedge \overline{\boldsymbol{y}} = (T_{\psi_1}(\boldsymbol{x}_{\psi_1}) \vee T_{\psi_2}(\boldsymbol{x}_{\psi_2})) \wedge \overline{\boldsymbol{y}}$, the first line becomes

$$(\psi_1(\boldsymbol{x}_{\psi_1}) \wedge \psi_2(\boldsymbol{x}_{\psi_2}) \wedge \boldsymbol{y}) \vee \left(\left(\overline{\psi_1(\overline{\boldsymbol{x}_{\psi_1}}) \wedge \psi_2(\overline{\boldsymbol{x}_{\psi_2}})}\right) \wedge \overline{\boldsymbol{y}}\right).$$

For the second line, we observe $\overline{T_\psi(\boldsymbol{x}_\psi)} \equiv \left(\overline{\psi(\boldsymbol{x}_\psi)} \vee \neg(\boldsymbol{y})\right) \wedge (\psi(\overline{\boldsymbol{x}_\psi}) \vee \neg(\overline{\boldsymbol{y}})) \wedge (\boldsymbol{y} \notin \{100, 010, 001\} \vee (\overline{\boldsymbol{x}_\psi} \wedge (\boldsymbol{y} = 001))) \wedge (\neg(\boldsymbol{x}_\psi) \vee \overline{(\boldsymbol{y} = 110)})$, thus $\overline{T_\psi(\boldsymbol{x}_\psi)} \wedge (\boldsymbol{y} = 001) \equiv \overline{\boldsymbol{x}_\psi} \wedge (\boldsymbol{y} = 001)$, and the second line becomes

$$\vee \left(\boldsymbol{y} \in \{100, 010, 001\} \wedge \neg (\overline{\boldsymbol{x}_{\psi_1}} \wedge \overline{\boldsymbol{x}_{\psi_2}} \wedge (\boldsymbol{y} = 001))\right).$$

Since $T_\psi(\boldsymbol{x}_\psi) \wedge (\boldsymbol{y} = 110) \equiv (\boldsymbol{x}_\psi \wedge (\boldsymbol{y} = 110))$ for any ψ, the third line becomes

$$\vee (\boldsymbol{x}_{\psi_1} \wedge \boldsymbol{x}_{\psi_2} \wedge (\boldsymbol{y} = 110)).$$

Now $T_\wedge [x_1/T_{\psi_1}, x_2/T_{\psi_2}]$ equals

$$T_{\psi_1 \wedge \psi_2} = (\psi_1(\boldsymbol{x}_{\psi_1}) \wedge \psi_2(\boldsymbol{x}_{\psi_2}) \wedge \boldsymbol{y}) \vee \left(\overline{\psi_1(\overline{\boldsymbol{x}_{\psi_1}}) \wedge \psi_2(\overline{\boldsymbol{x}_{\psi_2}})} \wedge \overline{\boldsymbol{y}}\right)$$

$$\vee \left(\boldsymbol{y} \in \{100, 010, 001\} \wedge \neg (\overline{\boldsymbol{x}_{\psi_1}} \wedge \overline{\boldsymbol{x}_{\psi_2}} \wedge (\boldsymbol{y} = 001))\right)$$

$$\vee (\boldsymbol{x}_{\psi_1} \wedge \boldsymbol{x}_{\psi_2} \wedge (\boldsymbol{y} = 110)).$$

2. This follows from 1. by Proposition 1.

3. By Lemma 2 there is an 1-reproducing $(n-3)$-ary function f with diameter of at least $2^{\lfloor \frac{n-3}{2} \rfloor}$. Let f be represented by a formula ϕ; then, T_ϕ represents an n-ary function of the same diameter in D_1. $\qquad \square$

Lemma 8. *If* $[B] \supseteq S_{02}^k$,

1. ST-BF-CONN(B) *and* BF-CONN(B) *are* PSPACE-*complete,*
2. ST-CIRC-CONN(B) *and* CIRC-CONN(B) *are* PSPACE-*complete,*
3. *for* $n \geq k + 4$, *there is an* n-*ary function* $f \in [B]$ *with diameter of at least* $2^{\lfloor \frac{n-k-2}{2} \rfloor}$.

Proof. 1. This proof is analogous to the previous one. For a 1-reproducing formula ψ over the n variables $x_1, \ldots, x_n$, we construct the formula $T_{\widetilde{\psi}} \in S_{02}^k$ with the additional variables y and $(z_1, \ldots, z_{k+1}) = \boldsymbol{z}$,

$$T_{\widetilde{\psi}} = (\psi \wedge y \wedge \overline{\boldsymbol{z}}) \vee \boldsymbol{z} \notin \{0 \cdots 0, 10 \cdots 0, 010 \cdots 0, \ldots, 0 \cdots 01\}.$$

$T_{\widetilde{\psi}}(\boldsymbol{x}, y, \boldsymbol{z})$ is 0-separating of degree k since all vectors that are no solutions of $T_{\widetilde{\psi}}$ end with a vector $\boldsymbol{b} \in \{0 \cdots 0, 10 \cdots 0, 010 \cdots 0, \ldots, 0 \cdots 01\} \subset \{0,1\}^{k+1}$ and thus any k of them have at least one common variable assigned 0. Also, $T_{\widetilde{\psi}}$ is 0-reproducing and still 1-reproducing.

Further, for any two solutions $\boldsymbol{s}$ and $\boldsymbol{t}$ of $\psi(\boldsymbol{x})$, $\boldsymbol{s}' = \boldsymbol{s} \cdot 1 \cdot 0 \cdots 0$ and $\boldsymbol{t}' = \boldsymbol{t} \cdot 1 \cdot 0 \cdots 0$ are solutions of $T_{\widetilde{\psi}}(\boldsymbol{x}, y, \boldsymbol{z})$ and are connected in $G(T_{\widetilde{\psi}})$ iff $\boldsymbol{s}$ and $\boldsymbol{t}$ are connected in $G(\psi)$.

But again, we have produced an additional connected component (consisting of all vectors not ending with $10 \cdots 0, 010 \cdots 0, \ldots, 0 \cdots 01$, or $0 \cdots 0$). To connect it to a component corresponding to one of ψ, we add $1 \cdots 1 \cdot 1 \cdot 10 \cdots 0$ as a solution,

$$T_\psi = (\psi \wedge y \wedge \overline{\boldsymbol{z}}) \vee \boldsymbol{z} \notin \{0 \cdots 0, 10 \cdots 0, 010 \cdots 0, \ldots, 0 \cdots 01\}$$
$$\vee (\boldsymbol{x} \wedge y \wedge (\boldsymbol{z} = 10 \cdots 0)).$$

Now $G(T_\psi)$ is connected iff $G(\psi)$ is connected.

Again we show that the algorithm TR works in this case. Here,

$$T_\wedge [x_1/T_{\psi_1}, x_2/T_{\psi_2}] = (T_{\psi_1}(\boldsymbol{x}_{\psi_1}) \wedge T_{\psi_2}(\boldsymbol{x}_{\psi_2}) \wedge y \wedge \overline{\boldsymbol{z}})$$
$$\vee \boldsymbol{z} \notin \{0 \cdots 0, 10 \cdots 0, 010 \cdots 0, \ldots, 0 \cdots 01\}$$
$$\vee (T_{\psi_1}(\boldsymbol{x}_{\psi_1}) \wedge T_{\psi_2}(\boldsymbol{x}_{\psi_2}) \wedge y \wedge (\boldsymbol{z} = 10 \cdots 0)),$$

which is equivalent to

$$T_{\psi_1 \wedge \psi_2} = (\psi_1(\boldsymbol{x}_{\psi_1}) \wedge \psi_2(\boldsymbol{x}_{\psi_2}) \wedge y \wedge \overline{\boldsymbol{z}})$$
$$\vee \boldsymbol{z} \notin \{0 \cdots 0, 10 \cdots 0, 010 \cdots 0, \ldots, 0 \cdots 01\}$$
$$\vee (\boldsymbol{x}_{\psi_1} \wedge \boldsymbol{x}_{\psi_2} \wedge y \wedge (\boldsymbol{z} = 10 \cdots 0)).$$

2. This follows from 1. by Proposition 1.

3. By Lemma 2 there is an 1-reproducing $(n - k - 2)$-ary function f with diameter of at least $2^{\lfloor \frac{n-k-2}{2} \rfloor}$. Let f be represented by a formula ϕ; then, T_ϕ represents an n-ary function of the same diameter in S_{02}^k.   □

This completes the proof of Theorem 1.

5   The Connectivity of Quantified Formulas

Definition 3. *A quantified B-formula ϕ (in prenex normal form) is an expression of the form*

$$Q_1 y_1 \cdots Q_m y_m \varphi(y_1, \ldots, y_m, x_1, \ldots x_n),$$

where φ is a B-formula, and $Q_1, \ldots, Q_m \in \{\exists, \forall\}$ are quantifiers. $x_1, \ldots, x_n$ are called the free variables of ϕ.

For quantified B-formulas, we define the connectivity problems

- QBF-CONN(B): Given a quantified B-formula ϕ, is $G(\phi)$ connected?
- ST-QBF-CONN(B): Given a quantified B-formula ϕ and two solutions s and t, is there a path from s to t in $G(\phi)$?

Theorem 2. *Let B be a finite set of Boolean functions.*

1. *If $B \subseteq \mathsf{M}$ or $B \subseteq \mathsf{L}$, then*
   (a) ST-QF-CONN(B) and QBF-CONN(B) are in P,
   (b) the diameter of every quantified B-formula is linear in the number of free variables.
2. *Otherwise,*
   (a) ST-QBF-CONN(B) and QBF-CONN(B) are PSPACE-complete,
   (b) there are quantified B-formulas with at most one quantifier such that their diameter is exponential in the number of free variables.

Proof. See the extended version of this paper [16].

Remark 1. An analog to Theorem 2 also holds for quantified circuits as defined in [12, Section 7].

6   Conclusions

While the classification for CSPs required an essential enhancement of Schaefer's framework and the introduction of new classes of CNF($\mathcal{S}$)-formulas, for B-formulas and B-circuits the connectivity issues fit entirely into Post's framework, although the proofs were quite novel, and made substantial use of Gopalan et al.'s results for 3-CNF-formulas.

As Gopalan et al. stated, we also believe that "connectivity properties of Boolean satisfiability merit study in their own right", which is substantiated by the recent interest in reconfiguration problems. Moreover, we imagine our results could aid the advancement of circuit based SAT solvers.

Acknowledgments. I am grateful to Heribert Vollmer for pointing me to these interesting themes.

References

1. Böhler, E., Creignou, N., Reith, S., Vollmer, H.: Playing with boolean blocks, part i: Posts lattice with applications to complexity theory. In: SIGACT News (2003)
2. Fu, Z., Malik, S.: Extracting logic circuit structure from conjunctive normal form descriptions. In: 20th International Conference on VLSI Design, Held Jointly with 6th International Conference on Embedded Systems, pp. 37–42. IEEE (2007)
3. Gopalan, P., Kolaitis, P.G., Maneva, E., Papadimitriou, C.H.: The connectivity of boolean satisfiability: Computational and structural dichotomies. SIAM J. Comput. 38(6), 2330–2355 (2009), http://dx.doi.org/10.1137/07070440X
4. Ito, T., Demaine, E.D., Harvey, N.J.A., Papadimitriou, C.H., Sideri, M., Uehara, R., Uno, Y.: On the complexity of reconfiguration problems. Theor. Comput. Sci. 412(12-14), 1054–1065 (2011), http://dx.doi.org/10.1016/j.tcs.2010.12.005
5. Kamiński, M., Medvedev, P., Milanič, M.: Shortest paths between shortest paths and independent sets. In: Iliopoulos, C.S., Smyth, W.F. (eds.) IWOCA 2010. LNCS, vol. 6460, pp. 56–67. Springer, Heidelberg (2011)
6. Lewis, H.R.: Satisfiability problems for propositional calculi. Mathematical Systems Theory 13(1), 45–53 (1979)
7. Makino, K., Tamaki, S., Yamamoto, M.: On the boolean connectivity problem for horn relations. In: Marques-Silva, J., Sakallah, K.A. (eds.) SAT 2007. LNCS, vol. 4501, pp. 187–200. Springer, Heidelberg (2007)
8. Maneva, E., Mossel, E., Wainwright, M.J.: A new look at survey propagation and its generalizations. Journal of the ACM (JACM) 54(4), 17 (2007)
9. Mézard, M., Mora, T., Zecchina, R.: Clustering of solutions in the random satisfiability problem. Physical Review Letters 94(19), 197205 (2005)
10. Michael, T.: On the applicability of post's lattice. Information Processing Letters 112(10), 386–391 (2012)
11. Post, E.L.: The Two-Valued Iterative Systems of Mathematical Logic(AM-5), vol. 5. Princeton University Press (1941)
12. Reith, S., Wagner, K.W.: The complexity of problems defined by Boolean circuits (2000)
13. Schaefer, T.J.: The complexity of satisfiability problems. In: STOC 1978, pp. 216–226 (1978)
14. Schnoor, H.: Algebraic techniques for satisfiability problems. Ph.D. thesis, Universität Hannover (2007)
15. Schwerdtfeger, K.W.: A computational trichotomy for connectivity of boolean satisfiability. ArXiv CoRR abs/1312.4524 (2013), extended version of a paper submitted to the JSAT Journal, http://arxiv.org/abs/1312.4524
16. Schwerdtfeger, K.W.: The connectivity of boolean satisfiability: Dichotomies for formulas and circuits. ArXiv CoRR abs/1312.6679 (2013), extended version of this paper, http://arxiv.org/abs/1312.6679
17. Vollmer, H.: Introduction to Circuit Complexity: A Uniform Approach. Springer-Verlag New York, Inc. (1999)
18. Wu, C.A., Lin, T.H., Lee, C.C., Huang, C.Y.R.: Qutesat: a robust circuit-based sat solver for complex circuit structure. In: Proceedings of the Conference on Design, Automation and Test in Europe, EDA Consortium, pp. 1313–1318 (2007)
19. Zverovich, I.E.: Characterizations of closed classes of boolean functions in terms of forbidden subfunctions and post classes. Discrete Appl. Math. 149(1-3), 200–218 (2005), http://dx.doi.org/10.1016/j.dam.2004.06.028

Randomized Communication Complexity of Approximating Kolmogorov Complexity

Nikolay Vereshchagin*

Moscow State University, Higher School of Economics, Yandex
ver@mech.math.msu.su

Abstract. The paper [Harry Buhrman, Michal Koucký, Nikolay Vereshchagin. Randomized Individual Communication Complexity. *IEEE Conference on Computational Complexity* 2008: 321-331] considered communication complexity of the following problem. Alice has a binary string x and Bob a binary string y, both of length n, and they want to compute or approximate Kolmogorov complexity $C(x|y)$ of x conditional to y. It is easy to show that deterministic communication complexity of approximating $C(x|y)$ with additive error α is at least $n - 2\alpha - O(1)$. The above referenced paper asks what is *randomized* communication complexity of this problem and shows that for r-round randomized protocols its communication complexity is at least $\Omega((n/\alpha)^{1/r})$. In this paper, for some positive ε, we show the lower bound $0.99n$ for (worst case) communication length of any randomized protocol that with probability at least 0.01 approximates $C(x|y)$ with additive error εn for all input pairs.

1 Introduction

Kolmogorov complexity of x conditional to y is defined as the minimal length of a program (for a universal machine) that given y as input prints x. Assume that Alice has x and Bob has y, which are strings of length n. Is there a communication protocol to transmit x to Bob (i.e. to compute the function $I(x,y) = x$) that communicates about $C(x|y)$ bits for all input pairs (x,y)?

The trivial upper bound for communication complexity of this problem is n (Alice sends her input to Bob). If Alice knew y, she could do better: she could find $C(x|y)$ bit program transforming y to x and send it to Bob. However, without any prior knowledge of y it seems impossible to solve the problem in about $C(x|y)$ communicated bits, and the paper [3] confirms this intuition for deterministic protocols. Moreover, for deterministic protocols even testing equality $x = y$ may require much more than $C(x|y)$ bits of communication. Indeed, for every deterministic protocol that tests equality there is an input pair (x,x) on which the protocol communicates at least n bits (see e.g. [5]). On the other hand, we have $C(x|x) = O(1)$.

---

* The work was in part supported by the RFBR grant 12-01-00864.

E.A. Hirsch et al. (Eds.): CSR 2014, LNCS 8476, pp. 365–374, 2014.
© Springer International Publishing Switzerland 2014

Surprisingly, the situation changes when we switch to randomized communication protocols. The paper [4] shows that for every positive ε there is a randomized communication protocol with public randomness that for all input pairs (x, y) communicates at most $C(x|y) + O(\sqrt{C(x|y)}) + \log(1/\varepsilon)$ bits and computes $I(x, y) = x$ with error probability at most ε. That protocol runs in $O(\sqrt{C(x|y)})$ rounds.

The paper [4] asks whether it is possible to reduce the number of rounds (keeping the communication close to $C(x|y)$) or to decrease the surplus term $O(\sqrt{C(x|y)})$ in communication length. Both questions are related to the communication complexity of approximating the conditional complexity $C(x|y)$. Indeed, assume that there is a randomized communication protocol that finds $C(x|y)$ with additive error α in r rounds and communicates at most l bits. Then the following randomized communication protocol computes $I(x, y) = x$ in $r + 1$ rounds with additional error ε and communicates at most $C(x|y) + l + \alpha + \log(1/\varepsilon)$ bits. Alice and Bob first run the given protocol to approximate $C(x|y)$. Assume that the protocol outputs an integer k. Then Alice communicates to Bob the value of a randomly chosen linear mapping $A : \{0,1\}^n \to \{0,1\}^{k+\alpha+\log(1/\varepsilon)}$ on her x. Bob finds any x' in the set $S = \{x' \mid C(x'|y) < k + \alpha\}$ such that $Ax' = Ax$ and outputs it (we consider protocols with public randomness thus Bob knows A). The additional error probability of this protocol is the probability of the event that S contains some $x' \neq x$ such that $Ax = Ax'$. By union bound this probability is at most $2^{k+\alpha}2^{-k-\alpha+\log\varepsilon} = \varepsilon$ (here $2^{k+\alpha}$ is an upper bound for the cardinality of S and $2^{-k-\alpha+\log\varepsilon}$ is the probability that $Ax' = Ax$ for any fixed $x' \neq x$).

The paper [4] shows that the worst case randomized communication complexity of approximating $C(x|y)$ with additive error α in r rounds is $\Omega((n/\alpha)^{1/r})$ and asks what happens when the number of rounds is not bounded. In this paper, we prove that for some positive ε every randomized protocol that for all input pairs with probability at least 0.01 computes $C(x|y)$ with additive error εn must communicate $0.99n$ bits for some input pair. That is, randomized communication complexity of approximating $C(x|y)$ is close to the trivial upper bound n unless the error is very bad (more than εn).

Actually, we prove more. In the strongest form, our result shows a lower bound for communication complexity of approximating the complexity of the pair $C(x, y|n)$ conditional to n, and not conditional complexity $C(x|y)$. Let us show that approximating $C(x, y|n)$ and $C(x|y)$ reduce to each other. By symmetry of information [6], we have

$$|C(x, y) - (C(y) + C(x|y))| \leqslant 4 \log n + O(1).$$

As Bob can find $C(y)$ privately[1] and transmit it to Alice in $\log n$ bits, approximating $C(x, y)$ and $C(x|y)$ with more than logarithmic additive error are almost

---

[1] Although $C(y)$ is not computable, Bob can do that, as we are using a non-uniform model of computation where the parties can just hard-wire a table containing $C(x)$ for all x of length up to $2n$.

equivalent. On the other hand,

$$|C(x,y) - C(x,y|n)| \leqslant 2\log n + O(1)$$

and hence approximating $C(x,y)$ and $C(x,y|n)$ with more than logarithmic additive error are also equivalent. More specifically, if a protocol approximates $C(x|y)$ with additive error α then it can approximate $C(x,y|n)$ with additive error $\alpha + 6\log n + O(1)$ by communicating extra $\log n$ bits, and the other way around.

We show that if a randomized protocol of depth d with shared randomness for every $(x,y) \in \{0,1\}^n \times \{0,1\}^n$ with probability at least p approximates $C(x,y|n)$ with additive error α, then

$$d \geqslant n - \log n - O(\alpha/p). \tag{1}$$

Moreover, our result holds for a weaker notion of *enumeration* in place of approximation. We say that a protocol e-enumerates a number if it outputs a list of e entries that contains that number.[2] Obviously, if a protocol approximates a function with additive error α then it is able to $2\alpha + 1$-enumerate that function. We show that the lower bound (1) holds also for the depth of any randomized protocol that with probability at least p for any input pair α-enumerates $C(x,y|n)$.

2 Preliminaries

All logarithms in this paper have the base 2.

2.1 Kolmogorov Complexity

Let U be a partial computable function that maps pairs of binary strings to binary strings. Kolmogorov complexity of a binary string x conditional to a binary string y with respect to U is defined as

$$C_U(x|y) = \min\{|p| \mid U(p,y) = x\}.$$

The notation $|p|$ refers to the length of p.

We call U *universal* or *optimal* if for any other partial computable function V there is a constant c such that

$$C_U(x|y) \leqslant C_V(x|y) + c$$

for all x, y.

By Solomonoff–Kolmogorov theorem universal partial computable functions exist [6]. We fix a universal U, drop the subscript U and call $C(x|y)$ *the Kolmogorov complexity of x conditional to y*. We call U also a "universal machine". If $U(p,y) = x$ we say that "program p outputs x on input y".

---

[2] The notion of enumeration has been studied in many contexts. In the context of communication complexity, it was first studied perhaps in [2].

Kolmogorov complexity of a string x is the minimal length of a program that prints x on the empty input Λ:

$$C(x) = C(x|\Lambda) = \min\{|p| \mid U(p, \Lambda) = x\}.$$

Kolmogorov complexity of other finite objects (like pairs of strings) is defined as follows: we fix a computable encoding of the objects in question by binary strings and declare Kolmogorov complexity of an object to be Kolmogorov complexity of its code.

For the properties of Kolmogorov complexity we refer to the textbook [6]. Actually, in this paper we do not need many of them. The first property we will need is an upper bound on the number of string of small complexity: for every y and k there are less than 2^k strings x with $C(x|y) < k$. We will use also the following obvious inequality $C(x) \leqslant |x| + O(1)$. Also we will use the inequality for the complexity $C(x, y)$ of the pair of strings x, y:

$$C(x, y) \leqslant 2C(x) + C(y) + O(1),$$

which is almost obvious: a short program to print the pair (x, y) can be identified by the shortest program to print x encoded in a prefix free way (the easiest prefix free encoding doubles the length) concatenated with the shortest program to print y. Finally, we will implicitly use the fact that algorithmic transformations do not increase complexity: $C(A(x)) \leqslant C(x) + O(1)$ for every algorithm A and all x (the constant $O(1)$ depends on A but not on x).

2.2   Communication Protocols

In this paper we use standard notions of a deterministic communication protocol and of a communication protocol with public randomness, as in the textbook [5]. Assume that Alice and Bob want to compute a function $f : X \times Y \to Z$ where the input $x \in X$ is given to Alice, and the input $y \in Y$ to Bob.

A deterministic communication protocol to compute such a function is identified by a rooted finite binary tree whose inner nodes are labeled with letters A (Alice) and B (Bob), labels indicate the turn to move. Additionally, each A-marked node is labeled by a function from X to $\{0, 1\}$ (different nodes may be labeled by different functions). This function identifies how the bit sent by Alice in her turn depends on her input. Similarly each B-marked node is labeled by a function from Y to $\{0, 1\}$. Each leaf of the tree is labeled by an element of Z (the output of the protocol).

Each node of the tree represents the state of the computation according to the protocol, which is the sequence of bits sent so far. The root is the initial state (no bits sent yet), the left son of a node u represents the state obtained after sending 0 in the state u and the right son of a node u represents the state obtained after sending 1 in the state u. When the current node is a leaf the computation halts, and the label of that leaf is considered as the result of the protocol, which should be equal to the value of the function f on the input pair.

The depth of the protocol tree is the worst case length of communication according to the protocol.

We will consider also randomized communication protocols. A randomized communication protocol of depth d with public randomness is a probability distribution $\mathcal{P}$ over deterministic communication protocols of depth d. We say that a randomized protocol $\mathcal{P}$ computes a function f with success probability p if for all input pairs (x, y) the protocol P drawn at random with respect to $\mathcal{P}$ computes $f(x, y)$ with probability at least p.

3 Results

3.1 Deterministic Protocols

Theorem 1. *If a deterministic protocol P computes $C(x|y)$ with additive error α then its depth d is at least $n - 2\alpha - O(1)$.*

Proof. Indeed, let $P(x, y)$ denote the output of P for the input pair (x, y). The protocol P defines a partition of the set $\{0, 1\}^n \times \{0, 1\}^n$ into at most 2^d rectangles[3] such that $P(x, y)$ is constant on every rectangle from the partition [5].

Let (y, y) be a diagonal input pair, $A \times B$ the rectangle in the partition containing it and k the value of P on that rectangle. As $C(y|y) = O(1)$, we have $k \leqslant \alpha + O(1)$. Since the rectangle $A \times B$ includes $A \times \{y\}$, we have $C(x|y) \leqslant 2\alpha + O(1)$ for all $x \in A$, which implies that $|A| \leqslant 2^{2\alpha + O(1)}$. Hence the number of diagonal pairs (y', y') in $A \times B$ is at most $2^{2\alpha + O(1)}$. As the total number of diagonal pairs is 2^n, it follows that the partition should have at least $2^{n - 2\alpha - O(1)}$ rectangles hence $d \geqslant n - 2\alpha - O(1)$.

3.2 Randomized Protocols

For randomized protocols it is much harder to derive lower bounds for communication complexity of our problem. For fixed number of rounds a lower bound was shown in [4].

Theorem 2 ([4]). *Assume that a randomized r round protocol with shared randomness for every $(x, y) \in \{0, 1\}^n \times \{0, 1\}^n$ communicates at most d bits and with probability at least $p > 1/2$ produces a number k such that $k \leqslant C(x|y) < k + \alpha$. Then $d \geqslant \Omega((n/\alpha)^{1/r})$. The constant in Ω-notation depends on r and p.*

We strengthen this theorem by removing the dependence of the lower bound on r. Our lower bound holds even for protocols whose success probability p may approach 0. Our main result shows that approximating $C(x, y|n)$ (and hence approximating $C(x|y)$) is hard for arbitrary randomized communication protocols.

---

[3] A rectangle is a set of the form $A \times B$.

Theorem 3. *Assume that a randomized protocol of depth d with shared randomness for every $(x, y) \in \{0, 1\}^n \times \{0, 1\}^n$ with probability at least p produces a list of α numbers containing $C(x, y|n)$. Then*

$$d \geqslant n - \log n - O(\alpha/p).$$

Corollary 1. *For some positive ε for all large enough n there is no randomized protocol of depth $0.99n$ that for all input pairs with probability at least 0.01 approximates $C(x, y|n)$ with additive error εn. The same statement holds for $C(x, y)$ and $C(x|y)$ in place of $C(x, y|n)$.*

Proof (Proof of Theorem 3). First notice that it suffices to prove the statement for $\alpha = 1$. Indeed, if a protocol computes a list with α entries containing $C(x, y|n)$ with probability p then a randomly chosen entry of the list equals $C(x, y|n)$ with probability p/α. Thus we will assume that $\alpha = 1$. In other words, we will consider protocols that compute $C(x, y|n)$ with success probability p.

Assume that there is a randomized protocol of depth d that computes $C(x, y|n)$ for every input pair (x, y) with success probability at least p. By Yao's principle [7], it follows that for any probability distribution μ on pairs $(x, y) \in \{0, 1\}^n \times \{0, 1\}^n$ there is a deterministic protocol of depth d that computes $C(x, y|n)$ on a fraction at least p of input pairs with respect to μ. Thus it suffices to find a distribution μ such that every deterministic protocol that computes $C(x, y|n)$ on a fraction at least p of input pairs with respect to μ has large depth.

To show that the constructed distribution μ has this property we will use a method similar to the discrepancy method [5]. More specifically, for the constructed distribution μ, for all rectangles $R \subset \{0, 1\}^n \times \{0, 1\}^n$ the following will hold: The fraction of pairs (with respect to μ) inside the rectangle that have any specific value of the function $C(x, y|n)$ is small compared to the size of the rectangle. The following lemma states that for such a μ every deterministic protocol of small depth is able to compute $C(x, y|n)$ only for a small fraction of input pairs. In that lemma μ is a probability distributions over $\{0, 1\}^n \times \{0, 1\}^n$ and f is any function from $\{0, 1\}^n \times \{0, 1\}^n$ into $\mathbb{N}$.

Lemma 1. *Assume that for every rectangle $R \subset \{0, 1\}^n \times \{0, 1\}^n$ and all $k \in \mathbb{N}$ we have*

$$\mu(\{(x, y) \in R \mid f(x, y) = k\}) \leqslant \varepsilon|R| + \delta.$$

Then every deterministic protocol of depth d computes f correctly on a fraction at most

$$\varepsilon 2^{2n} + \delta 2^d$$

of input pairs with respect to μ.

Proof. Fix a deterministic protocol P of depth d and call $P(x, y)$ its output on input pair (x, y). The protocol P defines a partition of the set $\{0, 1\}^n \times \{0, 1\}^n$ into at most 2^d rectangles such that $P(x, y)$ is constant on every rectangle from the partition [5]. The contribution of any rectangle R from the partition to the fraction of successful pairs equals

$$\mu(\{(x, y) \in R \mid f(x, y) = k\})$$

where k stands for the value of $P(x, y)$ on the rectangle. By the assumption this contribution is at most $\varepsilon|R| + \delta$. Summing up the contributions of all rectangles we obtain the upper bound $\varepsilon 2^{2n} + \delta 2^d$.

On the top level the construction of μ is the following. For some integer $l \leqslant n$, we construct a family of l distributions μ_i where $i = 2n - l + 1, \ldots, 2n$, with the following properties:

(1) $|C(x, y|n) - i| = O(1)$ for all pairs (x, y) in the support of μ_i;
(2) $\mu_i(R) \leqslant \varepsilon'|R| + \delta'$ for every rectangle $R \subset \{0, 1\}^n \times \{0, 1\}^n$.

Then we will let μ be the arithmetic mean of μ_i. The properties (1) and (2) imply that the assumptions of Lemma 1 are fulfilled for

$$\varepsilon = O\left(\frac{\varepsilon'}{l}\right) \text{ and } \delta = O\left(\frac{\delta'}{l}\right)$$

(for the function $f(x, y) = C(x, y|n)$). Indeed, for any k and for any rectangle R the μ-probability of the set

$$\{(x, y) \in R \mid C(x, y|n) = k\}$$

is the arithmetic mean of its μ_i-probabilities. By property (1) the μ_i-probability of this set is non-zero only when i is in the interval $[k - O(1); k + O(1)]$ and by property (2) for such i's it is at most $\varepsilon'|R| + \delta'$.

By Lemma 1 properties (1) and (2) imply that every deterministic protocol of depth d computes $C(x, y|n)$ correctly on a fraction at most

$$O\left(\frac{\varepsilon' 2^{2n} + \delta' 2^d}{l}\right)$$

of input pairs with respect to μ.

Is suffices to construct a large family of distributions such that properties (1) and (2) hold for small ε', δ'. To this end we will need the following combinatorial lemma.

Lemma 2. *For every $n \geqslant 1$ and every $3 < i \leqslant 2n$ there is a bipartite graph $G_{n,i}$ whose left and right nodes are all binary strings of length n, that has at least 2^{i-1} and at most 2^{i+1} edges and for every left set A and right set B with*

$$\log|A|, \log|B| > 2n - i + \log n + 4$$

the rectangle $A \times B$ has at most $|A \times B| \cdot 2^{i-2n+1}$ edges.

Let us finish the proof of the theorem assuming this lemma. Let $l \in [n; 2n)$ be an integer number to be chosen later. Apply Lemma 2 to all $i = 2n - l + 1, \ldots, 2n$. The number of edges $E_{n,i}$ in the resulting graph is between 2^{i-1} and 2^{i+1}. We may assume that the graph $G_{n,i}$ is computable given n, i (using brute force search we can find the first graph satisfying the lemma). Thus the Kolmogorov complexity of each edge in $G_{n,i}$ (conditional to n) is at most $i + O(1)$ (every edge can be identified by a its $i+1$ bit index). Remove from the graph all edges of complexity less than $i-2$. The number of removed edges is less that 2^{i-2} and hence the resulting graph has more than $2^{i-1} - 2^{i-2} = 2^{i-2}$ edges.

Let μ_i be the uniform probability distribution over the edges of $G_{n,i}$. The first property holds by construction. Let us show that the second property holds for some small ε', δ' for every rectangle $A \times B$. Assume first that both $\log|A|$ and $\log|B|$ are larger than $2n - i + \log n + 4$ (this bound comes from Lemma 2). The probability that a random edge from $G_{n,i}$ falls into $A \times B$ is at most the number of edges in $A \times B$ divided by the total number of edges in $G_{n,i}$. By Lemma 2 the number of edges in $A \times B$ is at most $|A \times B| \cdot 2^{i-2n+1}$ and $E_{n,i}$ is at least 2^{i-2}. Hence

$$\mu_i(A \times B) = O(|A \times B|/2^{2n}).$$

Otherwise either $|A|$, or $|B|$ is less than $2^{2n-i+\log n+4}$ and we use the trivial upper bound $|A \times B| \leqslant 2^n \times 2^{2n-i+\log n+4}$ for the number of edges of $G_{n,i}$ in $A \times B$ and the inequality $i > 2n - l$. We have

$$\mu_i(A \times B) \leqslant |A \times B|/2^{i-2} = O(2^{3n-2i+\log n})$$
$$= O(2^{2l-n+\log n}).$$

Thus the second property[4] holds for

$$\varepsilon' = O(2^{-2n}) \text{ and } \delta' = O(2^{2l-n+\log n}).$$

By Lemma 1 if a deterministic depth d protocol computes $C(x, y|n)$ on a fraction p of input pairs with respect to μ then

$$p \leqslant O\left(\frac{1 + 2^{d+2l-n+\log n}}{l}\right). \tag{2}$$

By Yao's principle Equation (2) also holds for the success probability of every depth d randomized protocol to compute $C(x, y|n)$. Now we have to choose l so that this inequality yields the best lower bound for d. A simple analysis reveals that an almost optimal choice of l is such that the exponent in the power of 2 in the right hand side of (2) is 0, that is $l = (n - d - \log n)/2$ (notice that if this l is negative then there is nothing to prove). Plugging such l in (2), we obtain

$$p \leqslant \frac{O(1)}{n - \log n - d}.$$

The statement of the theorem easily follows.

---

[4] The reader could wonder why we did not let μ_i be the uniform distribution over all pairs of Kolmogorov complexity about i. This distribution certainly satisfies the first property. However the second property is fulfilled only for $\varepsilon' = 2^{-i}$, which is much larger than 2^{-2n}. Indeed, let $R = A \times A$ where A is the set of all extensions of a fixed string of length $2n - i$ and complexity close to $2n - i$. Then complexity of almost all pairs in R is close to $(2n - i) + (i - n) + (i - n) = i$. Hence $\mu_i(R)$ is close to $|R|2^{-i}$.

It remains to prove Lemma 2. The lemma is proved by the probabilistic method. We will show that a randomly chosen graph has the desired properties with positive probability. The probability distribution over graphs is defined as follows. Every pair (left node, right node) is an edge of the graph with probability 2^{i-2n} and decisions for different pairs are independent.

We have to show that both requirements hold with probability more than one half. To this end we will use the Chernoff bound in the exponential form [1, Cor A.1.14]: for any independent random variables $T_1, \ldots, T_k$ with values 0,1 the probability that their sum T exceeds twice the expectation ET of T is less than $2^{-ET/4}$ and the probability that T is less than $ET/2$ is less than $2^{-ET/6}$.

The first requirement states that the number of edges in the graph is between 2^{i-1} and 2^{i+1}. The expected number of edges is 2^i. Hence by Chernoff bound[5] the probability that the requirement is not met is at most $2^{-2^i/4} + 2^{-2^i/6} < 1/2$, as $i \geqslant 4$.

The second requirement states that for all A, B of cardinality at least $2^{2n-i+\log n+4}$ the number of edges in $A \times B$ does not exceed twice its expectation. Fix a and b greater than $2^{2n-i+\log n+4} \geqslant 32$. Fix A and B of sizes a, b respectively. The expected number of edges that connect A and B is $ab2^{i-2n}$. Thus the probability that the number of edges between A and B exceeds its average two times is at most $2^{-ab2^{i-2n-2}}$. The number of possible A's of size a is at most 2na. Similarly, the number of possible B's of size b is at most 2^{nb}. By union bound, the probability that there are A and B of sizes a, b respectively, that violate the statement of the theorem is at most $2^{nb+na-ba2^{i-2n-2}}$. The exponent in this formula can be written as the sum of $b(n-a2^{i-2n-3})$ and $a(n-b2^{i-2n-3})$. The lower bound for $|A|, |B|$ was chosen so that both terms $n - a2^{i-2n-3}$ and $n - b2^{i-2n-3}$ be less than $-n$. By union bound the probability that there are A and B, that violate the statement of the theorem is at most

$$\sum_{b,a=32}^{2^n} 2^{-bn-an} = \sum_{b=32}^{2^n} 2^{-bn} \sum_{a=32}^{2^n} 2^{-an} < 1/2.$$

4   Open Problems and Acknowledgments

1. What is communication complexity of approximating $C(x|y)$ for quantum communication protocols?

2. Is it possible to drop the annoying $\log n$ term in the lower bound of Theorem 3?

3. Is it true that the depth of any randomized protocol which for every input pair (x, y) with probability at least p approximates $C(x, y)$ (or $C(x|y)$) with additive error α is also at least $n - O(\log n) - O(\alpha/p)$?

The author is grateful to anonymous referees for helpful suggestions.

---

[5] We could use here a weaker bound of large deviations.

References

1. Alon, N., Spencer, J.: The probabilistic method, 2nd edn. John Wiley & Sons (2000)
2. Ambainis, A., Buhrman, H., Gasarch, W.I., Kalyanasundaram, B., Torenvliet, L.: The communication complexity of enumeration, elimination and selection. Journal of Computer and System Sciences 63, 148–185 (2001)
3. Buhrman, H., Klauck, H., Vereshchagin, N.K., Vitányi, P.M.B.: Individual communication complexity. In: Diekert, V., Habib, M. (eds.) STACS 2004. LNCS, vol. 2996, pp. 19–30. Springer, Heidelberg (2004)
4. Buhrman, H., Koucký, M., Vereshchagin, N.: Randomized Individual Communication Complexity. In: IEEE Conference on Computational Complexity, pp. 321–331 (2008)
5. Kushilevitz, E., Nisan, N.: Communication Complexity. Cambridge University Press (1997)
6. Li, M., Vitányi, P.: An Introduction to Kolmogorov Complexity and its Applications. Springer (1997)
7. Yao, A.C.-C.: Probabilistic computations: Toward a unified measure of complexity. In: 18th Annual IEEE Symposium on Foundation of Computer Science, pp. 222–227 (1977)

Space Saving by Dynamic Algebraization

Martin Fürer* and Huiwen Yu

Department of Computer Science and Engineering
The Pennsylvania State University, University Park, PA, USA
{furer,hwyu}@cse.psu.edu

Abstract. Dynamic programming is widely used for exact computations based on tree decompositions of graphs. However, the space complexity is usually exponential in the treewidth. We study the problem of designing efficient dynamic programming algorithm based on tree decompositions in polynomial space. We show how to construct a tree decomposition and extend the algebraic techniques of Lokshtanov and Nederlof [18] such that the dynamic programming algorithm runs in time $O^*(2^h)$, where h is the maximum number of vertices in the union of bags on the root to leaf paths on a given tree decomposition, which is a parameter closely related to the tree-depth of a graph [21]. We apply our algorithm to the problem of counting perfect matchings on grids and show that it outperforms other polynomial-space solutions. We also apply the algorithm to other set covering and partitioning problems.

Keywords: Dynamic programming, tree decomposition, space-efficient algorithm, exponential time algorithms, zeta transform.

1   Introduction

Exact solutions to NP-hard problems typically adopt a branch-and-bound, inclusion/exclusion or dynamic programming framework. While algorithms based on branch-and-bound or inclusion/exclusion techniques [20] have shown to be both time and space efficient, one problem with dynamic programming is that for many NP-hard problems, it requires exponential space to store the computation table. As in practice programs usually run out of space before they run out of time [27], an exponential-space algorithm is considered not scalable. Lokshtanov and Nederlof [18] have recently shown that algebraic tools like the zeta transform and Möbius inversion [22,23] can be used to obtain space efficient dynamic programming under some circumstances. The idea is sometimes referred to as the coefficient extraction technique which also appears in [15,16].

The principle of space saving is best illustrated with the better known Fourier transform. Assume we want to compute a sequence of polynomial additions and multiplications modulo $x^n - 1$. We can either use a linear amount of storage and do many complicated convolution operations throughout, or we can start and end with the Fourier transforms and do the simpler component-wise

---

* Research supported in part by NSF Grant CCF-0964655 and CCF-1320814.

E.A. Hirsch et al. (Eds.): CSR 2014, LNCS 8476, pp. 375–388, 2014.

operations in between. Because we can handle one component after another, during the main computation, very little space is needed. This principle works for the zeta transform and subset convolution [3] as well.

In this paper, we study the problem of designing polynomial-space dynamic programming algorithms based on tree decompositions. Lokshtanov et al. [17] have also studied polynomial-space algorithms based on tree decomposition. They employ a divide and conquer approach. For a general introduction of tree decomposition, see the survey [6]. It is well-known that dynamic programming has wide applications and produces prominent results on efficient computations defined on path decomposition or tree decomposition in general [4]. Tree decomposition is very useful on low degree graphs as they are known to have a relatively low pathwidth [9]. For example, it is known that any degree 3 graph of n vertices has a path decomposition of pathwidth $\frac{n}{6}$. As a consequence, the minimum dominating set problem can be solved in time $O^*(3^{n/6})$[1], which is the best running time in this case [26]. However, the algorithm trades large space usage for fast running time.

To tackle the high space complexity issue, we extend the method of [18] in a novel way to problems based on tree decompositions. In contrast to [18], here we do not have a fixed ground set and cannot do the transformations only at the beginning and the end of the computation. The underlying set changes continuously, therefore a direct application on tree decomposition does not lead to an efficient algorithm. We introduce the new concept of zeta transforms for dynamic sets. Guided by a tree decomposition, the underlying set (of vertices in a bag) gradually changes. We adapt the transform so that it always corresponds to the current set of vertices. Herewith, we might greatly expand the applicability of the space saving method by algebraization.

We broadly explore problems which fit into this framework. Especially, we analyze the problem of counting perfect matchings on grids which is an interesting problem in statistical physics [12]. There is no previous theoretical analysis on the performance of any algorithm for counting perfect matchings on grids of dimension at least 3. We analyze two other natural types of polynomial-space algorithms, the branching algorithm and the dynamic programming algorithm based on path decomposition of a subgraph [14]. We show that our algorithm outperforms these two approaches. Our method is particularly useful when the treewidth of the graph is large. For example, grids, k-nearest-neighbor graphs [19] and low degree graphs are important graphs in practice with large treewidth. In these cases, the standard dynamic programming on tree decompositions requires exponential space.

The paper is organized as follows. In Section 2, we summarize the basis of tree decomposition and related techniques in [18]. In Section 3, we present the framework of our algorithm. In Section 4, we study the problem of counting perfect matchings on grids and extend our algorithmic framework to other problems.

---

[1] O^* notation hides the polynomial factors of the expression.

2 Preliminaries

2.1 Saving Space Using Algebraic Transformations

Lokshtanov and Nederlof [18] introduce algebraic techniques to solve three types of problems. The first technique is using discrete Fourier transforms (DFT) on problems of very large domains, e.g., for the subset sum problem. The second one is using Möbius and zeta transforms when recurrences used in dynamic programming can be formulated as subset convolutions, e.g., for the unweighted Steiner tree problem. The third one is to solve the minimization version of the second type of problems by combining the above transforms, e.g., for the traveling salesman problem. To the interest of this paper, we explain the techniques used in the second type of problems.

Given a universe V, let $\mathcal{R}$ be a ring and consider functions from 2^V to $\mathcal{R}$. Denote the collection of such functions by $\mathcal{R}[2^V]$. A singleton $f_A[X]$ is an element of $\mathcal{R}[2^V]$ which is zero unless $X = A$. The operator $\oplus$ is the pointwise addition and the operator $\odot$ is the pointwise multiplication. We first define some useful algebraic transforms.

The *zeta transform* of a function $f \in \mathcal{R}[2^V]$ is defined to be

$$\zeta f[Y] = \sum_{X \subseteq Y} f[X]. \tag{1}$$

The *Möbius transform/inversion* [22,23] of f is defined to be

$$\mu f[Y] = \sum_{X \subseteq Y} (-1)^{|Y \setminus X|} f[X]. \tag{2}$$

The Möbius transform is the inverse transform of the zeta transform, as they have the following relation [22,23]:

$$\mu(\zeta f)[X] = f[X]. \tag{3}$$

The high level idea of [18] is that, rather than directly computing $f[V]$ by storing exponentially many intermediate results $\{f[S]\}_{S \subseteq V}$, they compute the zeta transform of $f[S]$ using only polynomial space. $f[V]$ can be obtained by Möbius inversion (2) as $f[V] = \sum_{X \subseteq V} (-1)^{|V \setminus X|}(\zeta f)[X]$. Problems which can be solved in this manner have a common nature. They have recurrences which can be formulated by subset convolutions. The *subset convolution* [3] is defined to be

$$f *_{\mathcal{R}} g[X] = \sum_{X' \subseteq X} f(X')g(X \setminus X'). \tag{4}$$

To apply the zeta transform to $f *_{\mathcal{R}} g$, we need the *union product* [3] which is defined as

$$f *_u g[X] = \sum_{X_1 \bigcup X_2 = X} f(X_1)g(X_2). \tag{5}$$

The relation between the union product and the zeta transform is as follows [3]:

$$\zeta(f *_u g)[X] = (\zeta f) \odot (\zeta g)[X]. \tag{6}$$

In [18], functions over $(\mathcal{R}[2^V]; \oplus, *_\mathcal{R})$ are modeled by arithmetic circuits. Such a circuit is a directed acyclic graph where every node is either a singleton (constant gate), a $\oplus$ gate or a $*_\mathcal{R}$ gate. Given any circuit C over $(\mathcal{R}[2^V]; \oplus, *_\mathcal{R})$ which outputs f, every gate in C computing an output a from its inputs b, c is replaced by small circuits computing a relaxation $\{a^i\}_{i=1}^{|V|}$ of a from relaxations $\{b^i\}_{i=1}^{|V|}$ and $\{c^i\}_{i=1}^{|V|}$ of b and c respectively. (A *relaxation* of a function $f \in \mathcal{R}[2^V]$ is a sequence of functions $\{f^i : f^i \in \mathcal{R}[2^V], 0 \le i \le |V|\}$, such that $\forall i, X \subseteq V$, $f^i[X] = f[X]$ if $i = |X|$, $f^i[X] = 0$ if $i < |X|$, and $f^i[X]$ is an arbitrary value if $i > |X|$.) For a $\oplus$ gate, replace $a = b \oplus c$ by $a^i = b^i \oplus c^i$, for $0 \le i \le |V|$. For a $*_\mathcal{R}$ gate, replace $a = b *_\mathcal{R} c$ by $a^i = \sum_{j=0}^i b^j *_u c^{i-j}$, for $0 \le i \le |V|$. This new circuit C_1 over $(\mathcal{R}[2^V]; \oplus, *_u)$ is of size $O(|C| \cdot |V|)$ and outputs $f_{|V|}[V]$. The next step is to replace every $*_u$ gate by a gate $\odot$ and every constant gate a by ζa. It turns C_1 to a circuit C_2 over $(\mathcal{R}[2^V]; \oplus, \odot)$, such that for every gate $a \in C_1$, the corresponding gate in C_2 outputs ζa. Since additions and multiplications in C_2 are pointwise, C_2 can be viewed as $2^{|V|}$ disjoint circuits C^Y over $(\mathcal{R}[2^V]; +, \cdot)$ for every subset $Y \subseteq V$. The circuit C^Y outputs $(\zeta f)[Y]$. It is easy to see that the construction of every C^Y takes polynomial time.

As all problems of interest in this paper work on the integer domain $\mathbb{Z}$, we consider $\mathcal{R} = \mathbb{Z}$ and replace $*_\mathcal{R}$ by $*$ for simplicity. Assume $0 \le f[V] < m$ for some integer m, we can view the computation as on the finite ring $\mathbb{Z}_m$. Additions and multiplications can be implemented efficiently on $\mathbb{Z}_m$ (e.g., using the fast algorithm in [10] for multiplication).

Theorem 1 (Theorem 5.1 [18]). *Let C be a circuit over $(\mathbb{Z}[2^V]; \oplus, *)$ which outputs f. Let all constants in C be singletons and let $f[V] < m$ for some integer m. Then $f[V]$ can be computed in time $O^*(2^{|V|})$ and space $O(|V||C|\log m)$.*

2.2 Tree Decomposition

For any graph $G = (V, E)$, a *tree decomposition* of G is a tree $\mathcal{T} = (V_\mathcal{T}, E_\mathcal{T})$ such that every node x in $V_\mathcal{T}$ is associated with a set B_x (called the bag of x) of vertices in G and $\mathcal{T}$ has the following additional properties:

1. For any nodes x, y, and any node z belonging to the path connecting x and y in $\mathcal{T}$, $B_x \cap B_y \subseteq B_z$.
2. For any edge $e = \{u, v\} \in E$, there exists a node x such that $u, v \in B_x$.
3. $\cup_{x \in V_\mathcal{T}} B_x = V$.

The *width* of a tree decomposition $\mathcal{T}$ is $\max_{x \in V_\mathcal{T}} |B_x| - 1$. The *treewidth* of a graph G is the minimum width over all tree decompositions of G. We reserve the letter k for treewidth in the following context. Constructing a tree decomposition with minimum treewidth is an NP-hard problem. If the treewidth of a graph is bounded by a constant, a linear time algorithm for finding the minimum treewidth is known [5]. An $O(\log n)$ approximation algorithm of the treewidth is given in [7]. The result has been further improved to $O(\log k)$ in [8]. There are also a series of works studying constant approximation of treewidth k with running time exponential in k, see [5] and references therein.

To simplify the presentation of dynamic programming based on tree decomposition, an arbitrary tree decomposition is usually transformed into a *nice* tree decomposition which has the following additional properties. A node in a nice tree decomposition has at most 2 children. Let c be the only child of x or let c_1, c_2 be the two children of x. Any node x in a nice tree decomposition is of one of the following five types:

1. An *introduce vertex* node (introduce vertex v), where $B_x = B_c \cup \{v\}$.
2. An *introduce edge* node (introduce edge $e = \{u, v\}$), where $u, v \in B_x$ and $B_x = B_c$. We say that e is associated with x.
3. A *forget vertex* node (forget vertex v), where $B_x = B_c \setminus \{v\}$.
4. A *join* node, where x has two children and $B_x = B_{c_1} = B_{c_2}$.
5. A *leaf* node, a leaf of $\mathcal{T}$.

For any tree decomposition, a nice tree decomposition with the same treewidth can be constructed in polynomial time [13]. Notice that an introduce edge node is not a type of nodes in a common definition of a nice tree decomposition. We can create an introduce edge node after the two endpoints are introduced. We further transform every leaf node and the root to a node with an empty bag by adding a series of introduce nodes or forget nodes respectively.

3   Algorithmic Framework

We explain the algorithmic framework using the problem of counting perfect matchings based on tree decomposition as an example to help understand the recurrences. The result can be easily applied to other problems. A *perfect matching* in a graph $G = (V, E)$ is a collection of $|V|/2$ edges such that every vertex in G belongs to exactly one of these edges.

Consider a connected graph G and a nice tree decomposition $\mathcal{T}$ of treewidth k on G. Consider a function $f \in \mathbb{Z}[2^V]$. Assume that the recurrence for computing f on a join node can be formulated as a subset convolution, while on other types of tree nodes it is an addition or subtraction. We explain how to efficiently evaluate $f[V]$ on a nice tree decomposition by dynamic programming in polynomial space. Let $\mathcal{T}_x$ be the subtree rooted at x. Let T_x be the vertices contained in bags associated with nodes in $\mathcal{T}_x$ which are not in B_x. For any $X \subseteq B_x$, let Y_X be the union of X and T_x. For any $X \subseteq B_x$, let $f_x[X]$ be the number of perfect matchings in the subgraph Y_X with edges introduced in $\mathcal{T}_x$. As in the construction of Theorem 1, we first replace f_x by a relaxation $\{f_x^i\}_{0 \le i \le k+1}$ of f, where k is the treewidth. We then compute the zeta transform of f_x^i, for $0 \le i \le k + 1$. In the following context, we present only recurrences of f_x for all types of tree nodes except the join node where we need to use the relaxations. The recurrences of f_x based on f_c can be directly applied to their relaxations with the same index as in Theorem 1.

For any leaf node x, $(\zeta f_x)[\emptyset] = f_x[\emptyset]$ is a problem-dependent constant. In the case of the number of perfect matchings, $f_x[\emptyset] = 1$. For the root x, $(\zeta f_x)[\emptyset] = f_x[\emptyset] = f[V]$ which is the value of interest. For the other cases, consider an arbitrary subset $X \subseteq B_x$.

1. x is an introduce vertex node. If the introduced vertex v is not in X, $f_x[X] = f_c[X]$. If $v \in X$, in the case of the number of perfect matchings, v has no adjacent edges, hence $f_x[X] = 0$ (for other problems, $f_x[X]$ may equal to $f_c[X]$, which implies a similar recurrence). By definition of the zeta transform, if $v \in X$, we have $(\zeta f_x)[X] = \sum_{v \in X' \subseteq X} f_x[X'] + \sum_{v \notin X' \subseteq X} f_x[X'] = \sum_{v \notin X' \subseteq X} f_x[X']$. Therefore,

$$(\zeta f_x)[X] = \begin{cases} (\zeta f_c)[X] & v \notin X \\ (\zeta f_c)[X \setminus \{v\}] & v \in X \end{cases} \tag{7}$$

2. x is a forget vertex node. $f_x[X] = f_c[X \cup \{v\}]$ by definition.

$$(\zeta f_x)[X] = \sum_{X' \subseteq X} f_x[X'] = \sum_{X' \subseteq X} f_c[X' \cup \{v\}]$$
$$= (\zeta f_c)[X \cup \{v\}] - (\zeta f_c)[X]. \tag{8}$$

3. x is a join node with two children. By assumption, the computation of f_x on a join node can be formulated as a subset convolution. We have

$$f_x[X] = \sum_{X' \subseteq X} f_{c_1}[X'] f_{c_2}[X \setminus X'] = f_{c_1} * f_{c_2}[X]. \tag{9}$$

For the problem of counting perfect matchings, it is easy to verify that $f_x[X]$ can be computed using (9). Let $f_x^i = \sum_{j=0}^{i} f_{c_1}^j *_u f_{c_2}^{i-j}$. We can transform the computation to

$$(\zeta f_x^i)[X] = \sum_{j=0}^{i} (\zeta f_{c_1}^j)[X] \cdot (\zeta f_{c_2}^{i-j})[X], \text{ for } 0 \leq i \leq k+1. \tag{10}$$

4. x is an introduce edge node introducing $e = \{u, v\}$. The recurrence of f_x with respect to f_c is problem-dependent. Since the goal of the analysis of this case is to explain why we need to modify the construction of an introduce edge node, we consider only the recurrence for the counting perfect matchings problem. In this problem, if $e \not\subseteq X$, $f_x[X] = f_c[X]$, then $(\zeta f_x)[X] = (\zeta f_c)[X]$. If $e \subseteq X$, we can match u and v by e or not use e for matching, thus $f_x[X] = f_c[X] + f_c[X \setminus \{u, v\}]$. In this case, we have

$$(\zeta f_x)[X] = \sum_{e \subseteq X' \subseteq X} f_x[X'] + \sum_{e \not\subseteq X' \subseteq X} f_x[X'] = \sum_{e \subseteq X' \subseteq X} (f_c[X'] + f_c[X' \setminus \{u, v\}])$$

$$+ \sum_{e \not\subseteq X' \subseteq X} f_c[X'] = \sum_{X' \subseteq X} f_c[X'] + \sum_{e \subseteq X' \subseteq X} f(X' \setminus \{u, v\}).$$

Hence,

$$(\zeta f_x)[X] = \begin{cases} (\zeta f_c)[X] & e \nsubseteq X \\ (\zeta f_c)[X] + (\zeta f_c)[X \setminus \{u,v\}] & e \subseteq X \end{cases} \tag{11}$$

In cases 2 and 4, we see that the value of $(\zeta f_x)[X]$ depends on the values of ζf_c on two different subsets. We can visualize the computation along a path from a leaf to the root as a computation tree. This computation tree branches on introduce edge nodes and forget vertex nodes. Suppose along any path from the root to a leaf in $\mathcal{T}$, the maximum number of introduce edge nodes is m' and the maximum number of forget vertex nodes is h. To avoid exponentially large storage for keeping partial results in this computation tree, we compute along every path from a leaf to the root in this tree. This leads to an increase of the running time by a factor of $O(2^{m'+h})$, but the computation is in polynomial space (explained in detail later). As m' could be $\Omega(n)$, this could contribute a factor of $2^{\Omega(n)}$ to the time complexity. To reduce the running time, we eliminate the branching introduced by introduce edge nodes. On the other hand, the branching introduced by forget vertex nodes seems inevitable.

For any introduce edge node x which introduces an edge e and has a child c in the original nice tree decomposition $\mathcal{T}$, we add an auxiliary child c' of x, such that $B_{c'} = B_x$ and introduce the edge e at c'. c' is a special leaf which is not empty. We assume the evaluation of ζf on c' takes only polynomial time. For the counting perfect matchings problem, $f_{c'}[X] = 1$ only when $X = e$ or $X = \emptyset$, otherwise it is equal to 0. Then $(\zeta f_{c'})[X] = 2$ if $e \subseteq X$, otherwise $(\zeta f_{c'})[X] = 1$. We will verify that this assumption is valid for other problems considered in the following sections. We call x a *modified introduce edge* node and c' an *auxiliary leaf*. As the computation on x is the same as that on a join node, we do not talk about the computation on modified introduce edge nodes separately.

In cases 1 and 2, we observe that the addition operation is not a strictly pointwise addition as in Theorem 1. This is because in a tree decomposition, the set of vertices on every tree node might not be the same. However, there is a one-to-one correspondence from a set X in node x to a set X' in its child c. We call it a *relaxed pointwise addition* and denote it by $\oplus'$. Hence, f can be evaluated by a circuit C over $(\mathbb{Z}[2^V]; \oplus', *)$. We transform C to a circuit C_1 over $(\mathbb{Z}[2^V]; \oplus', *_u)$, then to C_2 over $(\mathbb{Z}[2^V]; \oplus', \odot)$, following constructions in Theorem 1.

In Theorem 1, C_2 can be viewed as $2^{|V|}$ disjoint circuits. In the case of tree decomposition, the computation makes branches on a forget node. Therefore, we cannot take C_2 as $O(2^k)$ disjoint circuits. Consider a subtree $\mathcal{T}_x$ of $\mathcal{T}$ where the root x is the only join node in the subtree. Take an arbitrary path from x to a leaf l and assume there are h' forget nodes along this path. We compute along every path of the computation tree expanded by the path from x to l, and sum up the result at the top. There are $2^{h'}$ computation paths which are independent. Hence we can view the computation as $2^{h'}$ disjoint circuits on $(\mathbb{Z}; +, \cdot)$. Assume the maximum number of forget nodes along any path from the root x to a leaf in $\mathcal{T}_x$ is h and there are n_l leaves, the total computation takes at most $n_l \cdot 2^h$ time and in polynomial space.

In general, we proceed the computation in an in-order depth-first traversal on a tree decomposition $\mathcal{T}$. Every time we hit a join node j, we need to complete all computations in the subtree rooted at j before going up. Suppose j_1, j_2 are the closest join nodes in two subtrees rooted at the children of j (if there is no other join node consider j_1 or j_2 to be empty). Assume there are at most h_j forget nodes between j, j_1 and j, j_2. Let T_x be the time to complete the computation of $(\zeta f_x)[X]$ at node x. We have $T_j \leq 2 \cdot 2^{h_j} \cdot \max\{T_{j_1}, T_{j_2}\}$). The modified edge node is a special type of join node. In this case, since one of its children c_1 is always a leaf, the running time only depends on the subtree rooted at c_2, thus similar to an introduce vertex node. Suppose there are n_j join nodes and let h be the maximum number of forget nodes along any path from the root to a leaf. By induction, it takes $2^{n_j} \cdot 2^h$ time to complete the computation on $\mathcal{T}$ and in polynomial space. Notice that 2^{n_j} is the number of leaves in $\mathcal{T}$, hence $2^{n_j} = O(|V| + |E|)$.

To summarize, we present the algorithm for the problem of counting perfect matchings based on a modified nice tree decomposition $\mathcal{T}$ in Algorithm 1.

Algorithm 1. Counting perfect matchings on a modified nice tree decomposition

---

    Input: a modified nice tree decomposition $\mathcal{T}$ with root r.
    return $(\zeta f)(r, \emptyset, 0)$.
    procedure $(\zeta f)(x, X, i)$. // $(\zeta f)(x, X, i)$ represents $(\zeta f_x^i)[X]$.
        if x is a leaf: **return** 1.
        if x is an auxiliary leaf: **return** 2 when $e \subseteq X$, otherwise 1.
        if x is an introduce vertex node: **return** $(\zeta f)(c, X, i)$ when $v \notin X$, or $(\zeta f)(c, X - \{v\}, i)$ when $v \in X$.
        if x is a forget vertex node: **return** $(\zeta f)(c, X \cup \{v\}, i) - (\zeta f)(c, X, i)$.
        if x is a join node: **return** $\sum_{j=0}^{i} (\zeta f)(c_1, X, j) \cdot (\zeta f)(c_2, X, i - j)$.
    end procedure

For any tree decomposition $\mathcal{T}$ of a graph G, we can transform it to a modified nice tree decomposition $\mathcal{T}'$ with the convention that the root has an empty bag. In this way, the parameter h, the maximum number of forget nodes along any path from the root to a leaf in $\mathcal{T}'$ is equal to the maximum size of the union of all bags along any path from the root to a leaf in $\mathcal{T}$. We directly tie this number h to the complexity of our algorithm. Let $h_m(G)$ be the minimum value of h for all tree decompositions of G. We show that $h_m(G)$ is closely related to a well-known parameter, the *tree-depth* of a graph [21].

Definition 1 (tree-depth [21]). *Given a rooted tree T with vertex set V, a closure of T, $clos(T)$ is a graph G with the same vertex V, and for any two vertices $x, y \in V$ such that x is an ancestor of y in T, there is a corresponding edge (x, y) in G. The tree-depth of T is the height of T. The tree-depth of a graph G, $td(G)$ is the minimum height of trees T such that $G \subseteq clos(T)$.*

Proposition 1. *For any connected graph G, $h_m(G) = td(G)$.*

Proof. For any tree decomposition of G, we first transform it to a modified nice tree decomposition $\mathcal{T}$. We contract $\mathcal{T}$ by deleting all nodes except the forget nodes. Let T_f be this contracted tree such that for every forget node in $\mathcal{T}$ which forgets a vertex x in G, the corresponding vertex in T_f is x. We have $G \subseteq clos(T_f)$. Therefore, $td(G) \le h$, here h is the maximum number of forget nodes along any path from the root to a leaf in $\mathcal{T}$.

For any tree T such that $G \subseteq clos(T)$, we construct a corresponding tree decomposition $\mathcal{T}$ of G such that, $\mathcal{T}$ is initialized to be T and every bag associated with the vertex x of T contains the vertex itself. For every vertex $x \in T$, we also put all ancestors of x in T into the bag associated with x. It is easy to verify that it is a valid tree decomposition of G. Therefore, the tree-depth of T, $td(T) \ge h_m(G)$. $\square$

In the following context, we also call the parameter h, the maximum size of the union of all bags along any path from the root to a leaf in a tree decomposition $\mathcal{T}$, the tree-depth of $\mathcal{T}$. Let k be the treewidth of G, it is shown in [21] that $td(G) \le (k+1) \log |V|$. Therefore, we also have $h_m(G) \le (k+1) \log |V|$. Moreover, it is obvious to have $h_m(G) \ge k + 1$.

Finally, we summarize the main result of this section in the following theorem.

Theorem 2. *Given any graph $G = (V, E)$ and tree decomposition $\mathcal{T}$ on G. Let f be a function evaluated by a circuit C over $(\mathbb{Z}[2^V]; \oplus', *)$ with constants being singletons. Assume $f[V] < m$ for integer m. We can compute $f[V]$ in time $O^*(((|V| + |E|)2^h)$ and in space $O(|V||C| \log m)$. Here h is the maximum size of the union of all bags along any path from the root to a leaf in $\mathcal{T}$.*

4   Counting Perfect Matchings

The problem of counting perfect matchings is $\sharp$P-complete. It has long been known that in a bipartite graph of size $2n$, counting perfect matchings takes $O^*(2^n)$ time using the inclusion and exclusion principle. A recent breakthrough [1] shows that the same running time is achievable for general graphs. For low degree graphs, improved results based on dynamic programming on path decomposition on a sufficiently large subgraph are known [2].

Counting perfect matchings on grids is an interesting problem in statistical physics [12]. The more generalized problem is the Monomer-Dimer problem [12], which essentially asks to compute the number of matchings of a specific size. We model the Monomer-Dimer problem as computing the matching polynomial problem . For grids in dimension 2, the pure Dimer (perfect matching) problem is polynomial-time tractable and an explicit expression of the solution is known [24]. We consider the problem of counting perfect matchings in cube/hypercube in Section 4.1. Results on counting perfect matchings in more general grids, computing the matching polynomial and applications to other set covering and partitioning problems are presented in Section 4.2.

4.1   Counting Perfect Matchings on Cube/Hypercube

We consider the case of counting perfect matchings on grids of dimension d, where $d \geq 3$ and the length of the grid is n in each dimension. We denote this grid by $G_d(n)$. To apply Algorithm 1, we first construct a balanced tree decomposition on $G_d(n)$ with the help of balanced separators. The balanced tree decomposition can easily be transformed into a modified nice tree decomposition.

Tree Decomposition Using Balanced Vertex Separators. We first explain how to construct a balanced tree decomposition using vertex separators of general graphs. An α-balanced vertex separator of a graph/subgraph G is a set of vertices $S \subseteq G$, such that after removing S, G is separated into two disjoint parts A and B with no edge between A and B, and $|A|, |B| \leq \alpha |G|$, where α is a constant in $(0, 1)$. Suppose we have an oracle to find an α-balanced vertex separator of a graph. We begin with creating the root of a tree decomposition $\mathcal{T}$ and associate the vertex separator S of the whole graph with the root. Consider a subtree $\mathcal{T}_x$ in $\mathcal{T}$ with the root x associated with a bag B_x. Denote the vertices belonging to nodes in $\mathcal{T}_x$ by V_x. Initially, $V_x = V$ and x is the root of $\mathcal{T}$. Suppose we have a vertex separator S_x which partitions V_x into two disjoint parts V_{c_1} and V_{c_2}. We create two children c_1, c_2 of x, such that the set of vertices belonging to $\mathcal{T}_{c_i}$ is $S_x \cup V_{c_i}$. Denote the set of vertices belonging to nodes in the path from x to the root of $\mathcal{T}$ by U_x, we define the bag B_{c_i} to be $S_x \cup (V_{c_i} \cap U_x)$, for $i = 1, 2$. It is easy to verify that this is a valid tree decomposition. Since V_x decreases by a factor of at least $1 - \alpha$ in each partition, the height of the tree is at most $\log_{\frac{1}{1-\alpha}} n$. To transform this decomposition into a modified nice tree decomposition, we only need to add a series of introduce vertex nodes, forget vertex nodes or modified introduce edge nodes between two originally adjacent nodes. We call this tree decomposition algorithm Algorithm 2.

We observe that after the transformation, the number of forget nodes from B_{c_i} to B_x is the size of the balanced vertex separator of V_x, i.e. $|S_x|$. Therefore, the number of forget nodes from the root to a leaf is the sum of the sizes of the balanced vertex separators used to construct this path in the tree decomposition.

A grid graph $G_d(n)$ has a nice symmetric structure. Denote the d dimensions by $x_1, x_2, ..., x_d$ and consider an arbitrary subgrid G'_d of $G_d(n)$ with length n'_i in dimension x_i. The hyperplane in G'_d which is perpendicular to x_i and cuts G'_d into halves can be used as a $1/2$-balanced vertex separator. We always cut the dimension with the longest length. If $n'_i = n'_{i+1}$, we choose to first cut the dimension x_i, then x_{i+1}.

To run Algorithm 2 on $G_d(n)$, we cut dimensions $x_1, x_2, ..., x_d$ consecutively with separators of size $\frac{1}{2^{i-1}} n^{d-1}$, for $i = 1, 2..., d$. Then we proceed with subgrids of length $n/2$ in every dimension. It is easy to see that the treewidth of this tree decomposition is $\frac{3}{2} n^{d-1}$. The tree-depth h of this tree decomposition is at most $\sum_{j=0}^{\infty} \sum_{i=0}^{d-1} \frac{1}{2^i} \cdot (\frac{1}{2^j} n)^{d-1}$, which is $\frac{2^d - 1}{2^{d-1} - 1} n^{d-1}$.

Lemma 1. *The treewidth of the tree decomposition $\mathcal{T}$ on $G_d(n)$ obtained by Algorithm 2 is $\frac{3}{2} n^{d-1}$. The tree-depth of $\mathcal{T}$ is at most $\frac{2^d - 1}{2^{d-1} - 1} n^{d-1}$.*

To apply Algorithm 1 to the problem of counting perfect matchings, we verify that $f[S] \leq \binom{|E|}{|V|/2} \leq |E|^{|V|/2}$ and all constants are singletons.

Theorem 3. *The problem of counting perfect matchings on grids of dimension d and uniform length n can be solved in time $O^*(2^{\frac{2^d-1}{2^{d-1}-1}n^{d-1}})$ and in polynomial space.*

To the best of our knowledge, there is no rigorous time complexity analysis of the counting perfect matchings problem in grids in the literature. To demonstrate the efficiency of Algorithm 1, we compare it to three other natural algorithms.

1. Dynamic programming based on path decomposition. A path decomposition is a special tree decomposition where the underlying tree is a path. A path decomposition with width $2n^{d-1}$ is obtained by putting all vertices with x_1 coordinate equal to j and $j+1$ into the bag of node j, for $j = 0, 1, ..., n-1$. A path decomposition with a smaller pathwidth of n^{d-1} can be obtained as follows. Construct n nodes $\{p_1, p_2, ..., p_n\}$ associated with a bag of vertices with x_1 coordinate equal to j, for $j = 0, 1, ..., n-1$. For any p_j, p_{j+1}, start from p_j, add a sequence of nodes by alternating between adding a vertex of $x_1 = j+1$ and deleting its neighbor with $x_1 = j$. The number of nodes increases by a factor of n^{d-1} than the first path decomposition. We run the standard dynamic programming on the second path decomposition. This algorithm runs in time $O^*(2^{n^{d-1}})$, however the space complexity is $O^*(2^{n^{d-1}})$. It is of no surprise that it has a better running time than Algorithm 1 due to an extra space usage. We remark that van Rooij et al. [25] give a dynamic programming algorithm for the counting perfect matching problem on any tree decomposition of treewidth k with running time $O^*(2^k)$ and space exponential to k.

2. Dynamic programming based on path decomposition on a subgrid. One way to obtain a polynomial space dynamic programming is to construct a low pathwidth decomposition on a sufficiently large subgraph. One can then run dynamic programming on this path decomposition and do an exhaustive enumeration on the remaining graph in a similar way as in [?]. To extract from $G_d(n)$ a subgrid of pathwidth $O(\log n)$ (notice that this is the maximum pathwidth for a polynomial space dynamic programming algorithm), we can delete a portion of vertices from $G_d(n)$ to turn a "cube"-shaped grid into a long "stripe" with $O(\log n)$ cross-section area. It is sufficient to remove $O(\frac{n^d}{(\log n)^{1/(d-1)}})$ vertices. This leads to a polynomial-space algorithm with running time $2^{O(\frac{n^d}{(\log n)^{1/(d-1)}})}$, which is worse than Algorithm 1.

3. Branching algorithm. A naive branching algorithm starting from any vertex in the grid could have time complexity $2^{O(n^d)}$ in the worst case. We analyze a branching algorithm with a careful selection of the starting point. The branching algorithm works by first finding a balanced separator S and partitioning the graph into $A \cup S \cup B$. The algorithm enumerates every subset $X \subseteq S$. A vertex in X either matches to vertices in A or to vertices in B while vertices in $S \setminus X$ are matched within S. Then the algorithm recurses on A and B. Let $T_d(n)$ be the running time of this branching algorithm on $G_d(n)$. We use the same balanced separator as in Algorithm 2. We have an upper bound of the running time as, $T_d(n) \leq 2T_d(\frac{n-|S|}{2})\sum_{X \subseteq S} 2^{|X|}T_{d-1}(|S \setminus X|)$. We can use any polynomial space algorithm to count perfect matchings on $S \setminus X$. For example using Algorithm 1, since the separator is of size $O(n^{d-1})$, we have $T_{d-1}(|S \setminus X|) = 2^{O(n^{d-2})}$. Therefore, $T_d(n) \leq 2T_d(\frac{n}{2}) \cdot 2^{o(n^{d-1})}\sum_{i=0}^{|S|}\binom{|S|}{i}2^i = 2T_d(\frac{n}{2}) \cdot 2^{o(n^{d-1})}3^{|S|}$. We get $T_d(n) = O^*(3^h)$, i.e. $O^*(3^{\frac{2^d-1}{2^d-1-1}n^{d-1}})$, which is worse than Algorithm 1. We remark that this branching algorithm can be viewed as a divide and conquer algorithm on balanced tree decomposition, which is similar as in [17].

4.2   Extensions

Counting Perfect Matchings on General Grids. Consider more general grids of dimension d with each dimension of length n_i, $1 \leq i \leq d$, which is at most n_m. We use Algorithm 2 to construct a balanced tree decomposition $\mathcal{T}$ of a general grid and obtain an upper bound of the tree-depth h of $\mathcal{T}$. The proof is omitted due to space constraint.

Lemma 2. *Given any grid of dimension d and volume $\mathcal{V}$. Using Algorithm 2, the tree-depth of this tree decomposition is at most $\frac{3d\mathcal{V}}{n_m}$.*

Based on Lemma 2, we give time complexity results of algorithms discussed in Section 4.1. First, h is the only parameter to the running time of Algorithm 1 and the branching algorithm. Algorithm 1 runs in time $O^*(2^{\frac{3d\mathcal{V}}{n_m}})$ and the branching algorithm runs in time $O^*(3^{\frac{3d\mathcal{V}}{n_m}})$. The dynamic programming algorithm based on path decomposition on a subgrid has a running time $2^{O(\frac{\mathcal{V}}{(\log n_m)^{1/(d-1)}})}$. Those three algorithms have polynomial space complexity. For constant d, Algorithm 1 has the best time complexity. For the dynamic programming algorithm based on path decomposition, it runs in time $O^*(2^{\frac{\mathcal{V}}{n_m}})$ but in exponential space.

Computing the Matching Polynomial. The matching polynomial of a graph G is defined to be $m[G, \lambda] = \sum_{i=0}^{|G|/2} m^i[G]\lambda^i$, where $m^i[G]$ is the number of matchings of size i in graph G. We put the coefficients of $m[G, \lambda]$ into a vector $\mathbf{m}[G]$. The problem is essentially to compute the coefficient vector $\mathbf{m}[G]$.

For every node x in a tree decomposition, let vector $\mathbf{m}_x[X]$ be the coefficient vector of the matching polynomial defined on Y_X. Notice that every entry of $\mathbf{m}_x[X]$ is at most $|E|^{|V|/2}$ and all constants are singletons. $\mathbf{m}_x^0[X] = 1$ and $\mathbf{m}_x^i[X] = 0$ for $i > |X|/2$. The case of x being a forget vertex node follows exactly from Algorithm 1. For any type of tree node x,

- x is a leaf node. $\mathbf{m}_x^i[\emptyset] = 1$ if $i = 0$, or 0 otherwise.
- x is an introduce vertex node. If $v \in X$, $\mathbf{m}_x^i[X] = \mathbf{m}_x^i[X \setminus \{v\}]$. Hence $(\zeta\mathbf{m}_c^i)[X] = 2(\zeta\mathbf{m}_c^i)[X \setminus \{v\}]$ if $v \in X$, or $(\zeta\mathbf{m}_c^i)[X] = (\zeta\mathbf{m}_c^i)[X]$ otherwise.
- x is an auxiliary leaf of a modified introduce edge node. $\mathbf{m}_x^i[X] = 1$ only when $u, v \in X$ and $i = 1$, or $i = 0$. Otherwise it is 0.
- x is a join node. $\mathbf{m}_x^i[X] = \sum_{X' \subseteq X}\sum_{j=0}^i \mathbf{m}_{c_1}^j[X']\mathbf{m}_{c_2}^{i-j}[X \setminus X']$.

Counting l-packings. Given a universe U of elements and a collection of subsets $\mathcal{S}$ on U, an l-packing is a collection of l disjoint sets. The l-packings problem can be solved in a similar way as computing the matching polynomial. Packing problems can be viewed as matching problems on hypergraphs. Tree decomposition on graphs can be generalized to tree decomposition on hypergraph, where we require every hyperedge to be assigned to a specific bag [11]. A hyperedge is introduced after all vertices covered by this edge are introduced.

Counting Dominating Sets, Counting Set Covers. The set cover problem is given a universe U of elements and a collection of sets $\mathcal{S}$ on U, find a subcollection of sets from $\mathcal{S}$ which covers the entire universe U. The dominating set problem is defined on a graph $G = (V, E)$. Let $U = V$, $\mathcal{S} = \{N[v]\}_{v \in V}$, where $N[v]$ is the union of the neighbors of v and v itself. The dominating set problem is to find a subset of vertices S from V such that $\bigcup_{v \in S} N[v]$ covers V.

The set cover problem can be viewed as a covering problem on a hypergraph, where one selects a collection of hyperedges which cover all vertices. The dominating set problem is then a special case of the set cover problem. If $\mathcal{S}$ is closed under subsets, a set cover can be viewed as a disjoint cover. We only consider the counting set covers problem. For any subset $X \subseteq B_x$, we define $h_x[X]$ to be the number of set covers of Y_X. We have $h_x[X] \le |U|^{|\mathcal{S}|}$, and all constants are singletons. We omit the recurrence for forget vertex nodes as we can directly apply recurrence (8) in Algorithm 1. For any node x, $h_x[\emptyset] = 1$.

- x is a leaf node. $h_x[\emptyset] = 1$.
  - x is an introduce vertex node. If $v \in X$, $h_x[X] = 0$. If $v \notin X$, $h_x[X] = h_c[X]$.
  - x is an auxiliary leaf of a modified introduce hyperedge node. $h_x[X] = 1$ when $X \subseteq e$, and $h_x[X] = 0$ otherwise.
  - x is a join node. $h_x[X] = \sum_{X' \subseteq X} h_{u_1}[X']h_{c_u}[X - X']$.

Finally, we point out that our framework has its limitations. First, it cannot be applied to problems where the computation on a join node cannot be formalized as a convolution. The maximum independent set problem is an example. Also it is not known if there is a way to adopt the framework to the Hamiltonian path problem, the counting l-path problems, and the unweighted Steiner tree problem. It seems that for theses problems we need a large storage space to record intermediate results. It is interesting to find more problems which fit in our framework.

References

1. Björklund, A.: Counting perfect matchings as fast as Ryser. In: SODA, pp. 914–921 (2012)
2. Björklund, A., Husfeldt, T.: Exact algorithms for exact satisfiability and number of perfect matchings. Algorithmica 52(2), 226–249 (2008)
3. Björklund, A., Husfeldt, T., Kaski, P., Koivisto, M.: Fourier meets Möbius: fast subset convolution. In: STOC, pp. 67–74 (2007)
4. Bodlaender, H.L.: Dynamic programming on graphs with bounded treewidth. In: Lepistö, T., Salomaa, A. (eds.) ICALP 1988. LNCS, vol. 317, pp. 105–118. Springer, Heidelberg (1988)
5. Bodlaender, H.L.: A linear time algorithm for finding tree-decompositions of small treewidth. In: STOC, pp. 226–234 (1993)

6. Bodlaender, H.L.: Discovering treewidth. In: Vojtáš, P., Bieliková, M., Charron-Bost, B., Sýkora, O. (eds.) SOFSEM 2005. LNCS, vol. 3381, pp. 1–16. Springer, Heidelberg (2005)
7. Bodlaender, H.L., Gilbert, J.R., Kloks, T., Hafsteinsson, H.: Approximating treewidth, pathwidth, and minimum elimination tree height. In: Schmidt, G., Berghammer, R. (eds.) WG 1991. LNCS, vol. 570, pp. 1–12. Springer, Heidelberg (1992)
8. Bouchitté, V., Kratsch, D., Müller, H., Todinca, I.: On treewidth approximations. Discrete Appl. Math. 136(2-3), 183–196 (2004)
9. Fomin, F.V., Gaspers, S., Saurabh, S., Stepanov, A.A.: On two techniques of combining branching and treewidth. Algorithmica 54(2), 181–207 (2009)
10. Fürer, M.: Faster integer multiplication. SIAM J. Comput. 39(3), 979–1005 (2009)
11. Gottlob, G., Leone, N., Scarcello, F.: Hypertree decompositions: A survey. In: Sgall, J., Pultr, A., Kolman, P. (eds.) MFCS 2001. LNCS, vol. 2136, pp. 37–57. Springer, Heidelberg (2001)
12. Kenyon, C., Randall, D., Sinclair, A.: Approximating the number of monomer-dimer coverings of a lattice. J. Stat. Phys. 83 (1996)
13. Kloks, T. (ed.): Treewidth. LNCS, vol. 842. Springer, Heidelberg (1994)
14. Kneis, J., Mölle, D., Richter, S., Rossmanith, P.: A bound on the pathwidth of sparse graphs with applications to exact algorithms. SIAM J. Discret. Math. 23(1), 407–427 (2009)
15. Koutis, I.: Faster algebraic algorithms for path and packing problems. In: ICALP, pp. 575–586 (2008)
16. Koutis, I., Williams, R.: Limits and applications of group algebras for parameterized problems. In: ICALP, pp. 653–664 (2009)
17. Lokshtanov, D., Mnich, M., Saurabh, S.: Planar k-path in subexponential time and polynomial space. In: Kolman, P., Kratochvíl, J. (eds.) WG 2011. LNCS, vol. 6986, pp. 262–270. Springer, Heidelberg (2011)
18. Lokshtanov, D., Nederlof, J.: Saving space by algebraization. In: STOC, pp. 321–330 (2010)
19. Miller, G.L., Teng, S.-H., Thurston, W., Vavasis, S.A.: Separators for sphere-packings and nearest neighbor graphs. J. ACM 44(1), 1–29 (1997)
20. Nederlof, J.: Fast polynomial-space algorithms using inclusion-exclusion. Algorithmica 65(4), 868–884 (2013)
21. Nešetřil, J., de Mendez, P.O.: Tree-depth, subgraph coloring and homomorphism bounds. Eur. J. Comb. 27(6), 1022–1041 (2006)
22. Rota, G.-C.: On the foundations of combinatorial theory. i. theory of möbius functions. Zeitschrift Wahrscheinlichkeitstheorie und Verwandte Gebiete 2(4), 340–368 (1964)
23. Stanley, R.P., Rota, G.C.: Enumerative Combinatorics, vol. 1. Cambridge University Press (2000)
24. Temperley, H.N.V., Fisher, M.: Dimer problem in statistical mechanics - an exact result. Philosophical Magazine 6, 1061–1063 (1961)
25. van Rooij, J.M.M., Bodlaender, H.L., Rossmanith, P.: Dynamic programming on tree decompositions using generalised fast subset convolution. In: Fiat, A., Sanders, P. (eds.) ESA 2009. LNCS, vol. 5757, pp. 566–577. Springer, Heidelberg (2009)
26. van Rooij, J.M.M., Nederlof, J., van Dijk, T.C.: Inclusion/Exclusion meets measure and conquer. In: Fiat, A., Sanders, P. (eds.) ESA 2009. LNCS, vol. 5757, pp. 554–565. Springer, Heidelberg (2009)
27. Woeginger, G.J.: Space and time complexity of exact algorithms: Some open problems (invited talk). In: 1st International Workshop on Parameterized and Exact Computation, pp. 281–290 (2004)

Author Index